IPT's PIPE TRADES HANDBOOK

by

ROBERT A. LE

Published by
IPT PUBLISHING AND TRAINING LTD.
BOX 9590, EDMONTON, ALBERTA, CANADA T6E 5X2
www.iptbooks.com
E-mail: info@iptbooks.com

Phone (780) 962-4548 Fax (780) 962-4819
Toll Free 1-888-808-6763

Printed by
grafikom.Speedfast, Edmonton, Alberta, Canada

IPT's PIPE TRADES HANDBOOK

First Printing, December 1991
Ninth Printing (revised edition), August 2006
Tenth Printing, May 2007
Eleventh Printing, February 2008

ISBN-13: 978-0-920855-18-8
ISBN-10: 0-920855-18-0

Acknowledgements

The author and publisher express their sincere appreciation to the following for their assistance in developing this publication:

Proofreading:
- Donald Miller, P.Eng. – Mechanical Engineering Technology, Northern Alberta Institute of Technology
- Brian Filax – Journeyman Pipefitter, Member of United Association Local 488

Technical Advice:
- David Cimesa (Hammond, Indiana) – President and Founder of Industrial Trades In-House Training, Member of ANSI, AWS, and United Association Local 597

Illustrations and Book Layout:
- Ian Holmes (Holmes Consulting), and Cindy Joly (Fine Lines Marketing & Communications Inc.). A well earned thank you to Ian and Cindy for their illustrations and work in laying out this book with its new format and revised text.
- Shawn Morgan Computer Graphics
- Ted Leach

SECTION ONE - PIPE DATA

SECTION TWO - TUBE DATA

SECTION FOUR - FITTINGS

SECTION FIVE - JOINTS

SECTION SIX - PIPE OFFSETS

SECTION TEN - PIPE WELDING

SECTION ELEVEN - RIGGING

SECTION TWELVE – APPENDICES

SECTION ONE

PIPE DATA

Pipe Manufacturing

Carbon steel pipe, wrought steel pipe, or "steel pipe", as it is most commonly referred to, is the material classification of pipe most often used in industry. Steel pipe, because of its high pressure and temperature ratings, ease of joining and overall durability, make it the pipe of choice for the majority of piping installations.

The three major methods of manufacturing steel pipe are referred to as:

- Type F, Furnace butt welded or Continuous welded pipe
- Type E, Electric resistance welded pipe
- Type S, Seamless pipe

Note: Another method of manufacturing steel pipe classified as lap welding has been replaced by the electric resistance welding method.

Both continuous welded and electric resistance welded pipe are made by shaping rolls of coiled steel called "skelp" into cylindrical forms, and welding along the longitudinal seam. In electric resistance welding, the edges of the skelp used to form the pipe are fused together by pressure and heat without the addition of filler metal. The heat is generated by the steel's resistance to an electrical current that is passed through the edges to be welded. The gradual forming of skelp and the electric resistance welding operation of the pipe is shown in illustration #1.

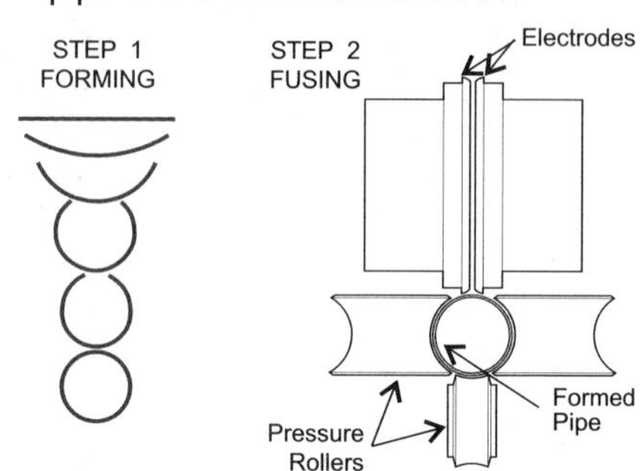

Illustration #1 - Forming and Electric Resistance Welding of Pipe

In the continuous butt welding process, skelp is first heated in a skelp heating furnace to a welding temperature of approximately 2300°F (1260°C). Immediately upon leaving the furnace, the skelp is run through forming rollers that form the skelp into a cylindrical shape. The hot edges of the pipe are squeezed together forming a welded seam.

Welding is performed by rollers that butt the two edges of the pipe together causing them to fuse, without the addition of new metal. In this method of manufacturing, continuous pipe making is possible because new rolls of skelp can be continuously added to the ends of used rolls without stopping the pipemaking process.

Illustration #2 displays the forming and welding rollers used in manufacturing of continuous welded pipe.

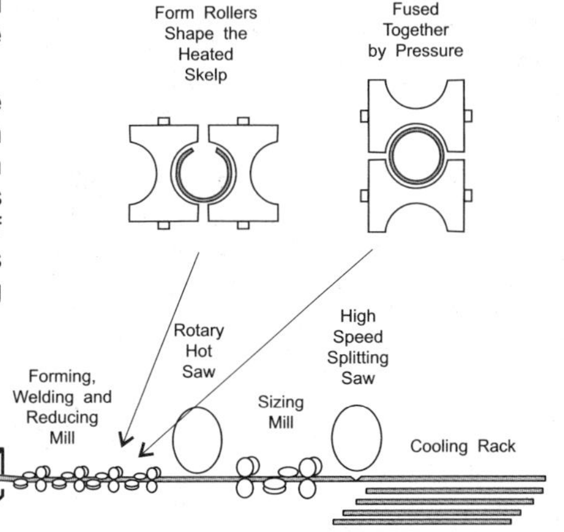

Illustration #2 - Manufacture of Continuous Butt-Welded Pipe

Seamless Pipe

There are two common methods of producing seamless pipe:

- Hot Rotary Piercing
- Extrusion Process Method

In the extrusion process, hot billets of metal are formed into seamless pipe by forcing the billets through an extrusion die and over a forming mandrel. This forms the metal into the desired seamless cylindrical shape.

Illustration #3 show an example of the extrusion process. The roughly formed pipe is then further reduced and sized in a reducing mill to the required dimensions.

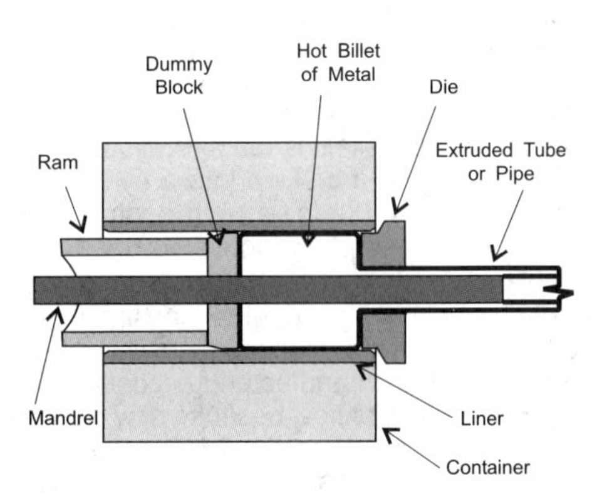

Illustration #3 - Extrusion of Seamless Pipe

Hot rotary piercing is the most common method of producing seamless pipe. In this process, heavy rollers rotate and advance hot billets of metal over a mandrel or piercing plug. The piercing plug produces the hole through the seamless cylindrical billets. Still further reduction, sizing, and shaping is carried out on the pipe by other mill rollers. Illustration #4A and #4B displays the piercing and mill processes on seamless pipe.

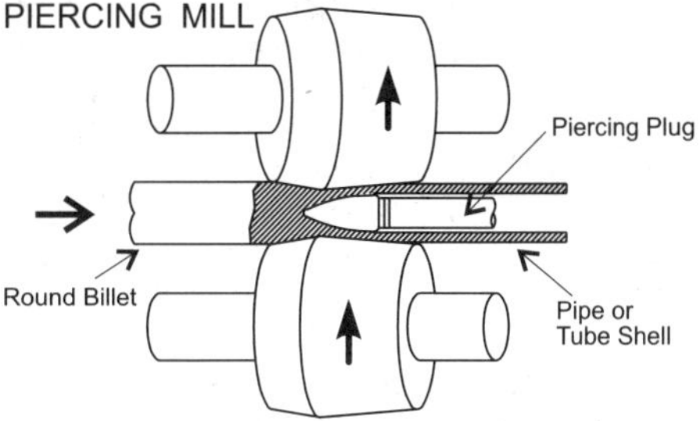

PIERCING MILL

Piercing Plug

Round Billet

Pipe or
Tube Shell

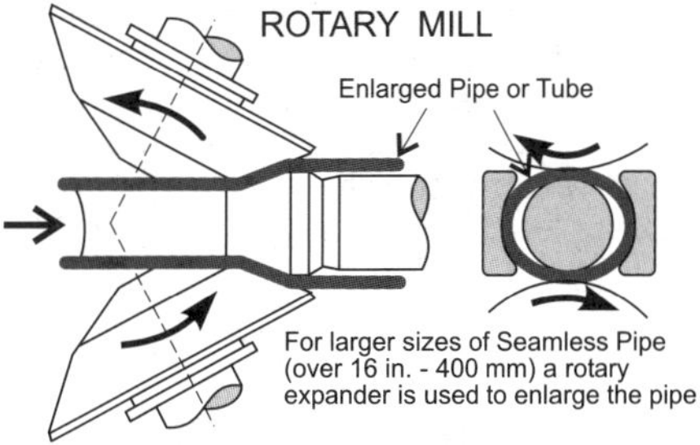

ROTARY MILL

Enlarged Pipe or Tube

Mandrell Rolling Mill
for Small Diameters

Plug Rolling Mill for
Medium Diameters

For larger sizes of Seamless Pipe
(over 16 in. - 400 mm) a rotary
expander is used to enlarge the pipe

Illustration #4A,B - Hot Rotary Piercing of Seamless Pipe

In addition to the previously mentioned methods of manufacturing, some larger sizes of steel pipe are electric fusion (arc) welded along the seams of the pipe. Usually the arc welding process is completed automatically using the submerged-arc welding process. Pipe that is electric fusion (arc) welded is manufactured by the straight seam method or by the spiral seam method.

The straight seam method of manufacturing shapes flat plate (skelp) into pipe and arc welds along the longitudinal seam. The spiral seam method of manufacturing takes skelp and spirally curls the plate into cylindrical shaped pipe. The welding then is done along the spiral seam of the pipe. See illustration #5 for examples of seamless and seam welded pipe.

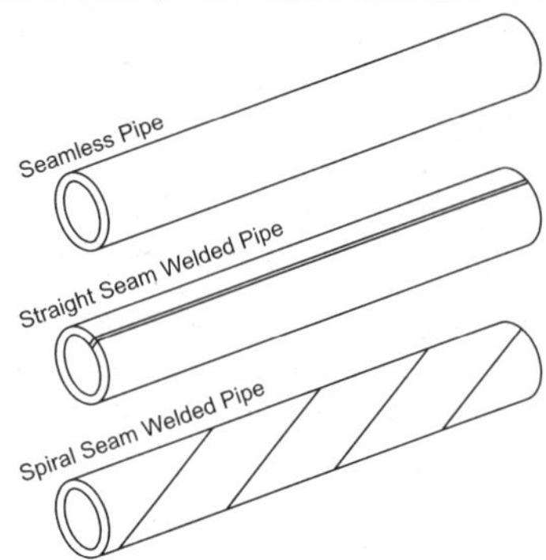

Illustration #5 - Seamless and Seam Welded Pipe

Pipe Standards and Specifications

Seamless and seam welded pipe and tube are manufactured to various pipe standards and specifications.

Some of the more common international and national standards and specifications for pipe and tube are listed in table #1.

National and International Pipe Standards & Specifications (Table #1)	
API-AMERICAN PETROLEUM INSTITUTE	
API 5L	Line Pipe
API 5LX	High Test Line Pipe
API 5LS	Spiral Weld Line Pipe
ASTM-AMERICAN SOCIETY FOR TESTING AND MATERIALS	
ASTM A53	Welded and Seamless Steel Pipe
ASTM A106	Seamless Carbon Steel Pipe for High Temperature Service
ASTM A120	Black and Hot Dipped Zinc Coating (Galvanized) Welded and Seamless Steel Pipe for Ordinary Use
ASTM A134	Electric-Fusion (ARC) Welded Steel Plate (Sizes 16 in. and Over)
ASTM A135	Electric-Resistance-Welded Steel Pipe
ASTM A139	Electric-Fusion (Arc)-Welded Steel Plate Pipe (Size 4 in. and Over)
ASTM A211	Spiral-Welded Steel or Iron Pipe
ASTM A312	Seamless and Welded Austenitic Stainless Steel Pipe

National and International Pipe Standards & Specifications (Table #1)	
ASTM A333	Seamless and Welded Steel Pipe for Low-Temperature Service
ASTM A335	Seamless Ferritic Alloy Steel Pipe for High-Temperature Service
ASTM A358	Electric-Fusion-Welded Austenitic Chromium-Nickel Alloy Steel Pipe for High-Temperature Service
ASTM A369	Carbon and Ferritic Alloy Steel Forged and Bored Pipe for HighTemperature Service
ASTM A376	Seamless Austenitic Steel Pipe for High-Temperature Central-Station Service
ASTM A381	Metal-Arc-Welded Steel Pipe for High-Pressure Systems
ASTM A405	Seamless Ferritic Alloy Steel Pipe Specially Heat Treated for HighTemperature Service
ASTM A409	Welded Large Diameter Austenitic Steel Pipe for Corrosive or HighTemperature Service
ASTM A430	Austenitic Steel Forged and Bored Pipe for High Temperature Service

Table #1 - National and International Pipe Standards & Specifications (part 1 of 5)

National and International Pipe Standards & Specifications (Table #1)	
ASTM A523	Plain End Seamless and Electric-Resistance-Welded Steel Pipe for High Pressure Pipe-Type Cable Circuits
ASTM A524	Seamless Carbon Steel Pipe for Process Piping
ASTM A530	General Requirements for Specialized Carbon and Alloy Steel Pipe
ASTM A671	Electric-Fusion-Welded Steel Pipe for Atmospheric and Lower Temperatures
ASTM A672	Electric-Fusion-Welded Steel Pipe for High-Pressure Service at Moderate Temperatures
ASTM A691	Carbon and Alloy Steel Pipe, Electric-Fusion-Welded for High-Pressure Service at High Temperatures
ASTM A714	High Strength Low-Alloy Welded and Seamless Steel Pipe
ASTM A731	Seamless and Welded Ferritic Stainless Steel Pipe

National and International Pipe Standards & Specifications (Table #1)	
ASTM A790	Seamless and Welded Ferritic/ Austenitic Stainless Steel Pipe
ASTM A795	Black and Hot-Dipped ZincCoated (Galvanized) Welded and Seamless Pipe for Fire Protection
AWWA-AMERICAN WATER WORKS ASSOCIATION	
AWWA C200	Steel Water Pipe, 6 Inches and Larger
AWWA C203	Standard for Coal-Tar Enamel Protective Coatings for Steel Water Pipe
AS-AUSTRALIAN STANDARD	
AS 1450	Steel Tubes for Mechanical Purposes
AS 1835	Seamless Steel Tubes for Pressure Purpose
AS 1836	Welded Steel Tubes for Pressure Purpose
BS-BRITISH STANDARDS	
BS 1387	Steel Tubes and Tubulars
BS 3059	Steel Boiler and Superheater Tubes

Table #1 - National and International Pipe Standards & Specifications (part 2 of 5)

National and International Pipe Standards & Specifications (Table #1)	
BS 3600	Dimensions and Masses per Unit Length of Welded and Seamless Steel Pipes and Tubes for Pressure Purposes: Metric Units
BS 3602	Steel Pipes and Tubes for Pressure Purposes Carbon Steel: High Temperature Duties
BS 3603	Steel Pipes and Tubes for Pressure Purposes Carbon and Alloy Steel: Low Temperature Duties
BS 3604	Steel Pipe and Tubes for Pressure Purposes Low and Medium-alloy Steel
BS 3605	Specification for Steel Pipe and Tubes for Pressure Purposes Austenitic Stainless Steel
CSA-CANADIAN STANDARD ASSOCIATION	
CSA Z245.1	Steel Line Pipe
CSA Z245.2	High Strength Steel Line Pipe 18 Inches and Larger in Diameter
CSA Z245.3	Low Strength Steel Line Pipe Less Than 18 Inches in Diameter

National and International Pipe Standards & Specifications (Table #1)	
CSA Z245.4	Low Strength Steel Line Pipe 18 Inches and Larger in Diameter
CSA Z245.5	High Strength Steel Line Pipe Less Than 18 Inches in Diameter
DIN-DEUTSCHE NORMEN (GERMANY)	
DIN 1615	Welded Circular Unalloyed Steel Tube not Subject to Special Requirements
DIN 1626	Welded Steel Pipes in Unalloyed and Low Alloy Steel for Supply Purposes
DIN 1629	Seamless Tubes in Unalloyed Steels
DIN 1630	High Performance Seamless Circular Unalloyed Steel Tubes
DIN 2440	Steel Tubes Medium-weight Suitable for Threading
DIN 2441	Steel Tubes Heavy-weight Suitable for Threading
DIN 17172	Steel Pipes for Pipelines for the Transport of Combustible Fluids and Gases
DIN 17175	Seamless Steel Tubes for Elevated Temperatures

Table #1 - National and International Pipe Standards & Specifications (part 3 of 5)

National and International Pipe Standards & Specifications (Table #1)	
DIN 17177	Electrical Resistance or Induction Welded Steel Tubes for Elevated Temperatures
GOST-USSR SPECIFICATIONS	
GOST 8731	Seamless Hot-Rolled Tubes Technical Requirements
GOST 8732	Seamless Hot-Rolled Steel Tubes Range
GOST 10704	Electric Welded Steel Tubes
GOST 10705	Specification for Delivery of Electric-Welded Steel Tubes of 8-530 mm in Diameter
ISO-INTERNATIONAL ORGANIZATION FOR STANDARDIZATION	
ISO 65	Steel Tubes Suitable for Screwing in accordance with International Standards IS07/1.
JIS-JAPANESE INDUSTRIAL STANDARDS	
JIS G3441	Alloy Steel Tubes for Machine Purpose
JIS G3442	Galvanized Steel Pipe for Water Service

National and International Pipe Standards & Specifications (Table #1)	
JIS G3443	Coating Steel Pipe for Water Service
JIS G3445	Carbon Steel Tubes for Machine and Structural Purpose
JIS G3446	Stainless Steel Tubes for Machine and Structural Purpose
JIS G3451	Deformed Pipe for Coating Steel Pipe for Water Service
JIS G3452	Carbon Steel Pipes for Ordinary Piping
JIS G3454	Carbon Steel Pipes for Pressure Services
JIS G3455	Carbon Steel Pipes for High Pressure Service
JIS G3457	Electric Arc-Welded Carbon Steel Pipe
JIS G3458	Alloy Steel Pipes
JIS G3459	Austenitic Stainless Steel Pipes
JIS G3460	Steel Pipes for Low-Temperature Service
JIS G3461	Carbon Steel Boiler and Heat-Exchanger Tubes

Table #1 - National and International Pipe Standards & Specifications (part 4 of 5)

National and International Pipe Standards & Specifications (Table #1)	
JIS G3462	Alloy Steel Boiler and Heat-Exchanger Tubes
JIS G3463	Stainless Steel Boiler and Heat-ExchangerTubes
JIS G3464	Steel Heat-Exchanger Tubes for Low Temperature Service
JIS G3473	Carbon Steel Tubes for Cylinder Barrels

National and International Pipe Standards & Specifications (Table #1)	
LR-LLOYD'S REGISTER OF SHIPPING	
	Pressure Pipes
	Boiler and Superheater Tubes
	Welded Pressure Pipes
	Ferritic, Steel Pressure Pipes for Low Temperature Service

Table #1 - National and International Pipe Standards & Specifications (part 5 of 5)

Note: Many of the standards and specifications correspond between countries and/or agencies.

Pipe Properties & Characteristics

Standards and/or specifications for pipe indicate pipe grade designations, intended use, testing practices, manufacturing methods, chemical properties and various physical characteristics of the pipe. Examples of chemical properties and physical characteristics of ASTM (A-53) and API (5L) specified pipe are shown in table #2.

Chemical Composition and Tensile Strength of Grade A and Grade B Steel Pipe							
Chemical Composition						**Tensile Properties**	
Designation	**Grade**	**C**	**Mn**	**P**	**S**	**min. yield strength psi (MPa)**	**min. tensile strength psi (MPa)**
ASTM A-53	A	0.25	0.95	0.05	0.06	30,000 (205)	48,000 (330)
	B	0.30	1.2	0.05	0.06	35,000 (240)	60,000 (415)
API 5L	A	0.22	0.9	0.04	0.05	30,000 (205)	48,000 (330)
	B	0.27	1.15	0.04	0.05	35,000 (240)	60,000 (415)
C = Carbon	Mn = Manganese		P = Phosphorous		S = Sulfur		

Table #2 - Grade A and B Pipe Chemical Composition

Even though steel pipe is available in various grades, the most common grades used are: Grade A and B, and the less common Grade C. It is important that the grade be specified for each application.

Even though Grade B may have a higher tensile strength than Grade A, Grade A pipe may be preferred in some applications. (See table #2 for tensile strengths). Grade A is preferred where close coiling or cold bending of pipe is required because of it's lower carbon content which makes it more ductile and less brittle.

Weights and Schedule Numbers

Steel pipe is produced in three weights or general wall thickness classifications:

- Standard (Std.)
- Extra Strong (XS) or Extra Heavy (XH).
- Double Extra Strong (XXS) or Double Extra Heavy (XXH).

Note: Designations of Strong and Heavy are interchangeable in the weight classifications.

Light wall, light weight or light gage pipe, as it may be referred to, is another weight classification sometimes given to steel pipe. This pipe classification is used extensively in many sprinkler installations and other applications where a thinner wall pipe may be preferred. The light wall pipe designation corresponds to schedule number 10 for steel pipe in most sizes.

Note: The weight classification denotes the wall thickness of the pipe.

For any pipe size, the outside diameter is constant and the inside diameter varies with the wall thickness. Pipe dies and fittings therefore remain the same for specific sizes of pipe, no matter what the weight.

Because of the variation in inside diameter, pipe sizes from 1/8 inch (6 mm) to 12 inches (300 mm) are designated by nominal inside diameter (ID), not by the actual inside diameter. Nominal sizes are referred to as Nominal Pipe Size (NPS) and less commonly, Iron Pipe Size (IPS). Pipe sizes over 12 inch (300 mm) are classified by actual outside diameter (OD).

Schedule Numbers

To further broaden the range of wall thicknesses for specific applications and various pressures, steel pipe is manufactured in assorted schedule numbers. These schedule numbers range from 10 through to 160 and are commercially available in schedules: 10, 20, 30, 40, 60, 80, 100, 120, 140, and 160.

Schedule Numbers & Pipe Weights

Some schedule numbers and weight classifications of steel pipe have the same wall thickness. Wall thicknesses for standard weight pipe and schedule 40 are the same for sizes 1/8 inch (6 mm) through to 10 inches (250 mm). All standard weight pipe sizes over 10 inches (250 mm) have a constant wall thickness of 3/8 inches (9.53 mm), whereas, schedule 40 pipe has a wall thickness that varies depending upon the particular size over 10 inches (250 mm).

Pipe sizes up to 8 inch (200 mm) in extra heavy and schedule 80 have identical dimensions. Over 8 inches (200 mm), extra heavy pipe has a constant wall thickness of 1/2 inch (12.7 mm), whereas the wall thickness for schedule 80 pipe varies depending on size.

There is no exact corresponding schedule number for double extra heavy pipe.

Generally, double extra heavy pipe up to 6 inches (150 mm) has a thicker wall than schedule 160 pipe. However, in sizes over 6 inches (150 mm) NPS, schedule number 160 becomes the thicker walled pipe. See illustration #6 for the relationship between wall thickness, schedule number and pipe weights.

General Piping Design

An approximation of the required schedule number, wall thickness, and maximum internal pressure for general design purposes can be determined by using the formulas:

1. $\text{Schedule Number} = 1000 x \dfrac{P}{S}$

2. $\text{Wall Thickness} = \dfrac{PxD}{2xS} + C$

3. $\text{Pressure} = 2xS \dfrac{(T - C)}{D}$

Nominal Pipe Size (NPS) 1 inch (25mm)

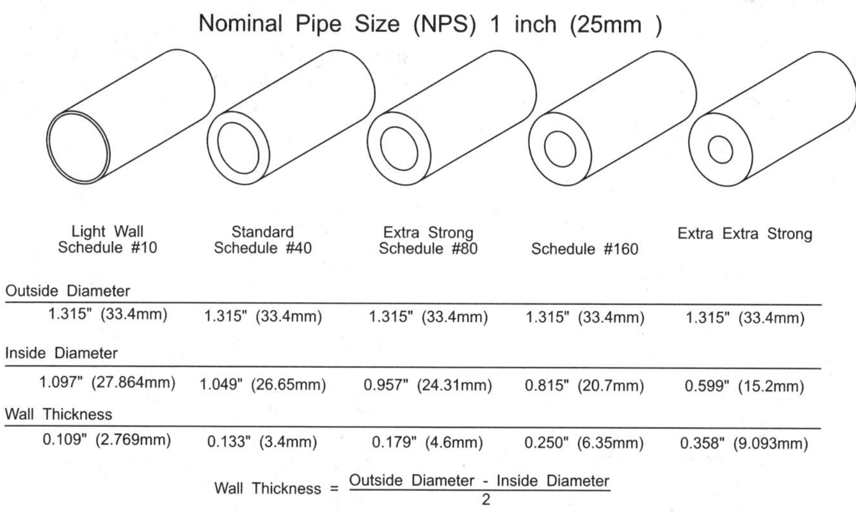

	Light Wall Schedule #10	Standard Schedule #40	Extra Strong Schedule #80	Schedule #160	Extra Extra Strong
Outside Diameter	1.315" (33.4mm)	1.315" (33.4mm)	1.315" (33.4mm)	1.315" (33.4mm)	1.315" (33.4mm)
Inside Diameter	1.097" (27.864mm)	1.049" (26.65mm)	0.957" (24.31mm)	0.815" (20.7mm)	0.599" (15.2mm)
Wall Thickness	0.109" (2.769mm)	0.133" (3.4mm)	0.179" (4.6mm)	0.250" (6.35mm)	0.358" (9.093mm)

$$\text{Wall Thickness} = \frac{\text{Outside Diameter} - \text{Inside Diameter}}{2}$$

Illustration #6 - Wall Thickness and Schedule Numbers

General Piping Design

The explanation below relates to the formulas on the previous page.

P = internal pressure in psi

S = maximum allowable stress value in psi

D = outside diameter of pipe

C = allowance for threading and grooving

T = wall thickness

General Allowance For Threading

Pipe Size (NPS)	Allowance
1/2"(15 mm) to 3/4"(20 mm)	.0571"(1.45 mm)
1" (25 mm) to 2"(50 mm)	.0696"(1.77 mm)
2 1/2"(65 mm) and larger	.1000"(2.54 mm)

Stress values may be found in manufacturers' publications and in various design specifications. See table #3 for S-values. It should be noted that the formula on page 14 is an approximate calculation only.

Other criteria such as corrosion, temperature, specific applications and various code regulations must also be taken into account before an exact figure can be determined for a particular job.

Pipe Sizes

Carbon steel pipe is commercially available in nominal pipe sizes ranging from 1/8 inch (6 mm) through to 42 inches (1050 mm). The following nominal pipe sizes are available within this range. Outside diameter (OD) designations are also given because of varying trade practices in specifying some line pipe by OD measurements. See table #4 for actual and nominal pipe sizes.

Note:

1. ***Metric nominal pipe sizes are based on International Organization for Standardization (ISO).***
2. ***For specific sizes, dimensions, weights, and schedule numbers of steel pipe refer to tables #5 (imperial) and #6 (metric).***

Maximum Allowable Stress (S-values) for Carbon Steel Pipe for ASTM and API Specifications, Grades and Manufacturing Methods Manufacturing Methods									
		Maximum Stress in 1000 PSI (MPa) for Various Pipe Temperatures							
Specification	Grade	- 20 to 100°F (-29 to 38°C)	200°F (93-C)	300°F (149°C)	400°F (204°C)	-20 to 650°F (-29 to 343°C)	700°F (371°C)	750°F (399°C)	800°F (427°C)
Seamless									
A53	A					12.0(82)	11.7(80)	10.7(73)	9.0(62)
	B					15.0(103)	14.4(99)	13.0(89)	10.8(74)
A106	A					12.0(82)	11.7(80)	10.7(73)	9.0(62)
	B					15.0(103)	14.4(99)	13.0(89)	10.8(74)
API 5L	A					12.0(82)	11.7(80)	10.7(73)	9.0(62)
	B					15.0(103)	14.4(99)	13.0(89)	10.8(74)
Continuous Welded									
A53						6.8(46)	6.5(44)	-	-
A120		6.4 (44)	6.3(43)	6.1(42)	5.8(39)				
Electric Resistance Welded									
A53	A					10.2(70)	9.9(68)	9.1(62)	7.7(52)
	B					12.8(88)	12.2(84)	11.0(75)	9.2(63)
A120	-	9.1(63)	9.0(62)	8.6(59)	8.3(57)				
API 5L	A					10.2(70)	9.9(68)	9.1(62)	7.7(52)
	B					12.8(88)	12.2(84)	11.0(75)	9.2(63)
Electric Fusion - Arc Welded									
API 5L	A					10.8(74)	10.5(72)	9.6(66)	8.1(56)
	B					13.5(93)	13.0(89)	11.7(80)	9.7(67)
A139	A					9.6(66)	7.4(51)	8.6(59)	-
	B					12.0(82)	11.5(79)	10.4(71)	-
Note: The S-Values include the longitudinal joint factor, which is the rated efficiency of the longitudinal welded seam for the various manufacturing methods.									

Table #3 - Stress Values for Carbon Steel Pipe

Pipe Sizes (Actual and Nominal)

Actual and Nominal Pipe Sizes			
Actual OD		Nominal Sizes	
inches	(mm)	inches	(mm)
0.405	10.3	1/8	6
0.540	13.7	1/4	8
0.675	17.1	3/8	10
0.840	21.3	1/2	15
1.050	26.7	3/4	20
1.315	33.4	1	25
1.660	42.2	1 1/4	32
1.900	48.3	1 1/2	40
2.375	60.3	2	50
2.875	73.0	2 1/2	65
3.500	88.9	3	80
4.000	101.6	3 1/2	90
4.500	114.3	4	100
5.563	141.3	5	125
6.625	168.3	6	150
8.625	219.1	8	200
10.750	273.1	10	250

Actual and Nominal Pipe Sizes			
Actual OD		Nominal Sizes	
inches	(mm)	inches	(mm)
12.750	323.9	12	300
14.000	355.6	14	350
16.000	406.4	16	400
18.000	457.0	18	450
20.000	508.0	20	500
22.000	559.0	22	550
24.000	610.0	24	600
26.000	660.0	26	650
28.000	711.0	28	700
30.000	762.0	30	750
32.000	813.0	32	800
34.000	864.0	34	850
36.000	914.0	36	900
38.000	965.0	38	950
40.000	1016.0	40	1000
42.000	1067.0	42	1050

Table #4 - Actual vs. Nominal Pipe Sizes

Steel Pipe Dimensions and Weights (Imperial Units)							
Weight Class	Sch. #	Inside Diameter	Wall Thick-ness	Outside Area	Inside Area	Plain End Weight	Water in Pipe
		inches	inches	ft^2/ft	ft^2/ft	lb/ft	lb/ft
		▼ NPS - 1/8		OD - 0.405 in. ▼			
STD	40	0.269	0.068	0.106	0.0705	0.244	0.025
XS	80	0.215	0.095	0.106	0.0563	0.314	0.016
		▼ NPS - 1/4		OD - 0.540 in. ▼			
STD	40	0.364	0.088	0.141	0.0955	0.424	0.045
XS	80	0.302	0.119	0.141	0.0794	0.535	0.031
		▼ NPS - 3/8		OD - 0.675 in. ▼			
STD	40	0.493	0.091	0.177	0.1295	0.567	0.083
XS	80	0.423	0.126	0.177	0.1106	0.738	0.061
		▼ NPS - 1/2		OD - 0.840 in. ▼			
STD	40	0.622	0.109	0.220	0.1637	0.850	0.132
XS	80	0.546	0.147	0.220	0.1433	1.087	0.101
-	160	0.464	0.188	0.220	0.1215	1.311	0.073
XXS	-	0.252	0.294	0.220	0.0660	1.714	0.022
		▼ NPS - 3/4		OD - 1.050 in. ▼			
STD	40	0.824	0.113	0.275	0.2168	1.130	0.230
XS	80	0.742	0.154	0.275	0.1948	1.473	0.187
-	160	0.612	0.219	0.275	0.1602	1.944	0.127
XXS	-	0.434	0.308	0.275	0.1137	2.440	0.063
		▼ NPS - 1		OD - 1.315 in. ▼			
STD	40	1.049	0.133	0.344	0.274	1.678	0.378
XS	80	0.957	0.179	0.344	0.252	2.171	0.311

Steel Pipe Dimensions and Weights (Imperial Units)							
Weight Class	Sch. #	Inside Diameter	Wall Thick-ness	Outside Area	Inside Area	Plain End Weight	Water in Pipe
		inches	inches	ft^2/ft	ft^2/ft	lb/ft	lb/ft
-	160	0.815	0.250	0.344	0.213	2.840	0.226
XXS	-	0.599	0.358	0.344	0.157	3.659	0.122
		▼ NPS - 1 1/4		OD - 1.660 in. ▼			
STD	40	1.380	0.140	0.434	0.362	2.272	0.647
XS	80	1.278	0.191	0.434	0.336	2.996	0.555
-	160	1.160	0.250	0.434	0.303	3.764	0.457
XXS	-	0.896	0.382	0.434	0.233	5.214	0.273
		▼ NPS - 1 1/2		OD - 1.900 in. ▼			
STD	40	1.610	0.145	0.497	0.421	2.717	0.882
XS	80	1.500	0.200	0.497	0.393	3.631	0.765
-	160	1.338	0.281	0.497	0.350	4.858	0.609
XXS	-	1.100	0.400	0.497	0.290	6.408	0.412
		▼ NPS - 2		OD - 2.375 in. ▼			
-	-	2.209	0.083	0.622	0.579	2.031	1.662
-	-	2.157	0.109	0.622	0.565	2.64	1.58
-	-	2.125	0.125	0.622	0.557	3.00	1.54
-	-	2.093	0.141	0.622	0.548	3.36	1.49
STD	40	2.067	0.154	0.622	0.540	3.65	1.45
		2.031	0.172	0.622	0.532	4.05	1.38
		1.999	0.188	0.622	0.524	4.39	1.36
XS	80	1.939	0.218	0.622	0.507	5.02	1.28
	-	1.875	0.250	0.622	0.492	5.67	1.20

Table #5 (part 1 of 14) - Imperial Steel Pipe Dimensions based on ANSI/ASME B-36.10M - 1985 Standard

Steel Pipe Dimensions and Weights (Imperial Units)

Weight Class	Sch. #	Inside Diameter	Wall Thick-ness	Outside Area	Inside Area	Plain End Weight	Water in Pipe
		inches	inches	ft²/ft	ft²/ft	lb/ft	lb/ft
	-	1.813	0.281	0.622	0.475	6.28	1.12
	160	1.687	0.344	0.622	0.442	7.46	0.97
XXS	-	1.503	0.436	0.622	0.393	9.03	0.77
▼ NPS - $2\frac{1}{2}$ OD - 2.875 in. ▼							
-	-	2.709	0.083	0.753	0.709	2.47	2.50
-	-	2.657	0.109	0.753	0.696	3.22	2.40
-	-	2.625	0.125	0.753	0.687	3.67	2.35
-	-	2.593	0.141	0.753	0.679	4.12	2.29
-	-	2.563	0.156	0.753	0.671	4.53	2.24
-	-	2.531	0.172	0.753	0.663	4.97	2.18
-	-	2.499	0.188	0.753	0.654	5.40	2.13
STD	40	2.469	0.203	0.753	0.646	5.79	2.07
		2.443	0.216	0.753	0.640	6.13	2.03
		2.375	0.250	0.753	0.622	7.01	1.92
XS	80	2.323	0.276	0.753	0.610	7.66	1.83
-	160	2.125	0.375	0.753	0.556	10.01	1.54
XXS	-	1.771	0.552	0.753	0.463	13.69	1.07
▼ NPS - 3 OD - 3.500 in. ▼							
-	-	3.334	0.083	0.916	0.873	3.03	3.78
-	-	3.282	0.109	0.916	0.860	3.95	3.67
-	-	3.250	0.125	0.916	0.851	4.51	3.60
-	-	3.218	0.141	0.916	0.843	5.06	3.53

Steel Pipe Dimensions and Weights (Imperial Units)

Weight Class	Sch. #	Inside Diameter	Wall Thick-ness	Outside Area	Inside Area	Plain End Weight	Water in Pipe
		inches	inches	ft²/ft	ft²/ft	lb/ft	lb/ft
-	-	3.188	0.156	0.916	0.835	5.57	3.46
-	-	3.156	0.172	0.916	0.827	6.11	3.39
-	-	3.124	0.188	0.916	0.818	6.65	3.32
STD	40	3.068	0.216	0.916	0.802	7.58	3.20
		3.000	0.250	0.916	0.785	8.68	3.06
		2.938	0.281	0.916	0.769	9.66	2.94
XS	80	2.900	0.300	0.916	0.761	10.25	2.86
-	160	2.624	0.438	0.916	0.681	14.32	2.34
XXS	-	2.300	0.600	0.916	0.601	18.58	1.80
▼ NPS - $3\frac{1}{2}$ OD - 4.000 in. ▼							
-	-	3.834	0.083	1.047	1.004	3.47	5.00
-	-	3.782	0.109	1.047	0.991	4.53	4.87
-	-	3.750	0.125	1.047	0.982	5.17	4.79
-	-	3.718	0.141	1.047	0.974	5.81	4.71
-	-	3.688	0.156	1.047	0.966	6.40	4.63
-	-	3.656	0.172	1.047	0.958	7.03	4.55
-	-	3.624	0.188	1.047	0.950	7.65	4.48
STD	40	3.548	0.226	1.047	0.929	9.11	4.28
		3.500	0.250	1.047	0.916	10.01	4.17
		3.438	0.281	1.047	0.900	11.16	4.02
XS	80	3.364	0.318	1.047	0.880	12.50	3.85

Table #5 (part 2 of 14) - Imperial Steel Pipe Dimensions based on ANSI/ASME B-36.10M - 1985 Standard

Steel Pipe Dimensions and Weights (Imperial Units)

Weight Class	Sch. #	Inside Diameter	Wall Thickness	Outside Area	Inside Area	Plain End Weight	Water in Pipe
		inches	inches	ft^2/ft	ft^2/ft	lb/ft	lb/ft
		▼ NPS - 4	OD - 4.500 in. ▼				
		4.334	0.083	1.178	1.135	3.92	6.39
		4.282	0.109	1.178	1.121	5.11	6.25
		4.250	0.125	1.178	1.113	5.84	6.15
		4.218	0.141	1.178	1.105	6.56	6.06
		4.188	0.156	1.178	1.096	7.24	5.97
		4.156	0.172	1.178	1.088	7.95	5.88
		4.124	0.188	1.178	1.082	8.66	5.80
		4.094	0.203	1.178	1.072	9.32	5.71
		4.062	0.219	1.178	1.063	10.01	5.62
STD	40	4.026	0.237	1.178	1.055	10.79	5.51
		4.000	0.250	1.178	1.049	11.35	5.45
		3.938	0.281	1.178	1.031	12.66	5.27
		3.876	0.312	1.178	1.013	13.96	5.12
XS	80	3.826	0.337	1.178	1.002	14.98	4.98
-	120	3.624	0.438	1.178	0.949	19.00	4.47
-	160	3.438	0.531	1.178	0.900	22.51	4.02
XXS	-	3.152	0.674	1.178	0.826	27.54	3.38
		▼ NPS - 5	OD - 5.563 in. ▼				
		5.397	0.083	1.456	1.413	4.86	9.92
		5.313	0.125	1.456	1.391	7.26	9.62
		5.251	0.156	1.456	1.375	9.01	9.39

Steel Pipe Dimensions and Weights (Imperial Units)

Weight Class	Sch. #	Inside Diameter	Wall Thickness	Outside Area	Inside Area	Plain End Weight	Water in Pipe
		inches	inches	ft^2/ft	ft^2/ft	lb/ft	lb/ft
		5.187	0.188	1.456	1.358	10.79	9.16
		5.125	0.219	1.456	1.342	12.50	8.94
		5.047	0.258	1.456	1.321	14.62	8.66
STD	40	5.001	0.281	1.456	1.309	15.85	6.52
		4.939	0.312	1.456	1.293	17.50	8.31
		4.875	0.344	1.456	1.276	19.17	8.09
XS	80	4.813	0.375	1.456	1.260	20.78	7.87
-	120	4.563	0.500	1.456	1.195	27.04	7.08
-	160	4.313	0.625	1.456	1.129	32.96	6.32
XXS	-	4.063	0.750	1.456	1.064	38.55	5.62
		▼ NPS - 6	OD - 6.625 in. ▼				
-	-	6.459	0.083	1.73	1.690	5.80	14.20
-	-	6.407	0.109	1.73	1.680	7.59	14.97
-	-	6.375	0.125	1.73	1.670	8.68	13.83
-	-	6.343	0.141	1.73	1.660	9.76	13.69
-	-	6.313	0.156	1.73	1.650	10.78	13.57
-	-	6.281	0.172	1.73	1.645	11.85	13.43
-	-	6.249	0.188	1.73	1.637	12.92	13.29
-	-	6.219	0.203	1.73	1.630	13.92	13.16
-	-	6.187	0.219	1.73	1.620	14.98	13.03
-	-	6.125	0.250	1.73	1.610	17.02	12.77
STD	40	6.065	0.280	1.73	1.590	18.97	12.52

Table #5 (part 3 of 14) - Imperial Steel Pipe Dimensions based on ANSI/ASME B-36.10M - 1985 Standard

Steel Pipe Dimensions and Weights (Imperial Units)

Weight Class	Sch. #	Inside Diameter	Wall Thick-ness	Outside Area	Inside Area	Plain End Weight	Water in Pipe
		inches	inches	ft^2/ft	ft^2/ft	lb/ft	lb/ft
		6.001	0.312	1.73	1.570	21.04	12.26
		5.937	0.344	1.73	1.550	23.08	12.00
		5.875	0.375	1.73	1.540	25.03	11.74
XS	80	5.761	0.432	1.73	1.510	28.57	11.30
	-	5.625	0.500	1.73	1.480	32.71	10.77
	120	5.501	0.562	1.73	1.470	36.39	10.30
	-	5.375	0.625	1.73	1.410	40.05	9.83
	160	5.187	0.719	1.73	1.360	45.35	9.16
	-	5.125	0.750	1.73	1.340	47.06	8.94
XXS	-	4.897	0.864	1.73	1.283	53.16	8.16
-	-	4.875	0.875	1.73	1.277	53.73	8.09
▼ NPS - 8		OD - 8.625 in. ▼					
-	-	8.375	0.125	2.26	2.19	11.35	23.87
-	-	8.313	0.156	2.26	2.18	14.11	23.52
-	-	8.249	0.188	2.26	2.16	16.94	23.16
-	-	8.219	0.203	2.26	2.15	18.26	22.99
-	-	8.187	0.219	2.26	2.15	19.66	22.81
-	20	8.125	0.250	2.26	2.13	22.36	22.47
-	30	8.071	0.277	2.26	2.12	24.70	22.77
-	-	8.001	0.312	2.26	2.10	27.70	21.79
STD	40	7.981	0.322	2.26	2.09	28.55	21.68
		7.937	0.344	2.26	2.08	30.42	21.44

Steel Pipe Dimensions and Weights (Imperial Units)

Weight Class	Sch. #	Inside Diameter	Wall Thick-ness	Outside Area	Inside Area	Plain End Weight	Water in Pipe
		inches	inches	ft^2/ft	ft^2/ft	lb/ft	lb/ft
		7.875	0.375	2.26	2.06	33.04	21.11
	60	7.813	0.406	2.26	2.04	35.64	20.78
	-	7.749	0.438	2.26	2.03	38.30	20.44
XS	80	7.625	0.500	2.26	2.01	43.39	19.79
	-	7.501	0.562	2.26	1.96	48.40	19.15
	100	7.437	0.594	2.26	1.95	50.95	18.83
	-	7.375	0.625	2.26	1.93	53.40	18.51
	120	7.187	0.719	2.26	1.88	60.71	17.58
	-	7.125	0.750	2.26	1.87	63.08	17.28
	140	7.001	0.812	2.26	1.83	67.76	16.68
XXS	-	6.875	0.875	2.26	1.80	72.42	16.09
	160	6.813	0.906	2.26	1.78	74.69	15.80
	-	6.625	1.000	2.26	1.74	81.44	14.94
▼ NPS - 10		OD - 10.750 in. ▼					
	-	10.438	0.156	2.814	2.732	17.65	37.08
	-	10.374	0.188	2.814	2.716	21.21	36.63
	-	10.344	0.203	2.814	2.708	22.87	36.42
	-	10.312	0.219	2.814	2.699	24.63	35.19
	20	10.250	0.250	2.814	2.683	28.04	35.76
	-	10.192	0.279	2.814	2.668	31.20	35.36
	30	10.136	0.307	2.814	2.653	34.24	34.97
	-	10.062	0.344	2.814	2.634	38.23	34.46

Table #5 (part 4 of 14) - Imperial Steel Pipe Dimensions based on ANSI/ASME B-36.10M - 1985 Standard

Steel Pipe Dimensions and Weights (Imperial Units)							
Weight Class	Sch. #	Inside Diameter	Wall Thick-ness	Outside Area	Inside Area	Plain End Weight	Water in Pipe
		inches	inches	ft²/ft	ft²/ft	lb/ft	lb/ft
STD	40	10.020	0.365	2.814	2.623	40.48	34.17
-	-	9.874	0.438	2.814	2.585	48.24	33.18
XS	60	9.750	0.500	2.814	2.552	54.74	32.36
	-	9.626	0.562	2.814	2.520	61.15	31.54
	80	9.562	0.594	2.814	2.503	64.43	31.12
	-	9.502	0.625	2.814	2.487	67.58	30.73
	100	9.312	0.719	2.814	2.438	77.03	29.51
	-	9.126	0.812	2.814	2.389	86.18	28.35
	120	9.062	0.844	2.814	2.372	89.29	27.95
	-	9.000	0.875	2.814	2.356	92.28	27.57
	-	8.874	0.938	2.814	2.323	98.30	26.80
XXS	140	8.750	1.000	2.814	2.291	104.13	26.06
	160	8.500	1.125	2.814	2.225	115.64	24.60
	-	8.250	1.250	2.814	2.160	126.83	23.17
▼ NPS - 12 OD - 12.750 in.▼							
	-	12.406	0.172	3.338	3.248	23.11	52.39
	-	12.374	0.188	3.338	3.239	25.22	52.12
	-	12.344	0.203	3.338	3.231	27.20	51.86
	-	12.312	0.219	3.338	3.223	29.31	51.60
	20	12.250	0.250	3.338	3.207	33.38	51.10
	-	12.188	0.281	3.338	3.191	37.42	50.56
	-	12.126	0.312	3.338	3.174	41.45	50.05

Steel Pipe Dimensions and Weights (Imperial Units)							
Weight Class	Sch. #	Inside Diameter	Wall Thick-ness	Outside Area	Inside Area	Plain End Weight	Water in Pipe
		inches	inches	ft²/ft	ft²/ft	lb/ft	lb/ft
	30	12.090	0.330	3.338	3.165	43.77	49.75
	-	12.062	0.344	3.338	3.158	45.58	49.52
STD	-	12.000	0.375	3.338	3.141	49.56	49.01
	40	11.938	0.406	3.338	3.125	53.52	48.51
	-	11.874	0.438	3.338	3.108	57.59	48.00
		11.750	0.500	3.338	3.076	65.42	47.00
	60	11.626	0.562	3.338	3.043	73.15	46.00
	-	11.500	0.625	3.338	3.010	80.93	45.00
	80	11.374	0.688	3.338	2.977	88.63	44.03
XS	-	11.250	0.750	3.338	2.945	96.12	43.08
	-	11.126	0.812	3.338	2.913	103.53	42.13
	100	11.062	0.844	3.338	2.896	107.32	41.65
	-	11.000	0.875	3.338	2.880	110.97	41.18
	-	10.874	0.938	3.338	2.847	118.33	40.25
	120	10.750	1.000	3.338	2.814	125.49	39.33
	-	10.626	1.062	3.338	2.782	132.57	38.43
	140	10.500	1.125	3.338	2.749	139.67	37.53
	-	10.250	1.250	3.338	2.683	153.53	35.76
XXS	160	10.126	1.312	3.338	2.651	160.27	34.90
▼ NPS - 14 OD - 14.0 in. ▼							
	-	13.624	0.188	3.67	3.57	27.73	63.18
	-	13.594	0.203	3.67	3.56	29.91	62.90

Table #5 (part 5 of 14) - Imperial Steel Pipe Dimensions based on ANSI/ASME B-36.10M - 1985 Standard

Steel Pipe Dimensions and Weights (Imperial Units)

Weight Class	Sch. #	Inside Diameter	Wall Thick-ness	Outside Area	Inside Area	Plain End Weight	Water in Pipe
		inches	inches	ft^2/ft	ft^2/ft	lb/ft	lb/ft
	-	13.580	0.210	3.67	3.56	30.93	62.77
	-	13.562	0.219	3.67	3.55	32.23	62.60
	10	13.500	0.250	3.67	3.53	36.71	62.03
	-	13.438	0.281	3.67	3.52	41.17	61.46
	20	13.376	0.312	3.67	3.50	45.61	60.90
	-	13.312	0.344	3.67	3.49	50.17	60.32
STD	30	13.250	0.375	3.67	3.47	54.57	59.76
	-	13.188	0.406	3.67	3.45	58.94	59.20
	40	13.124	0.438	3.67	3.44	63.44	58.62
	-	13.062	0.469	3.67	3.42	67.78	58.07
XS	-	13.000	0.500	3.67	3.40	72.09	57.52
-	-	12.876	0.562	3.67	3.37	80.66	56.43
-	60	12.812	0.594	3.67	3.35	85.05	55.87
-	-	12.750	0.625	3.67	3.34	89.28	55.33
-	-	12.624	0.688	3.67	3.30	97.81	54.24
-	80	12.500	0.750	3.67	3.27	106.13	53.18
	-	12.376	0.812	3.67	3.24	114.37	52.13
	-	12.250	0.875	3.67	3.21	122.65	51.08
	100	12.124	0.938	3.67	3.17	130.85	50.03
	-	12.000	1.000	3.67	3.14	138.84	49.01
	-	11.876	1.062	3.67	3.11	146.74	48.01
	120	11.812	1.094	3.67	3.10	150.79	47.49

Steel Pipe Dimensions and Weights (Imperial Units)

Weight Class	Sch. #	Inside Diameter	Wall Thick-ness	Outside Area	Inside Area	Plain End Weight	Water in Pipe
		inches	inches	ft^2/ft	ft^2/ft	lb/ft	lb/ft
	-	11.750	1.125	3.67	3.08	154.69	47.00
-	140	11.500	1.250	3.67	3.01	170.21	45.01
-	160	11.188	1.406	3.67	2.93	189.11	42.60
		10.000	2.000	3.67	2.62	256.32	34.04
		9.750	2.125	3.67	2.55	269.50	32.36
		9.600	2.200	3.67	2.51	277.25	31.37
		9.000	2.500	3.67	2.36	307.05	27.57
		▼ NPS - 16		OD - 16.0 in. ▼			
		15.624	0.188	4.19	4.09	31.75	83.09
		15.594	0.203	4.19	4.08	34.25	82.77
		15.562	0.219	4.19	4.07	36.91	82.43
	10	15.500	0.250	4.19	4.06	42.05	81.77
	-	15.438	0.281	4.19	4.04	47.17	81.12
	20	15.376	0.312	4.19	4.03	52.27	80.47
	-	15.312	0.344	4.19	4.01	57.52	79.80
STD	30	15.250	0.375	4.19	3.99	62.58	79.16
		15.188	0.406	4.19	3.98	67.62	78.51
		15.124	0.438	4.19	3.96	72.80	77.85
		15.062	0.469	4.19	3.94	77.79	77.22
XS	40	15.000	0.500	4.19	3.93	82.77	76.58
	-	14.876	0.562	4.19	3.89	92.66	75.32
	-	14.750	0.625	4.19	3.86	102.63	74.05

Table #5 (part 6 of 14) - Imperial Steel Pipe Dimensions based on ANSI/ASME B-36.10M - 1985 Standard

Steel Pipe Dimensions and Weights (Imperial Units)							
Weight Class	Sch. #	Inside Diameter	Wall Thick-ness	Outside Area	Inside Area	Plain End Weight	Water in Pipe
		inches	inches	ft²/ft	ft²/ft	lb/ft	lb/ft
	60	14.688	0.656	4.19	3.85	107.50	73.43
	-	14.624	0.688	4.19	3.83	112.51	72.79
	-	14.500	0.750	4.19	3.80	122.15	71.56
	-	14.376	0.812	4.19	3.76	131.71	70.34
	80	14.312	0.844	4.19	3.75	136.61	69.72
	-	14.250	0.875	4.19	3.73	141.34	69.12
	-	14.124	0.938	4.19	3.70	150.89	67.90
	-	14.000	1.000	4.19	3.67	160.20	66.71
	100	13.938	1.031	4.19	3.65	164.82	66.12
	-	13.876	1.062	4.19	3.63	169.43	65.54
	-	13.750	1.125	4.19	3.60	178.72	64.35
	-	13.624	1.188	4.19	3.57	187.93	63.18
	120	13.562	1.219	4.19	3.55	192.43	62.60
	-	13.500	1.250	4.19	3.53	196.91	62.03
-	140	13.124	1.438	4.19	3.44	223.64	58.62
-	160	12.812	1.594	4.19	3.35	245.25	55.87
▼ NPS - 18		OD - 18.0 in. ▼					
	-	17.624	0.188	4.71	4.61	35.76	105.72
	-	17.562	0.219	4.71	4.60	41.49	104.98
	-	17.500	0.250	4.71	4.58	47.39	104.24
	-	17.438	0.281	4.71	4.57	53.18	103.50
	20	17.376	0.312	4.71	4.55	58.94	102.77

Steel Pipe Dimensions and Weights (Imperial Units)							
Weight Class	Sch. #	Inside Diameter	Wall Thick-ness	Outside Area	Inside Area	Plain End Weight	Water in Pipe
		inches	inches	ft²/ft	ft²/ft	lb/ft	lb/ft
	-	17.312	0.344	4.71	4.53	64.87	102.01
STD	-	17.250	0.375	4.71	4.52	70.59	101.28
-	-	17.188	0.406	4.71	4.50	76.29	100.55
-	30	17.124	0.438	4.71	4.48	82.15	99.81
-	-	17.062	0.469	4.71	4.47	87.81	99.09
XS	-	17.000	0.500	4.71	4.45	93.45	98.37
	40	16.876	0.562	4.71	4.42	104.67	96.94
	-	16.750	0.625	4.71	4.39	115.98	95.49
	-	16.624	0.688	4.71	4.35	127.21	94.06
	60	16.500	0.750	4.71	4.32	138.17	92.67
	-	16.376	0.812	4.71	4.29	149.06	91.28
	-	16.250	0.875	4.71	4.25	160.03	89.88
	80	16.124	0.938	4.71	4.22	170.92	88.49
	-	16.000	1.000	4.71	4.19	181.56	87.13
	-	15.876	1.062	4.71	4.16	192.11	85.79
	-	15.750	1.125	4.71	4.12	202.75	84.43
	100	15.688	1.156	4.71	4.11	207.96	83.77
	-	15.624	1.188	4.71	4.09	213.31	83.09
	-	15.500	1.250	4.71	4.06	223.61	81.77
	120	15.250	1.375	4.71	3.99	244.14	79.16
	140	14.876	1.562	4.71	3.89	274.22	75.32
	160	14.438	1.781	4.71	3.78	308.50	70.95

Table #5 (part 7 of 14) - Imperial Steel Pipe Dimensions based on ANSI/ASME B-36.10M - 1985 Standard

Steel Pipe Dimensions and Weights (Imperial Units)

Weight Class	Sch. #	Inside Diameter	Wall Thick-ness	Outside Area	Inside Area	Plain End Weight	Water in Pipe
		inches	inches	ft^2/ft	ft^2/ft	lb/ft	lb/ft
		▼ NPS - 20		**OD - 20.0 in. ▼**			
	-	19.562	0.219	5.24	5.12	46.27	130.25
	10	19.500	0.250	5.24	5.11	52.73	129.43
	-	19.438	0.281	5.24	5.09	59.18	128.60
	-	19.376	0.312	5.24	5.07	65.60	127.78
	-	19.312	0.344	5.24	5.06	72.21	126.94
STD	20	19.250	0.375	5.24	5.04	78.60	126.13
		19.188	0.406	5.24	5.02	84.96	125.32
		19.124	0.438	5.24	5.01	91.51	124.48
		19.062	0.469	5.24	4.99	97.83	123.68
XS	30	19.000	0.500	5.24	4.97	104.13	122.87
	-	18.876	0.562	5.24	4.94	116.67	121.27
	40	18.812	0.594	5.24	4.92	123.11	120.45
	-	18.750	0.625	5.24	4.91	129.33	119.66
	-	18.624	0.688	5.24	4.88	141.90	118.81
	-	18.500	0.750	5.24	4.84	154.19	116.49
	60	18.376	0.812	5.24	4.81	166.40	114.93
	-	18.250	0.875	5.24	4.78	178.72	113.36
	-	18.124	0.938	5.24	4.74	190.96	111.80
	-	18.000	1.000	5.24	4.71	202.92	110.28
	80	17.938	1.031	5.24	4.70	208.87	109.52
	-	17.876	1.062	5.24	4.68	214.80	108.77
	-	17.750	1.125	5.24	4.65	226.78	107.24
	-	17.624	1.188	5.24	4.61	238.68	105.72
	-	17.500	1.250	5.24	4.58	250.31	104.24
	100	17.438	1.281	5.24	4.57	256.10	103.50
	-	17.376	1.312	5.24	4.55	261.86	102.77
	-	17.250	1.375	5.24	4.52	273.51	101.28
	120	17.000	1.500	5.24	4.45	296.37	98.37
	140	16.500	1.750	5.24	4.32	341.09	92.67
	160	16.062	1.969	5.24	4.21	379.17	87.81
		▼ NPS - 22		**OD - 22.0 in. ▼**			
	-	21.562	0.219	5.76	5.64	50.94	158.24
	10	21.500	0.250	5.76	5.63	58.07	157.34
	-	21.438	0.281	5.76	6.61	65.18	156.43
	-	21.376	0.312	5.76	5.60	72.27	155.53
	-	21.312	0.344	5.76	5.58	79.56	154.60
STD	20	21.250	0.375	5.76	5.56	86.61	153.70
		21.188	0.406	5.76	5.55	93.63	152.80
		21.124	0.438	5.76	5.53	100.86	151.88
		21.062	0.469	5.76	5.51	107.85	150.99
XS	30	21.000	0.500	5.76	5.50	114.81	150.10
		20.876	0.562	5.76	5.47	128.67	148.34
		20.750	0.625	5.76	5.43	142.68	146.55

Table #5 (part 8 of 14) - Imperial Steel Pipe Dimensions based on ANSI/ASME B-36.10M - 1985 Standard

Steel Pipe Dimensions and Weights (Imperial Units)

Weight Class	Sch. #	Inside Diameter	Wall Thickness	Outside Area	Inside Area	Plain End Weight	Water in Pipe
		inches	inches	ft²/ft	ft²/ft	lb/ft	lb/ft
		20.624	0.688	5.76	5.40	156.60	144.78
		20.500	0.750	5.76	5.37	170.21	143.04
		20.376	0.812	5.76	5.33	183.75	141.31
	60	20.250	0.875	5.76	5.30	197.41	139.57
	-	20.124	0.938	5.76	5.27	211.00	137.84
-	-	20.000	1.000	5.76	5.24	224.28	136.15
-	-	19.876	1.062	5.76	5.20	237.48	134.46
-	80	19.750	1.125	5.76	5.17	250.81	132.76
	-	19.624	1.188	5.76	5.14	264.06	131.08
	-	19.500	1.250	5.76	5.10	277.01	129.43
	-	19.376	1.312	5.76	5.07	289.88	127.78
	100	19.250	1.375	5.76	5.04	302.88	126.13
	-	19.124	1.438	5.76	5.01	315.79	124.48
	-	19.000	1.500	5.76	4.97	328.41	122.87
-	120	18.750	1.625	5.76	4.91	353.61	119.66
-	140	18.250	1.875	5.76	4.78	403.00	113.36
-	160	17.750	2.125	5.76	4.65	451.06	107.24
		▼ NPS - 24		**OD - 24.0 in. ▼**			
-	10	23.500	0.250	6.28	6.15	63.41	187.97
		23.438	0.281	6.28	6.14	71.18	186.98
		23.376	0.312	6.28	6.12	78.93	185.99
		23.312	0.344	6.28	6.10	86.91	184.97

Steel Pipe Dimensions and Weights (Imperial Units)

Weight Class	Sch. #	Inside Diameter	Wall Thickness	Outside Area	Inside Area	Plain End Weight	Water in Pipe
		inches	inches	ft²/ft	ft²/ft	lb/ft	lb/ft
STD	20	23.250	0.375	6.28	6.09	94.62	183.99
		23.188	0.406	6.28	6.07	102.31	183.01
		23.124	0.438	6.28	6.05	110.22	182.00
		23.062	0.469	6.28	6.04	117.86	181.03
XS	-	23.000	0.500	6.28	6.02	125.49	180.05
	30	22.876	0.562	6.28	5.99	140.68	178.11
	-	22.750	0.625	6.28	5.96	156.03	176.16
	40	22.624	0.688	6.28	5.92	171.29	174.22
	-	22.500	0.750	6.28	5.89	186.23	172.31
	-	22.376	0.812	6.28	5.86	201.09	170.41
	-	22.250	0.875	6.28	5.83	216.10	168.50
	-	22.124	0.938	6.28	5.79	231.03	166.60
	60	22.062	0.969	6.28	5.78	238.35	165.67
	-	22.000	1.000	6.28	5.76	245.64	164.74
	-	21.876	1.062	6.28	5.73	260.17	162.89
	-	21.750	1.125	6.28	5.69	274.84	161.02
	-	21.624	1.188	6.28	5.66	289.44	159.16
	80	21.562	1.219	6.28	5.64	296.58	158.24
	-	21.500	1.250	6.28	5.63	303.71	157.34
		21.376	1.312	6.28	5.60	317.91	155.53
		21.250	1.375	6.28	5.56	332.25	153.70
		21.124	1.438	6.28	5.53	346.50	151.88

Table #5 (part 9 of 14) - Imperial Steel Pipe Dimensions based on ANSI/ASME B-36.10M - 1985 Standard

Steel Pipe Dimensions and Weights (Imperial Units)

Weight Class	Sch. #	Inside Diameter	Wall Thick-ness	Outside Area	Inside Area	Plain End Weight	Water in Pipe
		inches	inches	ft²/ft	ft²/ft	lb/ft	lb/ft
		21.000	1.500	6.28	5.50	360.45	150.10
	100	20.938	1.531	6.28	5.48	367.39	149.22
	-	20.876	1.562	6.28	5.47	374.31	148.34
	120	20.376	1.812	6.28	5.33	429.39	141.31
	140	19.876	2.062	6.28	5.20	483.12	134.46
	160	19.312	2.344	6.28	5.06	542.13	126.94
▼ NPS - 26 OD - 26.0 in. ▼							
	-	25.500	0.250	6.81	6.68	68.75	221.32
	-	25.438	0.281	6.81	6.66	77.18	220.25
	10	25.376	0.312	6.81	6.64	85.60	219.18
	-	25.312	0.344	6.81	6.63	94.26	218.07
	-	25.250	0.375	6.81	6.61	102.63	217.01
	-	25.188	0.406	6.81	6.59	110.98	215.94
	-	25.124	0.438	6.81	6.58	119.57	214.85
	-	25.062	0.469	6.81	6.56	127.88	213.79
	20	25.000	0.500	6.81	6.54	136.17	212.73
		24.876	0.562	6.81	6.51	152.68	210.63
		24.750	0.625	6.81	6.48	169.38	208.50
		24.624	0.688	6.81	6.45	185.99	206.38
		24.500	0.750	6.81	6.41	202.25	204.31
		24.376	0.812	6.81	6.38	218.43	202.24

Steel Pipe Dimensions and Weights (Imperial Units)

Weight Class	Sch. #	Inside Diameter	Wall Thick-ness	Outside Area	Inside Area	Plain End Weight	Water in Pipe
		inches	inches	ft²/ft	ft²/ft	lb/ft	lb/ft
		24.250	0.875	6.81	6.35	234.79	200.16
		24.124	0.938	6.81	6.32	251.07	198.08
		24.000	1.000	6.81	6.28	267.00	196.05
▼ NPS - 28 OD - 28.0 in. ▼							
		27.500	0.250	7.33	7.20	74.09	257.40
		27.438	0.281	7.33	7.18	83.19	256.24
	10	27.376	0.312	7.33	7.17	92.26	255.09
	-	27.312	0.344	7.33	7.15	101.61	253.90
	-	27.250	0.375	7.33	7.13	110.64	252.74
	-	27.188	0.406	7.33	7.12	119.65	251.60
	-	27.124	0.438	7.33	7.10	128.93	250.41
	-	27.062	0.469	7.33	7.08	137.90	249.27
	20	27.000	0.500	7.33	7.07	146.85	248.13
	-	26.876	0.562	7.33	7.04	164.69	245.85
	30	26.750	0.625	7.33	7.00	182.73	243.55
	-	26.624	0.688	7.33	6.97	200.68	241.27
	-	26.500	0.750	7.33	6.94	218.27	239.02
	-	26.376	0.812	7.33	6.91	235.78	236.79
	-	26.250	0.875	7.33	6.87	253.48	243.54
	-	26.124	0.938	7.33	6.84	271.10	232.29
	-	26.000	1.000	7.33	6.81	288.36	230.09

Table #5 (part 10 of 14) - Imperial Steel Pipe Dimensions based on ANSI/ASME B-36.10M - 1985 Standard

Steel Pipe Dimensions and Weights (Imperial Units)

Weight Class	Sch. #	Inside Diameter	Wall Thick-ness	Outside Area	Inside Area	Plain End Weight	Water in Pipe
		inches	inches	ft²/ft	ft²/ft	lb/ft	lb/ft
▼ NPS - 30		OD - 30.0 in. ▼					
	-	29.500	0.250	7.85	7.72	79.43	296.21
	-	29.438	0.281	7.85	7.70	89.19	294.96
	10	29.376	0.312	7.85	7.69	98.93	293.72
		29.312	0.344	7.85	7.67	108.95	292.44
STD	-	29.250	0.375	7.85	7.66	118.65	291.21
-	-	29.188	0.406	7.85	7.64	128.32	289.97
-	-	29.124	0.438	7.85	7.62	138.20	288.70
-	-	29.062	0.469	7.85	7.61	147.92	287.48
XS	20	29.000	0.500	7.85	7.59	157.53	286.25
	-	28.876	0.562	7.85	7.56	176.69	283.81
	30	28.750	0.625	7.85	7.53	196.08	281.34
	-	28.624	0.688	7.85	7.49	215.38	278.88
	-	28.500	0.750	7.85	7.46	234.19	276.46
	-	28.376	0.812	7.85	7.43	253.12	274.06
	-	28.250	0.875	7.85	7.39	272.17	271.64
	-	28.124	0.938	7.85	7.36	291.14	269.22
	-	28.000	1.000	7.85	7.33	309.72	266.85
	-	27.876	1.062	7.85	7.30	328.22	264.49
	-	27.750	1.125	7.85	7.26	346.93	262.10
	-	27.624	1.188	7.85	7.23	365.56	259.73
	-	27.500	1.250	7.85	7.20	383.81	257.40

Steel Pipe Dimensions and Weights (Imperial Units)

Weight Class	Sch. #	Inside Diameter	Wall Thick-ness	Outside Area	Inside Area	Plain End Weight	Water in Pipe
		inches	inches	ft²/ft	ft²/ft	lb/ft	lb/ft
▼ NPS - 32		OD - 32.0 in. ▼					
	-	31.500	0.250	8.38	8.25	84.77	337.73
	-	31.438	0.281	8.38	8.23	95.19	336.40
	10	31.376	0.312	8.38	8.21	105.59	335.08
	-	31.312	0.344	8.38	8.20	116.30	333.71
STD	-	31.250	0.375	8.38	8.18	126.66	332.39
-	-	31.188	0.406	8.38	8.16	136.99	331.07
-	-	31.124	0.438	8.38	8.15	147.64	329.72
-	-	31.062	0.469	8.38	8.13	157.94	328.40
XS	20	31.000	0.500	8.38	8.11	168.21	327.09
	-	30.876	0.562	8.38	8.08	188.70	324.48
	30	30.750	0.625	8.38	8.05	209.43	321.84
	40	30.624	0.688	8.38	8.02	230.18	319.21
	-	30.500	0.750	8.38	7.98	250.31	316.63
	-	30.376	0.812	8.38	7.95	270.47	314.06
	-	30.250	0.875	8.38	7.92	290.86	311.46
	-	30.124	0.938	8.38	7.89	311.17	308.87
	-	30.000	1.000	8.38	7.85	331.08	306.33
	-	29.876	1.062	8.38	7.82	350.90	303.80
	-	29.750	1.125	8.38	7.79	370.96	301.25
	-	29.624	1.188	8.38	7.76	390.94	298.70
	-	29.500	1.250	8.38	7.72	410.51	296.21

Table #5 (part 11 of 14) - Imperial Steel Pipe Dimensions based on ANSI/ASME B-36.10M - 1985 Standard

Steel Pipe Dimensions and Weights (Imperial Units)

Weight Class	Sch. #	Inside Diameter	Wall Thick-ness	Outside Area	Inside Area	Plain End Weight	Water in Pipe
		inches	inches	ft²/ft	ft²/ft	lb/ft	lb/ft
		▼ NPS - 34		OD - 34.0 in. ▼			
	-	33.500	0.250	8.90	8.77	90.11	381.98
	-	33.438	0.281	8.90	8.75	101.19	380.57
	-	33.376	0.312	8.90	8.74	112.25	379.16
	-	33.312	0.344	8.90	8.72	123.65	377.70
	-	33.250	0.375	8.90	8.70	134.67	376.30
	-	33.188	0.406	8.90	8.69	145.67	374.90
	-	33.124	0.438	8.90	8.67	157.00	373.45
		33.062	0.469	8.90	8.66	167.95	372.06
		33.000	0.500	8.90	8.64	178.89	370.66
		32.876	0.562	8.90	8.61	200.70	367.88
		32.750	0.625	8.90	8.57	222.78	365.07
		32.624	0.688	8.90	8.54	244.77	362.26
		32.500	0.750	8.90	8.51	266.33	359.51
		32.376	0.812	6.90	8.48	387.61	356.78
		32.250	0.875	8.90	8.44	309.55	351.81
		32.124	0.938	8.90	8.41	331.21	351.24
		32.000	1.000	8.90	8.38	352.44	348.54
		31.876	1.062	8.90	8.35	373.59	345.84
		31.750	1.125	8.90	8.31	394.99	343.11
		31.624	1.188	8.90	8.28	416.31	340.39
		31.500	1.250	8.90	8.25	437.21	337.73

Steel Pipe Dimensions and Weights (Imperial Units)

Weight Class	Sch. #	Inside Diameter	Wall Thick-ness	Outside Area	Inside Area	Plain End Weight	Water in Pipe
		inches	inches	ft²/ft	ft²/ft	lb/ft	lb/ft
		▼ NPS - 36		OD - 36.0 in. ▼			
		35.500	0.250	9.42	9.29	95.45	428.95
		35.438	0.281	9.42	9.28	107.20	427.45
		35.376	0.312	9.42	9.26	118.92	425.96
		35.312	0.344	9.42	9.24	131.00	424.42
		35.250	0.375	9.42	9.23	142.68	422.93
		35.188	0.406	9.42	9.21	154.34	421.44
		35.124	0.438	9.42	9.19	166.35	419.91
		35.062	0.469	9.42	9.18	177.97	418.43
		35.000	0.500	9.42	9.16	189.57	416.95
		34.876	0.562	9.42	9.13	212.70	414.00
		34.750	0.625	9.42	9.10	236.13	411.02
		34.624	0.688	9.42	9.06	259.47	408.04
		34.500	0.750	9.42	9.03	282.35	405.12
		34.376	0.812	9.42	9.00	305.16	402.22
		34.250	0.875	9.42	8.97	328.24	399.27
		34.124	0.938	9.42	8.93	351.25	396.34
		34.000	1.000	9.42	8.90	373.80	393.47
		33.876	1.062	9.42	8.87	396.27	390.60
		33.750	1.125	9.42	8.84	419.02	387.70
		33.624	1.188	9.42	8.80	441.69	384.81
		33.500	1.250	9.42	8.77	463.91	381.98

Table #5 (part 12 of 14) - Imperial Steel Pipe Dimensions based on ANSI/ASME B-36.10M - 1985 Standard

Steel Pipe Dimensions and Weights (Imperial Units)

Weight Class	Sch. #	Inside Diameter	Wall Thick-ness	Outside Area	Inside Area	Plain End Weight	Water in Pipe
		inches	inches	ft²/ft	ft²/ft	lb/ft	lb/ft
▼ NPS - 38 OD - 38.0 in. ▼							
		37.376	0.312	9.95	9.79	125.58	475.48
		37.312	0.344	9.95	9.77	138.35	473.86
		37.250	0.375	9.95	9.75	150.69	472.28
		37.188	0.406	9.95	9.74	163.01	470.71
		37.124	0.438	9.95	9.72	175.71	469.09
		37.062	0.469	9.95	9.70	187.99	467.53
		37.000	0.500	9.95	9.69	200.25	465.96
		36.876	0.562	9.95	9.65	224.71	462.85
		36.750	0.625	9.95	9.62	249.48	459.69
		36.624	0.688	9.95	9.59	274.16	456.54
		36.500	0.750	9.95	9.56	298.37	453.46
		36.376	0.812	9.95	9.52	322.50	450.38
		36.250	0.875	9.95	9.49	346.93	447.27
		36.124	0.938	9.95	9.46	371.28	444.16
		36.000	1.000	9.95	9.42	395.16	441.12
		35.876	1.062	9.95	9.39	418.96	438.08
		35.750	1.125	9.95	9.36	443.05	435.01
		35.624	1.188	9.95	9.33	467.06	431.95
		35.500	1.250	9.95	9.29	490.61	428.95

Steel Pipe Dimensions and Weights (Imperial Units)

Weight Class	Sch. #	Inside Diameter	Wall Thick-ness	Outside Area	Inside Area	Plain End Weight	Water in Pipe
		inches	inches	ft²/ft	ft²/ft	lb/ft	lb/ft
▼ NPS - 40 OD - 40.0 in. ▼							
		39.376	0.312	10.47	10.31	132.25	527.73
		39.312	0.344	10.47	10.29	145.69	526.02
		39.250	0.375	10.47	10.28	158.70	524.36
		39.188	0.406	10.47	10.26	171.68	522.70
		39.124	0.438	10.47	10.24	185.06	521.00
		39.062	0.469	10.47	10.23	198.01	519.35
		39.000	0.500	10.47	10.21	210.93	517.70
		38.876	0.562	10.47	10.18	236.71	514.41
		38.750	0.625	10.47	10.14	262.83	511.08
		38.624	0.688	10.47	10.11	288.86	507.77
		38.500	0.750	10.47	10.08	314.39	504.51
		38.376	0.812	10.47	10.05	339.84	501.27
		38.250	0.875	10.47	10.01	365.62	497.98
		38.124	0.938	10.47	9.98	391.32	494.70
		38.000	1.000	10.47	9.95	416.52	491.49
		37.876	1.062	10.47	9.92	441.64	488.29
		37.750	1.125	10.47	9.88	467.08	485.05
		37.624	1.188	10.47	9.85	492.44	481.81
		37.500	1.250	10.47	9.82	517.31	478.64

Table #5 (part 13 of 14) - Imperial Steel Pipe Dimensions based on ANSI/ASME B-36.10M - 1985 Standard

Steel Pipe Dimensions and Weights (Imperial Units)							
Weight Class	Sch. #	Inside Diameter	Wall Thick-ness	Outside Area	Inside Area	Plain End Weight	Water in Pipe
		inches	inches	ft²/ft	ft²/ft	lb/ft	lb/ft
		▼ NPS - 42	OD - 42.0 in. ▼				
		41.312	0.344	11.00	10.82	153.04	580.90
		41.250	0.375	11.00	10.80	166.71	579.16
		41.188	0.406	11.00	10.78	180.35	577.42
		41.124	0.438	11.00	10.77	194.42	575.63
		41.062	0.469	11.00	10.75	208.03	573.89
		41.000	0.500	11.00	10.73	221.61	572.16
		40.876	0.562	11.00	10.70	248.72	568.70
		40.750	0.625	11.00	10.67	276.18	565.20
		40.624	0.688	11.00	10.64	303.55	561.71

Steel Pipe Dimensions and Weights (Imperial Units)							
Weight Class	Sch. #	Inside Diameter	Wall Thick-ness	Outside Area	Inside Area	Plain End Weight	Water in Pipe
		inches	inches	ft²/ft	ft²/ft	lb/ft	lb/ft
		40.500	0.750	11.00	10.60	330.41	558.29
		40.376	0.812	11.00	10.57	357.19	554.88
		40.250	0.875	11.00	10.54	384.31	551.42
		40.124	0.938	11.00	10.50	411.35	547.97
		40.000	1.000	11.00	10.47	437.88	544.59
		39.876	1.062	11.00	10.44	464.32	541.22
		39.750	1.125	11.00	10.41	491.11	538.80
		39.624	1.188	11.00	10.37	517.82	534.40
		39.500	1.250	11.00	10.34	544.01	531.06

Table #5 (part 14 of 14) - Imperial Steel Pipe Dimensions based on ANSI/ASME B-36.10M - 1985 Standard

Weight Class	Schedule #	Inside Diameter	Wall Thickness	Outside Area	Inside Area	Plain End Mass	Water Mass
		mm	mm	m^2/m	m^2/m	kg/m	kg/m
▼ NPS - 6		OD - 10.3mm ▼					
STD	40	6.84	1.73	0.0324	0.0215	0.37	0.037
XS	80	5.48	2.41	0.0324	0.0172	0.47	0.024
▼ NPS - 8		OD - 13.7mm ▼					
STD	40	9.22	2.24	0.0430	0.0290	0.63	0.067
XS	80	7.66	3.02	0.0430	0.0241	0.80	0.046
▼ NPS -10		OD - 17.1mm ▼					
STD	40	12.48	2.31	0.0537	0.0392	0.84	0.124
XS	80	10.70	3.20	0.0537	0.0336	1.10	0.091
▼ NPS - 15		OD - 21.3mm ▼					
STD	40	15.76	2.77	0.0669	0.0495	1.27	0.197
XS	80	13.84	3.73	0.0669	0.0435	1.62	0.150
-	160	11.74	4.78	0.0669	0.0369	1.95	0.109
XXS	-	6.36	7.47	0.0669	0.0200	2.55	0.033
▼ NPS - 20		OD - 26.7mm ▼					
STD	40	20.96	2.87	0.0839	0.0658	1.69	0.343
XS	80	18.88	3.91	0.0839	0.0593	2.20	0.279
-	160	15.58	5.56	0.0839	0.0489	2.90	0.189
XXS	-	11.06	7.82	0.0839	0.0347	3.64	0.094
▼ NPS - 25		OD - 33.4mm ▼					
STD	40	26.64	3.38	0.1049	0.0837	2.50	0.563
XS	80	24.30	4.55	0.1049	0.0763	3.24	0.463
-	160	20.70	6.35	0.1049	0.0650	4.24	0.337

Weight Class	Schedule #	Inside Diameter	Wall Thickness	Outside Area	Inside Area	Plain End Mass	Water Mass
		mm	mm	m^2/m	m^2/m	kg/m	kg/m
XXS	-	15.22	9.09	0.1049	0.0478	5.45	0.182
▼ NPS - 32		OD - 42.2mm ▼					
STD	40	35.08	3.56	0.1326	0.1102	3.39	0.964
XS	80	32.50	4.85	0.1326	0.1021	4.47	0.827
-	160	29.50	6.35	0.1326	0.0927	5.61	0.684
XXS	-	22.80	9.70	0.1326	0.0716	7.77	0.407
▼ NPS - 40		OD - 48.3mm ▼					
STD	40	40.94	3.68	0.1517	0.1286	4.05	1.314
XS	80	38.14	5.08	0.1517	0.1198	5.41	1.140
-	160	34.02	7.14	0.1517	0.1069	7.25	0.907
XXS	-	28.00	10.15	0.1517	0.0880	9.56	0.614
▼ NPS - 50		OD - 60.3mm ▼					
-	-	56.08	2.11	0.1894	0.1762	3.03	2.476
-	-	54.76	2.77	0.1894	0.1720	3.93	2.354
-	-	53.94	3.18	0.1894	0.1695	4.48	2.295
-	-	53.14	3.58	0.1894	0.1669	5.01	2.220
STD	40	52.48	3.91	0.1894	0.1649	5.44	2.161
		51.56	4.37	0.1894	0.1620	6.03	2.056
		50.74	4.78	0.1894	0.1594	6.54	2.026
XS	80	49.22	5.54	0.1894	0.1546	7.48	1.907
	-	47.60	6.35	0.1894	0.1495	8.45	1.788
	-	46.02	7.14	0.1894	0.1446	9.36	1.669
	160	42.82	8.74	0.1894	0.1345	11.11	1.445

Table #6 (part 1 of 13) - Metric Steel Pipe Dimensions based on ANSI/ASME B-36.10M - 1985 Standard

Steel Pipe Dimensions and Mass (Metric Units)

Weight Class	Schedule #	Inside Diameter	Wall Thickness	Outside Area	Inside Area	Plain End Mass	Water Mass
		mm	mm	m²/m	m²/m	kg/m	kg/m
XXS	-	38.16	11.07	0.1894	0.1199	13.44	1.147
▼ NPS - 65 OD - 73.0mm ▼							
-	-	68.78	2.11	0.2293	0.2161	3.69	3.73
-	-	67.46	2.77	0.2293	0.2119	4.80	3.58
-	-	66.64	3.18	0.2293	0.2094	5.48	3.50
-	-	65.84	3.58	0.2293	0.2068	6.13	3.41
-	-	65.08	3.96	0.2293	0.2045	6.74	3.34
-	-	64.26	4.37	0.2293	0.2019	7.40	3.25
-	-	63.44	4.78	0.2293	0.1993	8.04	3.17
STD	40	62.68	5.16	0.2293	0.1969	8.63	3.08
		62.02	5.49	0.2293	0.1948	9.14	3.02
		60.30	6.35	0.2293	0.1894	10.44	2.86
XS	80	58.98	7.01	0.2293	0.1853	11.41	2.73
-	160	53.94	9.53	0.2293	0.1695	14.92	2.29
XXS	-	44.96	14.02	0.2293	0.1412	20.39	1.59
▼ NPS - 80 OD - 88.9mm ▼							
-	-	84.68	2.11	2.7930	0.2660	4.52	5.63
-	-	83.36	2.77	0.2793	0.2619	5.88	5.47
-	-	82.54	3.18	0.2793	0.2593	6.72	5.36
-	-	81.74	3.58	0.2793	0.2568	7.53	5.26
-	-	80.98	3.96	0.2793	0.2544	8.29	5.16
-	-	80.16	4.37	0.2793	0.2518	9.11	5.05

Steel Pipe Dimensions and Mass (Metric Units)

Weight Class	Schedule #	Inside Diameter	Wall Thickness	Outside Area	Inside Area	Plain End Mass	Water Mass
		mm	mm	m²/m	m²/m	kg/m	kg/m
-	-	79.34	4.78	0.2793	0.2493	9.92	4.95
STD	40	77.92	5.49	0.2793	0.2448	11.29	4.77
		76.20	6.35	0.2793	0.2394	12.93	4.56
		74.62	7.14	0.2793	0.2344	14.40	4.38
XS	80	73.66	7.62	0.2793	0.2314	15.27	4.31
-	160	66.64	11.13	0.2793	0.2094	21.35	3.49
XXS	-	58.42	15.24	0.2793	0.1835	27.68	2.68
▼ NPS - 90 OD - 101.6mm ▼							
-	-	97.38	2.11	0.3192	0.3059	5.18	7.48
-	-	96.06	2.77	0.3192	0.3018	6.75	7.26
-	-	95.24	3.18	0.3192	0.2992	7.72	7.14
-	-	94.44	3.58	0.3192	0.2967	8.65	7.02
-	-	93.68	3.96	0.3192	0.2943	9.53	6.90
-	-	92.86	4.37	0.3192	0.2917	10.48	6.78
-	-	92.04	4.78	0.3192	0.2892	11.41	6.68
STD	40	90.12	5.74	0.3192	0.2831	13.57	6.38
		88.90	6.35	0.3192	0.2793	14.92	6.21
		87.32	7.14	0.3192	0.2743	16.63	5.99
XS	80	85.44	8.08	0.3192	0.2684	18.63	5.74
▼ NPS - 100 OD - 114.3mm ▼							
		110.08	2.11	0.3591	0.3458	5.84	9.52
		108.76	2.77	0.3591	0.3417	7.62	9.31
		107.94	3.18	0.3591	0.3391	8.71	9.16

Table #6 (part 2 of 13) - Metric Steel Pipe Dimensions based on ANSI/ASME B-36.10M - 1985 Standard

Steel Pipe Dimensions and Mass (Metric Units)

Weight Class	Schedule #	Inside Diameter	Wall Thickness	Outside Area	Inside Area	Plain End Mass	Water Mass
		mm	mm	m^2/m	m^2/m	kg/m	kg/m
		107.14	3.58	0.3591	0.3366	9.77	9.03
		106.38	3.96	0.3591	0.3342	10.78	8.90
		105.56	4.37	0.3591	0.3316	11.85	8.76
		104.74	4.78	0.3591	0.3291	12.91	8.64
		103.98	5.16	0.3591	0.3267	13.89	8.51
		103.18	5.56	0.3591	0.3242	14.91	8.37
STD	40	102.26	6.02	0.3591	0.3213	16.07	8.21
		101.60	6.35	0.3591	0.3192	16.90	8.12
		100.02	7.14	0.3591	0.3142	18.87	7.85
		98.46	7.92	0.3591	0.3093	20.78	7.63
XS	80	97.18	8.56	0.3591	0.3053	22.32	7.42
-	120	92.04	11.13	0.3591	0.2892	28.32	6.66
-	160	87.32	13.49	0.3591	0.2743	33.54	5.99
XXS	-	80.06	17.12	0.3591	0.2515	41.30	5.04
▼ NPS - 125		**OD - 141.3mm ▼**					
		137.08	2.11	0.4439	0.4307	7.24	14.78
		134.94	3.18	0.4439	0.4239	10.83	14.33
		133.38	3.96	0.3900	0.4190	13.41	13.99
		130.74	4.78	0.4439	0.4107	16.09	13.65
		130.18	5.56	0.4439	0.4090	18.61	13.32
STD	40	128.20	6.55	0.4439	0.4028	21.77	12.90
		127.02	7.14	0.4439	0.3990	23.62	12.69
		125.46	7.92	0.4439	0.3941	26.05	12.38

Steel Pipe Dimensions and Mass (Metric Units)

Weight Class	Schedule #	Inside Diameter	Wall Thickness	Outside Area	Inside Area	Plain End Mass	Water Mass
		mm	mm	m^2/m	m^2/m	kg/m	kg/m
		123.82	8.74	0.3900	0.3890	28.57	12.05
XS	80	122.24	9.53	0.4439	0.3840	30.97	11.73
-	120	115.90	12.70	0.4439	0.3641	40.28	10.55
-	160	109.54	15.88	0.4439	0.3441	49.11	9.42
XXS	-	103.20	19.05	0.4439	0.3242	57.43	8.37
▼ NPS - 150		**OD - 168.3mm ▼**					
-	-	164.08	2.11	0.5287	0.5155	8.65	21.16
-	-	162.76	2.77	0.5287	0.5113	11.31	22.31
-	-	161.94	3.18	0.5287	0.5088	12.95	20.61
-	-	161.14	3.58	0.5287	0.5062	14.54	20.40
-	-	160.38	3.96	0.5287	0.5038	16.05	20.22
-	-	159.56	4.37	0.5287	0.5013	17.67	20.01
-	-	158.74	4.78	0.5287	0.4987	19.27	19.80
-	-	157.98	5.16	0.2587	0.4963	20.76	19.61
-	-	157.18	5.56	0.5287	0.4938	22.31	19.41
-	-	155.60	6.35	0.5287	0.4888	25.36	19.03
STD	40	154.08	7.11	0.5287	0.4841	28.26	18.52
		152.46	7.92	0.5287	0.4790	31.32	18.27
		150.82	8.74	0.5287	0.4738	34.39	17.88
		149.24	9.53	0.5287	0.4689	37.31	17.49
XS	80	146.36	10.97	0.5287	0.4598	42.56	16.84
	-	142.90	12.70	0.5287	0.4489	48.73	16.05
	120	139.76	14.27	0.5287	0.4391	54.20	15.35

Table #6 (part 3 of 13) - Metric Steel Pipe Dimensions based on ANSI/ASME B-36.10M - 1985 Standard

Steel Pipe Dimensions and Mass (Metric Units)							
Weight Class	Schedule #	Inside Diameter	Wall Thickness	Outside Area	Inside Area	Plain End Mass	Water Mass
		mm	mm	m^2/m	m^2/m	kg/m	kg/m
	-	136.54	15.88	0.5287	0.4290	59.69	14.65
	160	131.78	18.26	0.5287	0.4140	67.56	13.65
	-	130.20	19.05	0.5287	0.4090	70.11	13.32
XXS	-	124.40	21.95	0.5287	0.3908	79.22	12.16
-	-	123.84	22.23	0.5287	0.3891	80.07	12.05
▼ NPS - 200 OD - 219.1mm ▼							
-	-	212.74	3.18	0.6883	0.6683	16.91	35.57
-	-	211.18	3.96	0.6883	0.6634	21.02	35.04
-	-	209.54	4.78	0.6883	0.6583	25.24	34.51
-	-	208.78	5.16	0.6883	0.6559	27.21	34.26
-	-	207.98	5.56	0.6883	0.6534	29.29	33.99
-	20	206.40	6.35	0.6883	0.6484	33.32	33.48
-	30	205.02	7.04	0.6883	0.6441	36.80	33.03
-	-	203.26	7.92	0.6883	0.6386	41.27	32.47
STD	40	202.74	8.18	0.6883	0.6369	42.54	32.30
		201.62	8.74	0.6883	0.6334	45.33	31.95
		200.04	9.53	0.6883	0.6284	49.23	31.45
	60	198.48	10.31	0.6883	0.6235	53.10	30.96
	-	196.84	11.13	0.6883	0.6184	57.07	30.46
	80	193.70	12.70	0.6883	0.6085	64.65	29.49
	-	190.56	14.27	0.6883	0.5987	72.12	28.53
	100	188.92	15.09	0.6883	0.5935	75.92	28.06

Steel Pipe Dimensions and Mass (Metric Units)							
Weight Class	Schedule #	Inside Diameter	Wall Thickness	Outside Area	Inside Area	Plain End Mass	Water Mass
		mm	mm	m^2/m	m^2/m	kg/m	kg/m
XS	-	187.34	15.88	0.6883	0.5885	79.57	27.58
	120	182.58	18.26	0.6883	0.5736	90.46	26.19
	-	181.00	19.05	0.6883	0.5686	93.99	25.75
	140	177.86	20.62	0.6883	0.5588	100.96	24.85
	-	174.64	22.23	0.6883	0.5486	107.91	23.97
	160	173.08	23.01	0.6883	0.5437	111.29	23.54
XXS	-	168.30	25.40	0.6883	0.5287	121.35	22.26
▼ NPS - 250 OD - 273.0mm ▼							
	-	265.08	3.96	0.8577	0.8328	26.30	55.25
	-	263.44	4.78	0.8577	0.8276	31.60	54.58
	-	262.68	5.16	0.8577	0.8252	34.08	54.27
	-	261.88	5.56	0.8577	0.8227	36.70	52.43
	20	260.03	6.35	0.8577	0.8169	41.78	53.28
	-	258.82	7.09	0.8577	0.8131	46.49	52.69
	30	257.40	7.80	0.8577	0.8086	51.02	52.11
	-	255.52	8.74	0.8577	0.8027	56.96	51.35
STD	40	254.46	9.27	0.8577	0.7994	60.32	50.91
-	-	250.74	11.13	0.8577	0.7877	71.88	49.44
XS	60	247.60	12.70	0.8577	0.7779	81.56	48.22
	-	244.46	14.27	0.8577	0.7680	91.11	46.99
	80	242.82	15.09	0.8577	0.7628	96.00	46.37
	-	241.24	15.88	0.8577	0.7579	100.69	45.79
	100	236.48	18.26	0.8577	0.7429	114.77	43.97

Table #6 (part 4 of 13) - Metric Steel Pipe Dimensions based on ANSI/ASME B-36.10M - 1985 Standard

Steel Pipe Dimensions and Mass (Metric Units)

Weight Class	Schedule #	Inside Diameter	Wall Thickness	Outside Area	Inside Area	Plain End Mass	Water Mass
		mm	mm	m^2/m	m^2/m	kg/m	kg/m
	-	231.76	20.62	0.8577	0.7281	128.41	42.24
	120	230.12	21.44	0.8577	0.7229	133.04	41.65
	-	228.54	22.23	0.8577	0.7180	137.50	41.08
	-	225.34	23.83	0.8577	0.7079	146.47	39.93
	140	222.20	25.40	0.8577	0.6981	155.15	38.83
XXS	160	215.84	28.58	0.8577	0.6781	172.30	36.65
		209.50	31.75	0.8577	0.6582	188.98	34.52
NPS - 300	**OD - 323.8mm**						
		315.06	4.37	1.0173	0.9898	34.43	78.06
		314.24	4.78	1.0173	0.9872	37.62	77.66
		313.48	5.16	1.0173	0.9848	40.56	77.27
		312.68	5.56	1.0173	0.9823	41.65	76.88
-	20	311.10	6.35	1.0173	0.9774	49.73	76.14
-	-	309.52	7.14	1.0173	0.9724	55.77	75.33
-	-	307.96	7.92	1.0173	0.9675	61.71	74.57
-	30	307.04	8.38	1.0173	0.9646	65.20	74.13
-	-	306.32	8.74	1.0173	0.9623	67.93	73.78
STD	-	304.74	9.53	1.0173	0.9574	73.88	73.02
	40	303.18	10.31	1.0173	0.9525	79.73	72.28
	-	301.54	11.13	1.0173	0.9473	85.84	71.52
		298.40	12.70	1.0173	0.9375	97.46	70.03
	60	295.26	14.27	1.0173	0.9276	108.96	68.54
	-	292.04	15.88	1.0173	0.9175	120.62	67.05

Steel Pipe Dimensions and Mass (Metric Units)

Weight Class	Schedule #	Inside Diameter	Wall Thickness	Outside Area	Inside Area	Plain End Mass	Water Mass
		mm	mm	m^2/m	m^2/m	kg/m	kg/m
	80	288.84	17.48	1.0173	0.9074	132.08	65.60
XS	-	285.70	19.05	1.0173	0.8976	143.21	64.19
	-	282.56	20.62	1.0173	0.8877	154.21	62.77
	100	280.92	21.44	1.0173	0.8825	159.91	62.06
	-	279.34	22.23	1.0173	0.8776	165.37	61.36
	-	276.14	23.83	1.0173	0.8675	176.33	59.97
	120	273.00	25.40	1.0173	0.8577	186.97	58.60
	-	269.86	26.97	1.0173	0.8478	197.48	57.26
	140	266.64	28.58	1.0173	0.8377	208.14	55.92
	-	260.30	31.75	1.0173	0.8178	228.74	53.28
XXS	160	257.16	33.32	1.0173	0.8079	238.76	52.00
▼ NPS - 350	**OD - 355.6mm ▼**						
	-	346.04	4.78	1.1172	1.0871	41.35	94.14
	-	345.28	5.16	1.1172	1.0847	44.59	93.72
	-	344.94	5.33	1.1172	1.0837	46.04	93.53
	-	344.48	5.56	1.1172	1.0822	47.99	93.27
	10	342.90	6.35	1.1172	1.0773	54.69	92.42
	-	341.32	7.14	1.1172	1.0723	61.35	91.58
	20	339.76	7.92	1.1172	1.0674	67.90	90.74
	-	338.12	8.74	1.1172	1.0622	74.76	89.88
STD	30	336.54	9.53	1.1172	1.0573	81.33	89.04
	-	334.98	10.31	1.1172	1.0524	87.79	88.21
	40	333.34	11.13	1.1172	1.0472	94.55	87.34

Table #6 (part 5 of 13) - Metric Steel Pipe Dimensions based on ANSI/ASME B-36.10M - 1985 Standard

Steel Pipe Dimensions and Mass (Metric Units)

Weight Class	Sche dule #	Inside Diameter	Wall Thick-ness	Outside Area	Inside Area	Plain End Mass	Water Mass
		mm	mm	m^2/m	m^2/m	kg/m	kg/m
	-	331.78	11.91	1.1172	1.0423	100.94	86.52
XS	-	330.20	12.10	1.1172	1.0374	107.39	85.70
-	-	327.06	14.27	1.1172	1.0275	120.11	84.08
-	60	325.42	15.09	1.1172	1.0223	126.71	83.25
-	-	323.84	15.88	1.1172	1.0174	133.03	82.44
-	-	320.64	17.48	1.1172	1.0073	145.75	80.82
	80	317.50	19.05	1.1172	0.9975	158.10	79.24
	-	314.36	20.62	1.1172	0.9876	170.33	77.67
	-	311.14	22.23	1.1172	0.9775	182.75	76.11
	100	307.94	23.83	1.1172	0.9674	194.96	74.54
	-	304.80	25.40	1.1172	0.9576	206.83	73.02
	-	301.66	26.97	1.1172	0.9477	218.57	71.58
	120	300.02	27.79	1.1172	0.9425	224.65	70.76
	-	298.44	28.58	1.1172	0.9376	230.48	70.03
	140	292.10	31.75	1.1172	0.9177	253.56	67.06
	160	284.18	35.71	1.1172	0.8928	281.70	63.47
	-	254.00	50.80	1.1172	0.7980	381.83	50.72
	-	247.64	53.98	1.1172	0.7780	401.50	48.22
	-	243.84	55.88	1.1172	0.7660	413.01	46.74
	-	228.60	63.50	1.1172	0.7182	457.40	41.00

Steel Pipe Dimensions and Mass (Metric Units)

Weight Class	Sche dule #	Inside Diameter	Wall Thick-ness	Outside Area	Inside Area	Plain End Mass	Water Mass
		mm	mm	m^2/m	m^2/m	kg/m	kg/m
▼ NPS - 400 OD - 406.4mm ▼							
	-	396.84	4.78	1.2767	1.2467	47.34	123.80
		396.08	5.16	1.2767	1.2443	51.06	123.33
		395.28	5.56	1.2767	1.2418	54.96	122.82
		393.70	6.35	1.2767	1.2368	62.64	121.84
		392.12	7.14	1.2767	1.2319	70.30	120.87
	20	390.56	7.92	1.2767	1.2270	77.83	119.90
	-	388.92	8.74	1.2767	1.2218	85.71	118.90
STD	30	387.34	9.53	1.2767	1.2169	93.27	117.95
		385.78	10.31	1.2767	1.2120	100.70	116.98
		384.14	11.13	1.2767	1.2068	108.49	116.00
		382.58	11.91	1.2767	1.2019	115.86	115.06
XS	40	381.00	12.70	1.2767	1.1969	23.30	114.10
	-	377.86	14.27	1.2767	1.1871	137.99	112.23
	-	374.64	15.88	1.2767	1.1770	152.93	110.33
	60	373.08	16.66	1.2767	1.1721	160.12	109.41
	-	371.44	17.48	1.2767	1.1670	167.65	108.46
	-	368.30	19.05	1.2767	1.1571	181.97	106.62
	-	365.16	20.62	1.2767	1.1472	196.16	104.81
	80	363.52	21.44	1.2767	1.1420	203.53	103.88

Table #6 (part 6 of 13) - Metric Steel Pipe Dimensions based on ANSI/ASME B-36.10M - 1985 Standard

Steel Pipe Dimensions and Mass (Metric Units)

Weight Class	Schedule #	Inside Diameter	Wall Thickness	Outside Area	Inside Area	Plain End Mass	Water Mass
		mm	mm	m²/m	m²/m	kg/m	kg/m
	-	361.94	22.23	1.2767	1.1371	210.60	102.98
	-	358.74	23.83	1.2767	1.1270	224.82	101.17
	-	355.60	25.40	1.2767	1.1172	238.64	99.40
	100	354.02	26.19	1.2767	1.1122	245.56	98.52
	-	352.46	26.97	1.2767	1.1073	252.35	97.65
	-	349.24	28.58	1.2767	1.0972	266.28	95.88
	-	346.04	30.18	1.2767	1.0872	280.00	94.14
	120	344.48	30.96	1.2767	1.0822	286.64	93.27
	-	342.90	31.75	1.2767	1.0773	293.33	92.42
	140	333.34	36.53	1.2767	1.0472	333.19	87.34
	160	325.42	40.49	1.2767	1.0223	365.35	83.25
▼ NPS - 450	OD - 457mm ▼						
	-	447.44	4.78	1.4357	1.4057	53.31	157.52
	-	445.88	5.56	1.4357	1.4008	61.90	156.42
	10	444.30	6.35	1.4357	1.3958	70.57	155.32
	-	442.72	7.14	1.4357	1.3908	79.21	154.22
	20	441.16	7.92	1.4357	1.3859	87.71	153.13
	-	439.52	8.74	1.4357	1.3808	96.61	151.99
STD	-	437.94	9.53	1.4357	1.3758	105.16	150.91
-	-	436.38	10.31	1.4357	1.3709	113.57	149.82
-	30	434.74	11.13	1.4357	1.3658	122.38	148.72
-	-	433.18	11.91	1.4357	1.3609	130.72	147.64
XS	-	431.60	12.70	1.4357	1.3559	139.15	146.57

Steel Pipe Dimensions and Mass (Metric Units)

Weight Class	Schedule #	Inside Diameter	Wall Thick-ness	Outside Area	Inside Area	Plain End Mass	Water Mass
		mm	mm	m²/m	m²/m	kg/m	kg/m
	40	428.46	14.27	1.4357	1.3460	155.80	144.44
	-	425.24	15.88	1.4357	1.3359	172.74	142.28
	-	422.04	17.48	1.4357	1.3259	189.46	140.15
	60	418.90	19.05	1.4357	1.3160	205.74	138.08
	-	415.76	20.62	1.4357	1.3062	221.89	136.01
	-	412.54	22.23	1.4357	1.2960	238.34	133.92
	80	409.34	23.83	1.4357	1.2860	254.55	131.49
	-	406.20	25.40	1.4357	1.2761	270.34	129.82
	-	403.06	26.97	1.4357	1.2663	286.00	127.83
	-	399.84	28.58	1.4357	1.2561	301.94	125.80
		398.28	29.36	1.4357	1.2512	309.62	124.82
	100	396.64	30.18	1.4357	1.2461	317.66	123.80
		393.50	31.75	1.4357	1.2362	332.95	121.84
	120	387.14	34.93	1.4357	1.2162	363.56	117.95
	140	377.66	39.67	1.4357	1.1865	408.26	112.23
	160	366.52	45.24	1.4357	1.1515	459.37	105.72
▼ NPS - 500	OD - 508mm ▼						
	-	496.88	5.56	1.5959	1.5610	68.89	194.07
	10	495.30	6.35	1.5959	1.5560	78.55	192.85
	-	493.72	7.14	1.5959	1.5511	88.19	191.61
	-	492.16	7.92	1.5959	1.5462	97.67	190.39
	-	490.52	8.74	1.5959	1.5410	107.60	189.14
	20	488.94	9.53	1.5959	1.5361	117.15	187.93

Table #6 (part 7 of 13) - Metric Steel Pipe Dimensions based on ANSI/ASME B-36.10M - 1985 Standard

Steel Pipe Dimensions and Mass (Metric Units)

Weight Class	Schedule #	Inside Diameter	Wall Thickness	Outside Area	Inside Area	Plain End Mass	Water Mass
		mm	mm	m^2/m	m^2/m	kg/m	kg/m
	-	487.38	10.31	1.5959	1.5312	126.53	186.73
	-	485.74	11.13	1.5959	1.5260	136.37	185.48
	-	484.18	11.91	1.5959	1.5211	145.70	184.28
	30	482.60	12.70	1.5959	1.5161	155.12	183.08
	-	479.46	14.27	1.5959	1.5063	173.74	180.69
	40	477.82	15.09	1.5959	1.5011	183.42	179.47
	-	476.24	15.88	1.5959	1.4962	192.71	178.29
-	-	473.04	17.48	1.5959	1.4861	211.44	177.03
-	-	469.90	19.05	1.5959	1.4762	229.70	173.57
-	60	466.76	20.62	1.5959	1.4664	247.83	171.25
	-	463.54	22.23	1.5959	1.4563	266.29	168.91
	-	460.34	23.83	1.5959	1.4462	284.52	166.58
	-	457.20	25.40	1.5959	1.4363	302.28	164.32
	80	455.62	26.19	1.5959	1.4314	311.17	163.18
	-	454.06	26.97	1.5959	1.4265	319.92	162.07
	-	450.84	28.58	1.5959	1.4164	337.89	159.79
	-	447.64	30.18	1.5959	1.4063	355.61	157.52
	-	444.50	31.75	1.5959	1.3964	372.88	155.32
	100	442.92	32.54	1.5959	1.3915	381.53	154.22
	-	441.36	33.32	1.5959	1.3866	390.03	153.13
	-	438.14	34.93	1.5959	1.3765	407.49	150.91
	120	431.80	38.10	1.5959	1.3565	441.49	146.57
	140	419.10	44.45	1.5959	1.3166	508.11	138.08
	160	407.98	50.01	1.5959	1.2817	564.81	130.84

Steel Pipe Dimensions and Mass (Metric Units)

Weight Class	Schedule #	Inside Diameter	Wall Thick-ness	Outside Area	Inside Area	Plain End Mass	Water Mass
		mm	mm	m^2/m	m^2/m	kg/m	kg/m
▼ NPS - 550 OD - 559mm ▼							
	-	547.88	5.56	1.7562	1.7212	75.88	235.78
	10	546.30	6.35	1.7562	1.7163	86.54	234.44
	-	544.72	7.14	1.7562	1.7113	97.17	233.08
	-	543.16	7.92	1.7562	1.7064	107.63	231.74
	-	541.52	8.74	1.7562	1.7012	118.60	230.35
STD	20	539.94	9.53	1.7562	1.6963	129.13	229.01
		538.38	10.31	1.7562	1.6914	139.50	227.67
		536.74	11.13	1.7562	1.6862	150.37	226.30
		535.18	11.91	1.7562	1.6813	160.68	224.98
XS	30	533.60	12.70	1.7562	1.6764	171.09	223.65
		530.46	14.27	1.7562	1.6665	191.69	221.03
		527.24	15.88	1.7562	1.6564	212.69	218.36
		524.04	17.48	1.7562	1.6463	233.43	215.72
		520.90	19.05	1.7562	1.6365	251.65	213.13
		517.76	20.62	1.7562	1.6266	273.76	210.55
	60	514.54	22.23	1.7562	1.6165	294.25	207.96
	-	511.34	23.83	1.7562	1.6064	314.49	205.38
-	-	508.20	25.40	1.7562	1.5966	334.23	202.86
-	-	505.06	26.97	1.7562	1.5867	353.84	200.35
-	80	501.84	28.58	1.7562	1.5766	373.83	197.81
	-	498.64	30.18	1.7562	1.5665	393.57	195.31
	-	495.50	31.75	1.7562	1.5567	412.81	192.85

Table #6 (part 8 of 13) - Metric Steel Pipe Dimensions based on ANSI/ASME B-36.10M - 1985 Standard

Steel Pipe Dimensions and Mass (Metric Units)							
Weight Class	Sche dule #	Inside Diameter	Wall Thickness	Outside Area	Inside Area	Plain End Mass	Water Mass
		mm	mm	m^2/m	m^2/m	kg/m	kg/m
	-	492.36	33.32	1.7562	1.5468	431.94	190.39
	100	489.14	34.93	1.7562	1.5367	451.42	187.93
	-	485.94	36.53	1.7562	1.5266	470.66	185.48
	-	482.80	38.10	1.7562	1.5168	489.41	183.08
	120	476.44	41.28	1.7562	1.4968	527.02	178.29
-	140	463.74	47.63	1.7562	1.4569	600.63	168.91
-	160	451.04	53.98	1.7562	1.4170	672.26	159.79
▼ NPS - 600 OD - 610mm ▼							
-	10	597.30	6.35	1.9164	1.8765	94.53	280.06
		595.72	7.14	1.9164	1.8715	106.15	278.60
		594.16	7.92	1.9164	1.8665	117.59	277.13
		592.52	8.74	1.9164	1.8615	129.59	275.61
STD	20	590.94	9.53	1.9164	1.8565	141.12	274.15
		589.38	10.31	1.9164	1.8516	152.47	272.68
		587.74	11.13	1.9164	1.8464	164.37	271.18
		586.18	11.91	1.9164	1.8415	175.66	269.73
XS	-	584.60	12.70	1.9164	1.8366	187.06	268.27
	30	581.46	14.27	1.9164	1.8267	209.64	265.40
	-	578.24	15.88	1.9164	1.8166	232.66	262.48
	40	575.04	17.48	1.9164	1.8065	255.41	259.59
	-	571.90	19.05	1.9164	1.7967	277.61	256.74
	-	568.76	20.62	1.9164	1.7868	299.69	253.93
	-	565.54	0.22	1.9164	1.7767	322.21	251.07

Steel Pipe Dimensions and Mass (Metric Units)							
Weight Class	Sche dule #	Inside Diameter	Wall Thick-ness	Outside Area	Inside Area	Plain End Mass	Water Mass
		mm	mm	m^2/m	m^2/m	kg/m	kg/m
	-	562.34	23.83	1.9164	1.7666	344.46	248.23
	60	560.78	24.61	1.9164	1.7617	355.26	246.85
	-	559.20	25.40	1.9164	1.7568	366.17	245.46
	-	556.06	26.97	1.9164	1.7486	387.76	242.71
	-	552.84	28.58	1.9164	1.7368	409.77	239.92
	-	549.64	30.18	1.9164	1.7267	431.52	237.15
	80	548.08	30.96	1.9164	1.7218	442.08	235.78
	-	546.50	31.75	1.9164	1.7169	452.74	234.44
		543.36	33.32	1.9164	1.7070	473.84	231.74
		540.14	34.93	1.9164	1.6969	495.35	229.01
		536.94	36.53	1.9164	1.6869	516.80	226.30
		533.80	38.10	1.9164	1.6770	537.33	223.65
	100	532.22	38.89	1.9164	1.6720	547.71	222.34
	-	530.66	39.67	1.9164	1.6671	557.43	221.03
	120	517.96	46.02	1.9164	1.6272	640.03	210.55
	140	505.26	52.37	1.9164	1.5873	720.15	200.35
	160	490.92	59.54	1.9164	1.5423	808.22	189.14
▼ NPS - 650 OD - 660mm ▼							
	-	647.30	6.35	2.0735	2.0336	102.36	329.77
	-	645.72	7.14	2.0735	2.0286	114.95	327.80
	10	644.16	7.92	2.0735	2.0237	127.36	326.58
	-	642.52	8.74	2.0735	2.0185	140.37	324.92
STD	-	640.94	9.53	2.0735	2.0136	152.87	323.34

Table #6 (part 9 of 13) - Metric Steel Pipe Dimensions based on ANSI/ASME B-36.10M - 1985 Standard

Steel Pipe Dimensions and Mass (Metric Units)

Weight Class	Schedule #	Inside Diameter	Wall Thickness	Outside Area	Inside Area	Plain End Mass	Water Mass
		mm	mm	m^2/m	m^2/m	kg/m	kg/m
-	-	639.38	10.31	2.0735	2.0087	165.18	321.75
-	-	637.74	11.13	2.0735	2.0035	178.09	320.13
-	-	636.18	11.91	2.0735	1.9986	190.34	318.55
XS	20	634.60	12.70	2.0735	1.9937	202.72	316.97
		631.46	14.27	2.0735	1.9838	227.23	313.84
		628.24	15.88	2.0735	1.9737	252.24	310.67
		625.04	17.48	2.0735	1.9636	276.96	307.51
		621.90	19.05	2.0735	1.9538	301.10	304.42
		618.76	20.62	2.0735	1.9439	325.12	301.34
		615.54	22.23	2.0735	1.9338	349.62	298.24
		612.34	23.83	2.0735	1.9237	373.84	295.14
		609.20	25.40	2.0735	1.9139	397.49	292.11
▼ NPS - 700		OD - 711mm ▼					
		698.30	6.35	2.2337	2.1938	110.34	383.53
		696.72	7.14	2.2337	2.1888	123.93	381.80
	10	695.16	7.92	2.2337	2.1839	137.32	380.08
	-	693.52	8.74	2.2337	2.1788	151.36	378.31
STD	-	691.94	9.53	2.2337	2.1738	164.85	376.58
-	-	690.38	10.31	2.2337	2.1689	178.15	374.88
-	-	688.74	11.13	2.2337	2.1637	192.09	373.11
-	-	687.18	11.91	2.2337	2.1588	205.32	371.41
XS	20	685.60	12.70	2.2337	2.1539	218.69	369.71
-	-	682.46	14.27	2.2337	2.1440	245.18	366.32

Steel Pipe Dimensions and Mass (Metric Units)

Weight Class	Schedule #	Inside Diameter	Wall Thickness	Outside Area	Inside Area	Plain End Mass	Water Mass
		mm	mm	m^2/m	m^2/m	kg/m	kg/m
-	30	679.24	15.88	2.2337	2.1339	271.21	362.89
		676.04	17.48	2.2337	2.1238	298.95	359.49
		672.90	19.05	2.2337	2.1140	325.06	356.14
		669.76	20.62	2.2337	2.1041	351.05	352.82
		666.54	22.23	2.2337	2.0940	377.58	349.46
		663.34	23.83	2.2337	2.0839	403.81	346.11
		660.20	25.40	2.2337	2.0741	429.44	342.83
▼ NPS - 750		OD - 762mm ▼					
		749.30	6.35	2.3939	2.3540	118.33	441.35
		747.72	7.14	2.3939	2.3490	132.91	439.49
		746.16	7.92	2.3939	2.3441	147.28	437.64
		744.52	8.74	2.3939	2.3390	162.35	435.74
STD	-	742.94	9.53	2.3939	2.3340	176.84	433.90
-	-	741.38	10.31	2.3939	2.3291	191.11	432.06
-	-	739.74	11.13	2.3939	2.3240	206.09	430.16
-	-	738.18	11.91	2.3939	2.3191	220.30	428.35
XS	20	736.60	12.70	2.3939	2.3141	234.67	426.51
	-	733.46	14.27	2.3939	2.3042	263.12	422.88
	30	730.24	15.88	2.3939	2.2941	292.18	419.20
	-	727.04	17.48	2.3939	2.2841	320.93	415.53
	-	723.90	19.05	2.3939	2.2742	349.02	411.93
	-	720.76	20.62	2.3939	2.2643	376.98	408.35
	-	717.54	22.23	2.3939	2.2542	405.54	404.74

Table #6 (part 10 of 13) - Metric Steel Pipe Dimensions based on ANSI/ASME B-36.10M - 1985 Standard

Steel Pipe Dimensions and Mass (Metric Units)

Weight Class	Schedule #	Inside Diameter	Wall Thickness	Outside Area	Inside Area	Plain End Mass	Water Mass
		mm	mm	m^2/m	m^2/m	kg/m	kg/m
	-	714.34	23.83	2.3939	2.2442	433.78	401.14
	-	711.20	25.40	2.3939	2.2343	461.38	397.61
	-	708.06	26.97	2.3939	2.2244	488.85	394.09
	-	704.84	28.58	2.3939	2.2143	516.90	390.53
	-	701.64	30.18	2.3939	2.2043	544.65	387.00
	-	698.50	31.75	2.3939	2.1944	571.75	383.63
		▼ NPS - 800	OD - 813mm ▼				
	-	800.30	6.35	2.5541	2.5142	126.31	503.22
	-	798.72	7.14	2.5541	2.5093	141.89	501.24
	10	797.16	7.92	2.5541	2.5044	157.24	499.27
	-	795.52	8.74	2.5541	2.4992	173.34	497.23
STD	-	793.94	9.53	2.5541	2.4942	188.82	495.26
-	-	792.38	10.31	2.5541	2.4893	204.08	493.29
-	-	790.74	11.13	2.5541	2.4842	220.08	491.28
-	-	789.18	11.91	2.5541	2.4793	235.28	489.32
XS	20	787.60	12.70	2.5541	2.4743	250.64	487.36
	-	784.46	14.27	2.5541	2.4645	281.07	483.48
	30	781.24	15.88	2.5541	2.4543	312.15	479.54
	40	778.04	17.48	2.5541	2.4443	342.91	475.62
	-	774.90	19.05	2.5541	2.4344	372.98	471.78
	-	771.76	20.62	2.5541	2.4246	402.92	467.95
	-	767.76	22.62	2.5541	2.4120	433.49	464.08
	-	765.34	23.83	2.5541	2.4044	463.75	460.22

Steel Pipe Dimensions and Mass (Metric Units)

Weight Class	Schedule #	Inside Diameter	Wall Thickness	Outside Area	Inside Area	Plain End Mass	Water Mass
		mm	mm	m^2/m	m^2/m	kg/m	kg/m
	-	762.20	25.40	2.5541	2.3945	493.32	456.43
	-	759.08	26.97	2.5541	2.3847	522.77	452.66
	-	755.84	28.58	2.5541	2.3745	552.85	448.86
	-	752.64	30.18	2.5541	2.3645	582.61	445.06
	-	749.50	31.75	2.5541	2.3546	611.68	441.35
		▼ NPS - 850	OD - 864mm ▼				
	-	851.30	6.35	2.7143	2.6744	134.30	569.15
	-	849.72	7.14	2.7143	2.6695	150.87	567.05
	10	848.16	7.92	2.7143	2.6646	167.20	564.95
	-	846.52	8.74	2.7143	2.6594	184.33	562.77
STD	-	844.94	9.53	2.7143	2.6545	200.31	560.69
-	-	843.38	10.31	2.7143	2.6496	217.05	558.60
-	-	841.74	11.13	2.7143	2.6444	234.08	556.44
		840.18	11.91	2.7143	2.6395	250.26	554.37
	20	838.60	12.70	2.7143	2.6345	266.61	552.29
	-	835.46	14.27	2.7143	2.6247	299.02	548.14
	30	832.24	15.88	2.7143	2.6146	332.12	543.95
	40	829.04	17.48	2.7143	2.6045	364.90	539.77
	-	825.90	19.05	2.7143	2.5946	396.93	535.67
	-	822.76	20.62	2.7143	2.5848	428.85	531.60
	-	819.54	22.23	2.7143	2.5747	461.45	524.20
	-	816.34	23.83	2.7143	2.5646	493.72	523.35
	-	813.20	25.40	2.7143	2.5547	525.27	519.32

Table #6 (part 11 of 13) - Metric Steel Pipe Dimensions based on ANSI/ASME B-36.10M - 1985 Standard

Steel Pipe Dimensions and Mass (Metric Units)							
Weight Class	Schedule #	Inside Diameter	Wall Thickness	Outside Area	Inside Area	Plain End Mass	Water Mass
		mm	mm	m^2/m	m^2/m	kg/m	kg/m
	-	810.06	26.97	2.7143	2.5449	556.69	515.30
	-	806.84	28.58	2.7143	2.5348	588.79	511.23
	-	803.64	30.18	2.7143	2.5247	620.56	507.18
XS	-	800.50	31.75	2.7143	2.5149	651.61	503.22
▼ NPS - 900 OD - 914mm ▼							
	-	901.30	6.35	2.8714	2.8352	142.13	639.14
	-	899.72	7.14	2.8714	2.8266	159.67	636.90
	10	898.16	7.92	2.8714	2.8217	176.96	634.68
	-	896.52	8.74	2.8714	2.8165	195.11	632.39
STD	-	894.94	9.53	2.8714	2.8115	212.56	630.17
-	-	893.38	10.31	2.8714	2.8066	229.76	627.95
-	-	891.74	11.13	2.8714	2.8015	247.31	625.67
-	-	890.18	11.91	2.8714	2.7966	264.94	623.46
XS	20	888.60	12.70	2.8714	2.7916	282.27	621.26
	-	885.46	14.27	2.8714	2.7818	316.11	616.86
	30	882.24	15.88	2.8714	2.7716	351.70	612.42
	-	879.04	17.48	2.8714	2.7616	386.45	607.98
	40	875.90	19.05	2.8714	2.7517	420.42	603.63
	-	872.76	20.62	2.8714	2.7419	454.27	599.31
	-	869.54	22.23	2.8714	2.7317	488.86	594.91
	-	866.34	23.83	2.8714	2.7217	523.11	590.55
	-	863.20	25.40	2.8714	2.7118	556.59	586.27
	-	860.06	26.97	2.8714	2.7020	589.95	581.99

Steel Pipe Dimensions and Mass (Metric Units)							
Weight Class	Schedule #	Inside Diameter	Wall Thickness	Outside Area	Inside Area	Plain End Mass	Water Mass
		mm	mm	m^2/m	m^2/m	kg/m	kg/m
	-	856.84	28.58	2.8714	2.6918	624.03	577.67
	-	853.64	30.18	2.8714	2.6818	657.77	573.37
	-	850.50	31.75	2.8714	2.6719	690.76	569.15
▼ NPS - 950 OD - 965mm ▼							
	-	949.16	7.92	3.0316	2.9819	186.92	708.47
	-	947.52	8.74	3.0316	2.9767	206.10	706.05
STD	-	945.94	9.53	3.0316	2.9718	224.54	703.70
-	-	944.38	10.31	3.0316	2.9669	242.72	701.36
-	-	942.74	11.13	3.0316	2.9617	261.80	698.94
-	-	941.18	11.91	3.0316	2.9568	279.92	696.62
XS	-	939.60	12.70	3.0316	2.9518	298.24	694.28
-	-	936.46	14.27	3.0316	2.9420	334.56	689.65
-	-	933.24	15.88	3.0316	2.9319	371.68	684.94
-	-	930.04	17.48	3.0316	2.9218	408.43	680.24
-	-	926.90	19.05	3.0316	2.9119	444.38	675.66
		923.76	20.62	3.0316	2.9021	480.21	671.07
		920.54	22.23	3.0316	2.8920	516.82	666.43
		917.34	23.83	3.0316	2.8819	553.08	661.80
		917.34	23.83	3.0316	2.8721	588.53	657.27
		911.06	26.97	3.0316	2.8622	621.87	652.74
		907.84	28.58	3.0316	2.8521	659.97	648.16
		904.64	30.18	3.0316	2.8420	695.73	643.61
		901.50	31.75	3.0316	2.8322	730.69	639.14

Table #6 (part 12 of 13) - Metric Steel Pipe Dimensions based on ANSI/ASME B-36.10M - 1985 Standard

Steel Pipe Dimensions and Mass (Metric Units)

Weight Class	Schedule #	Inside Diameter	Wall Thickness	Outside Area	Inside Area	Plain End Mass	Water Mass
		mm	mm	m^2/m	m^2/m	kg/m	kg/m
▼ NPS - 1000 OD - 1016mm ▼							
		1000.16	7.92	3.1919	3.1421	196.89	786.32
		998.52	8.74	3.1919	3.1370	217.09	783.77
STD	-	996.94	9.53	3.1919	3.1320	236.53	781.30
-	-	995.38	10.31	3.1919	3.1271	255.69	778.82
-	-	993.74	11.13	3.1919	3.1219	275.80	776.29
-	-	992.18	11.91	3.1919	3.1170	294.90	773.83
XS	-	990.60	12.70	3.1919	3.1121	314.22	771.37
-	-	987.46	14.27	3.1919	3.1022	352.51	766.47
-	-	984.24	15.88	3.1919	3.0921	391.65	761.51
-	-	981.04	17.48	3.1919	3.0820	430.42	756.58
-	-	977.90	19.05	3.1919	3.0722	468.34	751.72
-	-	974.76	20.62	3.1919	3.0623	506.14	746.89
-	-	971.54	22.23	3.1919	3.0522	544.78	741.99
-	-	968.34	23.83	3.1919	3.0421	583.05	737.10
-	-	965.20	25.40	3.1919	3.0323	620.48	732.32
-	-	962.06	26.97	3.1919	3.0224	657.78	727.55
-	-	958.84	28.58	3.1919	3.0123	695.92	722.72
-	-	955.64	30.18	3.1919	3.0022	733.68	717.90
-	-	952.50	31.75	3.1919	2.9924	770.62	713.17

Steel Pipe Dimensions and Mass (Metric Units)

Weight Class	Schedule #	Inside Diameter	Wall Thick-ness	Outside Area	Inside Area	Plain End Mass	Water Mass
		mm	mm	m^2/m	m^2/m	kg/m	kg/m
▼ NPS 1050 OD - 1067mm ▼							
-	-	1049.52	8.74	3.3521	3.2972	228.09	865.54
STD	-	1047.94	9.53	3.3521	3.2922	248.52	862.95
-	-	1046.38	10.31	3.3521	3.2873	268.66	860.36
-	-	1044.74	11.13	3.3521	3.2822	289.60	857.69
-	-	1043.18	11.91	3.3521	3.2773	309.88	655.10
XS	-	1041.60	12.70	3.3521	3.2723	330.19	852.52
-	-	1038.46	14.27	3.3521	3.2624	370.45	847.36
-	-	1035.24	15.88	3.3521	3.2523	411.62	842.15
-	-	1032.04	17.48	3.3521	3.2423	452.40	836.95
-	-	1028.90	19.05	3.3521	3.2324	492.30	831.85
-	-	1025.76	20.62	3.3521	3.2225	532.07	826.77
-	-	1022.54	22.23	3.3521	3.2124	572.73	821.62
-	-	1019.34	23.83	3.3521	3.2024	613.02	816.48
-	-	1016.20	25.40	3.3521	3.1925	652.42	811.00
-	-	1013.06	26.97	3.3521	3.1826	691.70	806.42
-	-	1009.84	28.58	3.3521	3.1725	731.86	802.81
-	-	1006.64	30.18	3.3521	3.1625	771.64	796.26
-	-	1003.50	31.75	3.3521	3.1526	810.55	791.20

Table #6 (part 13 of 13) - Metric Steel Pipe Dimensions based on ANSI/ASME B-36.10M - 1985 Standard

Pipe Lengths

Generally, steel pipe when specified, is supplied in 21 ft. (6.4 m) lengths, +/- 1/2 inch (12.7 mm). It is also available in random lengths, 16 ft. to 22 ft. (4.88 m to 6.71 m), and double random lengths, 22 ft. (6.71 m) minimum, with a minimum average length per order of 35 ft. (10.67 m).

Pipe End Finish

The ends of pipe are commercially available in the following end finishes:

- Threaded and Coupled
- Threaded without Couplings
- Plain End - cut square
- Beveled for Welding - 30 degree bevel with a 1/16" (1.6 mm) land.
- Grooved End - cut or rolled for mechanical couplings

The preceding end finishes are shown in illustration #7.

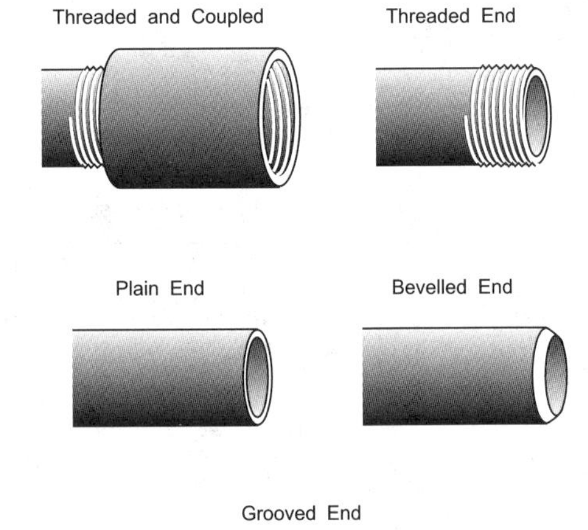

Illustration #7 - Five Common Pipe End Finishes

When ends are threaded and coupled, only one end of each length is supplied with a coupling. In sizes 4 in. (100 mm) and larger, it is common practice to supply the unprotected thread with a thread protector. The pipe couplings supplied for standard weight pipe 2 inches (50 mm) and under are parallel threaded (straight-tapped).

Couplings supplied for extra strong and double extra strong pipe are taper threaded (taper-tapped) and recessed for all sizes. Illustrations #8A and #8B display cross sectional drawings of both a straight-tapped merchant coupling and a taper-tapped line coupling used on pipe ends.

Supplied on pipe sizes
2 in. (50 mm) and smaller

Straight Tapped Coupling

Taper Pipe Thread

Joints made with straight tapped couplings
are inferior to those with taper tapped couplings

Illustration #8A - Straight Tapped Coupling

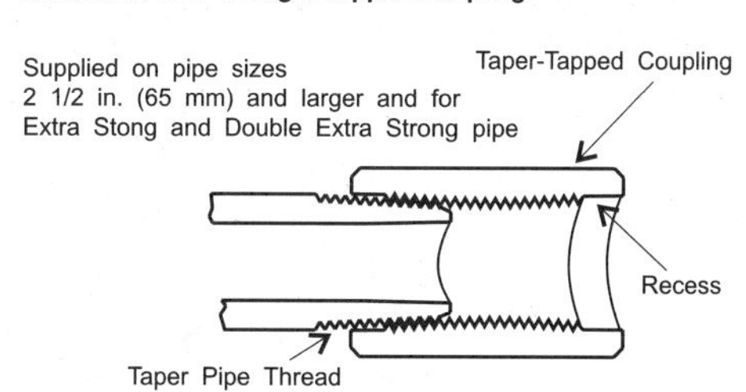

Recesses in coupling facilitate pipe alignment when
making-up threaded joints of larger sizes and heavier pipe

Illustration #8B - Taper Tapped and Recessed Coupling

Pipe Identification and Marking

Steel pipe is available in numerous manufacturing classifications, grades, weights, schedule numbers, sizes, and lengths . For this reason it is important to be able to interpret steel pipe's identification markings. Steel pipe can be identified by paint stencil, or stamped markings on the pipe itself or by a tag attached to smaller sized bundles of pipe. ***The American Society for Testing and Materials (ASTM)*** require pipe made to their specifications to be labeled with:

- Manufacturer's Name (trademark or brand may be used).
- Pipe manufacturing method:
  F = Furnace butt welded, continuous welded
  E = Electric resistance welded
  S = Seamless
- Weight of Pipe: Std., XS., or XXS.
- Specification Number: ASTM specification number.
- Length: units pipe was ordered in.

In addition to the above marking, certain ASTM pipe classifications may require pipe to be marked with:

- Grade of pipe.
- Hydrostatic Test Pressure or:
  - NH when not tested
  - Schedule Number
  - S for supplementary requirements
  - Pipe size

An example of ASTM marking on pipe is shown in illustration #9.

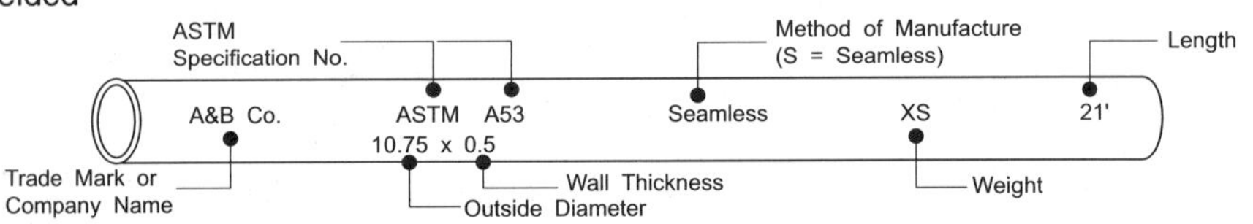

Illustration #9 - Typical ASTM Pipe Marking

The American Petroleum Institute (API) require pipe conforming to their specifications to be marked with the following:

- Manufacturer's Name or Mark
- API's Specification Number
- Diameter: either the nominal or outside diameter in inches
- Weight per Foot
- Grade: API's grades use numbers follow ing letters to indicate minimum yield strength
- Process of Manufacture:
  S = Seamless
  E = Welded, except butt-welded
  F = Butt-welded
  SW = Spiral welded
- Type of Steel:
  E - Electric-furnace steel
  R - Rephosphorized steel (class 11) no marking required for open hearth or basic-oxygen steel
- Heat Treatment:
  - HN-Normalized or normalized/tempered
  - HS-Subcritical stress relieved
  - HA-Subcritical age hardened
  - HQ-Quench and tempered
- Test Pressure: if pressure is higher than in tables

An example of API marking on pipe is shown in illustration #10.

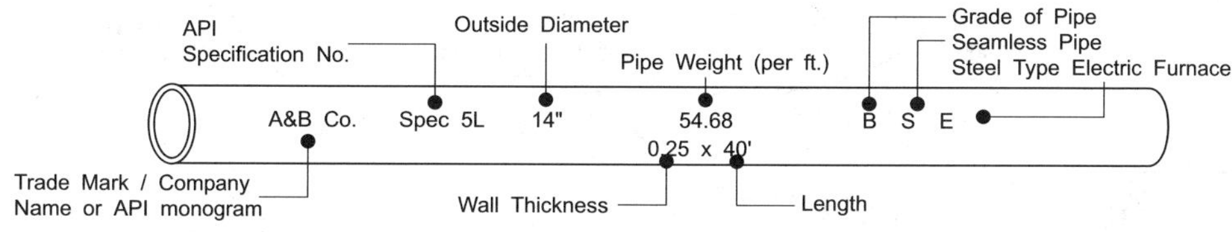

Illustration #10 - Typical API Pipe Marking

The Canadian Standards Association (CSA) require pipe made to their specifications to be marked with the following:

- Manufacturer's Name or Mark.
- Specified Outside Diameter, in millimetres.
- Specified Wall Thickness, in millimetres.
- Grade of Pipe (CSA Z245.1 Grades range from 172 to grade 550 inclusive).
- Sour Service: if pipe is intended for this use it must be labeled with the symbol SS.
- Process of Manufacture:
  E-Electric welded or submerged arc welded pipe
  F-Butt welded pipe
- Type of Steel:
  E - electric furnace steel
  Open hearth or oxygen steel: no mark required.
- Heat Treatment: if required.
- Length: in metres to two decimal places.
- Hydrostatic Test Pressure: if applicable.

An example of CSA marking on pipe is shown in illustration #11.

Note: Pipe marking may vary depending upon specific pipe specification requirements within each standard.

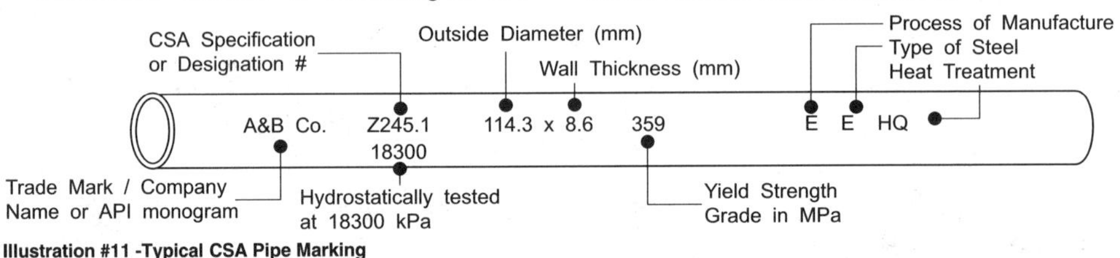

Illustration #11 -Typical CSA Pipe Marking

Steel Pipe Finishing Processes

The standard finish given to steel pipe is a lacquer coating that is intended to prevent corrosion in transit. This pipe is usually referred to as black iron pipe. Other common surface finishes include:

- Bare Metal
- Pickled only
- Pickled and Oiled
- Galvanized

Galvanizing

Zinc is one of the most common types of corrosion protective coatings given to steel pipe. Galvanized pipe is carbon steel pipe which has been coated with zinc both on the inside and outside of the pipe.

The zinc is usually applied by a hot-dipping process where the pipe is submerged in a molten bath of zinc. The standard weight of zinc applied to the pipe is a minimum average of 1.8 oz. of zinc per square ft. of pipe (.55 kg per square m).

Corrosion Protective/ Lined and Coated Pipe

To enhance pipe's corrosion resistance, and to modify flow characteristics, pipe can be lined and/or coated with a variety of materials. Table #7 shows some corrosion protective materials available for pipe. It is by no means comprehensive, but is intended to give an overview of some of the materials that may be utilized. Even though most of the materials listed are used to prevent corrosion, some are also used to improve flow characteristics within the piping system.

Corrosion Protective Coatings and Linings

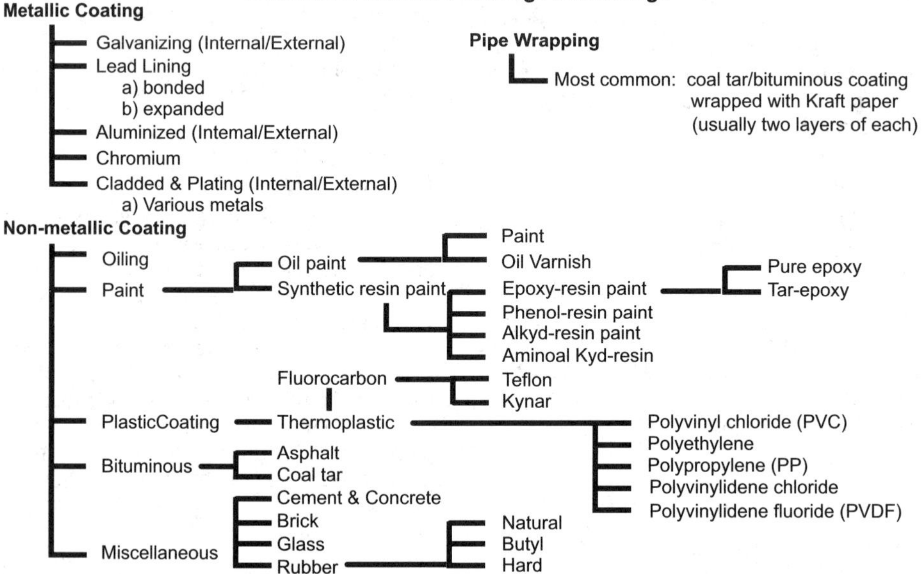

Table #7 - Pipe Corrosion Protective Coatings

Wrought Iron Pipe

The material classification of wrought iron pipe should not be confused with being the same type of pipe or pipe material classification as wrought steel pipe. Wrought steel pipe is another term given to carbon steel pipe because of the method of manufacturing. Wrought in this sense means to work or to be formed as compared to casting.

Wrought iron pipe primarily consists of refined iron and iron silicate (slag) which gives the pipe a tough fibrous structure. The iron silicate (approximately 1% to 3% by weight) was believed to give the pipe an added resistance to corrosion. However, steel pipe has generally proven to be equal to wrought iron in withstanding corrosion and is less expensive to produce.

Because of these factors, wrought iron pipe is no longer commercially produced, however it is still encountered in existing installations.

Distinguishing between wrought iron pipe and steel pipe can be made by wrought iron's somewhat rougher surface and its thicker pipe wall. Since wrought iron is less dense than steel but made to the same outside diameter and weight per ft., its wall thickness is slightly larger to compensate for the lighter density.

Stainless Steel Pipe

The principal alloy elements in stainless steel pipe are chromium and nickel, which are added to steel to give it a high resistance to corrosion and heat. Chromium forms a thin oxide layer on pipe which prevents further oxidation and corrosion of the pipe. The major effects of nickel are to increase the pipe's toughness, strength and resistance to heat. The amounts of these two elements vary depending on the type of stainless steel, ranging from 4% to 27% for chromium and 0% to 22% for nickel.

Stainless Classifications: The American Iron and Steel Institute (AISI) uses the following standard type numbers to identify the three major types of stainless steel:

- Austenitic (AISI types 200 and 300 series)
- Ferritic (AISI type 400 series)
- Martensitic (AISI types 400 and 500 series)

The majority of stainless steel pipe is made from Austenitic stainless steel which provides maximum corrosion resistance, strength and ductility.

To distinguish between the two most common series of stainless steel (300 series and 400 series) use a magnet. Series 400 is magnetic while series 300 is non-magnetic.

Stainless Steel Pipe Dimensions

Wall thickness dimensions for stainless steel pipe, as a rule coincide to steel pipe dimensions. However, there are some exceptions: 12 inch (300 mm) schedule 40S stainless pipe has a slightly thinner wall than schedule 40 steel pipe and 10 inch (250 mm) and 12 inch (300 mm) schedule 80S stainless both have thinner walls than the corresponding sizes in schedule 80 steel pipe.

Stainless steel pipe is also available in schedule 5S and I0S virtually throughout the size ranges available.

Note: The suffix S in schedule numbers distinguish stainless steel numbers from carbon steel schedule numbers.

All dimensions for stainless steel pipe are shown in tables #8 and #9.

Stainless Steel Pipe Dimensions and Weight (Imperial Units)

Nominal Pipe Size	Sch. #	O.D.	I.D.	Wall Thick-ness	Nominal Weight Plain End	Nominal Weight Waterfill
Inches		Inches	Inches	Inches	lb/ft	lb/ft
1/8	10S	0.405	0.307	0.049	0.19	0.032
	40S	0.405	0.269	0.068	0.24	0.025
	80S	0.405	0.215	0.095	0.31	0.016
1/4	10S	0.540	0.410	0.065	0.33	0.057
	40S	0.540	0.364	0.088	0.42	0.045
	80S	0.540	0.302	0.119	0.54	0.031
3/8	10S	0.675	0.545	0.065	0.42	0.101
	40S	0.675	0.493	0.091	0.57	0.083
	80S	0.675	0.423	0.126	0.74	0.061
1/2	5S	0.840	0.710	0.065	0.54	0.172
	10S	0.840	0.674	0.083	0.67	0.155
	40S	0.840	0.622	0.109	0.85	0.132
	80S	0.840	0.546	0.147	1.09	0.102
3/4	5S	1.050	0.920	0.065	0.69	0.288
	10S	1.050	0.884	0.083	0.86	0.266
	40S	1.050	0.824	0.113	1.13	0.231
	80S	1.050	0.742	0.154	1.47	0.188
1	5S	1.315	1.185	0.065	0.87	0.478
	10S	1.315	1.095	0.109	1.40	0.409
	40S	1.315	1.045	0.133	1.68	0.375
	80S	1.315	0.957	0.179	2.17	0.312

Stainless Steel Pipe Dimensions and Weight (Imperial Units)

Nominal Pipe Size	Sch. #	O.D.	I.D.	Wall Thick-ness	Nominal Weight Plain End	Nominal Weight Waterfill
Inches		Inches	Inches	Inches	lb/ft	lb/ft
1 1/4	5S	1.660	1.530	0.065	1.11	0.797
	10S	1.660	1.442	0.109	1.81	0.708
	40S	1.660	1.380	0.140	2.27	0.649
	80S	1.660	1.278	0.191	3.00	0.555
1 1/2	5S	1.900	1.770	0.065	1.28	1.066
	10S	1.900	1.682	0.109	2.09	0.963
	40S	1.900	1.610	0.145	2.72	0.882
	80S	1.900	1.500	0.200	3.63	0.765
2	5S	2.375	2.245	0.065	1.61	1.72
	10S	2.375	2.157	0.109	2.64	1.58
	40S	2.375	2.067	0.154	3.65	1.45
	80S	2.375	1.939	0.218	5.02	1.28
2 1/2	5S	2.875	2.709	0.083	2.48	2.50
	10S	2.875	2.635	0.120	3.53	2.36
	40S	2.875	2.469	0.203	5.79	2.07
	80S	2.875	2.323	0.276	7.66	1.87
3	5S	3.500	3.334	0.083	3.03	3.78
	10S	3.500	3.260	0.120	4.33	3.62
	40S	3.500	3.068	0.216	7.58	3.20
	80S	3.500	2.900	0.300	10.25	2.86

Table #8A -Stainless Pipe Dimensions (Imperial Units)

Stainless Steel Pipe Dimensions and Weight (Imperial Units)						
Nominal Pipe Size	Sch. #	O.D.	I.D.	Wall Thick-ness	Nominal Weight Plain End	Nominal Weight Waterfill
Inches		Inches	Inches	Inches	lb/ft	lb/ft
	5S	4.000	3.834	0.083	3.48	5.00
$3^1/_2$	10S	4.000	3.760	0.120	4.97	4.81
	40S	4.000	3.548	0.266	9.11	4.29
	80S	4.000	3.364	0.318	12.51	3.84
	5S	4.500	4.334	0.083	3.92	6.39
4	10S	4.500	4.260	0.120	5.61	6.18
	40S	4.500	4.026	0.237	10.79	5.50
	80S	4.500	3.826	0.337	14.98	4.98
	5S	5.563	5.345	0.109	6.36	9.72
5	10S	5.563	5.295	0.134	7.77	9.54
	40S	5.563	5.041	0.258	14.62	8.67
	80S	5.563	4.813	0.375	20.78	7.88
	5S	6.625	6.407	0.109	7.60	13.97
6	10S	6.625	6.357	0.134	9.29	13.75
	40S	6.625	6.065	0.280	18.97	12.51
	80S	6.625	5.761	0.432	28.57	11.29
	5S	8.625	8.407	0.109	9.93	24.06
8	10S	8.625	8.329	0.148	13.40	23.61
	40S	8.625	7.981	0.322	28.55	21.70
	80S	8.625	7.625	0.500	43.39	19.78
	5S	10.750	10.482	0.134	15.19	37.39
10	10S	10.750	10.420	0.165	18.65	36.95
	40S	10.750	10.020	0.365	40.48	34.20
	80S	10.750	9.750	0.500	54.74	32.35

Table #8B -Stainless Pipe Dimensions (Imperial Units)

Stainless Steel Pipe Dimensions and Weight (Imperial Units)						
Nominal Pipe Size	Sch. #	O.D.	I.D.	Wall Thick-ness	Nominal Weight Plain End	Nominal Weight Waterfill
Inches		Inches	Inches	Inches	lb/ft	lb/ft
	5S	12.750	12.438	0.156	20.98	52.65
12	10S	12.750	12.390	0.180	24.17	52.25
	40S	12.750	12.000	0.375	49.56	49.00
	80S	12.750	11.750	0.500	65.42	46.92
14	5S	14.000	13.688	0.156	23.07	63.77
	10S	14.000	13.624	0.188	27.73	63.17
16	5S	16.000	15.670	0.165	27.90	83.57
	10S	16.000	15.624	0.188	31.75	83.08
18	5S	18.000	7.670	0.165	31.43	106.00
	10S	18.000	17.624	0.188	35.76	105.70
20	5S	20.000	19.624	0.188	39.78	131.00
	10S	20.000	19.564	0.218	46.06	130.20
22	5S	22.000	21.624	0.188	43.80	159.10
	10S	22.000	21.564	0.218	50.71	158.20
24	5S	24.000	23.564	0.218	55.37	188.90
	10S	24.000	23.500	0.250	63.41	187.90
30	5S	30.000	29.500	0.250	79.43	296.30
	10S	30.000	29.376	0.312	98.93	293.70

Note: Because of the various types of stainless steel available nominal weight (mass) based on carbon steel has been used.
For general correction factor multiply by:
0.99 for AISI 400 series stainless steel
1.02 for AISI 300 series stainless steel

Stainless Steel Pipe Dimensions and Mass (Metric Units)						
Nominal Pipe Size	Sch. #	O.D.	I.D.	Wall Thick-ness	Nominal Mass Plain End	Nominal Mass Waterfill
mm		mm	mm	mm	kg/m	kg/m
6	10S	10.3	7.82	1.24	0.28	0.048
	40S	10.3	6.84	1.73	0.37	0.037
	80S	10.3	5.48	2.41	0.47	0.030
8	10S	13.7	10.40	1.65	0.49	0.085
	40S	13.7	9.22	2.24	0.63	0.067
	80S	13.7	7.66	3.02	0.80	0.046
10	10S	17.1	13.80	1.65	0.63	0.150
	40S	17.1	12.48	2.31	0.84	0.124
	80S	17.1	10.70	3.20	1.10	0.091
15	5S	21.3	18.00	1.65	0.80	0.256
	10S	21.3	17.08	2.11	1.00	0.231
	40S	21.3	15.76	2.77	1.27	0.197
	80S	21.3	13.84	3.73	1.62	0.152
20	5S	26.7	23.40	1.65	1.03	0.429
	10S	26.7	22.48	2.11	1.28	0.396
	40S	26.7	20.96	2.87	1.69	0.344
	80S	26.7	18-88	3.91	2.20	0.280
25	5S	33.4	30.10	1.65	1.30	0.712
	10S	33.4	27.86	2.77	2.09	0.609
	40S	33.4	26.64	3.38	2.50	0.559
	80S	33.4	24.30	4.55	3.24	0.465

Stainless Steel Pipe Dimensions and Mass (Metric Units)						
Nominal Pipe Size	Sch. #	O.D.	I.D.	Wall Thick-ness	Nominal Mass Plain End	Nominal Mass Waterfill
mm		mm	mm	mm	kg/m	kg/m
32	5S	42.2	38.90	1.65	1.65	1.187
	10S	42.2	36.66	2.77	2.70	1.055
	40S	42.2	35.08	3.56	3.39	0.967
	80S	42.2	32.50	4.85	4.47	0.827
40	5S	48.3	45.00	1.65	1.91	1.588
	10S	48.3	42.76	2.77	3.11	1.434
	40S	48.3	40.94	3.68	4.05	1.314
	80S	48.3	38.14	5.08	5.41	1.139
50	5S	60.3	57.00	1.65	2.40	2.562
	10S	60.3	54.76	2.77	3.93	2.353
	40S	60.3	52.48	3.91	5.44	2.160
	80S	60.3	49.22	5.54	7.48	1.907
65	5S	73.0	68.78	2.11	3.69	3.724
	10S	73.0	66.90	3.05	5.26	3.515
	40S	73.0	62.68	5.16	8.63	3.083
	80S	73.0	58.98	7.01	11.41	2.785
80	5S	88.9	84.68	2.11	4.51	5.630
	10S	88.9	82.80	3.05	6.45	5.392
	40S	88.9	77.92	5.49	11.29	4.766
	80S	88.9	73.66	7.62	15.27	4.260
90	5S	101.6	97.38	2.11	5.18	7.448
	10S	101.6	95.50	3.05	7.40	7.165
	40S	101.6	90.12	5.74	13.57	6.390
	80S	101.6	85.44	8.08	18.63	5.720

Table #9A - Stainless Pipe Dimensions (Metric)

Stainless Steel Pipe Dimensions and Mass (Metric Units)						
Nominal Pipe Size	Sch. #	O.D.	I.D.	Wall Thick-ness	Nominal Mass Plain End	Nominal Mass Waterfill
mm		mm	mm	mm	kg/m	kg/m
	5 S	114.3	110.06	2.11	5.84	9.518
100	10S	114.3	108.20	3.05	8.36	9.205
	40S	114.3	102.26	6.02	16.07	8.192
	80S	114.3	97.18	8.56	22.32	7.418
	5S	141.3	135.76	2.77	9.47	14.478
125	10S	141.3	134.50	3.40	11.57	14.210
	40S	141.3	128.20	6.55	21.77	12.914
	80S	141.3	122.24	9.53	30.97	11.737
	5S	168.3	162.76	2.77	11.32	20-808
150	10S	168.3	161.50	3.40	13.84	20.481
	40S	168.3	154.08	7.11	28.26	18.634
	80S	168.3	146.36	10.97	42.56	16.816
	5S	219.1	213.56	2.77	14.79	35.838
200	10S	219.1	211.58	3.76	19.96	35.167
	40S	219.1	202.74	8.18	42.55	32.322
	80S	219.1	193.70	12.70	64.64	29.462
	5S	273.1	266.3	3.40	22.63	55.692
	10S	273.1	264.72	4.19	27.78	55.037
250	40S	273.1	254.56	9.27	60.31	50.941
	80S	273.1	247.70	12.70	96.01	48.185
	5S	323.9	315.98	3.96	31.25	78.422
300	10S	323.9	314.76	4.57	36.00	77.826
	40S	321.9	304.84	9.53	73.88	72.986
	80S	323.9	298.50	12.70	132.08	69.887

Stainless Steel Pipe Dimensions and Mass (Metric Units)						
Nominal Pipe Size	Sch. #	O.D.	I.D.	Wall Thick-ness	Nominal Mass Plain End	Nominal Mass Waterfill
mm		mm	mm	mm	kg/m	kg/m
350	5S	355.6	347.68	3.96	34.36	94.985
	10S	355.6	346.04	4.78	41.30	94.092
400	5S	406.4	398.02	4.19	41.56	124.478
	10S	406.4	396.84	4.78	47.29	123.748
450	5S	457	448.62	4.19	46.81	158.260
	10S	457	447.44	4.78	53.26	157.440
500	5S	508	498.44	4.78	59.25	195.125
	10S	508	496.92	5.54	68.61	193.933
550	5S	559	549.44	4.78	65.24	236.980
	10S	559	547.92	5.54	75.53	235.639
600	5S	610	604.46	5.54	82.47	281.367
	10S	610	597.30	6.35	94.45	279.877
750	5S	762	749.30	6.35	118.31	441.339
	10S	762	746.16	7.92	147.36	437.466

Note: Because of the various types of stainless steel available nominal mass (weight) based on carbon steel has been used. For a general correction factor multiply by:
- 0.99 for AISI 400 series stainless steel
- 1.02 for AISI 300 stainless steel

Table #9B - Stainless Pipe Dimensions (Metric)

Copper & Brass Pipe

Copper and copper alloy pipe offer excellent resistance to corrosion in potable and non-potable water piping, boiler feed water lines, and other similar systems. The copper alloy used most often in these systems is red brass. Red brass has a chemical composition of approximately 84.0% to 86.0% copper with the remainder of the alloy being zinc.

Both copper and red brass pipe are produced in two general weights or wall thickness classifications:

- Regular weight
- Extra-strong weight

The pipe dimensions within these two classifications conform to standardized pipe size dimensions (see table #10A, B for actual dimensions and weights).

Copper and red brass pipe is available in seamless 12 ft. (3.66 m) lengths, with a possible tolerance of +/- 1/2 inch (13 mm) per length. Sizes range from 1/8 inch (6 mm) to 12 inches (300 mm) NPS.

REGULAR - Copper & Red Brass Pipe Dimensions and Weights (Mass)									
Imperial Units					Metric Units				
Nominal Pipe Size	O.D.	Wall Thick-ness	Copper Weight	Brass Weight	Nominal Pipe Size	O.D.	Wall Thick-ness	Copper Mass	Brass Mass
inches	inches	inches	lb/ft	lb/ft	mm	mm	mm	kg/m	kg/m
1/8	0.405	0.062	0.259	0.53	6	10.3	1.57	0.385	0.376
1/4	0.540	0.082	0.457	0.447	8	13.7	2.08	0.680	0.665
3/8	0.675	0.090	0.641	0.627	10	17.1	2.29	0.954	0.933
1/2	0.840	0.107	0.955	0.934	15	21.3	2.72	1.42	1.39
3/4	1.050	0.114	1.30	1.27	20	26.7	2.90	1.93	1.89
1	1.315	0.126	1.82	1.78	25	33.4	3.20	2.71	2.65
1 1/4	1.660	0.146	2.69	2.63	32	42.2	3.71	4.00	3.91
1 1/2	1.900	0.150	3.20	3.13	40	48.3	3.81	4.76	4.66
2	2.375	0.156	4.22	4.12	50	60.3	3.96	6.28	6.13
2 1/2	2.875	0.187	6.12	5.99	65	73.0	4.75	9.11	8.91
3	3.500	0.219	8.76	8.56	80	88.9	5.56	13.0	12.7
3 1/2	4.000	0.250	11.4	11.2	90	102	6.35	17.0	16.7
4	4.500	0.250	12.9	12.7	100	114	6.35	19.2	18.9
5	5.562	0.250	16.2	15.8	125	141	6.35	24.1	23.5
6	6.625	0.25	19.4	19.0	150	168	6.35	28.9	28.3
8	8.625	0.3120	31.6	30.9	200	219	7.92	47.0	46.0
10	10.750	0.365	46.2	45.2	250	273	9.27	68.7	67.3
12	12.750	0.375	56.5	55.3	300	324	9.52	84.1	82.3

Table #10A - Copper and Red Brass Pipe Dimensions - Regular

EXTRA STRONG -Copper and Red Brass Pipe Dimensions and Weights (Mass)									
Imperial Units					Metric Units				
Nominal Pipe Size	O.D.	Wall Thick-ness	Copper Weight	Brass Weight	Nominal Pipe Size	O.D.	Wall Thick-ness	Copper Mass	Brass Mass
inches	inches	inches	lb/ft	lb/ft	mm	mm	mm	kg/m	kg/m
1/8	0.405	0.100	0.371	0.363	6	10.3	2.54	0.552	0.540
1/4	0.540	0.123	0.625	0.611	8	13.7	3.12	0.930	0.909
3/8	0.675	0.127	0.847	0.829	10	17.1	3.23	1.26	1.23
1/2	0.840	0.149	1.25	1.23	15	21.3	3.78	1.86	1.83
3/4	1.050	0.157	1.71	1.67	20	26.7	3.99	2.54	2.48
1	1.315	0.182	2.51	2.46	25	33.4	4.62	3.73	3.66
1 1/4	1.660	0.194	3.46	3.39	32	42.2	4.93	5.15	5.04
1 1/2	1.900	0.203	4.19	4.10	40	48.3	5.16	6.23	6.10
2	2.375	0.221	5.80	5.67	50	60.3	5.61	8.63	8.44
2 1/2	2.875	0.280	8.85	8.66	65	73.0	7.11	13.2	12.9
3	3.500	0.304	11.8	11.6	80	88.9	7.72	17.6	17.0
3 1/2	4.000	0.321	14.4	14.1	90	102	8.15	21.4	21.0
4	4.500	0.341	17.3	16.9	100	114	8.66	25.7	25.1
5	5.562	0.375	23.7	23.2	125	141	9.52	35.3	34.5
6	6.625	0.437	32.9	32.2	150	168	11.1	49.0	47.9
8	8.625	0.500	49.5	48.4	200	219	12.7	73.7	72.0
10	10.750	0.500	62.4	61.1	250	273	12.7	92.9	90.9

Table #10B - Copper and Red Brass Pipe Dimensions - Extra Strong

Cast Iron Pipe

Cast iron pipe is an iron alloy containing between 2% to 4% carbon and approximately 1% to 3% silicon. This chemical composition may vary depending on type and specific properties needed in the pipe.

Casting of pipe is accomplished by pouring molten hot iron alloy into either a stationary mold form or into the more commonly used centrifugal mold. Centrifugal pipe cast molding uses centrifugal force (produced by the rotating mold) to evenly distribute the molten metal throughout the pipe casting.

Illustration #12 shows an example of a centrifugal pipe casting mold.

The four principal types of cast iron available for pipe and fitting manufacturing are:

- White cast iron
- Ductile iron
- Gray cast iron
- Malleable iron

These types of cast iron are manufactured into two major cast iron pipe classifications:

- Cast Iron Pressure Pipe
- Cast Iron Soil Pipe

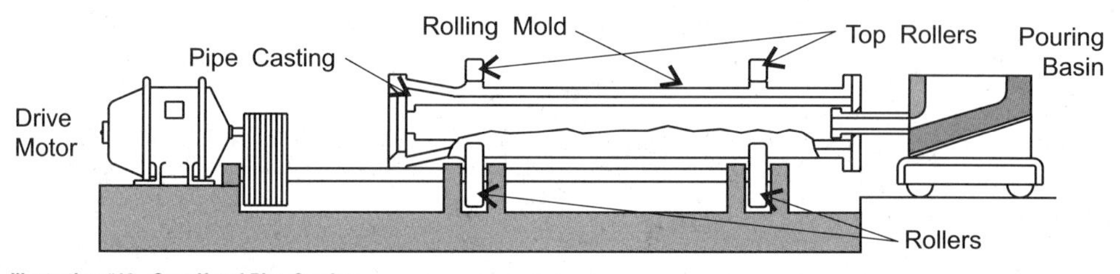

Illustration #12 - Centrifugal Pipe Casting

Cast Iron Pressure Pipe

Cast iron pressure pipe was commonly used for carrying gas, water, and other liquid services under pressure. It is now being rapidly replaced by various types of plastic pipe. Ductile and gray cast iron fittings and ductile pipe are the two predominant cast iron types used in cast iron pressure systems. Ductile iron is essentially made by adding magnesium to molten cast iron. Magnesium changes most of the free carbon into a modular form which gives it added strength while still having the corrosion resistance of cast iron.

Pressure cast iron can be specified by weight, class, wall thickness and/or pressure rating. Because the majority of pressure cast iron is intended for underground service, common standards usually incorporate loading and laying conditions in trenching. American National Standards Institute (ANSI) and American Water Works Association (AWWA) standards cover the majority of manufacturing specifications for ductile and gray cast iron fittings and ductile pipe (gray cast iron pressure pipe is no longer commercially being manufactured).

Cast Iron Soil Pipe

Due to building code regulations, cast iron soil pipe is commonly used in commercial buildings for DWV (drainage, waste, and vent). There are three styles of soil pipe manufactured: single hub, double hub, and hubless pipe. All three styles of pipe are usually made from gray cast iron. American Society for Testing and Materials (ASTM) standards categorize cast iron soil pipe into two weights, extra heavy and service weight pipe.

Tables #11A and #11B give specific dimensions for extra heavy and service weight soil pipe based on ASTM standards. The Canadian Standards Association (CSA) uses only standard grade (CAN3-B70-M86) to designate their cast iron pipe. Dimensions based on CSA standards are presented in table #11C.

Sizes for soil pipe range from 2 inches (50mm) to 15 inches (380mm) nominal inside diameter for all sizes. Soil pipe is available in 5 ft. (1.5 m) and 10 ft. (3.0 m) lengths which are referred to as laying lengths. Illustration #13 indicates laying lengths of all three styles of soil pipe.

Silicon Iron Pipe

Cast iron soil pipe and fittings can be manufactured with a special high silicon composition (often referred to as Duriron), that is intended for services requiring high corrosion resistance.

A typical use of this type of pipe would be for laboratory drainage and waste.

Silicon in this type of pipe makes up approximately 14.2% to 14.75% of the chemical composition.

When used with hub and spigot joints, it is recommended that acid proof rope packing be used.

Note: If M.J. couplings (Mechanical Joints) are used, special acid resistant sleeves or coatings on the elastomer sleeve should be used.

DOUBLE HUB PIPE

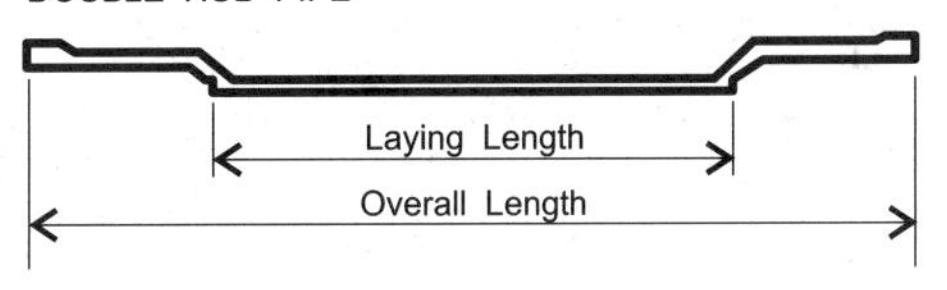

SINGLE HUB PIPE

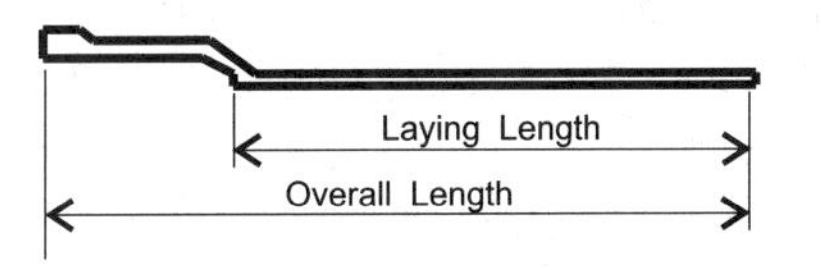

HUBLESS PIPE

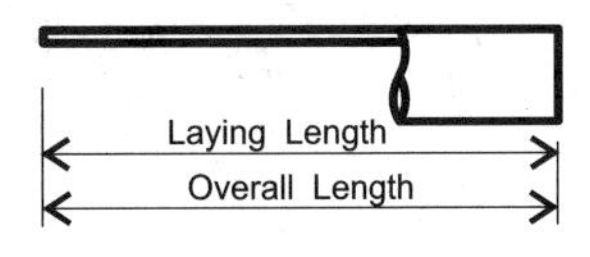

Illustration #13 - Laying Lengths for Cast Iron Soil Pipe

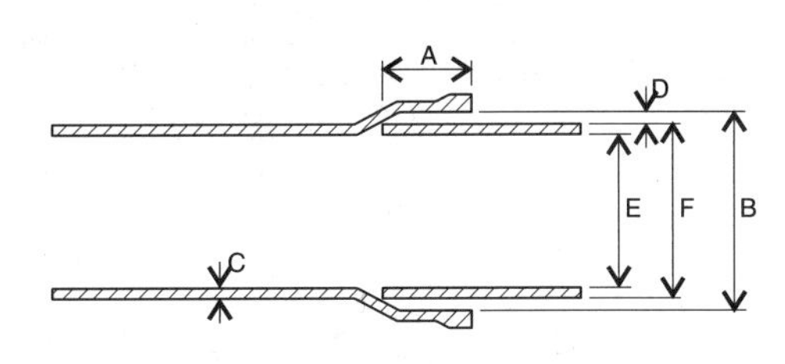

LEGEND

A = Hub Length
B = Hub I.D.
C = Wall Thickness
D = Caulking Width
E = Barrel I.D.
F = Barrel O.D.

Table #11A,B,C Diagram - Hub, Spigot and Pipe Dimensions

EXTRA HEAVY Cast Iron Soil Pipe Dimensions Based on ASTM Standards							
Nominal Pipe Size	(C) Min. Wall Thickness	(C) Nominal Wall Thickness	(F) O.D.	(E) Nominal I.D.	(A) Min. Hub Length	(B) Hub I.D.	(D) Caulking Width
Imperial Units (inches)							
2	0.12	0.19	2.38	2.00	2.50	3.06	0.35
3	0.18	0.25	3.50	3.00	2.75	4.19	0.35
4	0.18	0.25	4.50	4.00	3.00	5.19	0.35
5	0.18	0.25	5.50	5.00	3.00	6.19	0.35
6	0.18	0.25	6.50	6.00	3.00	7.19	0.35
8	0.25	0.31	8.62	8.00	3.50	9.50	0.44
10	0.31	0.37	10.75	10.00	3.50	11.62	0.44
12	0.31	0.37	12.75	12.00	4.25	13.75	0.50
15	0.37	0.44	15.88	15.00	4.25	17.00	0.56
Metric Units (mm)							
50	3.05	4.83	60.45	50.8	63.50	77.72	8.89
75	4.57	6.35	88.90	76.2	69.85	106.43	8.89
100	4.57	6.35	114.90	101.6	76.20	131.83	8.89
125	4.57	635	139.70	127.0	76.20	157.23	8.89
150	4.57	6.35	165.10	152.4	76.20	182.63	8.89
200	6.35	7.87	218.95	203.2	88.90	241.30	11.18
250	7.87	9.40	273.05	254.0	88.90	295.15	11.18
300	7.87	9.40	323.85	304.8	107.95	249.25	12.70
375	9.40	11.18	403.35	381.0	107.95	431.80	14.22

Table #11A - Extra Heavy Cast Iron Soil Pipe

SERVICE WEIGHT Cast Iron Soil Pipe Dimensions Based on ASTM Standards							
Nominal Pipe Size	(C) Min. Wall Thickness	(C) Nominal Wall Thickness	(F) O.D.	(E) Nominal I.D.	(A) Min. Hub Length	(B) Hub I.D.	(D) Caulking Width
Imperial Units (inches)							
2	0.12	0.17	2.250	2.08	2.50	2.94	0.320
3	0.13	0.17	3.250	3.08	2.75	3.94	0.320
4	0.14	0.18	4.250	4.07	3.00	4.94	0.320
5	0.15	0.19	5.250	5.06	3.00	5.94	0.320
6	0.16	0.20	6.250	6.05	3.00	6.94	0.320
8	0.17	0.22	8.375	8.16	3.50	9.25	0.440
10	0.21	0.26	10.500	10.24	3.50	11.38	0.440
12	0.22	0.28	12.500	12.22	4.25	13.50	0.500
15	0.25	0.30	15.375	15.08	4.25	16.75	0.565
Metric Units (mm)							
50	3.05	4.32	57.15	52.83	63.50	74.68	8.13
75	3.30	4.32	82.55	78.23	69.85	100.08	8.13
100	3.56	4.57	107-95	103.38	76.20	125.48	8.13
125	3.81	4.83	133.35	128.52	76.20	150.88	8.13
150	4.06	5.08	158.75	153.67	76.20	176.28	8.13
200	4.32	5.59	212.73	207.26	88.90	234.95	11.18
250	5.33	6.60	266.70	260.10	88.90	289.05	11.18
300	5.59	7.11	317.50	310.39	107.95	342.90	12.70
375	6.35	7.62	390.53	383.03	107.95	425.45	14.35

Table #11B - Service Weight Cast Iron Soil Pipe

Cast Iron Soil Pipe Dimensions Based on CSA (CAN3-B70-M86) Standards							
Nominal Pipe Size	(C) Min. Wall Thickness	(E) Minimum I.D.	(F) Minimum O.D.	(F) Maximum O.D.	(A) Min. Hub Length	(B) Hub I.D.	(D) Caulking Width
Imperial Units (Inches)							
1½	0.118	1.398	1.811	1.969			
2	0.118	1.882	2.244	2.441	2.244	2.937	0.252
3	0.150	2.874	3.268	3.445	2.244	4.173	0.299
4	0.181	3.882	4.291	4.488	2.244	5.197	0.311
5	0.201	4.843	5.315	5.472	2.480	6.063	0.319
6	0.209	5.827	6.299	6.535	2.480	7.126	0.330
8	0.252	7.756	8.386	8.622	2.756	9.488	0.339
10	0.280	9.685	10.512	10.669	2.756	11.575	0.358
12	0.300	11.693	12.520	12.677	2.992	13.819	0.370
15	0.339	14.567	15.630	15.827	3.228	17.047	0.421
Metric Units (mm)							
40	3.0	35.5	46	50			
50	3.0	47.8	57	62	57	74.6	6.4
75	3.8	73.0	83	87.5	57	106	7.6
100	4.6	98.6	109	114	57	132	7.9
125	5.1	123	135	139	63	154	8.1
150	5.3	148	160	166	63	181	8.4
200	6.4	197	213	219	70	241	8.6
250	7.1	246	267	271	70	294	9.1
300	7.6	297	318	322	76	351	9.4
375	8.6	370	397	402	82	433	10.7

Table #11C - CSA Standards for Cast Iron Soil Pipe

Thermoplastic Pipe

Plastic pipe can be divided into two separate material categories. These are:

- Thermoplastic
- Thermoset.

Thermoplastic materials can be repeatedly heated to a liquid state, cooled and reformed without any chemical change to the plastic. Pipe made from thermoplastic becomes soft when heated, with a decrease in working pressure and tensile strength. When cooled, it becomes hard and brittle with a decrease in its impact resistance.

The following common plastic piping materials are classified as thermoplastics:

- Acrylonitrile Butadiene Styrene (ABS)
- Polyvinyl Chloride (PVC)
- Chlorinated Polyvinyl Chloride (CPVC)
- Polybutylene (PB)
- Polyethylene (PE)
- Polypropylene (PP)
- Polyvinylidine Fluoride (PVDF)

Thermoset Plastic Pipe

Thermoset plastic, once cured, can not be reheated without effecting the chemical make up of the plastic. Plastic pipe made from thermoset material is permanently rigid and encompasses all reinforced thermosetting resin pipes (RTRP). This type of piping is often referred to as fiberglass pipe or fiberglass reinforced pipe (FRP). The following are examples of thermoset plastic pipe:

- Glass Reinforced Epoxy
- Glass Reinforced Polyester
- Glass Reinforced Vinylestes
- Glass Reinforced Furan

Laminated and Composition Plastic Pipe

In addition to the two separate material categories of plastic pipe, there is a third type that combines a thermoplastic liner with a thermoset outer structure. This type of plastic pipe is referred to as dual laminated (DL).

It combines the advantages of thermoplastic with the rigid structural properties of a reinforced thermosetting resin.

Metals can also be used with plastics to improve physical properties. Plastic metal composition pipe uses a metal core laminated between interior and exterior layers of thermoplastic. The result is a plastic composition pipe that has the durability of plastic and the strength of metal.

Plastic Standards and Ratings

Most plastic pipe used in the North American market is manufactured to standards set out by ASTM, CSA, and/or NSF (National Sanitation Foundation).

Plastic pipe is pressure designated (aside from sewer drainage grade) by schedule number, pressure rating at a given temperature and/or given a standard dimension ratio (SDR).

The standard dimension ratio (SDR) is the average outside diameter of the pipe divided by the minimum wall thickness.

Examples of thermoplastic pressure ratings and comparable SDR numbers are shown in table #12A. Maximum operating pressure rates for schedule 40 and 80 thermoplastics are given in tables #12B and #12C. Temperature correction factors are given in table #12D.

Thermoplastic Pressure Ratings and Comparable SDR Numbers								
PVC & CPVC			Polybutylene			Polyethylene (3408)		
PSI	kPa	SDR No.	PSI	kPa	SDR No.	PSI	kPa	SDR No.
315	2172	13.5	160	1103	13.5	160	1103	9
250	1724	17	125	862	17	130	862	13.5
200	1380	21	100	690	21	100	690	17
160	1103	26	80	552	26	80	552	21
125	862	32.5	45	310	45.5	65	448	26
100	690	41	-	-	-	50	345	32.5
63	435	63	-	-	-	-	-	-
Note: Pressure ratings are taken at 73°F (23°C).								

Table #12A - Thermoplastic Ratings and SDR Numbers

Plastic Pipe Maximum Pressure Rating (psi at 73°F)									
Nominal Pipe Size	Schedule 40 PVC & CPVC	Schedule 80 PVC & CPVC			Schedule 80 Poly-propylene	Schedule 80 Poly-ethylene	Schedule 80 PVDF		DL & RTRP
Inches	Socket End	Socket End	Threaded End	Flanged	Heat Fusion Joint	Heat Fusion Joint	Heat Fusion Joint	Threaded End	Filament Wound Socket
1/2	600	850	420	150	218	-	580	290	150
3/4	480	690	340	150	218	176	470	235	150
1	450	630	320	150	174	164	430	215	150
1 1/4	370	520	260	150	174	134	370	185	150
1 1/2	330	470	240	150	174	121	320	160	150
2	280	400	200	150	162	101	275	135	150
2 1/2	300	420	210	150	-	-	300	150	150
3	260	370	190	150	130	96	260	130	150
4	220	320	160	150	130	81	220	110	150
6	180	280	-	150	Low Pressure or Drainage Applications Only	64	180	90	150
8	160	250	-	150		57	-	-	150
10	140	230	-	150		51	-	-	150
12	130	230	-	150		48	-	-	150
14	130	220	-	150		-	-	-	150
16-24	130	220	-	150			-	-	150

Table #12B - Plastic Pipe Pressure Ratings (psi)

Plastic Pipe Maximum Pressure Rating (kPa at 23°C)									
Nominal Pipe Size	Schedule 40 PVC & CPVC	Schedule 80 PVC & CPVC			Schedule 80 Poly-propylene	Schedule 80 Poly-ethylene	Schedule 80 PVDF		DL & RTRP
mm	Socket End	Socket End	Threaded End	Flanged	Heat Fusion Joint	Heat Fusion Joint	Heat Fusion Joint	Threaded End	Filament Wound Socket
15	4140	5860	2896	1035	1503	-	4000	2000	1035
20	3300	4760	2345	1035	1503	1214	3240	1620	1035
25	3100	4340	2210	1035	1200	1131	2965	1482	1035
32	2500	3590	1793	1035	1200	924	2550	1276	1035
40	2280	3240	1655	1035	1117	835	2210	1100	1035
50	1930	2760	1380	1035	-	697	1896	930	1035
65	2070	2900	1448	1035	897	-	2070	1035	1035
80	1790	2550	1310	1035	897	662	1790	900	1035
100	1520	2210	1100	1035	Low Pressure or Drainage Applications Only	559	1520	760	1035
150	1240	1930		1035		442	1240	620	1035
200	1100	1720		1035		393	-	-	1035
250	970	1590		1035		352	-	-	*1035
300	900	1520		1035		331	-	-	*1035
350	900	1520		1035		-	-	-	*1035
400-600	900	1520		1035		-	-	-	*1035

Note: 1. Metric Pressureratings are rounded off.
2. * indicates pressures only apply to RTRP.
3. All pressures based on water service.
4. For higher temperatures multiply the pressure rating by factor given in the temperature correction table.

Table #12C - Plastic Pipe Pressure Ratings (kPa)

Temperature Correction Factors For Commonly Used Thermoplastics						
Operating Temperature		Factors				
°F	°C	PVC	CPVC	PP	PE	PVDF
70	21	1.00	1.00	1.00	1.00	1.00
80	27	0.90	0.96	0.97	0.95	0.95
90	32	0.75	0.92	0.91	0.88	0.87
100	38	0.62	0.85	0.85	0.82	0.80
110	43	0.50	0.77	0.80	0.76	0.75
115	46	0.45	0.74	0.77	0.72	0.71
120	49	0.40	0.70	0.75	0.69	0.68
125	52	0.35	0.66	0.71	0.66	0.66
130	54	0.30	0.62	0.68	0.63	0.62
140	60	0.22	0.55	0.65	N.R.	0.58
150	66	N.R.	0.47	0.57	N.R.	0.52
160	71	N.R.	0.40	0.50	N.R.	0.49
170	77	N.R.	0.32	0.26	N.R.	0.45
180	82	N.R.	0.25	N.R.	N.R.	0.42
200	93	N.R.	0.18	N.R.	N.R.	0.36
210	99	N.R.	0.15	N.R.	N.R.	0.33
240	116	N.R.	N.R.	N.R.	N.R.	0.25
280	138	N.R.	N.R.	N.R.	N.R.	0.18

Note: 1. N.R. - Not Recommended
2. Multiply maximum pressure rating by correction factor to determine pressure rating at given temperature

Table #12D - Thermoplastic Temperature Correction

Plastic Pipe Selection and Usage

Acrylonitrile Butadiene Styrene (ABS).

ABS pipe and fittings are colored black for schedule 40 and blue-gray for pressure ratings 145, 180, and 230 psi (1000, 1240, & 1590 kPa). Airline, one of the only plastic pipe designs approved for compressed air service, is light blue in color.

Most ABS pipe is used in drainage, waste and vent (DWV) housing applications, but it does have limited use in well casing, electrical and communication conduit and industrial chemical services.

The methods used to join ABS are: solvent cementing, threading, grooved joints, and flanging.

It is supplied in rigid lengths and has a maximum operating temperature of 180 degrees F (82 degrees C).

Polyvinyl Chloride (PVC)

PVC pipe and fittings are colored gray for schedule 40, 80 and 120. AWWA categories of PVC are white, blue or green in color. All other pressure and sewer grades of pipe are white.

PVC is the most widely used type of plastic pipe material with use in: drainage, waste and vent (DWV); building sewers; well casing; irrigation systems; chilled water piping; industrial applications; water service and transmission lines.

The methods used to join PVC are: solvent cementing, O-rings, threading, grooved joints, and flanging.

PVC is supplied in rigid lengths and has a maximum operating temperature of 140 degrees F (60 degrees C).

Chlorinated Polyvinyl Chloride (CPVC)

CPVC pipe and fittings are colored purple-gray for schedule 40 and 80. Copper tubing equivalent sizes are beige in color, and SDR 13.5 rated pipe is orange.

Applications include: hot and cold water distribution, sprinkler systems (UL & FM approved), industrial applications and for services requiring higher temperature than PVC.

The methods used to join CPVC are: solvent cementing, threading, grooved joints, and flanging.

CPVC is supplied in rigid lengths and has a maximum operating temperature of 210 degrees F (99 degrees C).

Polyethylene (PE)

PE gas service piping and fittings are colored orange, beige or black. All other SDR and classes are produced in the color black.

Most PE piping is used for gas and water distribution and service. Other usage include: irrigation systems and industrial applications.

The methods used to join PE piping are: butt and socket heat fusion, O-rings, insert fittings, threading and flanged connections.

PE pipe is normally supplied in coils up to 3 inch (75 mm), and in rigid lengths over this size. The maximum operating temperature is 160 degrees F (71 degrees C).

Polypropylene (PP)

PP pipe and fittings are colored black or light blue for schedule 40 and 80. All other pressure ratings are colored white.

It is used in industrial applications, laboratory waste and pure water systems.

The methods used to join PP piping are: butt and socket heat fusion, electrical resistance heat fusion and mechanical couplings.

PP pipe is supplied in rigid lengths and has a maximum operating temperature of 180 degrees F (82 degrees C).

Polybutylene (PB)

All SDR ratings and classes of PB are colored black or blue.

Applications include: hot and cold water distribution, hydronic heating, industrial applications and sprinkler systems (UL & FM approved).

The methods used to join PB piping are: butt and socket heat fusion, O-rings, insert fittings, crimped joints, threaded and flanged.

PB pipe is supplied in rigid lengths or coils. The maximum operating temperature is 210 degrees F (99 degrees C).

Polyvinylidine Fluoride (PVDF)

PVDF pipe and fittings are colored red or natural for schedule 80 and natural or white for all other classes or ratings.

PVDF is used in highly corrosive and chemical services including: wet and dry chloride, bromine, pure water and halogens.

The methods used to join PVDF include: butt and socket heat fusion, grooved, threaded and flanged connections.

It is supplied in rigid lengths and has a maximum operating temperature of 280 degrees F (138 degrees C).

Reinforced Thermosetting Resin Pipes (RTRP).

RTRP pipe and fittings are supplied in various colors depending on the manufacturer. It is used in all types of industrial and commercial applications.

The methods used to join RTR piping include: butt, bell and spigot, adhesive bonding, flanged and threaded connections.

It is supplied in rigid lengths and has maximum operating temperatures of:

- Glass Reinforced Epoxy 300 degrees F (149 degrees C).
- Glass Reinforced Polyester 225 degrees F (107 degrees C).
- Glass Reinforced Vinylestes 250 degrees F (121 degrees C).
- Glass Reinforced Furan 300 degrees F (149 degrees C).

Solvent Cementing

Solvent cementing is the most common method used to join thermoplastic (ABS, PVC, and CPVC) pipe and fittings. The following give a brief description of the steps involved in the assembling of a solvent cement joint:

1. Cut the pipe squarely with a miter box and hand saw or with a plastic pipe cutter.
2. Remove all burrs and ridges from the pipe end. ***Ridges or raised beads on the pipe will have a tendency to wipe away the cement when fitting the joint together.***
3. Wipe the end of the pipe and socket of the fitting to remove any dirt, moisture or grease.
4. Select the appropriate applicator for the size of pipe used. See table #13.

APPROPRIATE APPLICATOR BRUSH SIZE					
Nominal Pipe Size		Maximum Width		Minimum Length	
Inches	mm	Inches	mm	Inches	mm
1 to $1^1/_4$	25 to 32	1	25	$1^1/_2$	40
$1^1/_2$ to 2	40 to 50	$1^1/_2$	40	2	50
3	80	$2^1/_2$	65	3	80
4	100	3	80	$3^1/_2$	90
6	150	5	125	$5^1/_2$	140
8	200	6	150	6	150

Table #13 - Applicator Brush Sizes

5. The joining surfaces must be softened by the use of primer, cement or a combination of both primer and cement.

Note: Primer is not required on ABS.

6. Apply sufficient cement to pipe and fitting to fill the gap space in the joint. See illustration #14.
7. Assemble the pipe and fitting while the cement is still wet and fluid. Twist the pipe slightly while assembling, and when bottomed, hold for approximately 30 seconds to prevent push out from the tapered fitting.
8. Wipe off any excess cement from the assembled joint. Handle the joint with care during set time. Table #14 gives initial set times for various pipe sizes and temperatures.

Initial Set and Cure Items							
Pipe Diameter		Temperature Range					
		60° - 100°F 15° - 40°C		40° - 60°F 5° - 15°C		0° - 40°F -20° - +5°C	
Inches	(mm)	Set	Cure	Set	Cure	Set	Cure
$^{1}/_{2}$" to 1$^{1}/_{4}$"	(15 to 32)	15 min	1 - 6 hr	1 hr	2 - 12 hr	3 hr	8 - 48 hr
1$^{1}/_{2}$" to 3"	(40 to 80)	30 min	2 - 12 hr	2 hr	4 - 24 hr	6 hr	16 - 96 hr
3$^{1}/_{2}$" to 8"	(90 to 200)	I hr	6 to 24 hr	4 hr	12 - 48 hr	12 hr	48 - 192 hr
10" to 14"	(250 to 350)	2 hr	24 hr	8 hr	72 hr	24 hr	192 hr
16" to 24"	(400 to 600)	4 hr	48 - 72 hr	16 hr	120 hr	48 hr	240- 336 hr

Note: 1. Initial set time indicates joints will withstand normal installation and handling stresses.
2. Cure times indicates required time before testing or before line pressure can be applied.
3. 50% more cure time is required in damp or humid conditions.

Table #14 - Set and Cure Times

Solvent Cementing

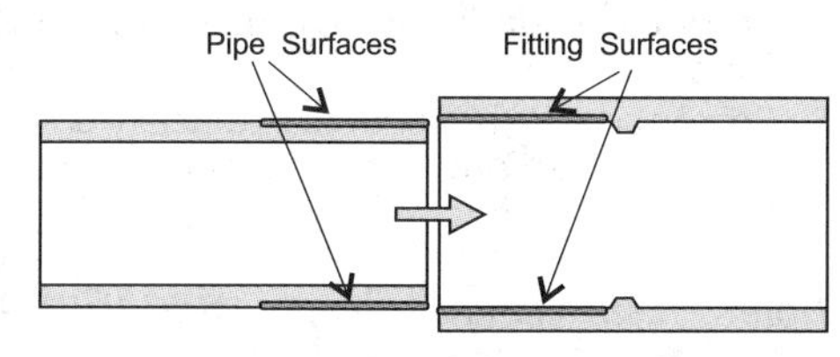

Illustration #14 - Fitting Surfaces for Plastic Pipe

Threading Plastic

Note: When threading plastic pipe, use only Schedule 80 or heavier pipe, and after threading the pressure rating must be reduced by 50 percent.

1. Use pipe dies designed for plastic pipe (recommended front rake angle of 5 to 10 degrees).
2. Do not over tighten the vise used to hold the pipe and, if needed, insert an aluminum or wooden plug in the pipe end to maintain roundness.
3. Cutting oil is not needed. A small amount of oil periodically applied to the chasers is adequate.
4. Only use Teflon tape or other approved joining compound to assemble the joint (do not use an oil base compound or Teflon paste).
5. Tighten by hand. Then an additional 2 turns with a strap wrench is usually sufficient, this is approximately 150 inch/pounds torque (17 Newton-metres).

Grooved Plastic Joints

1. Check with the pipe supplier for the recommended grooving method (roll groove or cut groove). Maximum operating pressure will depend on the grooving method used.
2. Groove dimensions and coupling assembly coincide to methods used with standard steel pipe.
3. Use grooved couplings recommended for plastic use.
4. The following PVC pipes may be grooved:

- SDR 26 - sizes 6 inches to 12 inches (150 mm to 300 mm)
- SDR 21 - sizes 4 inches to 12 inches (100 mm to 300 mm)
- Sch 40 - maximum size 8 inches (200 mm)
- Sch 80

Flanged Plastic Joints

1. Align bolt holes and flange faces without putting stress on the flange or piping.
2. Place a full flat faced soft gasket of approximately 1/8 in. (3 mm) between the flange faces.
3. Tighten bolts in a proper diametrically opposed pattern and final torque to values given in table #15.

Note: Plastic flanges comply with ANSI 150 lb. steel flange dimensions.

RECOMMENDED PLASTIC FLANGE TORQUE VALUES			
Flange Size		Torque	
Inches	millimetres	Ft. lbs.	Newton metres
$^1/_2$ to $1^1/_2$	15 to 40	15	21
2 to 4	50 to 100	30	41
6 to 8	150 to 200	50	68
10	250	70	95
12 to 24	300 to 600	100	136

Table #15 - Plastic Flange Torque Values

INSERTION DISTANCE FOR AWWA C-900 PVC PIPE			
Nominal Pipe Size		Insertion Depth	
Inches	millimetres	Inches	millimetres
4	100	4.02	102
6	150	5.30	135
8	200	5.70	145
10	250	7.70	195
12	300	8.20	210

Table #16 - Plastic Pipe Insertion Distances

O-Ring or Gasket Joints for Underground Service

1. Wipe clean the spigot end, bell end and gasket groove of the pipe or fitting to be joined.
2. Inspect (making sure the correct gasket is used) and insert the gasket in bell groove. An easy way of inserting the gasket is to bend the gasket into a heart shape and then place into the bell groove.
3. Lubricate the pipe end that will be inserted into the bell. If the pipe is field cut chamfer the end of the pipe (approx. 15 degrees) with chamfering tool, coarse file or rasp.
4. Align and then push the end of the pipe into the bell up to the appropriate distance or reference mark, see illustration #15. Reference mark distances are shown in table #16.

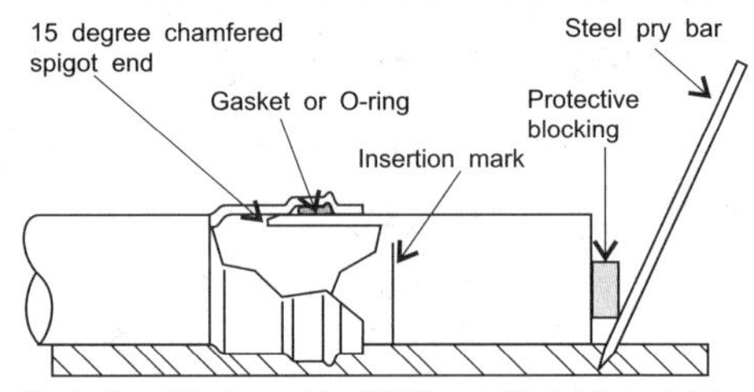

Illustration #15 - Assembly of O-Ring or Gasket Type Joining

Joining RTR Pipe

The two standard methods of joining RTR pipe is by butt and strap adhesive bonding or bell and spigot adhesive bonding.

Butt and strap joints are made by butting two pieces of pipe together and then wrapping the joint with layers of resin saturated glass mat.

Bell and spigot adhesive bonding uses tapered fittings and epoxy to make the joint. Pipe ends are sanded and fitted to the taper fittings before an adhesive is applied to the pipe and fitting.

SECTION TWO

TUBING DATA

Tube vs. Pipe

Tube differs from pipe in that it does not have the more liberal tolerances for inside diameter, outside diameter, wall thickness and nominal sizes given to pipe. ***Pipe sizes up to 12 inches (300 mm) are designated by nominal sizes which are smaller than the outside diameter of the pipe; where as, in most cases, tube sizes are identical to the outside diameter of the tube.*** Tubing is classified into three major types: structural tube, mechanical tube, and pressure tube.

Structural Tube

Structural tube is used in construction of such things as: building frameworks, roadway median barriers, bridge structures and for other general structural applications. It is available not only in round tubing shapes, but also in rectangular, square and other special structural shapes as needed. Illustration #16 shows some of the more common tube shapes available.

ASTM standards cover ferrous and non ferrous structural tube in both welded and seamless forms. Sizing of structural tube is specified by actual outside diameter and wall thickness. Maximum sizes normally extend up to 24 inches (609.4 mm) for round tubing with wall thickness up to 1 inch (25.4 mm). Structural tubing with other dimensions may be furnished providing they meet ASTM or equivalent specifications.

Mechanical Tube

Mechanical tube is utilized in a variety of mechanical and structural applications ***and like structural tube, it is not intended to carry fluids or gases under pressure***. Because mechanical tubing is usually manufactured for specific applications needing particular mechanical and chemical properties, only limited standards are covered by ASTM or other agencies. Sizes and dimensions are usually determined by established end usage or customer needs.

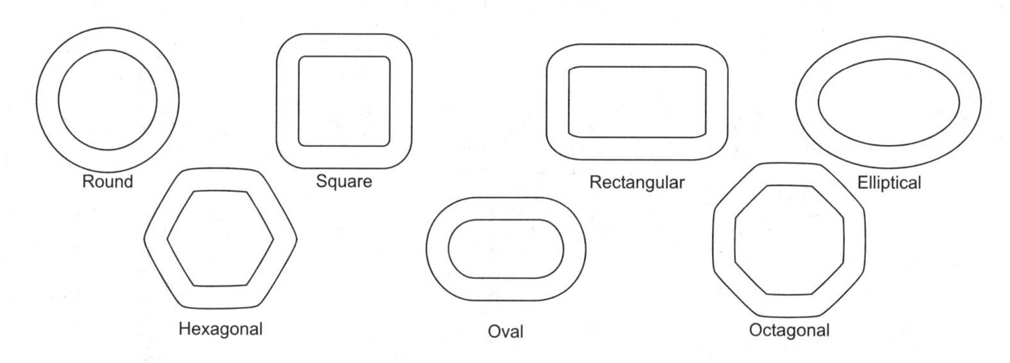

Illustration #16 - Structural Tube Shapes

Pressure Tube

Pressure tube is the type of tubing used most often in the piping industry. It is designed to carry fluids under pressure. Sizing is customarily designated by the tube's actual outside diameter and wall thickness or tube gage given in either:

Fractions of an inch, decimals of an inch, millimetres or by wire gage (usually in the Birmingham Iron Wire Gage/"BWG").

Table #17 gives wall thickness equivalents for BWG, decimals of an inch and millimetres.

Tubing Wall Thickness in Birmingham Wire Gage, Decimal of an Inch and Millimetres		
BWG	Decimal of an Inch	mm
30	0.012	0.3048
29	0.013	0.3302
28	0.014	0.3556
27	0.016	0.4064
26	0.018	0.4572
25	0.020	0.5080
24	0.022	0.5588
23	0.025	0.6350
22	0.028	0.7112
21	0,032	0.8128

Tubing Wall Thickness in Birmingham Wire Gage, Decimal of an Inch and Millimetres		
BWG	Decimal of an Inch	mm
20	0.035	0.8890
19	0.042	1.0668
18	0.049	1.2446
17	0.058	1.4732
16	0.065	1.6510
15	0.072	1.8288
14	0.083	2.1082
13	0.095	2.4130
12	0.109	2.7686
11	0.120	3.0480

Tubing Wall Thickness in Birmingham Wire Gage, Decimal of an Inch and Millimetres		
BWG	Decimal of an Inch	mm
10	0.134	3.4036
9	0.148	3.7592
8	0.165	4.1910
7	0.180	4.5720
6	0.203	5.1562
5	0.220	5.5880
4	0.238	6.0452
3	0.259	6.5786
2	0,284	7.2136
1	0.300	7.6200

Table #17 - BWG Tube Wall Thicknes

Pressure Tube

Tubing is available in several material classifications and in sub-groupings under each material heading. The following are the general material classifications most often given to pressure tubing:

- Aluminum Tubing
- Special Application Tubing
- Copper Tubing
- Special Alloy Tubing
- Carbon Steel Tubing
- Plastic Tubing
- Stainless Steel Tubing

Aluminum Tubing

Aluminum tubing is available in both welded and seamless forms in various alloys, tempers, and wall thicknesses. Pure aluminum is seldom used for manufacturing tube, but is alloyed with other metals to improve its physical properties.

Most tube acquired is produced to the general purpose ASTM B-210 standards for drawn seamless aluminum alloy tube. This tube is supplied in sizes from 1/8 inch (3.175 mm) to 12 inches (304.8 mm) outside diameter and is available in straight lengths and coils.

Coils are supplied in annealed temper only with maximum wall thickness not exceeding 0.083 inches (2.11 mm). Coiled rolls are usually supplied in 50 ft. (15 m) and 100 ft. (30.5 m) lengths, but lengths up to and exceeding 500 ft. (153.5 m) are available.

Straight tube lengths are usually available in 12 ft.(3.66 m) lengths but, may be special ordered up to 50 ft. (15 m) or more. Wall thickness for straight aluminum alloy tube can range between 0.01 inches (0.25 mm) to 0.50 inches (12.50 mm). Table #18 gives common commercial sizes, wall thicknesses, and pressure ratings for general purpose aluminum alloy tubing up to 1 inch (25 mm).

Aluminum Tubing - Suggested Allowable Working Pressure and Wall Thickness
General Purpose ASTM B-210 or Equivalent (Aluminum Alloy 6061 -76 Annealed Seamless Tubing)

O.D. in.	Wall Thickness Inches								
	0.020	0.028	0.035	0.049	0.065	0.083	0.095	0.109	0.120
	Pressure in P.S.I.								
1/8	3854	5597							
3/16	2447	3560	4578	6605					
1/4	1796	2583	3308	4830	6563				
5/16	1418	2027	2583	3770	5156				
3/8		1670	2121	3066	4221	5523			
1/2		1229	1554	2232	3045	4022	4673		
5/8			1229	1754	2384	3119	3633		
3/4			1019	1449	1953	2552	2961	3455	3854
7/8			867	1229	1659	2153	2499	2909	3234
1			756	1071	1439	1869	2163	2510	2783

O.D. mm	Wall Thickness millimetres								
	0.508	0.711	0.884	1.245	1.651	2.108	2.415	2.769	3.048
	Pressure in MPa								
3.18	26.5	38.6							
4.76	16.9	24.5	31.6	45.5					
6.35	12.4	17.8	22.8	33.3	45.2				
7.94	9.8	14.0	17.8	26.0	35.5				
9.53		11.5	14.6	21.1	29.1	38.0			
12.70		8.5	10.7	15.4	21.0	27.7	32.2		
15.88			8.5	12.1	16.4	21.5	25.0		
19.05			7.0	10.0	13.4	17.6	20.4	23.8	26.5
22.23			6.0	8.5	11.4	14.8	17.2	20.0	22.3
25.40			5.2	7.4	9.9	12.9	14.9	17.3	19.2

Table #18 - Aluminum Tub Wall Dimensions

NOTES:

1. Table based on allowable stress of 10,500 psi (72.4 MPa) at -20°F(-29°C) to 100°F (39°C)

2. Safety factor of 4 used in table

Copper Tube

There are numerous types of copper and copper alloy tubing being manufactured. However, the greater part of copper tube used in the piping industry is manufactured from 99.90 percent pure copper (minimum). This copper tube can be classified into two general groupings.

One type of tubing designation has tubing measured by its approximate inside diameter and is often referred to as plumbing tube. This designation consists of tube types: K, L, M, pressure tube and DWV, a non-pressure type of tubing.

The other type has tubing which is designated by the outside diameter and wall thickness measurements. It comprises of "ACR" (air conditioning and refrigeration) tubing and "GP" (general purpose) tubing.

Seamless Copper Tube Types K, L, M

Type K, L, and M seamless copper tube is manufactured to ASTM B-88 specifications for seamless water tube.

Nominal I.D. sizes are used to designate tubing sizes for the imperial or inch system, with actual outside diameters of the tube being 1/8 (0.125) inches (3 mm) larger than the nominal tube size.

Metric Version: ASTM B-88M, the metric version of the B-88 standard, re-classifies the tube into types A, B,and C classifications. The metric standard also designates the tube size by its actual outside diameter, not its nominal size. All other dimensions and properties of the tube remain comparable to the imperial specifications.

Tube Types K, L, M

Copper tubes K, L, and M are available in either annealed (soft) temper copper, or drawn (hard) temper copper.

Soft copper tube is commercially available in types K and L, however, type M soft can be supplied. Soft copper is normally supplied in coils ranging from 40 ft. (12.2 m) to 100 ft. (30.5 m) lengths. Coils are manufactured in nominal sizes ranging from 1/4 inch (8 mm) through to 2 inches (50 mm).

Hard copper (types K, L, and M) is supplied in straight lengths of 12 ft. (3.66 m) or 20 ft. (6.10 m). Nominal sizes of copper water tube range from 1/4 inch (6 mm) through to 12 inches (308 mm), with wall thickness varying as per type classification. Tables #19A, #19B, and #19C give specific size and dimensions for type K, L, and M respectively.

Oxygen Tube: Copper tube that is used for oxygen lines and other medical gases is often referred to as oxygen tube. This tube is basically type K or L tube that has been specially cleaned and capped. The cleaning is a safety measure to prevent contamination and possible spontaneous combustion with organic oils or impurities in the tube.

Color Coding of Copper

Color coding of copper tubing is applied to help distinguish between the various types of tube.

Type K copper has the heaviest wall thickness and is color coded green. Type L is the intermediate wall thickness tube and is color coded blue. Type M is the thinnest wall thickness of pressure tube and is color coded red. DWV, a non-pressure copper tube, is color coded yellow. It should be noted that annealed copper is not color coded in any types.

Plumbing Tube Color Code

Type K = Green	Type M = Red
Type L = Blue	Type DWV = Yellow

Type K Copper Tube Dimensions given in Inches and Millimetres									
Nominal Size	O.D.	I.D.	Wall Thick-ness	Approx. weight	Nominal Size	O.D.	I.D.	Wall Thick-ness	Approx. Mass
in.	in.	in.	in.	lb/ft	mm	mm	mm	mm	kg/m
1/4	0.375	0.305	0.035	0.145	6	9.53	7.75	0.89	0.216
3/8	0.500	0.402	0.049	0.269	10	12.70	10.21	1.24	0.400
1/2	0.625	0.527	0.049	0.344	12	15.88	13.39	1.24	0.512
5/8	0.750	0.652	0.049	0.418	15	19.05	16.56	1.24	0.622
3/4	0.875	0.745	0.065	0.641	20	22.23	18.92	1.65	0.954
1	1.125	0.995	0.065	0.839	25	28.58	25.27	1.65	1.25
1 1/4	1.375	1.245	0.065	1.04	32	34.93	31.62	1.65	1.55
1 1/2	1.625	1.481	0.072	1.36	40	41.28	37.62	1.83	2.02
2	2.125	1.959	0.083	2.06	50	53.98	49.76	2.11	3.06
2 1/2	2.625	2.435	0.095	2.93	65	66.68	61.85	2.41	4.36
3	3.125	2.907	0.109	4.00	80	79.38	73.84	2.77	5.95
3 1/2	3.625	3.385	0.120	5.12	90	92.08	85.98	3.05	7.62
4	4.125	3.857	0.134	6.51	100	104.78	97.97	3.40	9.69
5	5.125	4.805	0.160	9.67	125	130.18	122.05	4.02	14.39
6	6.125	5.741	0.192	13.9	150	155.58	145.82	4.88	20.68
8	8.125	7.583	0.271	25.9	200	206.38	192.61	6.88	38.53
10	10.125	9.449	0.338	40.3	250	257.18	240.00	8.59	59.96
12	12.125	11.315	0.405	57.8	300	307.98	287.40	10.29	85.00

Notes: 1. Dimensions based on ASTM Standard B-88.
2. Use for underground gas, oil and water lines and for plumbing and heating systems above ground. Has thick wall. Furnished in hard and soft copper. Can be bent cold. Weight: .323 lbs. per cu. in.

Table #19A - Type K Copper Tube

Type L Copper Tube Dimensions given in Inches and Millimetres									
Nominal Size	O.D.	I.D.	Wall Thickness	Approx. weight	Nominal Size	O.D.	I.D.	Wall Thick-ness	Approx. Mass
in.	in.	in.	in.	lb/ft	mm	mm	mm	mm	kg/m
1/4	0.375	0.315	0.030	0.126	6	9.53	8.00	0.76	0.187
3/8	0.500	0.430	0.035	0.198	10	12.70	10.92	0.89	0.295
1/2	0.625	0.545	0.040	0.285	12	15.88	13.84	1.02	0.424
5/8	0.750	0.666	0.042	0.362	15	19.05	16.92	1.07	0.539
3/4	0.875	0.785	0.045	0.455	20	22.23	19.94	1.14	0.677
1	1.125	1.025	0.050	0.655	25	28.58	26.04	1.27	0.976
1 1/4	1.375	1.265	0.055	0.884	32	34.93	32.13	1.40	1.320
1 1/2	1.625	1.505	0.060	1.14	40	41.28	38.23	1.52	1.70
2	2.125	1.985	0.070	1.75	50	53.98	50.42	1.78	2.60
2 1/2	2.625	2.465	0.080	2.48	65	66.68	62.61	2.03	3.69
3	3.125	2.945	0.090	3.33	80	79.38	74.80	2.29	4.96
3 1/2	3.625	3.425	0.100	4.29	90	92.08	86.00	2.54	6.38
4	4.125	3.905	0.110	5.38	100	104.78	99.19	2.79	8.01
5	5.125	4.875	0.125	7.61	125	130.18	123.83	3.18	11.32
6	6.125	5.845	0.140	10.2	150	155.58	148.46	3.56	15.18
8	8.125	7.725	0.200	19.3	200	206.38	196.22	5.08	28.71
10	10.125	9.625	0.250	30.1	250	257.18	244.48	6.35	44.78
12	12A25	11.565	0.280	40.4	300	307.98	293.75	7.11	60.11

Notes: 1. Dimensions based on ASTM Standard B-88.
2. Use for plumbing and heating systems. Medium wall thickness. Furnished in hard and soft copper. Can be bent cold. Weight: .323 lbs. per cu. in.

Table #19B - Type L Copper Tube

Type M Copper Tube Dimensions given in Inches and Millimetres									
Nominal Size	O.D.	I.D.	Wall Thickness	Approx. weight	Nominal Size	O.D.	I.D.	Wall Thickness	Approx. Mass
in.	in.	in.	in.	lb/ft	mm	mm	mm	mm	kg/m
3/8	0.500	0.450	0.025	0.145	10	12.70	11.43	0.64	0.216
1/2	0.625	0.569	0.028	0.204	12	15.88	14.45	0.71	0.304
3/4	0.875	0.811	0.032	0.328	20	22.23	20.60	0.81	0.488
1	1.125	1.055	0.035	0.465	25	28.58	26.80	0.89	0.692
1 1/4	1.375	1.291	0.042	0.682	32	34.93	32.79	1.07	1.01
1 1/2	1.625	1.527	0.049	0.940	40	41.28	38.79	1.24	1.40
2	2.125	2.009	0.058	1.46	50	53.98	51.03	1.47	2.17
2 1/2	2.625	2.495	0.065	2.03	65	66.68	63.37	1.65	3.02
3	3.125	2.981	0.072	2.68	80	79.38	75.72	1.83	3.99
3 1/2	3.625	3.459	0.083	3.58	90	92.08	87.86	2.11	5.33
4	4.125	3.935	0.095	4.66	100	104.78	99.95	2.41	6.93
5	5.125	4.907	0.109	6.66	125	130.18	124.64	2.77	9.91
6	6.125	5.881	0.122	8.92	150	155.58	149.38	3.10	13.27
8	8.125	7.785	0.17	16.5	200	206.38	197.74	4.32	24.55
10	10.125	9.701	0.212	25.6	250	257.18	246.41	5.38	38.09
12	12.125	11.617	0.254	36.7	300	307.98	295.07	6.45	54.60

Notes: 1. Dimensions based on ASTM Standard B-88.
2. Applications include water distribution systems, solar, heating and sprinkler systems.
3. Type M is available in hard straight lengths and not normally available in annealed temper.

Table #19C - Type M Copper Tube

DWV Drainage Tube

Type DWV copper drainage tube is another type of copper tube which is classified by its approximate inside diameter. Specifications for the tube are covered in the ASTM B-306 standard for copper drainage tube. ***DWV tube is intended for use in drainage, waste, and vent applications, which the "DWV" abbreviated letters stand for.*** The tube has a color code of yellow and is manufactured in 12 ft. (3.66 m) and 20 ft. (6.10 m) straight hard temper lengths only. The wall thickness of DWV is thinner than types K, L, or M tubing. See table #20 for specific sizes and wall thickness for type DWV copper tube.

Air Conditioning and Refrigeration Tube and General Purpose Copper Tube

ACR (air conditioning and refrigeration) and GP (general purpose) copper tubing are two types of tube which are measured and designated by outside diameters and wall thickness.

ACR Tube: Seamless copper tube designed for air conditioning and refrigeration field service. ACR tube is covered under ASTM specification B-280. The tube is good for carrying most commercial refrigerants (except ammonia).

ACR tube is available in 12 ft. (3.66 m) or 20 ft. (6.10 m) straight hard temper lengths and standard 50 ft. (15.2 m) soft temper coils.

Hard temper lengths of tube are color coded blue and identified with the ACR inscription. The tube differs from other copper tube in that it is thoroughly cleaned, degreased, dehydrated, and capped prior to delivery (tube may be supplied nitrogen charged).

Another point of deviation from other copper tube is that annealed temper tube dimensions differ in size designations. Table #21 gives sizes and dimensions of annealed and hard temper ACR tube.

Type DWV Copper Tube Dimensions given in Inches and Millimetres									
Nominal Size	**O.D.**	**I.D.**	**Wall Thickness**	**Approx. Weight**	**Nominal Size**	**O.D.**	**I.D.**	**Wall Thickness**	**Approx. Mass**
inches	inches	inches	inches	lb/ft	mm	mm	mm	mm	kg/m
1 1/4	1.375	1.295	0.040	0.650	32	34.93	32.89	1.02	0.967
1 1/2	1.625	1.541	0.042	0.809	40	41.28	39.14	1.07	1.20
2	2.125	2.041	0.042	1.07	50	53.98	51.84	1.07	1.59
3	3.125	3.030	0.045	1.69	80	79.38	76.96	1.14	2.51
4	4.125	4.009	0.058	2.87	100	104.78	101.83	1.47	4.27
5	5.125	4.981	0.072	4.43	125	130.18	126.52	1.83	6.59
6	6.125	5.959	0.083	6.10	150	155.58	151.36	2.11	9.08
8	8.125	7.907	0.109	10.6	200	206.38	200.84	2.77	15.8

Notes:
1. Dimensions based on ASTM Standard B-306.
2. Used for drainage, waste and vent plumbing applications and other non-pressure uses
3. Furnished in hard temper lengths only and joined with solder fittings
4. Millimetre dimensions are calculated by multiplying inches by 25.4 and rounding off to two places of decimals.
5. Mass (kg/m) is calculated by multiplying lbs/ft by 1.49 and rounding off to two places of decimals.

Table #20 - Type DWV Copper Tube

Note: Trade practice in some areas often uses the terms Tube or Tubing to distinguish between inside diameter (ID) and outside diameter (OD) copper designations. "Tubing" is used to indicate exact OD copper (ACR & GP tubing) and "Tube" for approximate ID specified copper (types K, L, M & DWV). This is not an official way to designate copper tube and should be avoided. If there is any possibility of confusion, state exact dimensions needed for copper tube/tubing.

Air Conditioning and Refrigeration (ACR) - Copper Tube Dimensions given in Inches & mm										
Available Temper	Size	O.D.	I.D.	Wall Thickness	Approx. Weight	Size	O.D.	I.D.	Wall Thickness	Approx. Mass
form	in.	in.	in.	in.	lb/ft	mm	mm	mm	mm	kg/m
C	1/8	0.125	0.065	0.030	0.0347	3.18	3.18	1.65	0.76	0.0516
C	3/16	0.188	0.128	0.030	0.0575	4.78	4.78	3.25	0.76	0.086
C	1/4	0.250	0.190	0.030	0.0804	6.35	6.35	4.83	0.76	0.120
C	5/16	0.312	0.248	0.032	0.109	7.92	7.92	6.30	0.81	0.162
S	3/8	0.375	0.315	0.030	0.126	9.53	9.53	8.00	0.76	0.187
C	3/8	0.375	0.311	0.032	0.134	9.53	9.53	7.90	0.81	0.199
C	1/2	0.500	0.436	0.032	0.182	12.70	12.70	11.07	0.81	0.271
S	1/2	0.500	0.430	0.035	0.198	12.70	12.70	10.92	0.89	0.295
C	5/8	0.625	0.555	0.035	0.251	15.88	15.88	14.10	0.89	0.373
S	5/8	0.625	0.545	0.040	0.285	15.88	15.88	13.84	1.02	0.424
B	3/4	0.750	0.666	0.042	0.362	19.05	19.05	16.91	1.07	0.539
B	7/8	0.875	0.785	0.045	0.455	22.23	22.23	19.94	1.14	0.677
B	1 1/8	1.125	1.025	0.050	0.655	28.58	28.58	26.04	1.27	0.975
B	1 3/8	1.375	1.265	0.055	0.884	34.93	34.93	32.13	1.40	1.32
S	1 5/8	1.625	1.505	0.060	1.14	41.28	41.28	38.23	1.52	1.70
S	2 1/8	2.125	1.985	0.070	1.75	53.98	53.98	50.42	1.78	2.60
S	2 5/8	2.625	2.465	0.080	2.48	66.68	66.68	62.61	2.03	3.69
S	3 1/8	3.125	2.945	0.090	3.33	79.38	79.38	74.80	2.29	4.96
S	3 5/8	3.625	3.425	0.100	4.29	92.08	92.08	86.00	2.54	6.38
S	4 1/8	4.125	3.905	0.110	5.38	104.78	104.78	99.19	2.79	8.01

Note: Based on ASTM Standard B-280. mm = inch x 25.4 kg/m = lbs/ft x 1.49
C = Coiled lengths, soft annealed temper; S = Straight lengths, hand drawn general purpose; B=Both forms

Table #21 - ACR Copper Tube

General Purpose (GP) Copper Tube - Commercial Sizes and Wall Thickness									
O.D. Inches	Wall Thickness, Inches								
1/8	0.028	0.032	0.035						
3/16	0.028	0.032	0.035	0.049					
1/4	0.028	0.032	0.035	0.049	0.065				
5/16		0.032	0.035	0.049	0.065				
3/8		0.032	0.035	0.049	0.065	0.083			
1/2			0.035	0.049	0.065	0.083			
5/8			0.035	0.049	0.065	0.083	0.095		
3/4			0.035	0.049	0.065	0.083	0.095	0.109	
7/8			0.035	0.049	0.065	0.083	0.095	0.109	
1			0.035	0.049	0.065	0.083	0.095	0.109	0.120

O.D. mm	Wall Thickness, Millimetres								
3.18	0.71	0.81	0.89						
4.76	0.71	0.81	0.89	1.25					
6.35	0.71	0.81	0.89	1.25	1.65				
7.94		0.81	0.89	1.25	1.65				
9.53		0.81	0.89	1.25	1.65	2.11			
12.70			0.89	1.25	1.65	2.11			
15.88			0.89	1.25	1.65	2.11	2.41		
19.05			0.89	1.25	1.65	2.11	2.41	2.77	
22.23			0.89	1.25	1.65	2.11	2.41	2.77	
25.40			0.89	1.25	1.65	2.11	2.41	2.77	3.05

Note: Table based on ASTM Standard B-75 or equivalent

Table #22 - GP Copper Tube Wall Thickness

General Purpose (GP) Copper Tube

GP tubing is the classification of seamless copper tube designed for general engineering purposes. ASTM specification B-75 covers specific requirements while ASTM B-251 and B-251 M (Metric) standards cover common requirements such as: lengths, wall thickness and other general dimensions and specifications.

See table #22 for common GP tube sizes and wall thickness dimensions.

Carbon Steel and Stainless Steel Tube

Carbon steel and stainless steel pressure tubing are used for various applications and these applications can be classified into general groupings. The following are the major grouping classifications for both carbon steel and stainless steel pressure tubing:

- Boiler and superheater tubing
- Heat exchanger and condenser tubing
- Still tubing
- General purpose tubing

See tables #23 and #24 for common tube characteristics.

Boiler & Superheater Tubes: Seamless and welded boiler and superheater tubes are available in both hot rolled and cold drawn tube for various pressure and temperature applications. The tube is sized by outside diameter and minimum wall thickness. Boiler and superheater tubes are made to various ASTM specifications depending on type of operations and designated use.

Heat Exchanger and Condenser Tubes: Tubing under this classification is used for heat exchangers, condensers, and similar units where the tube is used to transfer heat from one medium to another. Tubing is designated by outside diameter, minimum wall thickness, and frequently, the exact tubing length. A typical shell and tube heat exchanger with a removable tube bundle is shown in illustration #17.

General Tube Dimensions and Weight (Imperial)							
O.D.	I.D.	Wall Thick-ness	BWG	Cross Sec-tion	Ext. Sur-face	Int. Sur-face	Weight
inch	inch	inch	gage	sq.in.	ft^2/ft	ft^2/ft	lb/ft
1/4	0.194	0.028	22	0.0295	0.0655	0.0508	0.066
	0.206	0.022	24	0.0333	0.0655	0.0539	0.054
	0.214	0.018	26	0.0360	0.0655	0.0560	0.045
	0.218	0.016	27	0.0373	0.0655	0.0570	0.040
3/8	0.277	0.049	18	0.0603	0.0982	0.0725	0.171
	0.305	0.035	20	0.0731	0.0982	0.0798	0.127
	0.319	0.028	22	0.0799	0.0982	0.0835	0.104
	0.331	0.022	24	0.0860	0.0982	0.0867	0.083
1/2	0.370	0.065	16	0.1075	0.1309	0.0969	0.302
	0.402	0.049	18	0.1269	0.1309	0.1052	0.236
	0.430	0.035	20	0.1452	0.1309	0.1126	0.174
	0.444	0.028	22	0.1548	0.1309	0.1162	0.141
5/8	0.407	0.109	12	0.1301	0.1636	0.1066	0.602
	0.435	0.095	13	0.1486	0.1636	0.1139	0.537
	0.459	0.083	14	0.1655	0.1636	0.1202	0.479
	0.481	0.072	15	0.1817	0.1636	0.1259	0.425
	0.495	0.065	16	0.1924	0.1636	0.1296	0.388
	0.509	0.058	17	0.2035	0.1636	0.1333	0.350
	0.527	0.049	18	0.2181	0.1636	0.1380	0.303
	0.541	0.042	19	0.2298	0.1636	0.1416	0.262
	0.555	0.035	20	0.2419	0.1636	0.1453	0.221

General Tube Dimensions and Weight (Imperial)							
O.D.	I.D.	Wall Thick-ness	BWG	Cross Sec-tion	Ext. Sur-face	Int. Sur-face	Weight
inch	inch	inch	gage	sq.in.	ft^2/ft	ft^2/ft	lb/ft
3/4	0.482	0.134	10	0.1825	0.1963	0.1262	0.884
	0.510	0.120	11	0.2043	0.1963	0.1335	0.809
	0.532	0.109	12	0.2223	0.1963	0.1393	0.748
	0.560	0.095	13	0.2463	0.1963	0.1466	0.666
	0.584	0.083	14	0.2679	0.1963	0.1529	0.592
	0.606	0.072	15	0.2884	0.1963	0.1587	0.520
	0.620	0.065	16	0.3019	0.1963	0.1623	0.476
	0.634	0.058	17	0.3157	0.1963	0.1660	0.428
	0.652	0.049	18	0.3339	0.1963	0.1707	0.367
	0.680	0.035	20	0.3632	0.1963	0.1780	0.269
1	0.670	0.165	8	0.3526	0.2618	0.1754	1.462
	0.732	0.134	10	0.4208	0.2618	0.1916	1.237
	0.760	0.120	11	0.4536	0.2618	0.1990	1.129
	0.782	0.109	12	0.4803	0.2618	0.2047	1.037
	0.810	0.095	13	0.5153	0.2618	0.2121	0.918
	0.834	0.083	14	0.5463	0.2618	0.2183	0.813
	0.856	0.072	15	0.5755	0.2618	0.2241	0.714
	0.870	0.065	16	0.5945	0.2618	0.2278	0.649
	0.902	0.049	18	0.6390	0.2618	0.2361	0.496
	0.930	0.035	20	0.6793	0.2618	0.2435	0.360

Table #23 - General Tube Dimensions and Weight (Imperial)

General Tube Dimensions and Weight (Imperial)							
O.D.	I.D.	Wall Thick-ness	BWG	Cross Sec-tion	Ext. Sur-face	Int. Sur-face	Weight
inch	inch	inch	gage	sq.in.	ft^2/ft	ft^2/ft	lb/ft
1 1/4	0.890	0.180	7	0.6221	0.3272	0.2330	2.057
	0.920	0.165	8	0.6648	0.3272	0.2409	1.921
	0.982	0.134	10	0.7574	0.3272	0.2571	1.598
	1.010	0.120	11	0.8012	0.3272	0.2644	1.448
	1.032	0.109	12	0.8365	0.3272	0.2702	1.329
	1.060	0.095	13	0.8825	0.3272	0.2775	1.173
	1.084	0.083	14	0.9229	0.3272	0.2838	1.033
	1.120	0.065	16	0.9852	0.3272	0.2932	0.823
	1.152	0.049	18	1.0420	0.3272	0.3016	0.629
	1.180	0.035	20	1.0940	0.3272	0.3089	0.456

General Tube Dimensions and Weight (Imperial)							
O.D.	I.D.	Wall Thick-ness	BWG	Cross Sec-tion	Ext. Sur-face	Int. Sur-face	Weight
inch	inch	inch	gage	sq.in.	ft^2/ft	ft^2/ft	lb/ft
1 1/2	1.232	0.134	10	1.1920	0.3927	0.3225	1.955
	1.282	0.109	12	1.2910	0.3927	0.3356	1.618
	1.334	0.083	14	1.3960	0.3927	0.3492	1.258
	1.370	0.065	16	1.4740	0.3927	0.3587	0.996
2	1.760	0.120	11	2.4330	0.5236	0.4608	2.211
	1.782	0.109	12	2.4940	0.5236	0.4665	2.201
	1.810	0.095	13	2.5730	0.5236	0.4739	1.934
	1.834	0.083	14	2.6420	0.5236	0.4801	1.699

Table #23 (cont'd) - General Tube Dimensions and Weight (Imperial)

Note: Table weights based on carbon steel tube.
To establish other metal weights multiply by:

Aluminum. 0.35
Titanium. 0.58
A.I.S.I. 400 Series Stainless Steels 0.99
A.I.S.I. 300 Series Stainless Steels 1.02
Aluminum Bronze. 1.04
Aluminum Brass. 1.06
Nickel-Chrome-Iron 1.07
Admiralty.. . . . . 1.09
Nickel and Nickel-Copper. 1.13
Copper and Cupro-Nickels. 1.14

General Tube Dimensions and Weight (Metric)							
O.D.	I.D.	Wall Thickness	BWG	Cross Section	Ext. Surface	Int. Surface	Mass
mm	mm	mm	gage	cm^2	m^2/m	m^2/m	kg/m
6.35	4.93	0.71	22	0.190	0.020	0.015	0.098
	5.23	0.56	24	0.215	0.020	0.016	0.080
	5.44	0.46	26	0.232	0.020	0.017	0.067
	5.54	0.41	27	0.241	0.020	0.017	0.060
9.53	7.04	1.25	18	0.389	0.030	0.022	0.255
	7.75	0.89	20	0.472	0.030	0.024	0.189
	8.10	0.71	22	0.515	0.030	0.025	0.155
	8.41	0.56	24	0.555	0.030	0.026	0.124
12.70	9.40	1.65	16	0.694	0.040	0.030	0.450
	10.21	1.25	18	0.819	0.040	0.032	0.351
	10.92	0.89	20	0.937	0.040	0.034	0.260
	11.28	0.71	22	0.999	0.040	0.035	0.210
15.88	10.34	2.77	12	0.839	0.050	0.032	0.897
	11.05	2.41	13	0.959	0.050	0.035	0.800
	11.66	2.11	14	1.068	0.050	0.037	0.714
	12.22	1.83	15	1.172	0.050	0.038	0.633
	12.57	1.65	16	1.241	0.050	0.040	0.578
	12.93	1.47	17	1.313	0.050	0.041	0.522
	13.39	1.25	18	1.407	0.050	0.042	0.451
	13.74	1.07	19	1.483	0.050	0.043	0.390
	14.10	0.89	20	1.561	0.050	0.044	0.329

General Tube Dimensions and Weight (Metric)							
O.D.	I.D.	Wall Thickness	BWG	Cross Section	Ext. Surface	Int. Surface	Mass
mm	mm	mm	gage	cm^2	m^2/m	m^2/m	kg/m
19.05	12.24	3.40	10	1.177	0.060	0.038	1.317
	12.95	3.05	11	1.318	0.060	0.041	1.205
	13.51	2.77	12	1.434	0.060	0.042	1.115
	14.22	2.41	13	1.589	0.060	0.045	0.992
	14.83	2.11	14	1.728	0.060	0.047	0.882
	15.39	1.83	15	1.861	0.060	0.048	0.775
	15.75	1.65	16	1.948	0.060	0.049	0.709
	16.10	1.47	17	2.037	0.060	0.051	0.638
	16.56	1.25	18	2.154	0.060	0.052	0.547
	17.27	0.89	20	2.343	0.060	0.052	0.401
25.4	17.02	4.19	8	2.275	0.08	0.053	2.462
	18.59	3.40	10	2.715	0.08	0.058	1.843
	19.30	3.05	11	2.926	0.08	0.061	1.682
	19.86	2.77	12	3.099	0.08	0.062	1.545
	20.57	2.41	13	3.325	0.08	0.065	1.368
	21.18	2.11	14	3.525	0.08	0.067	1.211
	21.74	1.83	15	3.713	0.08	0.068	1.064
	22.10	1.65	16	3.835	0.08	0.069	0.967
	22.91	1.25	18	4.123	0.08	0.072	0.739
	23.62	0.89	20	4.383	0.08	0.074	0.536

Table #24 - General Tube Dimensions and Weight (Metric)

General Tube Dimensions and Weight (Metric)

O.D.	I.D.	Wall Thick-ness	BWG	Cross Sec-tion	Ext. Sur-face	Int. Sur-face	Mass
mm	mm	mm	gage	cm^2	m^2/m	m^2/m	kg/m
31.75	22.61	4.57	7	4.014	0.10	0.071	1.326
	23.37	4.19	8	4.289	0.10	0.073	1.371
	24.94	3.40	10	4.886	0.10	0.078	1.463
	25.65	3.05	11	5.169	0.10	0.081	1.505
	26.21	2.77	12	5.397	0.10	0.082	1.538
	26.92	2.41	13	5.694	0.10	0.085	1.579
	27.53	2.11	14	5.954	0.10	0.087	1.615
	28.45	1.65	16	6.356	0.10	0.089	1.669
	29.26	1.25	18	6.723	0.10	0.092	1.716
	29.97	0.89	20	7.058	0.10	0.094	1.758

Table #24(cont'd) - General Tube Dimensions and Weight (Metric)

General Tube Dimensions and Weight (Metric)

O.D.	I.D.	Wall Thick-ness	BWG	Cross Sec-tion	Ext. Sur-face	Int. Sur-face	Mass
mm	mm	mm	gage	cm^2	m^2/m	m^2/m	kg/m
38.1	31.29	3.40	10	7.690	0.12	0.098	1.836
	32.56	2.77	12	8.329	0.12	0.102	1.910
	33.88	2.11	14	9.019	0.12	0.106	1.988
	34.80	1.65	16	9.490	0.12	0.109	2.041
50.8	44.70	3.05	11	15.697	0.16	0.140	2.622
	45.26	2.77	12	16.090	0.16	0.142	2.655
	45.97	2.41	13	16.600	0.16	0.144	2.697
	46.58	2.11	14	17.045	0.16	0.146	2.733

Note: Mass based on carbon steel tube. To establish other metal weights multiply by:

Aluminum. 0.35
Titanium. 0.58
A. I.S.I. 400 Series Stainless Steels. 0.99
A. I.S.I. 300 Series Stainless Steels. 1.02
Aluminum Bronze. 1.04
Aluminum Brass. 1.06
Nickel-Chrome- Iron. 1.07
Admiralty. 1.09
Nickel and Nickel-Copper. 1.13
Copper and Cupro-Nickels. 1.14

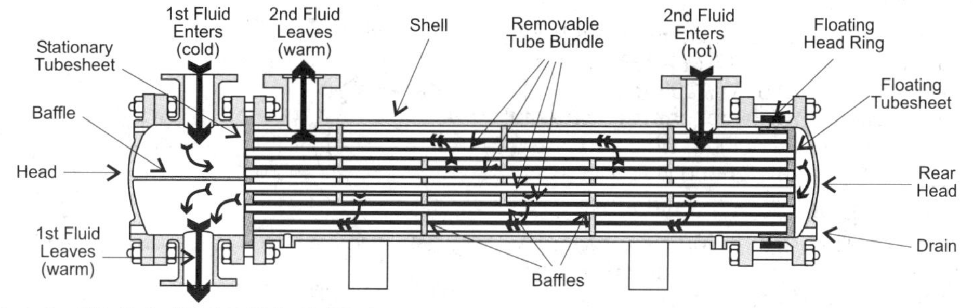

Illustration #17 - Typical Shell and Tube Heat Exchanger

Still Tubing: Still tubing is designed for process refinery type heater applications where the tube is subject to external furnace temperatures higher than the fluid or vapors contained internally in the tube. It is usually supplied in seamless, hot-finished or cold drawn types for various ranges of temperature and pressure. Like other types of tube, it is sized by outside diameter and minimum wall thickness.

General Purpose Tube: Tube classified under this heading is used for applications that require general service types of tubing. ASTM A179 and SAE J524b (Society of Automotive Engineers) standards are usually the two major carbon steel tube specifications used under this classification. The tubes are low carbon seamless tubes which are cold-drawn and in the case of the SAE tube, are annealed for bending and flaring.

General Purpose

Stainless steel tubing for general service is covered under either ASTM A268 for ferritic stainless steel or ASTM A269 for austenitic stainless steel. These standards cover both welded and seamless tube, which are sized by the outside diameter and wall thickness.

Special Application Tube

Tubing classified in this group is used for applications where environmental conditions or installation requirements require a specific type of tube or tube arrangement.

Copper Brazed Steel Tube: This type of tubing is used in the automotive, refrigeration and stove industries. The tube is manufactured from steel strips coated with copper and formed into tubing. It is then brazed to form a complete tube. Sizes are limited from 3/16 inches (4.76 mm) to 5/8 inches (15.88 mm) OD. Illustration #18 details two types of copper-brazed steel tubing identified in ASTM A 254-84 standards.

Double Wall 360 Degree Brazed Tubing

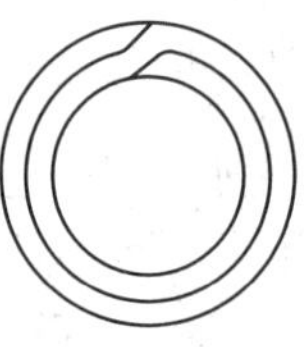

Single-Strip Type

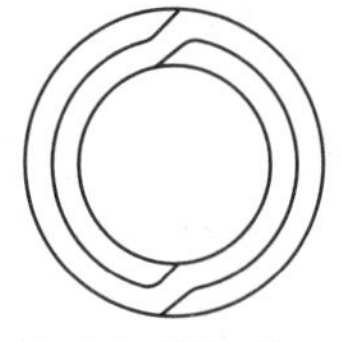

Double-Strip Type

Illustration #18 - Copper Brazed Steel Tubing

Bundled Metallic or Plastic Tube: These tube bundles can also be classified as special application tubing. The bundles can incorporate either metallic or plastic tubes encased in plastic or steel armor protective jackets. Tubing bundles are primarly used for pneumatic measurement and control signal instrumentation lines. They provide protection for individual tubes and faster field installation. Two typical tube bundles are shown in illustration #19.

Bundled Metallic or Plastic Tube

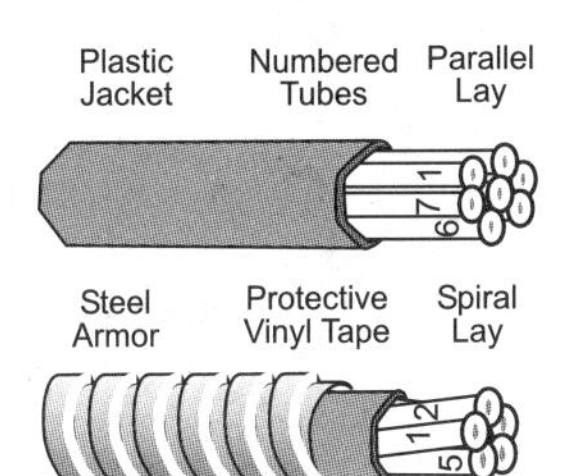

illustration #19 -Typical Bundled Tube

Special Alloy Tube

Service conditions that require tube outside of the scope of carbon steel or stainless steel specifications can usually be accommodated by one of the special alloy tubes. The following are commonly available tube alloys:

Monel	Inconels, Incoloys
Cupro Nickel	Nickel
Brass Alloys	Chrome Moly
Hastelloy	Stainless Steel Alloys
Titanium	Molybdenum

Plastic Tube

Because of plastic's good corrosion resistance, ease of handling and low cost of manufacturing, it is becoming commonly used. Typical applications of plastic tubing range from instrumentation air to beverage and pharmaceutical lines. The more common types of plastic tubing used are: Nylon, Polyethylene, Polypropylene, Teflon and PVC.

Working pressure for these plastics range up to 600 psi (4100 kPa) with temperature maximums as high as 400°F (204°C). Suggested allowable working pressure and wall thickness dimensions for nylon and polyethylene tube are given in tables #25 and #26.

Nylon Tubing Suggested Allowable Working Pressure and Wall Thickness											
O.D. in.	Wall Thickness, Inches					O.D. mm	Wall Thickness, Millimetres				
	0.030	0.040	0.062	0.080	0.125		0.762	1.14	1.57	2.03	3.18
	Pressure in P.S.I.						Pressure in MPa				
3/16	530	790				4.76	3.65	5.45			
1/4	400	590	810			6.35	2.76	4.07	5.58		
3/8	260	390	540	700		9.53	1.79	2.69	3.72	4.83	
1/2		300	410	530	820	12.70		2.07	2.83	3.65	5.65
3/4			270	350	550	19.05			1.86	2.41	3.79
1			200	260	410	25.4			1.38	1.79	2.83

Table #25 - Nylon Tube Dimensions

Polyethylene Tubing Suggested Allowable Working Pressure and Wall Thickness									
O.D. in.	Wall Thickness, Inches				O.D. mm	Wall Thickness, Millimetres			
	0.045	0.062	0.090	0.125		1.14	1.57	2.29	3.18
	Pressure in P.S.I.					Pressure in MPa			
3/16	190				4.76	1.31			
1/4	140	195			6.35	0.966	1.34		
3/8	95	130	190		9.53	0.655	0.896	1.31	
1/2	70	100	140	200	12.7	0.483	0.689	0.966	1.38
3/4	45	65	95	130	19.05	0.310	0.448	0.655	0.986
1		50	70	100	25.4		0.345	0.483	0.689
Note: Pressure is determined at room temperature									

Table #26 - Polyethylene Tube Dimensions

Tube Bending

Tube benders range from the most simple spring types shown in illustration #20A to the diverse mechanical types used on larger sizes of tubing. Most smaller sizes of tubing can bs easily and accurately bent by hand, using the appropriate type of hand tubing benders. Bending of such tube is not limited to annealed copper and aluminum, but can be performed readily on annealed carbon steel and stainless steel tubing.

Maximum tubing size for hand benders varies depending on tube material, temper, wall thickness, and the style of benders used. Table #27A gives minimum and maximum wall thickness for bending tube.

Compression Tube Bender

This instruction section on bending will deal with a hand lever type tube bender often referred to as a compression tube bender, see illustration #20B.

Tube size dimensions for this type of bender usually range from 3/16 of an inch (4.76 mm) through to 3/4 inch (19 mm) outside diameter.

Bender Part Identification

All parts identified in the following description of benders are shown in illustration #20B.

1. "Bending Wheel" shown provides the circular form the tubing rotates around to make the actual bend. It is marked off in degrees of a half circle.
2. "Stationary Handle" supports the bending wheel while bending.
3. "Pull Handle" pulls the tube around the bending wheel to designated bend needed.
4. "Placement Link" indicated where to place the measured mark on the tubing that is to be bent.
5. "Fastener" holds tubing in place while bending.

Tube Bending

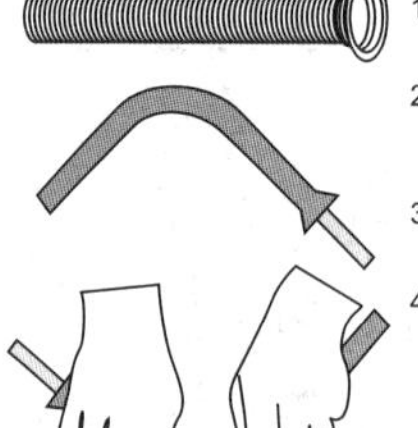

1. Select bending spring that just slips over the tube.
2. Slip the center of the spring over the center of the desired bend.
3. Hold bender in both hands and slowly make the bend.
4. Grasp flat end of the spring and remove from the tubing.

Illustration #20A - Spring Tube Benders

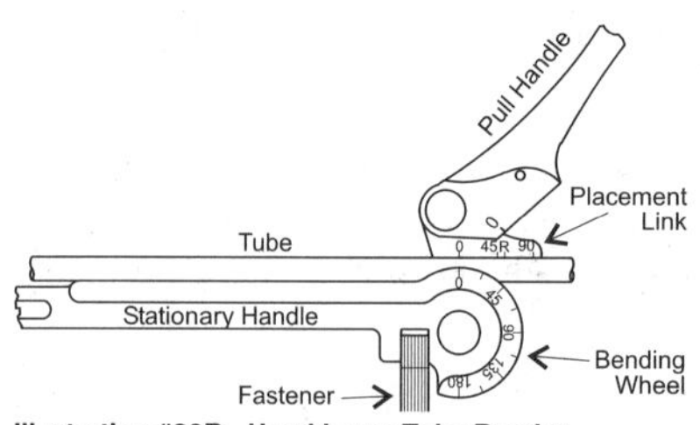

Illustration #20B - Hand Lever Tube Bender

Installing Fittings Near Bends

When installing fittings close to tube bends, the distorted tube section (caused by bending) must not enter the tube fitting. Table #27B and #27C give the necessary straight tube lengths. See illustration #21.

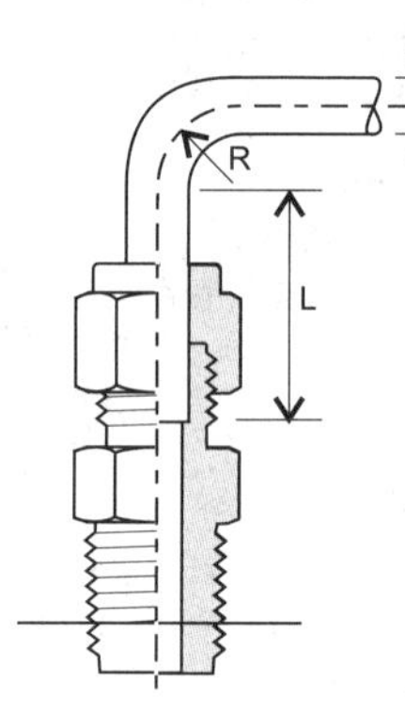

T Tube Outside Diameter

R Radius of Tubing Bend as Required or Minimum Allowed for Specified Wall Thickness and Tube Size as Recommended by the Bender Manufacturer.

L Straight Tube Length Required from End of Tube to Beginning of Bend

Illustration #21 - Fittings Near Tube Bends

Recommended Maximum and Minimum Wall Thickness for Tube Bending													
Tube O.D.		Min. Wall Without Flattening		Recommended Maximum Wall Thickness						Radius To Tube Centering		Gain	
				Copper		Aluminum		Steel					
in.	mm	in.	mm	in.	mm	in.	mm	in.	mm	in.	mm	in.	mm
1/8	3.2	0.012	0.3	-	-	-		0.032	0.8	3/8	9.5	0.17	4.3
3/16	4.8	0.020	0.5	-	-	-		0.032	0.8	7/16	11.1	0.19	4.8
1/4	6.4	0.028	0.7	0.083	2.1	0.083	2.1	0.032	0.8	9/16	14.3	0.29	7.4
5/16	7.9	0.032	0.8	0.095	2.4	0.083	2.1	0.035	0.9	11/16	17.5	0.29	7.4
3/8	9.5	0.032	0.8	0.109	2.8	0.083	2.1	0.049	1.2	15/16	23.8	0.50	12.7
1/2	12.7	0.042	1.1	0.109	2.8	0.083	2.1	0.049	1.2	1 1/2	37.5	0.65	16.5
5/8	15.9	0.042	1.1	0.109	2.8	0.095	2.4	0.049	1.2	2	75.0	0.86	21.8
3/4	19.0	0.049	1.3	0.109	2.8	0.095	2.4	0.065	1.7	2 1/2	93.8	1.08	27.4
7/8	22.2	0.049	1.3	0.109	2.8	0.109	2.8	0.065	1.7	3	93.8	1.29	32.8
1	25.4	0.065	1.7	0.109	2.8	0.109	2.8	0.065	1.7	3 1/2	87.5	1.62	41.1

Table #27A - Recommended Thickness for Bending,

T Tube O.D.	L - Straight Tube Length	
	A	B
in.	in.	in.
1/16	1/2	13/32
1/8	23/32	19/32
3/16	3/4	5/8
1/4	13/16	11/16
5/16	7/8	23/32
3/8	1 5/16	3/4
1/2	1 3/16	31/32
5/8	1 1/4	1 1/32
3/4	1 1/4	1 1/32
7/8	1 5/16	1 3/32
1	1 1/2	1 9/32

T Tube O.D.	L- Straight Tube Length	
	A	B
mm	mm	mm
3	19	16
6	21	17
8	22	18
10	25	20
12	29	24
14	31	25
16	32	25
18	32	25
20	33	26
22	33	27
25	40	33

Table #27B - Straight Tube Lengths

Notes:

1. R=Radius of tube bend as recommended by bender manufacturer
2. Dimensions in column A represent recommended straight tube length. Dimensions in column B are to be used when an absolute minimum straight tube length is necessary.
3. Metric tube fittings ARE NOT INTERCHANGEABLE with fractional sizes and conversions

Bending Steps

1. 90 Degree Bend (illustration #21A)

a. Measure and mark tube to the desired length; measuring from the end of the tube to the center of the 90 degree bend and mark with a pencil or fine point felt marker. If more than one bend is required, measure from the center line of the previous bend to the center line of the required 90 degree bend.

b. Place the tubing in the bender with the measured end of the tubing along the stationary handle (from left to right).

c. Position the mark at which to bend the tube directly in line with the 90 degree notch on the placement link. Secure the fastener latch onto the tube (making sure the "0" notch on both the bending wheel and placement link are in line).

d. While moving the pull handle towards the stationary handle, pull down smoothly until the "0" notch on the placement link lines up with the 90 degree on the bending wheel.

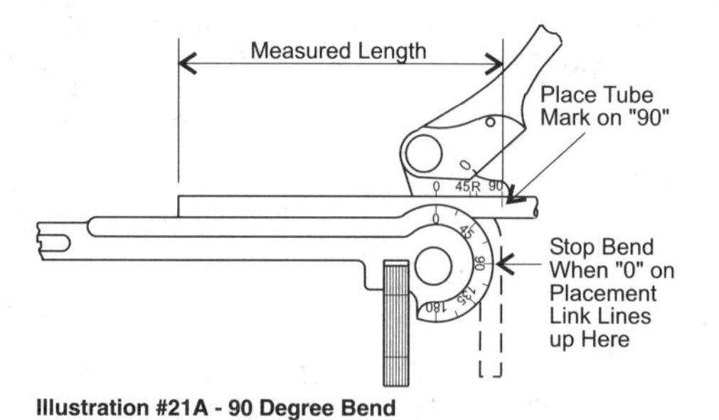

Illustration #21A - 90 Degree Bend

2. Reverse 90 Degree Bend (illustration #21B)

The "R" indicator notch on the placement link is used when the end of the tube from which the bend is measured from is reversed. That is, the tube is positioned in the bender from the opposite or reverse side of the stationary handle. All of the aforementioned bending procedures remain the same except the mark on the tube is placed on the "R" notch on the placement link.

Bending Steps

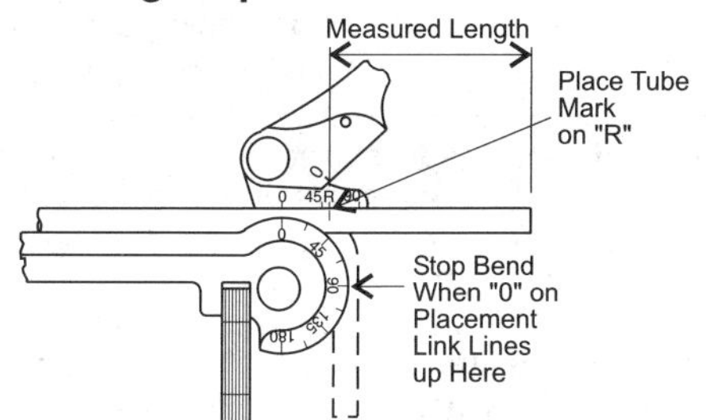

Illustration #21B - Reverse 90 Degree Bend

3. 45 Degree Bend (illustration #21C)

a. Measure and mark tube to the needed length, measure from the end of the tube to the center mark of the required 45 degree bend. If more than one bend is required, measure from the center line of the first bend to the center of the second 45 degree bend and mark.

b. Tubing can be placed in either direction (left or right) from the form handle.

c. Position the mark at which to bend the tube directly in line with the 45 degree bend notch on the placement link. Close the fastener onto the tube (making sure the "0" notch is lined up on both the bending wheel and the placement link).

d. While moving the pull handle towards the stationary handle, pull down smoothly on the pull handle till the "0" notch on the placement link lines up with the 45 degree on the bending wheel.

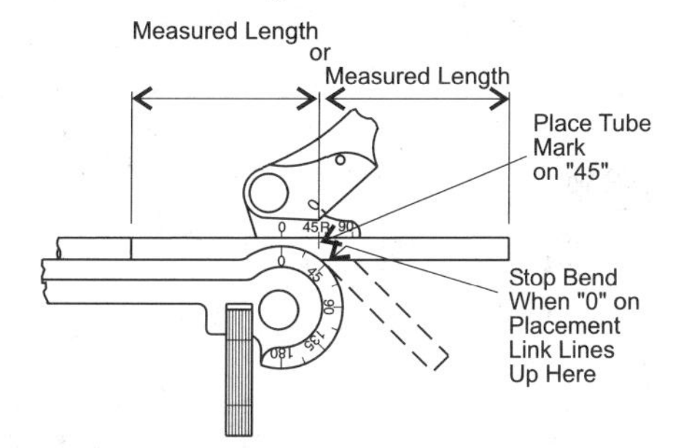

Illustration #21C - 45 Degree Bend

Tube Gain (Bending Length Increase)

When tube bending, the actual bend made does not follow the absolute profile of a straight line lay out, but short cuts the angles.

This short cutting of the corners gives additional length to the remaining straight tube. The length acquired is commonly referred to as gain. Illustration #22 shows this short cutting of the straight angle by the tube.

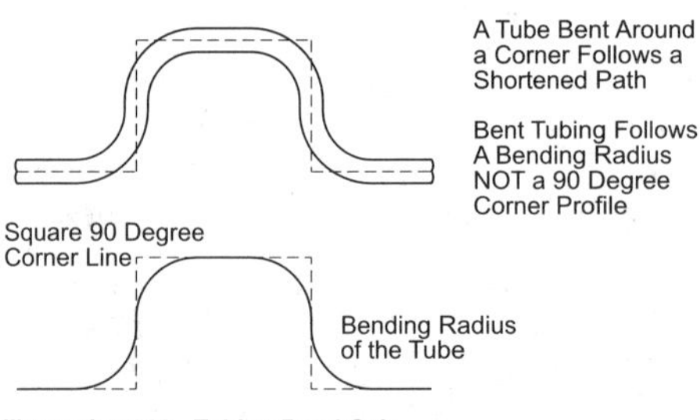

Illustration #22 - Tubing Bend Gain

Gain for 90 degree bends is approximately equal to the O.D. of the tube being bent.

Table #28 gives exact gain lengths for various sizes of tube and different bending radii. If bends are made correctly by positioning the mark to be bent on the proper placement link notch (either 90 degree or R), and in multibending situations, measuring each bend from the previous bend, the gain appears on the straight length of tube opposite the measured length.

Bending in this manner ensures correct end to center or center to center measurements in bending.

Tube Gain Calculation Table							
Tube Size		Common Bender Radius		90° Bend Gain		45° Bend Gain	
in	mm	in	mm	in	mm	in	mm
1/8	3.2	3/8	9.5	0.16	4.1	0.02	0.4
3/16	4.8	7/16	11.1	0.19	4.8	0.02	0.5
1/4	6.4	9/16	14.3	0.24	6.1	0.02	0.6
1/4	6.4	3/4	19.0	0.32	8.2	0.03	0.8
5/16	7.9	11/16	17.5	0.29	7.5	0.03	0.8
3/8	9.5	15/16	23.8	0.40	10.2	0.04	1.0
3/8	9.5	1 1/8	28.6	0.48	12.3	0.05	1.2
1/2	12.7	1 1/2	38.1	0.64	16.3	0.06	1.6
5/8	15.9	1 7/8	47.6	0.80	20.4	0.08	2.0
5/8	15.9	2	50.8	0.86	21.8	0.09	2.2
3/4	19.1	2 1/4	57.2	0.97	24.5	0.10	2.5
3/4	19.1	2 1/2	63.5	1.07	27.2	0.11	2.7
7/8	22.2	2 5/8	66.7	1.13	28.6	0.11	2.9
7/8	22.2	3	76.2	1.29	32.7	0.13	3.3
1	25.4	3	76.2	1.29	32.7	0.13	3.3
1	25.4	3 1/2	88.9	1.50	38.1	0.15	3.8
1 1/4	31.8	3 3/4	95.3	1.61	40.9	0.16	4.1
1 1/2	38.1	4 1/2	114.3	1.93	49.0	0.19	4.9
1 1/2	38.1	5	127.0	2.15	54.5	0.22	5.5
2	50.8	8	203.2	3.43	87.2	0.35	8.7

Table #28 - Tube Bend Gain

SECTION THREE

VALVES

Basic Valve Types

There are numerous valve types, styles, sizes, and shapes available for use in industry. Even though there are dozens of valve varieties to choose from, the primary purpose of valves remains the same; that is to stop or start flow, or to regulate flow.

Regulation of flow includes: throttling, prevention of flow reversal, and relieving or regulating pressure within a system.

Selection of valves for a system is based on the valve's intended service and design function. There are eight basic valve designs available:

- Gate
- Globe
- Check
- Diaphragm
- Ball
- Butterfly
- Plug
- Relief

Gate Valves

Gate valves are used for on-off service and are designed to operate fully open or fully closed.

Because of excessive vibration and wear created in partially closed gates, the valves are not intended for throttling or flow regulation. Gate valves are available in solid wedge, flexible wedge, split wedge and double disc styles. A typical gate valve and its major parts are shown in illustration #24.

See tables #32 and #34 through #37 of this section for sizes and dimensions.

Wedge type gate valves have a tapered wedge that wedges between two tapered seats when the valve is closed. The solid wedge design (shown in illustration #25) is widely used and is suitable for air, gas, oil, steam, and water service. Flexible wedge gate valves are used in services that have a tendency to bind the solid wedge design due to excessive variations in temperatures.

Gate Valves

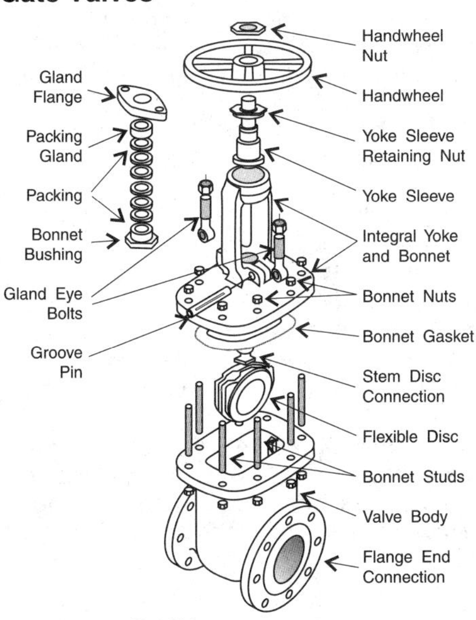

Illustration #24 - Gate Valve

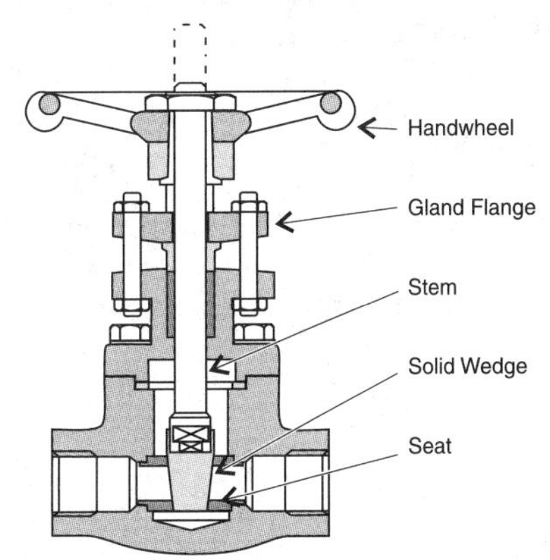

Illustration #25 - Solid Wedge Gate Valve

The design of the flexible wedge (shown in illustration #26) provides good seating characteristics (opening and closing) for a wide range of temperatures while providing positive shutoff.

Gate Valves

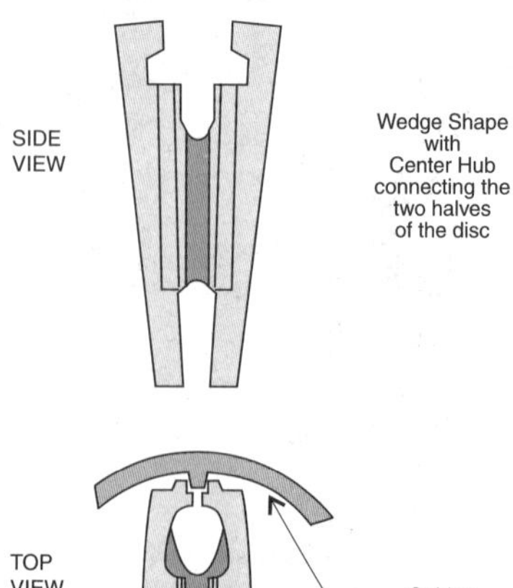

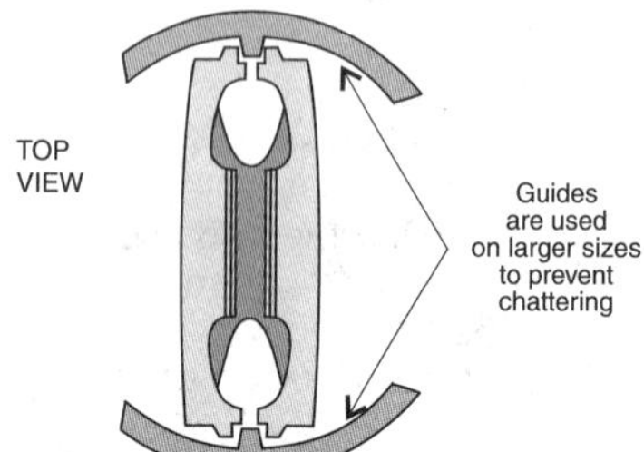

Illustration #26 - Flexible Wedge Disc

Split wedge and double disc gate valves have discs or wedges that are made of two pieces. In the split wedge design, (shown in illustration #27) the last turn of the hand-wheel forces the two discs against the tapered seats. The double disc gate valve has parallel discs and seats. Closure of the double disc is accomplished by a spreader or wedge which forces the parallel discs against the seats. The double disc gate valve is shown in illustration #28.

Friction, which causes wear on seats and discs, is kept to a minimum in both the split wedge and double disc gate valves because the seat and disc are in contact only on closing the valve.

When installing either the split wedge or double disc gate valve, the stem of the valve must be vertical (valve upright) which insures the discs do not jam apart before closing.

Gate Valves

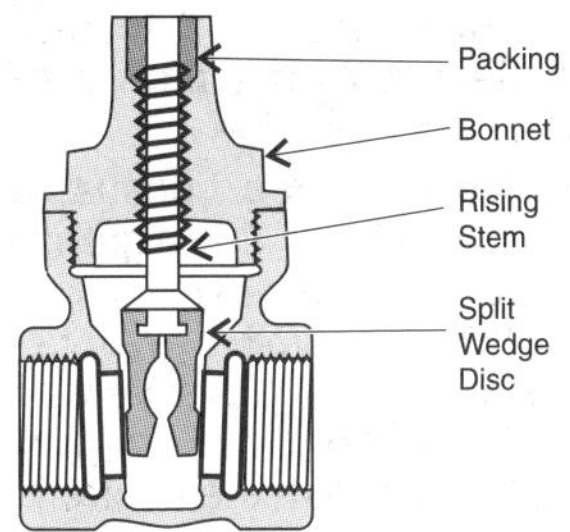

Illustration #27 - Split Wedge Gate Valve

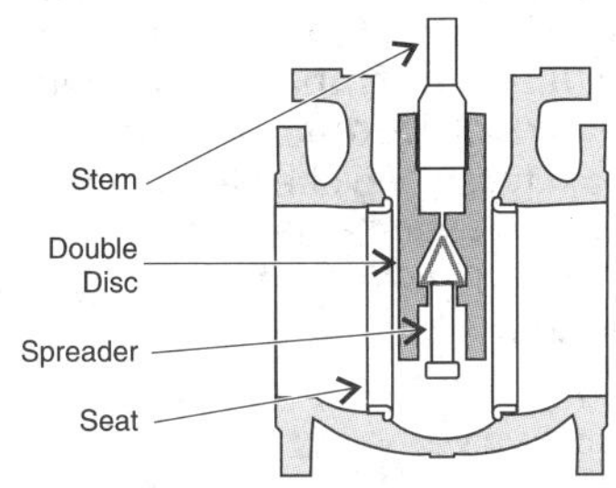

illustration #28 - Double Disc Gate Valve

Globe Valves

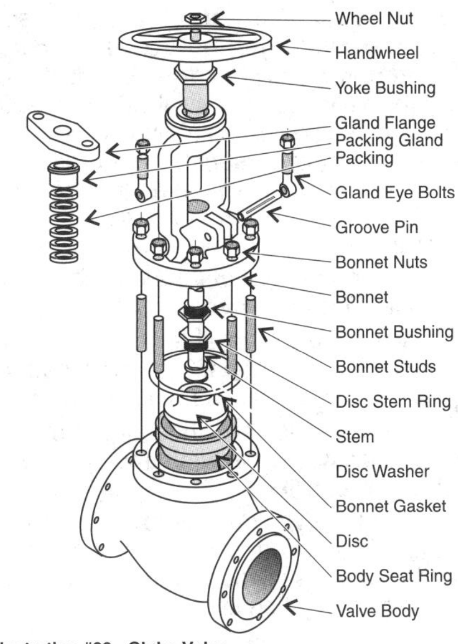

Illustration #29 - Globe Valve

Globe Valves

Globe valves, unlike gate valves, are used in applications requiring frequent operation and/or throttling of flow. The design of the globe valve (shown in illustration #29) keeps seat erosion to a minimum, while making it an easy valve to service. When flow begins in the globe valve design, the disc moves completely away from the seat thus minimizing erosion and wire drawing.

See tables #33 and #38 through #41 of this section for sizes and dimensions.

Globe valves are available in three body styles: angle body, Y-pattern and T-pattern body or straightway style (most common).

All three body styles are suited for throttling, but each has its own flow characteristic and service applicability.

Angle body valves provide for a 90 degree change of direction (see illustration #30) which in some installations saves space, material, and installation time.

The inner design of the angle body valve offers less flow restriction than the conventional T-pattern, but more than the Y-pattern globe valve.

Y-pattern globe valves, because of the angle of the stem (45 or 60 degrees from the run), give very little flow restriction. Illustration #31 shows a typical Y-pattern globe valve. The Y-pattern globe valve is ideally suited for applications requiring almost full flow in a valve, but with the characteristics of a globe valve.

Applications for the Y-pattern globe valve include boiler blow-offs and services where mud, grit and/or viscous fluids may be encountered.

The main types of globe valves according to their seat arrangements are:

- conventional disc
- plug disc
- composition disc
- needle valve

Globe Valves

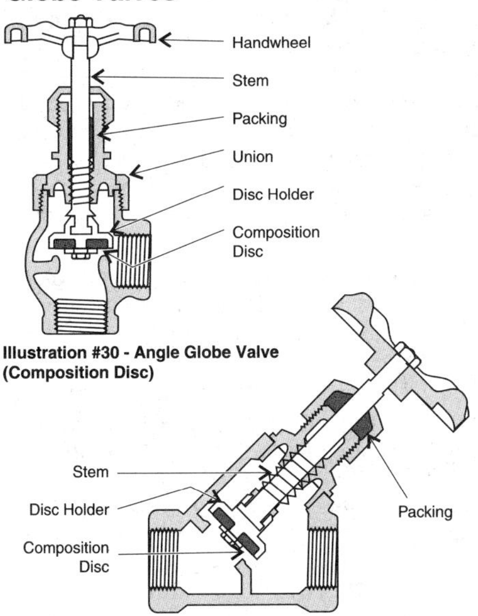

Illustration #30 - Angle Globe Valve (Composition Disc)

Illustration #31 - Y Pattern Globe Valve (Composition Disc)

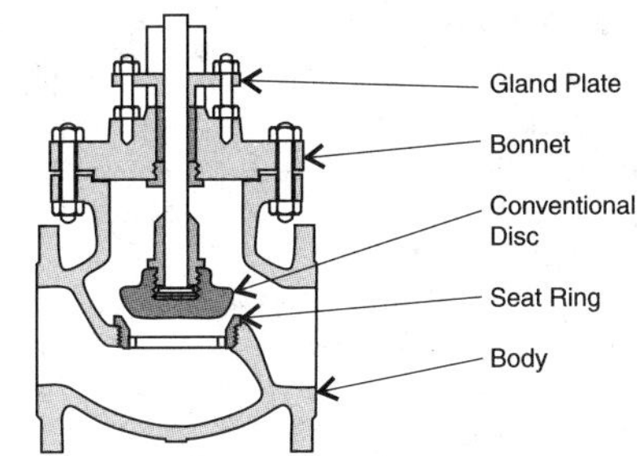

Illustration #32 - Plug Disc Globe Valve

Conventional disc globe valves use a short tapered disc which fits into a matched tapered seat for closure and throttling (shown in illustration #32). When flow conditions have a tendency to deposit or coke on valve seats, the conventional design disc valve is often preferred. The narrow disc used in the conventional disc will usually break through deposits, giving positive seating, rather than packing the deposits.

Globe Valves

The plug disc differs from the conventional disc in that the plug disc and seat arrangement are longer and more tapered. The longer area of the plug and seat give the plug valve maximum resistance to flow induced erosion. An example of a plug globe valve is shown in illustration #33.

The composition disc globe valve has a flat composition disc that fits flat against a seat rather than into a seat. This arrangement can be seen in illustration #34.

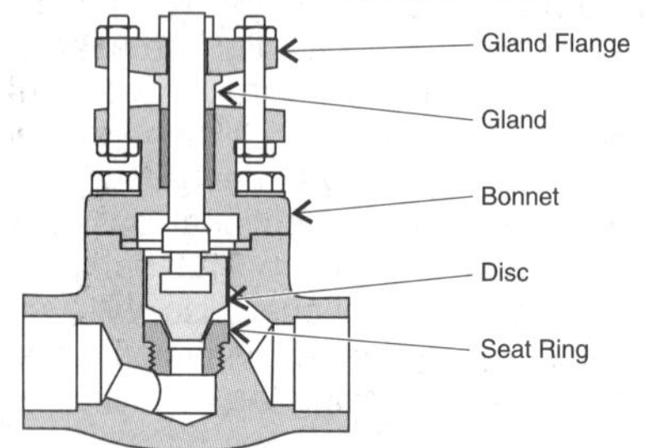

Illustration #33 - Plug Disc Globe Valve

Various composition materials can be used in the disc holder depending on the intended service of the valve. By changing the composition disc material, the valve can be changed from one service to another. Because of the softer nature of certain compositions used in the disc, foreign matter usually inbeds in the disc rather than causing leakage or scoring of the seat.

The needle valve is another type of globe valve which is used for accurate throttling in high pressure and/or high temperature service.

Needle valves are designed for small diameter lines requiring fine throttling of gases, steam, oil, water, or any other light liquid.

The valve consists of a sharp pointed (needle shaped) stem that controls flow through the seat. Illustration #35 shows a typical needle valve.

Globe Valves

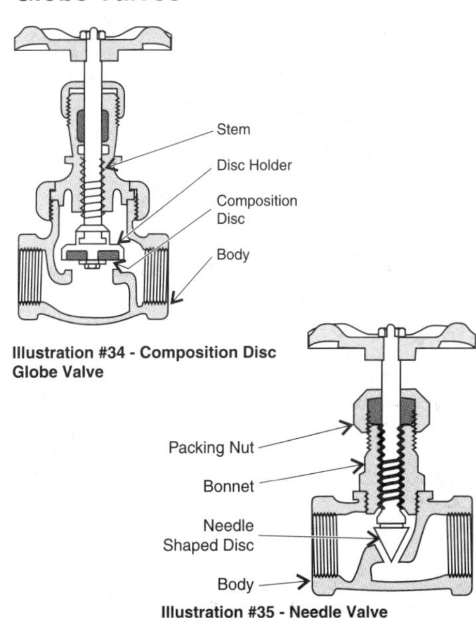

Illustration #34 - Composition Disc Globe Valve

Illustration #35 - Needle Valve

Check Valves

Check valves are used to prevent flow reverse in piping systems and connected equipment. The two most common design forms of check valves are:

- swing check
- lift check

The swing check (shown in illustration #36) consists of a hinged disc that swings open when flow starts in the desired direction and swings closed in flow reversal situations. Because of this swinging action of the disc, it is important to install all swing check valves so that the disc closes positively by gravity. When fully open, the swing check offers less flow resistance than the lift check valve design.

In flow situations where shock closure and/or disc chatter are encountered, other varieties of swing check valves can be used to help minimize the problems.

See tables #38 through #41 of this section for sizes and dimensions.

Check Valves

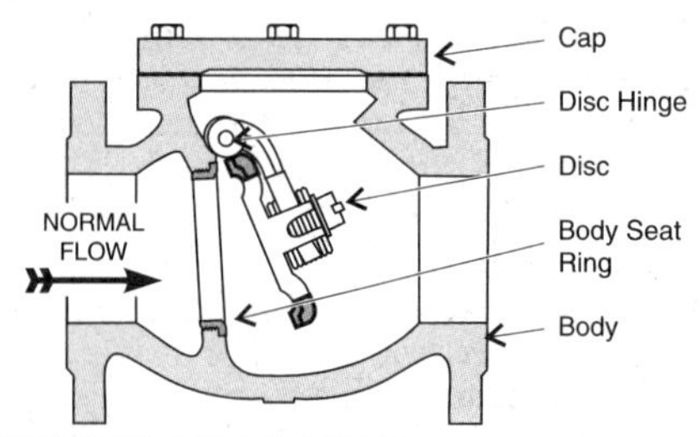

Illustration #36 - Swing Check Valve

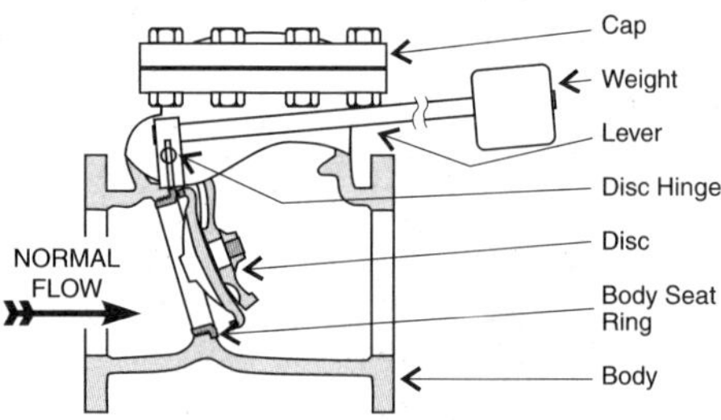

Illustration #37A - Lever & Weight Swing Check Valve

Swing check valves with outside lever and weight arrangements (shown in illustration #37A) or spring loaded discs can facilitate immediate closure in flow reverse. This immediate closure minimizes the possible damages of shock and disc chatter in systems. The tilting disc swing check valve is another type of swing check valve that is used to help prevent slamming (see illustration #37B).

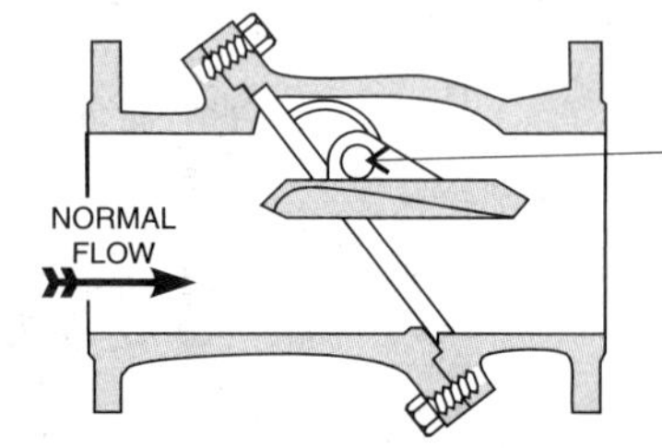

Illustration #37B - Tilting Check Valve

Check Valves

The lift check valve design (shown in illustration #38A) is used in line situations where pressure drop is not considered critical. The flow pattern through the valve corresponds to that of the globe valve. Lift check valves are available in horizontal and vertical designs.

Vertical design check valves (shown in illustration #38B) are for use in up-feed vertical lines only and will not work in the inverted or in the horizontal position. It is important when installing any lift check valve that the disc or ball lifts vertically when in operation.

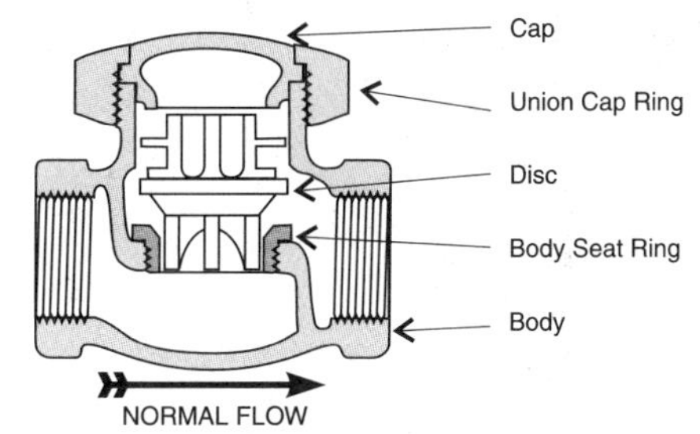

Illustration #38A - Horizontal Lift Check Valve

By-Pass Valves

In larger valving situations where high pressure and/or high temperatures are encountered, a smaller by-pass valve is often installed to equalize pressure and/or to allow the pipe down stream to warm up before the larger valve is opened. Illustration #39A shows a typical by-pass arrangement, while illustration #39B gives standard by-pass and auxiliary connection locations and sizes.

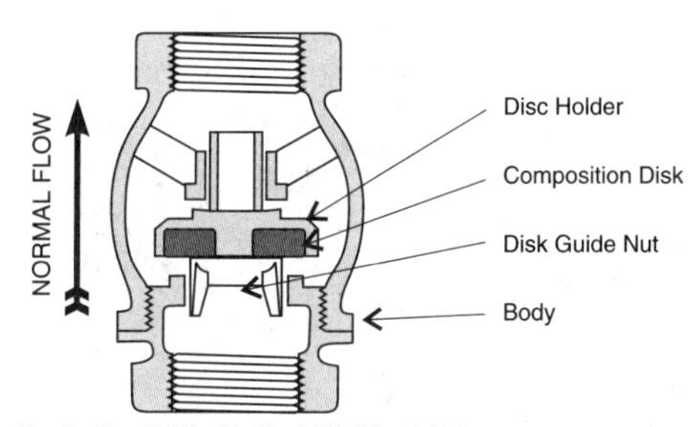

Illustration #38B - Vertical Lift Check Valve

By-Pass Valves

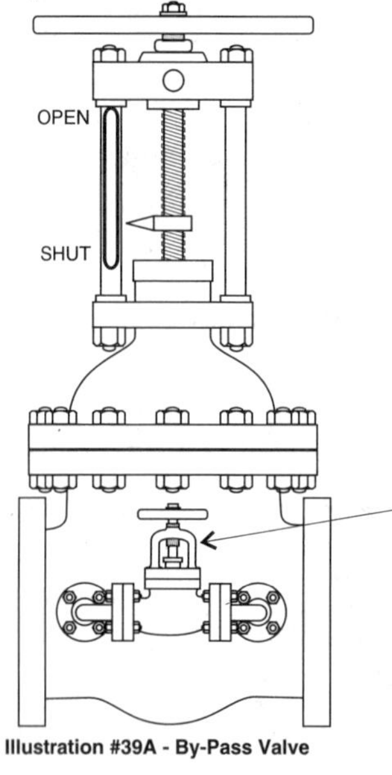

Illustration #39A - By-Pass Valve

Diaphragm Valves

Diaphragm valves (sometimes referred to as saunders valves) are designed to control flow in corrosive services where line content could adversely affect valve components. Other applications for diaphragm valves are in services where contamination from outside sources can not be tolerated; for example, the pharmaceutical and food industry.

The diaphragm valves differ from other valves in that the body of the valve and line content is sealed off from all moving parts of the valve by a flexible diaphragm. This flexible diaphragm seal prevents stem packing leakage of line content and flow contamination by packing lubricants.

Even though there are many variations in diaphragm valve design, most can be classified as either:

- weir type
- straightway type.

By-Pass Valves

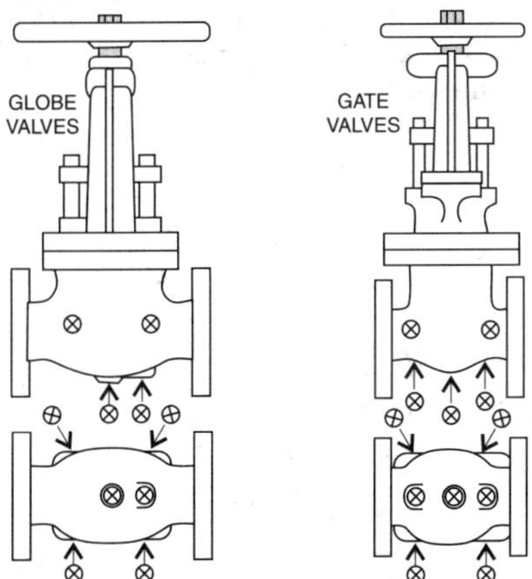

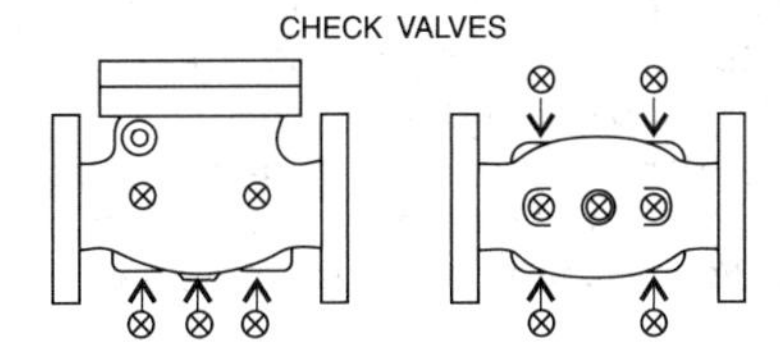

Recommended Sizes of By-Pass Valve Piping

Main Valve	2-4" 50-100mm	5-8" 125-200mm	10-24" 250-600mm
By-Pass	1/2" 15mm	3/4" 20mm	1" 25mm

Illustration #39B - By-Pass and Auxiliary Connections

Diaphragm Valves

Weir type diaphragm valves are the most common type of diaphragm valve used. Illustration #40A shows a typical weir valve along with its major components. The weir diaphragm incorporates a raised section (weir) half way through the valve which acts as a closure point for the flexible diaphragm. Because of the way the weir is formed in the body, diaphragm movement is shortened, which in turn prolongs diaphragm life and reduces overall maintenance.

Straightway diaphragm valves have no weir incorporated in the valve design. Illustration #40B shows a typical straightway valve. This design gives the valve an uninterrupted passageway suited for flows which are viscous or contain solids.

There are many types of diaphragm materials available (see table #29) depending on service and temperature conditions.

Because of longer diaphragm movement needed in the straightway valve compared to the weir type, material selection is limited for the straightway type.

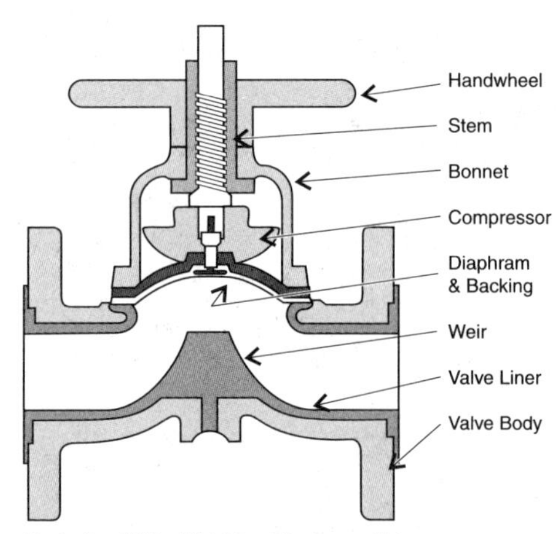

Illustration #40A - Weir Type Diaphragm Valve

Diaphragm Valves

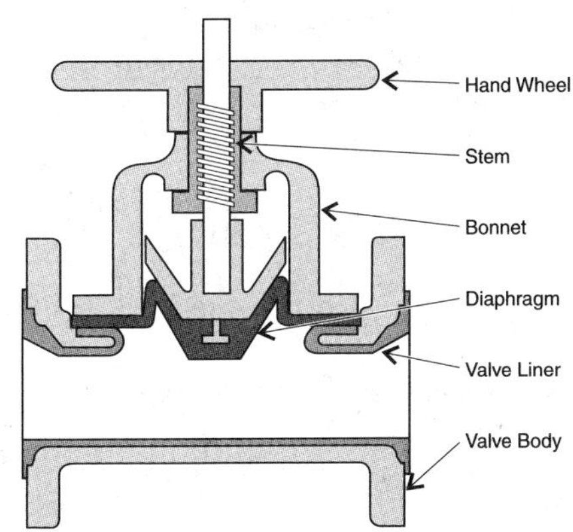

Illustration #40B - Straightway Type Diaphragm Valve

Pinch Valves

The pinch valve, like the diaphragm valve, uses a flexible diaphragm in the closing and opening of the valve. Pinch valves use a flexible hollow sleeve which is pinched closed to stop flow by manual or power methods. A simple air operated pinch valve is shown in illustration #41.

Pinch valves are ideally suited in services which carry suspended matter, slurries and solid powder flows.

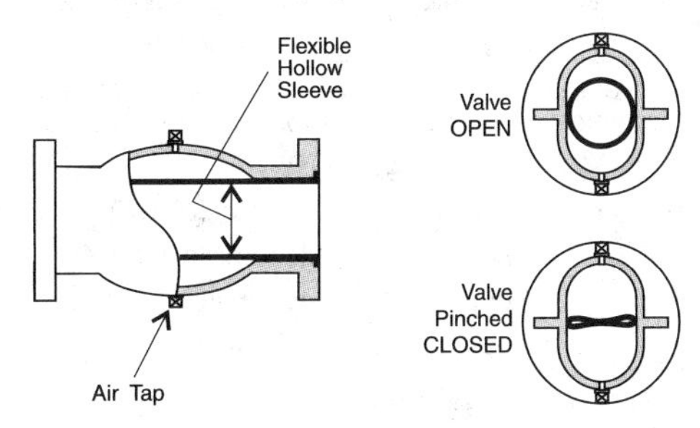

Illustration #41 - Air Operated Pinch Valve

COMMON MATERIALS USED FOR DIAPHRAGMS IN DIAPHRAGM VALVES						
Service Application	Material	Temperature Range			
		Temp. °F		Temp. °C	
		Min	Max	Min	Max
water systems	Natural Rubber	-30	180	-34	82
chemical, air & oil	Neoprene	-30	200	-34	93
oxidizing services	Hypalon	0	200	-18	93
food & beverage	White chlorinated butyl	-10	200	-23	93
fatty acids, chemical & gases	Black chlorinated butyl	0	200	-18	93
chemical, air & oil	Buna N	10	180	-12	82

Note: Full valve body liners of plastic are often used in diaphragm valves to enhance all around chemical resistance. The most common plastic lining materials are:

polypropylene (PP)
polyvinylidene fluoride (PVDF)
fluorinated ethylene propylene (FEP)
ethylene tetrafluorethylene (ETFE)
perfluoro alkoxy (PFA)

Table #29 - Diaphragm Material

Ball Valves

The ball valve, as the name indicates, contains a ball shaped plug within a valve body which regulates flow. The ball has a circular hole or flow way through its center and when turned one quarter turn, the flow stops. Ball valves come in three general patterns:

- venturi port
- full port
- reduced port design (sometimes referred to as regular port).

The port patterns indicate the inside diameter of the ball flow-way.

Ball valves may also be classified by their body style. Body styles come in one piece and multi-piece bodies.

A typical reduced (regular) port ball valve with a multi-piece body is shown in illustration #42.

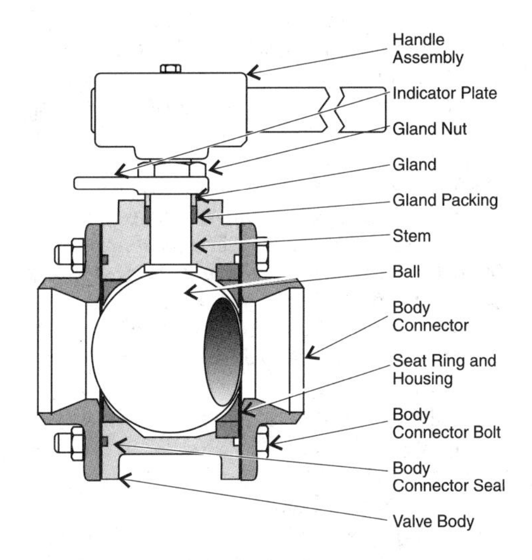

Illustration #42 - Reduced Port Ball Valve

Butterfly Valves

The butterfly valve offers a simple design that is lightweight, compact, and inexpensive, particularly in larger valve sizes. It consists of a flat circular disc hinged in its center, which closes or fully opens with a quarter turn. Seating for the disc is supplied by metal seats or resilient types of material like elastomer and plastics. ***Because of the advances in seating material, butterfly valves have found general acceptance in the oil, gas, chemical, water, and process fields. The valve is often used in place of a gate valve, but has the added advantage of flow regulation.***

Butterfly valves are available in two basic body types:

- wafer type
- double flanged type

The wafer type (shown in illustration #43) is mounted between two flanges and is held in place by flange bolts.

In situations where dismantling of equipment or lines may require disconnecting of one of the holding flanges, a lug wafer butterfly valve can be used. The lug wafer butterfly valve (shown in illustration #44) has tapped lugs through which the flange bolts are screwed and hold the valve in place, even when one flange is removed. In most wafer butterfly valves, the elastomer seal also acts as its own flange gasket.

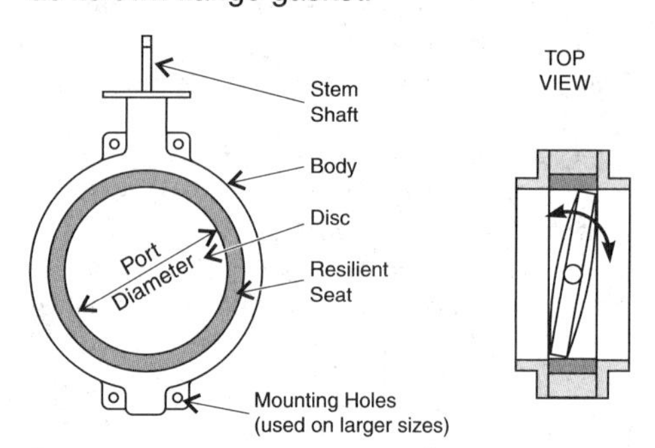

Illustration #43 - Wafer Type Butterfly Valve

Butterfly Valves

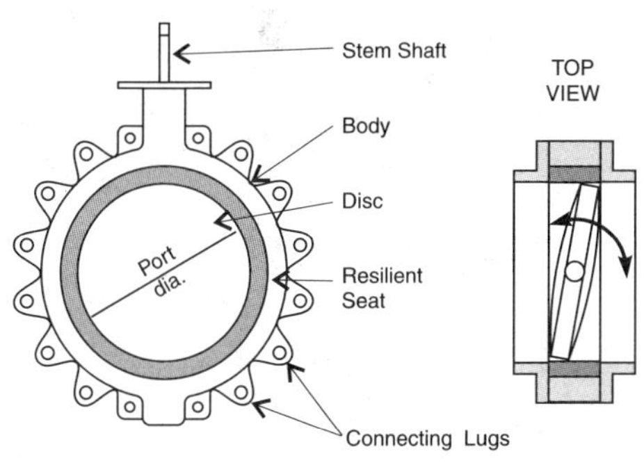

Illustration #44 - Lug-Wafer Butterfly Valve

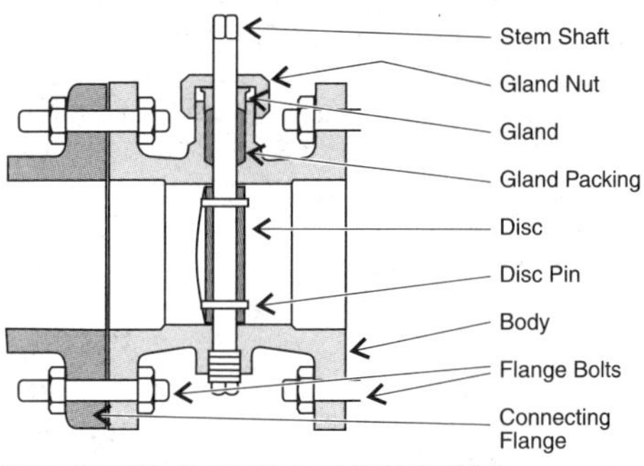

Illustration #45 - Double Flanged Butterfly Valve

The ***double flanged butterfly*** (shown in illustration #45) incorporates two flange ends which are bolted individually into the pipework or equipment flanges. Gaskets are used between the valve ends and connecting flanges.

Plug Valves

Plug valves (also known as cocks) consist of a tapered or parallel sided plug which can be rotated a quarter turn within a valve body. The quarter turn gives full closure or fully open operation of the valve. There are two basic types of plug valves available:

- lubricated plug
- non-lubricated plug

One variation of the lubricated plug is shown in illustration #46.

The lubricated plug differs from the non-lubricated plug in that it provides a means to lubricate the seating surfaces of the valve. This lubrication helps to eliminate valve seizing while still providing a positive seal. Lubricated plugs should not be used if flow contamination may be a problem.

The plug valve design also lends itself to multiport valve arrangements as shown in illustration #47.

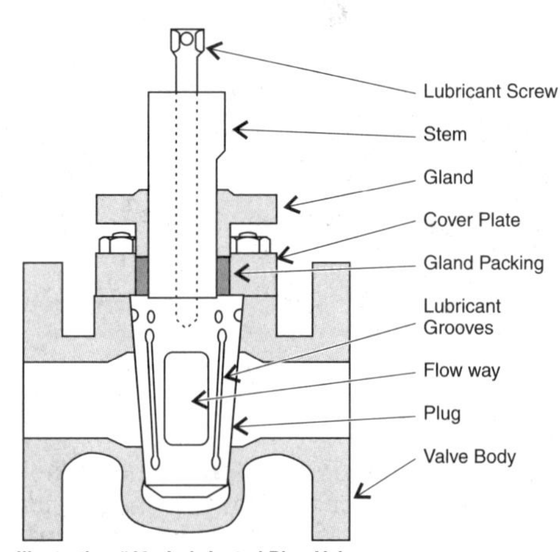

Illustration #46 - Lubricated Plug Valve

Plug Valve (Multi Port)

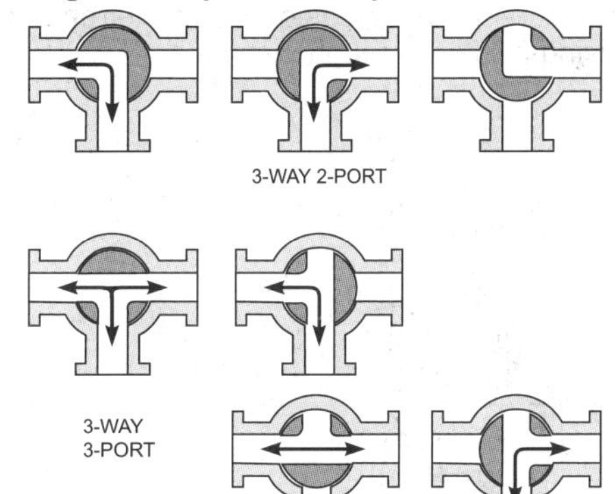

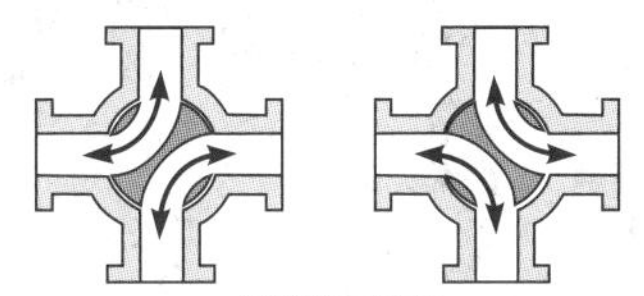

Illustration #47 - Multi-Port Plug Valve

Relief and Safety Valves

These valves automatically relieve excess pressure in pressure vessels or piping systems. Illustrations #48 and #49 show a relief valve and a safety valve (often referred to as a pop-safety valve).

The relief valve is used in applications where full discharge of pressure is not required to relieve pressure excesses.

The opening of the valve is proportional to the increase of pressure above the valve's set point. Usually, relief valves are specified in services where non compressible fluids are used and no likelihood of explosion will be encountered by overpressure.

Relief & Safety Valves

Safety valves, on the other hand, are used for relief of compressible fluids and gases which may cause an explosion when overpressured.

Cap
Spring Adjusting Screw
Spring
Valve Body
Valve Stem
Discharge
Disc
Disc Seat
Disc Guide
Inlet

Illustration #48 - Relief Valve

Safety valves or pop safety valves relieve pressure by fully opening (popping full open), giving full flow through the valve.

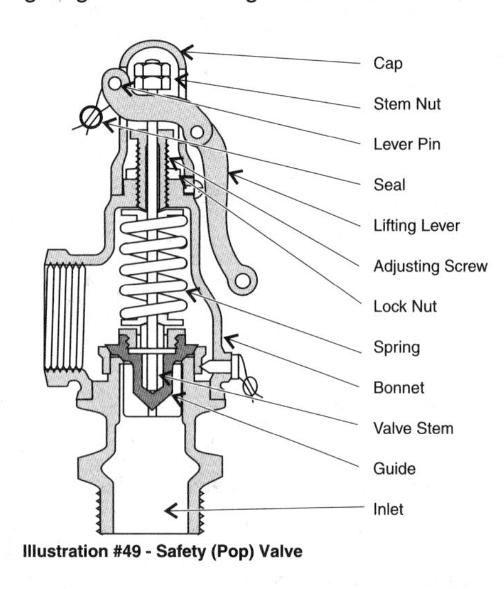

Illustration #49 - Safety (Pop) Valve

Variations in General Features of Valves

The features of most valves vary in:

- specific end connections
- bonnet assembly methods
- stem design selections

End Connections

The principal end connections for valves are:

a. ***Threaded End*** - usually a female tapered pipe thread used on valves under 6 inches (150 mm).

b. ***Welded End*** - used primarily on steel and steel alloy valves and can be either socket or butt welded. Socket welding is usually limited to valve sizes 2 inches (50 mm) and smaller. Typical valve socket weld depths and diameters are given in table #30.

c. ***Flanged End*** - found on larger valves with the most common flange end connections being the flat face, raised face and ring type.

d. ***Compression End*** - used for connecting to tubing and smaller sizes of pipe.

e. ***Flared End*** - used in connecting to tubing systems under 2 inches (50 mm).

f. ***Solder End*** - found on copper and copper alloy low pressure valves.

g. ***Hub End*** - used on cast iron valves for water supply and sewage systems.

Bonnet Assembly Methods

The primary function of a valve bonnet is to provide for a pressure tight transition assembly from the valve body to the valve stem. The four most common bonnet assemblies are:

1. threaded (illustration #50)
2. union (illustration #51)
3. bolted (illustration #52)
4. pressure seal type (illustration #53)

Two less common bonnet types are:

5. Welded bonnet, used in high temperature or pressure applications (illustration #54)
6. Clip type bonnet, a utility valve which is easy to dismantle and reassemble (illustration #55)

SOCKET WELD DIMENSION							
NOMINAL		SOCKET BORE DIAMETER				SOCKET DEPTH	
PIPE SIZE		Inches		Millimetres		Minimum	
Inch	mm	Max	Min	Max	Min	Inch	mm
1/4	8	0.565	0.555	14.35	14.10	0.38	9.6
3/8	10	0.700	0.690	17.78	17.53	0.38	9.6
1/2	15	0.865	0.855	21.97	21.72	0.38	9.6
3/4	20	1.075	1.065	27.30	27.05	0.50	12.7
1	25	1.340	1.330	34.04	33.78	0.50	12.7
1 1/4	32	1.685	1.675	42.80	42.54	0.50	12.7
1 1/2	40	1.925	1.915	48.90	48.64	0.50	12.7
2	50	2.416	2.406	61.37	61.11	0.62	15.8

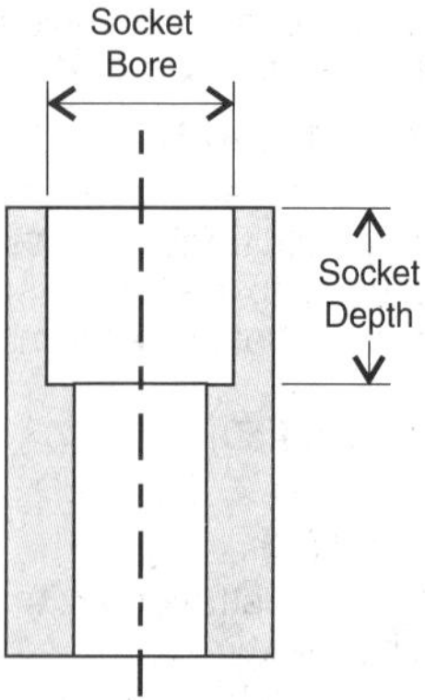

NOTE: Socket weld dimension based on ANSI B16.11

Table #30 - Socket Weld Dimensions

Bonnet Assembly Methods

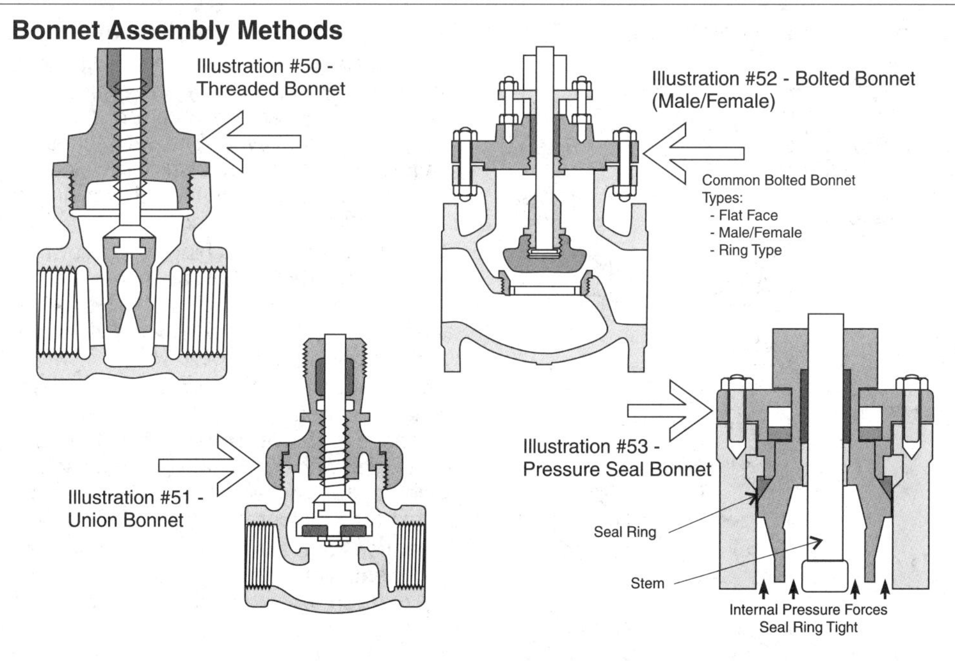

Bonnet Assembly Methods

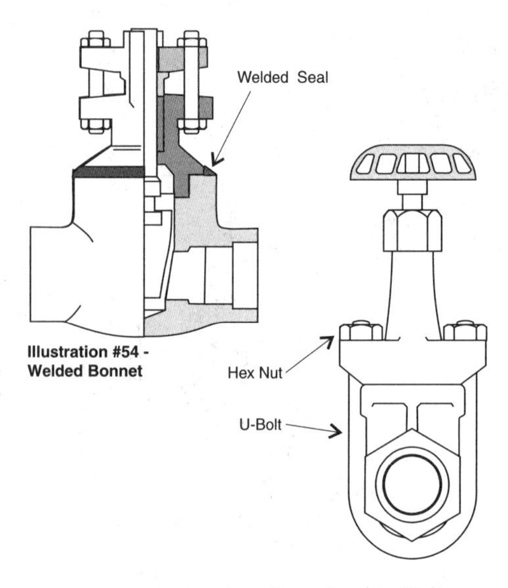

Illustration #54 - Welded Bonnet

Illustration #55 - Clip Bonnet

Special Bonnet Design Valves:

Bellows sealed valves are used in situations where no stem leakage can be allowed because of the hazardous nature of the substance being conveyed. The valve consists of a flexible metal bellows which is welded to the stem on one end and the valve casing or bonnet on the opposite end.

Three methods of attaching the bellows to the valve are shown in illustration #56A. To prevent damage to the bellows, anti-stem rotation devices are used in the valves to inhibit torque or twisting of the bellows by the stem.

Extended bonnet valves are used when temperature extremes could affect the stem packing of the valve. The bonnet of the valve is extended with the stuffing box and packing placed on top of the bonnet well away from the extreme temperature zone of the valve.

Illustration #56B shows an example of a cryogenic gate valve which is suited for liquified air, oxygen or other cold temperature service.

Special Bonnet Design Valves

Three Typical Locations for Welding the Bellows to Stem & Bonnet of a Valve

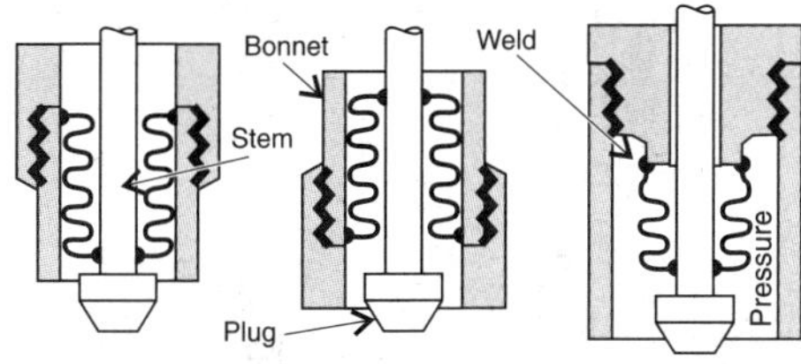

Illustration #56A -Bellows Seal Attachment

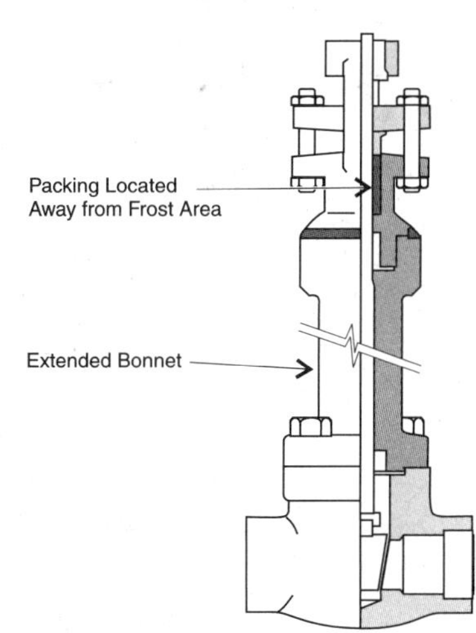

illustration #56B -Cryogenic Service Bonnet Gate Valve

Stem Design

There are four basic stem mechanism designs available for valves:

1. inside screw rising stem (illustration #57)
2. inside screw non-rising stem (ill. #58)
3. outside screw rising stem (ill. #59)
4. quick opening sliding stem (ill. #60)

The outside screw and rising stem design is often referred to as an OS&Y valve (outside screw and yoke).

Stem Designs

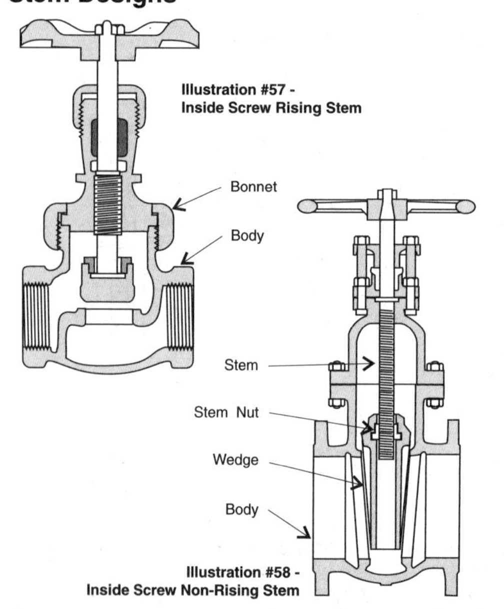

Illustration #57 - Inside Screw Rising Stem

Illustration #58 - Inside Screw Non-Rising Stem

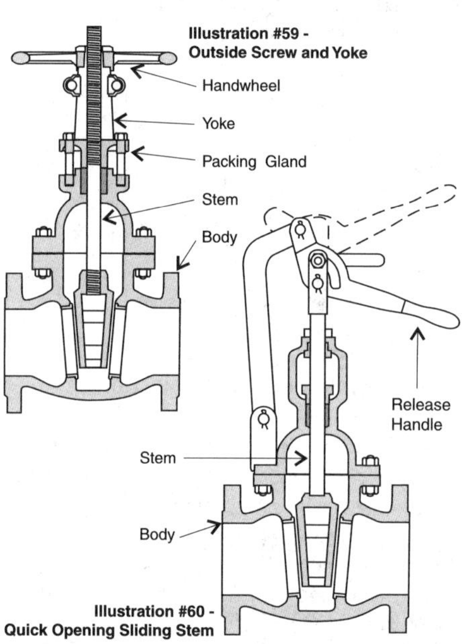

Illustration #59 - Outside Screw and Yoke

Illustration #60 - Quick Opening Sliding Stem

Valve ID, Marking & Symbols

Because valves are available in a multitude of styles, sizes, materials, and rating designations, it is important to be able to distinguish between specific valve categories. Most valve manufacturers mark their valve bodies and/or use an identification tag or plate to assist in proper identification of valves. ***Valve identification may include all or part of the following information (depending on valve classification). The Manufacturers Standardization Society of the Valve and Fittings Industry (MSS) specify the following marking requirements for valves:***

Manufacturer Name, Trademark, or Symbol:

- ***Rating Designation:*** this may include appropriate pressure rating class designations or specific maximum pressure temperature designations of the valve.

The following letters and designations correspond to Cold Working Pressure, (CWP) under normal ambient temperatures ranging between - 20° to 100°F (-29° to 36°C):

a. WO . . . Water & Oil Pressure
b. OWG . . Oil, Water & Gas Pressure
c. WOG . . Water, Oil & Gas Pressure
d. GLP . . . Gas & Liquid Pressure
e. WWP . . Working Water Pressure
f. W. . . . . Water Pressure

The following letters and designations correspond to Steam Working Pressure (SWP):

a. S. . . . . = Steam
b. SP . . . . = Steam Pressure
c. WSP . . = Working Steam Pressure

Example of valve designation rating:

Maximum 200 psig (SWP) service → 200 = S ← Steam

Maximum 400 psig (CWP) service → 400 - WOG ← Water, Oil and Gas

Valve ID, Marking & Symbols

NOTE:

1. When SI (metric) units are required, pressure is given in "Bar" units (1 bar = 100,000 Newtons per square metre, or 1 bar = 14.5 psi.
2. When SI (metric) units are required, temperature is given in degrees Celsius (°C).

- ***Material Designation:*** materials used in the construction of the body and bonnet of the valve are identified. Forged and fabricated valves are usually identified with ASTM specification numbers and grade identification symbols. Table #31 gives common symbols for metallic and nonmetallic materials used in valve construction.
- ***Melt Identification:*** cast valves are marked with the melt numbers or melt identification symbols.
- ***Valve Trim Identification:*** when stem, disc (includes plug, ball, and gate) or seat face material differs from body material, it is identified.
- ***Size Designation:*** given in nominal pipe size.
- ***Identification of Threaded End:*** when the threaded end connection of a valve is other than American National, Standard Pipe Thread or American National Standard Hose Thread, it is identified by the type of thread used.
- ***Identification of Ring-Joint Face:*** standard ring joints complying with ANSI standard B16.20 or API standard 6A are marked with an "R ".
- ***Special Function Requirements:*** manufacturers may place catalog numbers, dates, reference numbers or any additional marking to distinguish the product.

SYMBOLS USED FOR METALLIC AND NON-METALLIC MATERIALS

METALLIC SYMBOLS

Material	Symbol
All Iron	AI
Aluminum	AL
Brass	BRS
Bronze	BRZ
Carbon Steel	CS
Cast Iron	CI
Chromium	CR
Copper-Nickel	CUN
Ductile Iron	DI
Forged Steel	FS
Hardfacing	HF
Integral Seat	INT
Malleable Iron	MI
Molybdenum	MO
Monel Metal	M
Nickel-Copper	NI CU
Nickel-Iron	NI
Soft Metal	SM
Stainless Steel	SS
Steel, 13 Chromium	CR13
Steel, 18 Chromium	CR18
Steel, 28 Chromium	CR28
Steel, 18-8	18-8
Steel, 18-8 Molybdenum	18-8SMO
Steel, 18-8 Columbium	18-8SCB
Surface Hardened	SH

NON-METALLIC SYMBOLS

Material	Symbol
Asbestos	ASB
Butadiene Rubber	BR
Butyl Rubber	HR
Chlorinated polyvinyl chloride	CPVC
Chloroprene or Neoprene	CR
Chlorosulfonated Polyethylene	CSM
Chlorotrifluoreoethylene	CIFE
Ethylene-Propylene Rubber	EPR
Fluorocarbon or Viton Rubber	FPM
Fluorinated Ethylene Propylene	FEP
Isoprene Rubber	IR
Natural Rubber	NR
Nitrile or Buna N	NBR
Nylon	NYL
Polyacrylic Rubber	ACM
Poly Vinyl Chloride	PVC
Polypropylene	PP
Polyvinylidene	PVDF
Silicone Rubber	SI
Styrene Butadiene	SBR
Tetrafluoroethylene	TFE
Teflon	TEF
Thermoplastic	T-PLAS
Thermosetting	T-SET

Table #31 - Valve Material Symbols

Valve Installation

1. Follow manufacturers specification for any specific installation procedures that may be required with the valve.
2. Check the valve specifications and/or ID to ensure they match with the valve needed in a specific application.
3. Make sure pipe scale, metal chips, welding slag and any other foreign materials are removed from the piping system or equipment before installing the valve.
4. If foreign particles (especially abrasives) may be present in the line, install a strainer upstream from the valve.
5. Check valve operation before installing the valve, making sure any packing or shipping material is removed.
6. Install valve in correct position. The valve stem should be in the vertical position with the stem of the valve pointing up. This is particularly true for split wedge and double disc gate valves. See illustration #61.
7. When installing check valves, make sure that discs or balls in lift checks lift vertically, and that in swing check valves the disc closes positively with gravity.
8. Valve flow direction design must correspond to system flow direction. Follow flow direction arrows or inlet/outlet stampings on valves when installing.

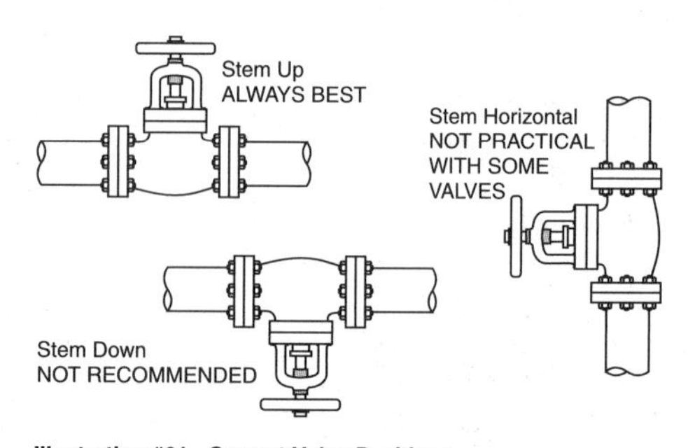

Illustration #61 - Correct Valve Positions

9. Ensure adequate space is available so that the valve can be operated and maintained. See illustration #62.

10. Flanges, piping, and equipment must line up square and true with the valve without putting undue stress on valve. See illustration #63.

NOTES:

Dimensions to center of valve handwheel for vertical valves, stem center line for horizontal valves.

Use chain operators on valve sizes over 1 1/2 in. (40 mm) located 6'-6" to 20' (2 to 6 m) above platforms or floors.

Chains should extend to within 3' (1 m) from floor level.

If a railing is present, horizontal valves should be placed 5' to 5'-6" (1.5 to 1.7 m) above the floor.

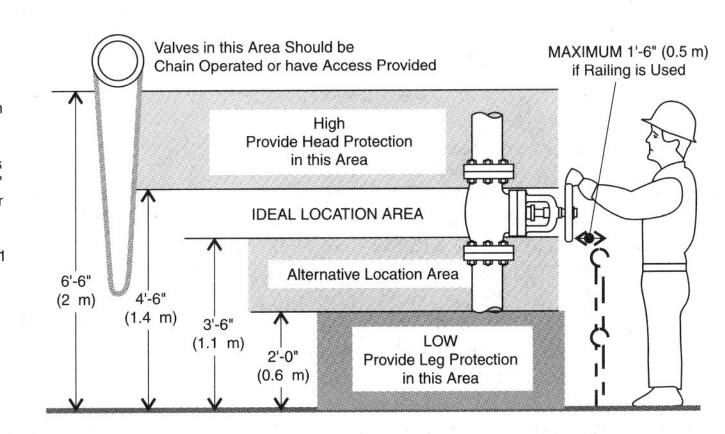

Illustration #62 - Valve Locations

Valve Installation

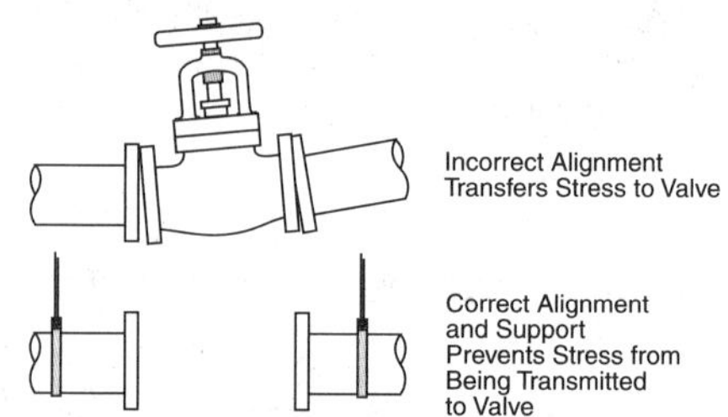

Illustration #63 - Align Piping and Valve

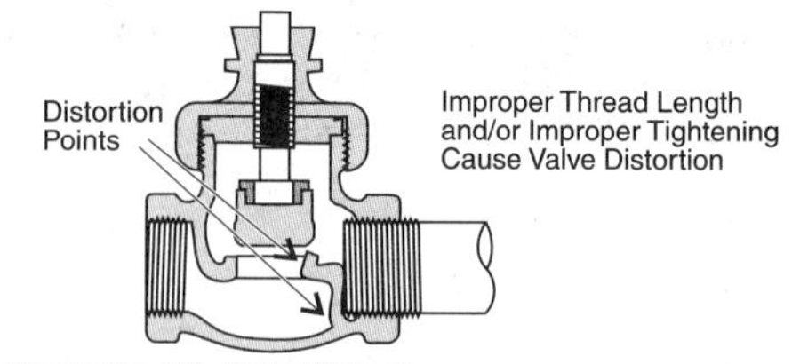

Illustration #64 - Valve Distortion

11. Appropriate space must be provided for the valve to ensure the valve fits into the piping when installed. Standard valve dimensions are given in tables #32 - #42.
12. Do not over tighten or extend pipe threads too far into a valve because of the possibility of damage and distortion of valve or parts. See illustration #64.
13. Use proper flange make-up procedure by tightening flange bolts in cross over order. See illustration #65.
14. Wrench placement should be positioned on the end of the valve nearest to the joint being made up. See illustration #66.
15. Do not place valves in locations where they may be subject to accidental damage or mistreatment.
16. Valves must be properly supported to prevent the weight of the valve from being transferred to piping or equipment. See Appendices for hanger information.

Valve Installation

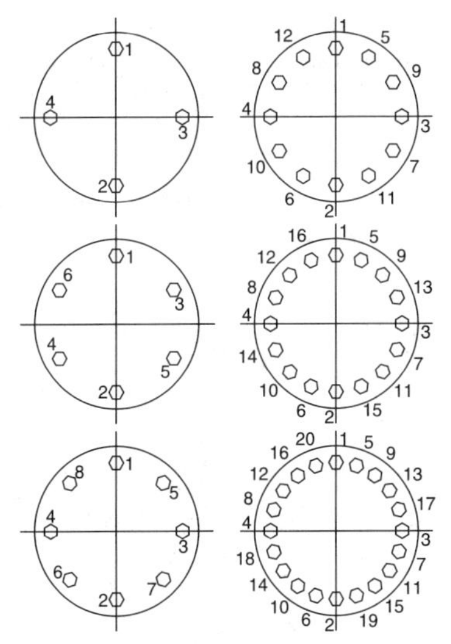

NOTE: Numbers indicate tightening sequence

Illustration #65 - Valve Flange Tightening

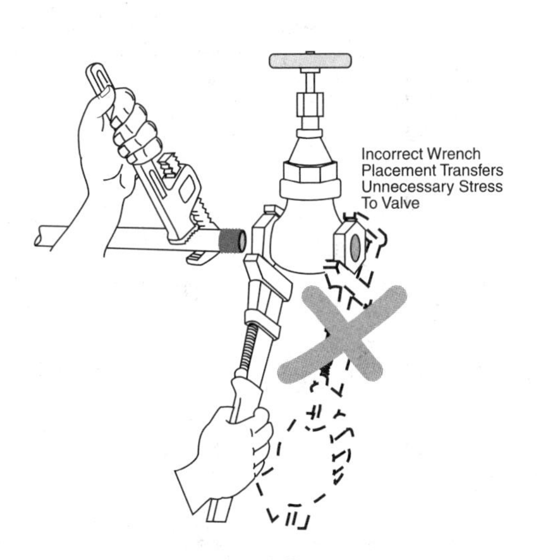

Illustration #66 - Proper Wrench Placement

Cast Iron Gate Valves/Dimensions Given in Inches and Millimetres (mm) taken from Face to Face of the Valve

CAST IRON GATE VALVES / ANSI CLASS 125, 250 & 800

Nominal Pipe Size		Flat Face/Class 125				Raised Face/Class 250				Raised Face/Class 800			
		Solid Wedge		Double Disc		Solid Wedge		Double Disc		Solid Wedge		Double Disc	
Inch	mm	Inch	mm	Inch	mm	Inch	mm	Inch	mm	Inch	mm	Inch	mm
1/4	8	-	-	-	-	-	-	-	-	-	-	-	-
3/8	10	-	-	-	-	-	-	-	-	-	-	-	-
1/2	15	-	-	-	-	-	-	-	-	-	-	-	-
3/4	20	-	-	-	-	-	-	-	-	-	-	-	-
1	25	-	-	-	-	-	-	-	-	-	-	-	-
1 1/4	32	-	-	-	-	-	-	-	-	-	-	-	-
1 1/2	40	-	-	-	-	-	-	-	-	-	-	-	-
2	50	7.00	177.8	7.00	177.8	8.50	215.9	8.50	215.9	11.50	292.1	11.50	292.1
2 1/2	65	7.50	190.5	7.50	190.5	9.50	241.3	9.50	241.3	13.00	330.2	13.00	330.2
3	80	8.00	203.2	8.00	203.2	11.12	282.5	11.12	282.5	14.00	355.6	14.00	355.6
4	100	9.00	228.6	9.00	228.6	12.00	304.8	12.00	304.8	17.00	431.8	17.00	431.8
5	125	10.00	254.0	10.00	254.0	15.00	381.0	15.00	381.0	-	-	-	-
6	150	10.50	266.7	10.50	266.7	15.88	403.4	15.88	403.4	22.00	558.8	22.00	558.8
8	200	11.50	292.1	11.50	292.1	16.50	419.1	16.50	419.1	26.00	660.4	26.00	660.4
10	250	13.00	330.2	13.00	330.2	18.00	457.2	18.00	457.2	31.00	787.4	31.00	787.4
12	300	14.00	355.6	14.00	355.6	19.75	501.7	19.75	501.7	33.00	838.2	33.00	838.2
14	350	15.00	381.0	-	-	22.50	571.5	22.50	571.5	-	-	-	-
16	400	16.00	406.4	-	-	24.00	609.6	24.00	609.6	-	-	-	-
18	450	17.00	431.8	-	-	26.00	660.4	26.00	660.4	-	-	-	-
20	500	18.00	457.2	-	-	28.00	711.2	28.00	711.2	-	-	-	-
22	550	-	-	-	-	-	-	-	-	-	-	-	-
24	600	20.00	508.0			31.00	787.4	31.00	787.4	-	-	-	-
26	650	-	-	-	-	-							
28	700	-	-	-	-	-							
30	750	-	-	-	-	-							
32	800	-	-	-	-	-							
34	850	-	-	-	-	-							
36	900	-	-	-	-	-							

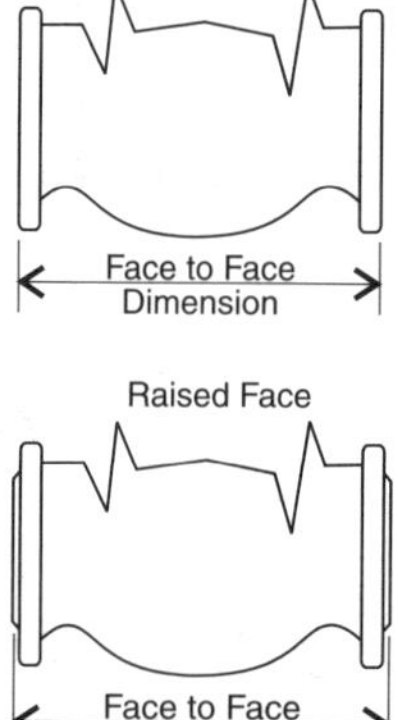

NOTES:
1. (-) indicates valve size is not shown in ASME/ANSI standards but may be commercially available.
2. Millimetre dimensions are rounded off to first decimal point.
3. Dimensions based on ASME/ANSI B16.10 Standard.

Table #32 - Cast Iron Gate Valve

Cast Iron Globe Valves/Dimensions Given in Inches and Millimetres (mm) taken from Face to Face for Straightway Valves and Center to Face for Angle Valves

CAST IRON GLOBE VALVES/ANSI CLASS 125 & 250

Nominal Pipe Size		Flat Face/Class 125						Raised Face/Class 250					
		Straightway		Angle Globe		Control-Style		Straightway		Angle Globe		Control-Style	
Inch.	mm	Inch.	mm	Inch.	mm	Inch.	mm	Inch.	mm	Inch.	mm	Inch.	mm
1/4	8	-	-	-	-	-	-	-	-	-	-	-	-
3/8	10	-	-	-	-	-	-	-	-	-	-	-	-
1/2	15	-	-	-	-	-	-	-	-	-	-	7.50	190.5
3/4	20	-	-	-	-	-	-	-	-	-	-	7.62	193.5
1	25	-	-	-	-	7.25	190.5	-	-	-	-	7.75	196.9
1 1/4	32	-	-	-	-			-	-	-	-	-	-
1 1/2	40	-	-	-	-	8.75	222.3	-	-	-	-	9.25	235.0
2	50	8.00	203.2	4.00	101.6	10.00	254.0	10.50	266.7	5.25	133.4	10.50	266.7
2 1/2	65	8.50	215.9	4.25	107.0	10.88	276.4	11.50	292.1	5.75	146.1	11.50	292.1
3	80	9.50	241.3	4.75	120.7	11.75	298.5	12.50	317.5	6.25	158.8	12.50	317.5
4	100	11.50	292.1	5.75	146.1	13.88	352.6	14.00	355.6	7.00	177.8	14.50	368.3
5	125	13.00	330.2	6.50	165.1	-	-	15.75	400.1	7.88	200.2	-	
6	150	14.00	355.6	7.00	177.8	17.75	450.9	17.50	444.5	8.75	222.3	18.62	473.0
8	200	19.50	495.3	9.75	247.7	21.38	543.1	21.00	533.4	10.50	266.7	22.38	568.5
10	250	24.50	622.3	12.25	311.1	26.50	673.1	24.50	622.3	12.25	311.2	27.88	708.2
12	300	27.50	698.5	13.75	349.3	29.00	736.6	28.00	711.2	14.00	355.6	30.50	774.7
14	350	31.00	787.4	15.50	393.7	35.00	889.0					36.50	927.1
16	400	36.00	914.4	18.00	457.2	40.00	1016.0					41.62	1057.2
18	450												
20	500												
22	550												
24	600												
26	650												
28	700												
30	750												
32	800												
36	900												

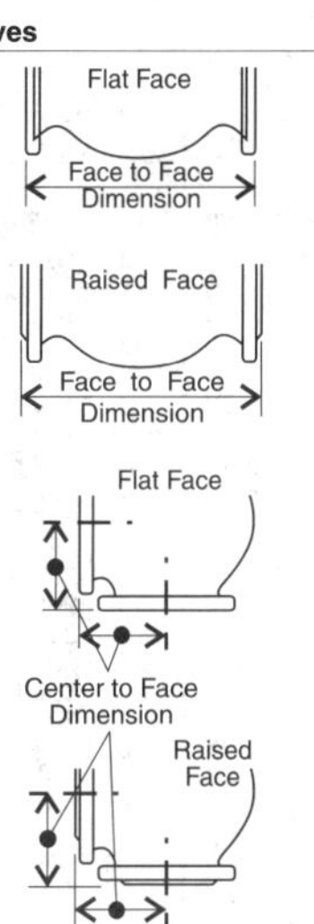

NOTES: 1. (-) indicates valve size is not shown in ASME/ANSI standards but may be commercially available. 2. Millimetre dimensions are rounded off to first decimal point. 3. Dimensions based on ASME/ANSI B16.10 Standard. 4. Dimensions for standard straightway lift and swing check valves are interchangeable with straightway Globe valve dimensions (excluding 16" (400mm) vertical swing checks). 5. Dimensions for standard angle lift check valves are interchangeable with angle globe valve dimensions.

Table #33 - Cast Iron Globe Valve

Steel Gate Valves/Dimensions Given in Inches and Millimetres (mm) taken from Face to Face and End to End of Valves																	
STEEL GATE VALVES / ANSI CLASS 150 & 300																	
Nominal Pipe Size		Raised Face/Class 150				Beveled End/Class 150				Raised Face/Class 300				Beveled End/Class 300			
		Solid Wedge		Double Disc		Solid Wedge		Double Disc		Solid Wedge		Double Disc		Solid Wedge		Double Disc	
Inch	mm	Inch.	mm	Inch.	mm	Inch.	mm	Inch.	mm	Inch.	mm	Inch.	mm	Inch.	mm	Inch.	mm
1/4	8	4.00	101.6	4.00	101.6	4.00	101.6	4.00	101.6	-	-	-	-	-	-	-	-
3/8	10	4.00	101.6	4.00	101.6	4.00	101.6	4.00	101.6	-	-	-	-	-	-	-	-
1/2	15	4.25	108.0	4.25	108.0	4.25	108.0	4.25	108.0	5.50	139.7	-	-	5.50	139.7	-	-
3/4	20	4.62	117.4	4.62	117.4	4.62	117.4	4.62	117.4	6.00	152.4	-	-	6.00	152.4	-	-
1	25	5.00	127.0	5.00	127.0	5.00	127.0	5.00	127.0	6.50	165.1	-	-	6.50	165.1	-	-
1 1/4	32	5.50	139.7	5.50	139.7	5.50	139.7	5.50	139.7	7.00	177.8	-	-	7.00	177.8	-	-
1 1/2	40	6.50	165.1	6.50	165.1	6.50	165.1	6.50	165.1	7.50	190.5	7.50	190.5	7.50	190.5	7.50	190.5
2	50	7.00	177.8	7.00	177.8	8.50	215.9	8.50	215.9	8.50	215.9	8.50	215.9	8.50	215.9	8.50	215.9
2 1/2	65	7.50	190.5	7.50	190.5	9.50	241.3	9.50	241.3	9.50	241.3	9.50	241.3	9.50	241.3	9.50	241.3
3	80	8.00	203.2	8.00	203.2	11.12	282.5	11.12	282.5	11.12	282.5	11.12	282.5	11.12	282.5	11.12	282.5
4	100	9.00	228.6	9.00	228.6	12.00	304.8	12.00	304.8	12.00	304.8	12.00	304.8	12.00	304.8	12.00	304.8
5	125	10.00	254.0	10.00	254.0	15.00	381.0	15.00	381.0	15.00	381.0	15.00	381.0	15.00	381.0	15.00	381.0
6	150	10.50	266.7	10.50	266.7	15.88	403.4	15.88	403.4	15.88	403.4	15.88	403.4	15.88	403.4	15.88	403.4
8	200	11.50	292.1	11.50	292.1	16.50	419.1	16.50	419.1	16.50	419.1	16.50	419.1	16.50	419.1	16.50	419.1
10	250	13.00	330.2	13.00	330.2	18.00	457.2	18.00	457.2	18.00	457.2	18.00	457.2	18.00	457.2	18.00	457.2
12	300	14.00	355.6	14.00	355.6	19.75	501.7	19.75	501.7	19.75	501.7	19.75	501.7	19.75	501.7	19.75	501.7
14	350	15.00	381.0	15.00	381.0	22.50	571.5	22.50	571.5	30.00	762.0	30.00	762.0	30.00	762.0	30.00	762.0
16	400	16.00	406.4	16.00	406.4	24.00	609.6	24.00	609.6	33.00	838.2	33.00	838.2	33.00	838.2	33.00	838.2
18	450	17.00	431.8	17.00	431.8	26.00	660.4	26.00	660.4	36.00	914.4	36.00	914.4	36.00	914.4	36.00	914.4
20	500	18.00	457.2	18.00	457.2	28.00	711.2	28.00	711.2	39.00	990.6	39.00	990.6	39.00	990.6	39.00	990.6
22	550	-	-	-	-	30.00	762.0	30.00	762.0	43.00	1092.2	43.00	1092.2	43.00	1092.2	43.00	1092.2
24	600	20.00	508.0	20.00	508.0	32.00	812.8	32.00	812.8	45.00	1143.0	45.00	1143.0	45.00	1143.0	45.00	1143.0
26	650	22.00	558.8	22.00	558.8	-	-	34.00	863.6	49.00	1244.6	49.00	1244.6	49.00	1244.6	49.00	1244.6
28	700	24.00	609.6	24.00	609.6	-	-	36.00	914.4	53.00	1346.2	53.00	1346.2	53.00	1346.2	53.00	1346.2
30	750	24.00	609.6	24.00	609.6	-	-	36.00	914.4	55.00	1397.0	55.00	1397.0	55.00	1397.0	55.00	1397.0
32	800	-	-	-	-	-	-	38.00	965.2	60.00	1524.0	60.00	1524.0	60.00	1524.0	60.00	1524.0
34	850	-	-	-	-	-	-	40.00	1016	64.00	1625.6	64.00	1625.6	64.00	1625.6	64.00	1625.6
36	900	28.00	711.2	28.00	711.2	-	-	40.00	1016	68.00	1727.2	68.00	1727.2	68.00	1727.2	68.00	1727.2

Table #34 - Steel Gate Valve (150 & 300)

Steel Gate Valves/Dimensions Given in Inches and Millimetres (mm) taken from Face to Face and End to End of Valves															
STEEL GATE VALVES / ANSI CLASS 400 & 600															
Nominal Pipe Size		Raised Face/Class 400 & Beveled End/Class 400				Raised Face/Class 600				Beveled End/Class 600					
		Solid Wedge		Double Disc		Solid Wedge		Double Disc		Solid Wedge		Double Disc		Short Pattern	
Inch	mm	Inch	mm	Inch	mm	Inch	mm	Inch	mm	Inch	mm	Inch	mm	Inch	mm
1/2	15	6.50	165.1	-	-	6.50	165.1	-	-	6.50	165.1	-	-	-	-
3/4	20	7.50	190.5	-	-	7.50	190.5	-	-	7.50	190.5	-	-	-	-
1	25	8.50	215.9	8.50	215.9	8.50	215.9	8.50	215.9	8.50	215.9	8.50	215.9	5.25	133.4
1 1/4	32	9.00	228.6	9.00	228.6	9.00	228.6	9.00	228.6	9.00	228.6	9.00	228.6	5.75	146.1
1 1/2	40	9.50	241.3	9.50	241.3	9.50	241.3	9.50	241.3	9.50	241.3	9.50	241.3	6.00	152.4
2	50	11.50	292.1	11.50	292.1	11.50	292.1	11.50	292.1	11.50	292.1	11.50	292.1	7.00	177.8
2 1/2	65	13.00	330.2	13.00	330.2	13.00	330.2	13.00	330.2	13.00	330.2	13.00	330.2	8.50	215.9
3	80	14.00	355.6	14.00	355.6	14.00	355.6	14.00	355.6	14.00	355.6	14.00	355.6	10.00	254.0
4	100	16.00	406.4	16.00	406.4	17.00	431.8	17.00	431.8	17.00	431.8	17.00	431.8	12.00	304.8
5	125	18.00	457.2	18.00	457.2	20.00	508.0	20.00	508.0	20.00	508.0	20.00	508.0	15.00	381.0
6	150	19.50	495.3	19.50	495.3	22.00	558.8	22.00	558.8	22.00	558.8	22.00	558.8	18.00	457.2
8	200	23.50	596.9	23.50	596.9	26.00	660.4	26.00	660.4	26.00	660.4	26.00	660.4	23.00	584.2
10	250	26.50	673.1	26.50	673.1	31.00	787.4	31.00	787.4	31.00	787.4	31.00	787.4	28.00	711.2
12	300	30.00	762.0	30.00	762.0	33.00	838.2	33.00	838.2	33.00	838.2	33.00	838.2	32.00	812.8
14	350	32.50	825.5	32.50	825.5	35.00	889.0	35.00	889.0	35.00	889.0	35.00	889.0	35.00	889.0
16	400	35.50	901.7	35.50	901.7	39.00	990.6	39.00	990.6	39.00	990.6	39.00	990.6	39.00	990.6
18	450	38.50	977.9	38.50	977.9	43.00	1092.2	43.00	1092.2	43.00	1092.2	43.00	1092.2	43.00	1092.2
20	500	41.50	1054.1	41.50	1054.1	47.00	1193.8	47.00	1193.8	47.00	1193.8	47.00	1193.8	47.00	1193.8
22	550	45.00	1143.0	45.00	1143.0	51.00	1295.4	51.00	1295.4	51.00	1295.4	51.00	1295.4	-	-
24	600	48.50	1231.9	48.50	1231.9	55.00	1397.0	55.00	1397.0	55.00	1397.0	55.00	1397.0	55.00	1397.0
26	650	-	-	51.50	1308.1	57.00	1447.8	57.00	1447.8	57.00	1447.8	57.00	1447.8	-	-
28	700	-	-	55.00	1397.0	61.00	1549.4	61.00	1549.4	61.00	1549.4	61.00	1549.4	-	-
30	750	-	-	60.00	1524.0	65.00	-	65.00	1651.0	65.00	1651.0	65.00	1651.0	-	-
32	800	-	-	65.00	1651.0	-	-	70.00	1778.0	-	-	70.00	1778.0	-	-
34	850	-	-	70.00	1778.0	-	-	76.00	1930.4	-	-	76.00	1930.4	-	-
36	900	-	-	74.00	1879.6	-	-	82.00	2082.8	-	-	82.00	2082.8	-	-

Table #35 - Steel Gate Valve (400 & 600)

Steel Gate Valves/Dimensions Given in Inches and Millimetres (mm) taken from Face to Face and End to End of Valves														
STEEL GATE VALVES / ANSI CLASS 900 & 1500														
Nominal Pipe Size		Raised Face/Class 900 & Beveled End/Class 900				Beveled End/ Class 900		Raised Face/Class 1500 & Beveled End/Class 1500						
		Solid Wedge		Double Disc		Short Pattern		Solid Wedge		Double Disc		Short Pattern		
Inch	mm	Inch	mm	Inch	mm	Inch	mm	Inch	mm	Inch	mm	Inch	mm	
3/4	20	-	-	-	-	-	-	-	-	-	-	-	-	
1	25	10.00	254.0	-	-	5.50	139.7	10.00	254.0	-	-	5.50	139.7	
1 1/4	32	11.00	279.4	-	-	6.50	165.1	11.00	279.4	-	-	6.50	165.1	
1 1/2	40	12.00	304.8	-	-	7.00	177.8	12.00	304.8	-	-	7.00	177.8	
2	50	14.50	368.3	14.50	368.3	8.50	215.9	14.50	368.3	14.50	368.3	8.50	215.9	
2 1/2	65	16.50	419.1	16.50	419.1	10.00	254.0	16.50	419.1	16.50	419.1	10.00	254.0	
3	80	15.00	381.0	15.00	381.0	12.00	304.8	18.50	469.9	18.50	469.9	12.00	304.8	
4	100	18.00	457.2	18.00	457.2	14.00	355.6	21.50	546.1	21.50	546.1	16.00	406.4	
5	125	22.00	558.8	22.00	558.8	17.00	431.8	26.50	673.1	26.50	673.1	19.00	482.6	
6	150	24.00	609.6	24.00	609.6	20.00	508.0	27.75	704.9	27.75	704.9	22.00	558.8	
8	200	29.00	736.6	29.00	736.6	26.00	660.4	32.75	831.9	32.75	831.9	28.00	711.2	
10	250	33.00	838.2	33.00	838.2	31.00	787.4	39.00	990.6	39.00	990.6	34.00	863.6	
12	300	38.00	965.2	38.00	965.2	36.00	914.4	44.50	1130.3	44.50	1130.3	39.00	990.6	
14	350	40.50	1028.7	40.50	1028.7	39.00	990.6	49.50	1257.3	49.50	1257.3	42.00	1066.8	
16	400	44.50	1130.3	44.50	1130.3	43.00	1092.2	54.50	1384.3	54.50	1384.3	47.00	1193.8	
18	450	48.00	1219.2	48.00	1219.2	-	-	60.50	1536.7	60.50	1536.7	53.00	1346.2	
20	500	52.00	1320.8	52.00	1320.8	-	-	65.50	1663.7	65.50	1663.7	58.00	1473.2	
22	550	-	-	-	-	-	-	-	-	-	-	-	-	
24	600	61.00	1549.4	61.00	1549.4	-	-	76.50	1943.1	76.50	1943.1	-	-	

Table #36 - Steel Gate Valve (900 & 1500)

Steel Gate Valves/Dimensions Given in Inches and Millimetres (mm) taken from Face to Face and End to End of Valves													
STEEL GATE VALVES / ANSI CLASS 2500													
Nominal Pipe Size		Raised Face/Class 2500				Beveled End/Class 2500							
		Solid Wedge		Double Disc		Solid Wedge		Double Disc		Short Pattern			
Inch	mm	Inch	mm	Inch	mm	Inch	mm	Inch	mm	Inch	mm		
1/2	15	10.38	263.7	-	-	10.38	263.7	-	-	-	-		
3/4	20	10.75	273.1	-	-	10.75	273.1	-	-	-	-		
1	25	12.12	307.9	-	-	12.12	307.9	-	-	7.31	185.7		
1 1/4	32	13.75	349.3	-	-	13.75	349.3	-	-	9.12	231.7		
1 1/2	40	15.12	384.1	-	-	15.12	384.1	-	-	9.12	231.7		
2	50	17.75	450.9	17.75	450.9	17.75	450.9	17.75	450.9	11.00	279.4		
2 1/2	65	20.00	508.0	20.00	508.0	20.00	508.0	20.00	508.0	13.00	330.2		
3	80	22.75	577.9	22.75	577.9	22.75	577.9	22.75	577.9	14.50	368.3		
4	100	26.50	673.1	26.50	673.1	26.50	673.1	26.50	673.1	18.00	457.2		
5	125	31.25	793.8	31.25	793.8	31.25	793.8	31.25	793.8	21.00	533.4		
6	150	36.00	914.4	36.00	914.4	36.00	914.4	36.00	914.4	24.00	609.6		
8	200	40.25	1022.4	40.25	1022.4	40.25	1022.4	40.25	1022.4	30.00	762.0		
10	250	50.00	1270.0	50.00	1270.0	50.00	1270.0	50.00	1270.0	36.00	914.4		
12	300	56.00	1422.4	56.00	1422.4	56.00	1422.4	56.00	1422.4	41.00	1041.4		
14	350	-	-	-	-	-	-	-	-	44.00	1117.6		
16	400	-	-	-	-	-	-	-	-	49.00	1244.6		
18	450	-	-	-	-	-	-	-	-	55.00	1397.0		

NOTE:
1. (-) indicates valve size is not shown in ASME/ANSI standards, but may be commercially available.
2. Millimeter dimensions are rounded off to the first decimal point.
3. Dimensions based on ASME/ANSI B16.10 standard.
4. Short pattern dimensions apply to pressure seal or flange less bonnet valves (option on bolted bonnets.)

Table #37 - Steel Gate Valve (2500)

Steel Globe and Check Valves/Dimensions Given in Inches and Millimetres (mm) taken from Face to Face and End to End for Straightway Valves and Center to Face and Center to End for Angle Valves

STEEL GLOBE AND CHECK VALVES / ANSI CLASS 150 & 300

Nominal Pipe Size		Raised Face and Beveled End/Class 150										Raised Face and Beveled End/Class 300							
		Straight-way Globe		Angle Globe		Y-Pattern Globe		Control Valve Globe (*)		Swing Check		Straight-way Globe		Angle Globe		Control Valve Globe (*)		Swing Check	
In.	mm	Inch	mm	Inch	mm	Inch	mm	Inch	mm	Inch	mm	Inch	mm	Inch	mm	Inch	mm	Inch	mm
1/4	8	4.00	101.6	2.00	50.8	-		-	-	4.00	101.6	-	-	-	-	-	-	-	-
3/8	10	4.00	101.6	2.00	50.8	-		-	-	4.00	101.6	-	-	-	-	-	-	-	-
1/2	15	4.25	108.0	2.25	57.2	5.50	139.7	-	-	4.25	108.0	6.00	152.4	3.00	76.2	7.50	190.5	-	-
3/4	20	4.62	117.4	2.50	63.5	6.00	152.4	-	-	4.62	117.4	7.00	177.8	3.50	88.9	7.62	193.6	-	-
1	25	5.00	127.0	2.75	69.9	6.50	165.1	7.25	184.2	5.00	127.0	8.00	203.2	4.00	101.6	7.75	196.9	8.50	215.9
1 1/4	32	5.50	139.7	3.00	76.2	7.25	184.2	-	-	5.50	139.7	8.50	215.9	4.25	108.0	-	-	9.00	228.6
1 1/2	40	6.50	165.1	3.25	82.6	8.00	203.2	8.75	222.3	6.50	165.1	9.00	228.6	4.50	114.3	9.25	235.0	9.50	241.3
2	50	8.00	203.2	4.00	101.6	9.00	228.6	10.00	254.0	8.00	203.2	10.50	266.7	5.25	133.4	10.50	266.7	10.50	266.7
2 1/2	65	8.50	215.9	4.25	108.0	11.00	279.4	10.88	276.4	8.50	215.9	11.50	292.1	5.75	146.1	11.50	292.1	11.50	292.1
3	80	9.50	241.3	4.75	120.7	12.50	317.5	11.75	298.5	9.50	241.3	12.50	317.5	6.25	158.8	12.50	321.3	12.50	317.5
4	100	11.50	292.1	5.75	146.1	14.50	368.3	13.88	352.6	11.50	292.1	14.00	355.6	7.00	177.8	14.50	368.3	14.00	355.6
5	125	14.00	355.6	7.00	177.8	-	-	-	-	13.00	355.6	15.75	400.1	7.88	200.2	-	-	15.75	400.1
6	150	16.00	406.4	8.00	203.2	18.50	469.9	17.75	450.9	14.00	406.4	17.50	444.5	8.75	222.3	18.62	473.0	17.50	444.5
8	200	19.50	495.3	9.75	247.7	23.50	596.9	21.38	543.1	19.50	495.3	22.00	558.8	11.00	279.4	22.38	568.5	21.00	533.4
10	250	24.50	622.3	12.25	311.2	26.50	673.1	26.50	673.1	24.50	622.3	24.50	622.3	12.25	311.2	27.88	708.2	24.50	622.3
12	300	27.50	698.5	13.75	349.3	30.50	774.7	29.00	736.6	27.50	698.5	28.00	711.2	14.00	355.6	30.50	774.7	28.00	711.2
14	350	31.00	787.4	15.50	393.7	-	-	35.00	889.0	31.50	787.4	-	-	-	-	36.50	927.1	33.00	838.2
16	400	36.00	914.4	18.00	457.2	-	-	40.00	1016.0	34.00	914.4	-	-	-	-	41.62	1057.2	34.00	863.6
18	450	-	-	-	-	-	-	-	-	38.50	977.9	-	-	-	-	-	-	38.50	977.9
20	500	-	-	-	-	-	-	-	-	38.50	977.9	-	-	-	-	-	-	40.00	1016.0
22	550	-	-	-	-	-	-	-	-	42.00	1066.8	-	-	-	-	-	-	44.00	1117.6
24	600	-	-	-	-	-	-	-	-	51.00	1295.4	-	-	-	-	-	-	53.00	1346.2
26	650	-	-	-	-	-	-	-	-	51.00	1295.4	-	-	-	-	-	-	53.00	1346.2
28	700	-	-	-	-	-	-	-	-	57.00	1447.8	-	-	-	-	-	-	59.00	1498.6
30	750	-	-	-	-	-	-	-	-	60.00	1524.0	-	-	-	-	-	-	62.75	1593.9
32	800	-	-	-	-	-	-	-	-	-	-	-	-	-	-	-	-	-	-
34	850	-	-	-	-	-	-	-	-	-	-	-	-	-	-	-	-	-	-
36	900	-	-	-	-	-	-	-	-	77.00	1955.8	-	-	-	-	-	-	82.00	2082.8

NOTE: Control valve dimensions pertain to raised face flange valves only.

Table #38 - Steel Globe / Check Valve (150 & 300)

Steel Globe and Check Valves/Dimensions Given in Inches and Millimetres (mm) taken from Face to Face and End to End for Straightway Valves and Center to Face and Center to End for Angle Valves													
								Long Pattern					
Nominal Pipe Size		Raised Face & Beveled End / Class 400						Raised Face & Beveled End / Class 600					
		Straightway Globe		Angle Globe		Swing Check		Straightway Globe		Angle Globe		Swing Check	
Inch	mm	Inch	mm	Inch	mm	Inch	mm	Inch	mm	Inch	mm	Inch	mm
1/2	15	6.50	165.1	3.25	82.6	6.50	165.1	6.50	165.1	3.25	82.6	6.50	165.1
3/4	20	7.50	190.5	3.75	95.3	7.50	190.5	7.50	190.5	3.75	95.3	7.50	190.5
1	25	8.50	215.9	4.25	108.0	8.50	215.9	8.50	215.9	4.25	108.0	8.50	215.9
1 1/4	32	9.00	228.6	4.50	114.3	9.00	228.6	9.00	228.6	4.50	114.3	9.00	228.6
1 1/2	40	9.50	241.3	4.75	120.7	9.50	241.3	9.50	241.3	4.75	120.7	9.50	241.3
2	50	11.50	292.1	5.75	146.1	11.50	292.1	11.50	292.1	5.75	146.1	11.50	292.1
2 1/2	65	13.00	330.2	6.50	165.1	13.00	330.2	13.00	330.2	6.50	165.1	13.00	330.2
3	80	14.00	355.6	7.00	177.8	14.00	355.6	14.00	355.6	7.00	177.8	14.00	355.6
4	100	16.00	406.4	8.00	203.2	16.00	406.4	17.00	431.8	8.50	215.9	17.00	431.8
5	125	18.00	457.2	9.00	228.6	18.00	457.2	20.00	508.0	10.00	254.0	20.00	508.0
6	150	19.50	495.3	9.75	247.7	19.50	495.3	22.00	558.8	11.00	279.4	22.00	558.8
8	200	23.50	596.9	11.75	298.5	23.50	596.9	26.00	660.4	13.00	330.2	26.00	660.4
10	250	26.50	673.1	13.25	336.6	26.50	673.1	31.00	787.4	15.50	393.7	31.00	787.4
12	300	30.00	762.0	15.00	381.0	30.00	762.0	33.00	838.2	16.50	419.1	33.00	838.2
14	350	-	-	-	-	35.00	889.0	-	-	-	-	35.00	889.0
16	400	-	-	-	-	35.50	901.7	-	-	-	-	39.00	990.6
18	450	-	-	-	-	40.00	1016.0	-	-	-	-	43.00	1092.2
20	500	-	-	-	-	41.50	1054.1	-	-	-	-	47.00	1193.8
22	550	-	-	-	-	45.00	1143.0	-	-	-	-	51.00	1295.4
24	600	-	-	-	-	55.00	1397.0	-	-	-	-	55.00	1397.0
26	650	-	-	-	-	55.00	1397.0	-	-	-	-	57.00	1447.8
28	700	-	-	-	-	63.00	1600.2	-	-	-	-	63.00	1600.2
30	750	-	-	-	-	65.00	1651.0	-	-	-	-	65.00	1651.0
32	800	-	-	-	-	-	-	-	-	-	-	-	-
34	850	-	-	-	-	-	-	-	-	-	-	-	-
36	900	-	-	-	-	82.00	2082.8	-	-	-	-	82.00	2082.8

Table #39A - Steel Globe/Check Valve (400 & 600)

Steel Globe and Check Valves/Dimensions Given in Inches and Millimetres taken from Face to Face and End to End for Straightway Valves and Center to Face and Center to End for Angle Valves									
Nominal Pipe Size		Short Pattern							
		Beveled End / Class 600							
		Straightway Globe		Angle Globe		Swing Check		Control Valve Globe	
Inch	mm	Inch	mm	Inch	mm	Inch	mm	Inch	mm
1/2	15	-	-	-	-	-		8.00	203.2
3/4	20	-	-	-	-	-		8.12	206.3
1	25	5.25	133.4	-	-	5.25	133.4	8.25	209.6
1 1/4	32	5.75	146.1	-	-	5.75	146.1	-	-
1 1/2	40	6.00	152.4	-	-	6.00	152.4	9.88	251.0
2	50	7.00	177.8	4.25	108.0	7.00	177.8	11.25	285.8
2 1/2	65	8.50	215.9	5.00	127.0	8.50	215.9	12.25	311.2
3	80	10.00	254.0	6.00	152.4	10.00	254.0	13.25	336.6
4	100	12.00	304.8	7.00	177.8	12.00	304.8	15.50	393.7
5	125	15.00	381.0	8.50	215.9	15.00	381.0	-	-
6	150	18.00	457.2	10.00	254.0	18.00	457.2	20.00	508.0
8	200	23.00	584.2	-	-	23.00	584.2	24.00	609.6
10	250	28.00	711.2	-	-	28.00	711.2	29.62	752.4
12	300	32.00	812.8	-	-	32.00	812.2	32.25	819.2
14	350	-	-	-	-	-	-	38.25	971.6
16	400	-	-	-	-	-	-	43.62	1108.0
18	450	-	-	-	-	-	-	-	-
20	500	-	-	-	-	-	-	-	-
22	550	-	-	-	-	-	-	-	-
24	600	-	-	-	-	-	-	-	-
26	650	-	-	-	-	-	-	-	-
28	700	-	-	-	-	-	-	-	-
30	750	-	-	-	-	-	-	-	-
32	800	-	-	-	-	-	-	-	-
34	850	-	-	-	-	-	-	-	-
36	900	-	-	-	-	-	-	-	-

Table #39B - Steel Globe/Check Valve (600)

Steel Globe and Check Valves/Dimensions Given in Inches and Millimetres (mm) taken from Face to Face and End to End for Straightway Valves and Center to Face and Center to End for Angle Valves													
Steel Globe and Check Valves / ANSI Class 900													
Nominal Pipe Size		Long Pattern						Short Pattern					
		Raised Face & Beveled End / Class 900						Beveled End / Class 900					
		Straightway Globe		Angle Globe		Swing Check		Straightway Globe		Angle Globe		Swing Check	
Inch	mm	Inch	mm	Inch	mm	Inch	mm	Inch	mm	Inch	mm	Inch	mm
1/2	15	-	-	-	-	-	-	-	-	-	-	-	-
3/4	20	9.00	228.6	4.50	114.3	9.00	228.6	-	-	-	-	-	-
1	25	10.00	254.0	5.00	127.0	10.00	254.0	-	-	-	-	-	-
1 1/4	32	11.00	279.4	5.50	139.7	11.00	279.4	-	-	-	-	-	-
1 1/2	40	12.00	304.8	6.00	152.4	12.00	304.8	-	-	-	-	-	-
2	50	14.50	368.3	7.25	184.2	14.50	368.3	-	-	-	-	-	-
2 1/2	65	16.50	419.1	8.25	209.6	16.50	419.1	10.00	254.0	-	-	10.00	254.0
3	80	15.00	381.0	7.50	190.5	15.00	381.0	12.00	304.8	6.00	152.4	12.00	304.8
4	100	18.00	457.2	9.00	228.6	18.00	457.2	14.00	355.6	7.00	177.8	14.00	355.6
5	125	22.00	558.8	11.00	279.4	22.00	558.8	17.00	431.8	8.50	215.9	17.00	431.8
6	150	24.00	609.6	12.00	304.8	24.00	609.6	20.00	508.0	10.00	254.0	20.00	508.0
8	200	29.00	736.6	14.50	368.3	29.00	736.6	26.00	660.4	13.00	330.2	26.00	660.4
10	250	33.00	838.2	16.50	419.1	33.00	838.2	31.00	787.4	15.50	393.7	31.00	787.4
12	300	38.00	965.2	19.00	482.6	38.00	965.2	36.00	914.4	18.00	457.2	36.00	914.4
14	350	40.50	1028.7	20.25	514.4	40.50	1028.7	39.00	990.6	19.50	495.3	39.00	990.6
16	400	-	-	26.00	660.4	44.50	1130.3	43.00	1092.2	-	-	43.00	1092.2
18	450	-	-	29.00	736.6	48.00	1219.2	-	-	-	-	-	-
20	500	-	-	32.50	825.5	52.00	1320.8	-	-	-	-	-	-
22	550	-	-	-	-	-	-	-	-	-	-	-	-
24	600	-	-	39.00	990.6	61.00	1549.4	-	-	-	-	-	-

Table #40A - Steel Globe/Check Valve (900)

Steel Globe and Check Valves/Dimensions Given in Inches and Millimetres taken from Face to Face and End to End for Straightway Valves and Center to Face and Center to End for Angle Valves													
Nominal Pipe Size		Steel Globe and Check Valves / ANSI Class 1500											
		Long Pattern						Short Pattern					
		Raised Face and Beveled End / Class 1500						Beveled End / Class 1500					
		Straightway Globe		Angle Globe		Swing Check		Straightway Globe		Swing Check			
Inch	mm	Inch	mm	Inch	mm	Inch	mm	Inch	mm	Inch	mm		
1/2	15	8.50	215.9	4.25	108.0	-	-	-	-	-	-		
3/4	20	9.00	228.6	4.50	114.3	9.00	228.6	-	-	-	-		
1	25	10.00	254.0	5.00	127.0	10.00	254.0	-	-	-	-		
1 1/4	32	11.00	279.4	5.50	139.7	11.00	279.4	-	-	-	-		
1 1/2	40	12.00	304.8	6.00	152.4	12.00	304.8	-	-	-	-		
2	50	14.50	368.3	7.25	184.2	14.50	368.3	8.50	215.9	8.50	215.9		
2 1/2	65	16.50	419.1	8.25	209.6	16.50	419.1	10.00	254.0	10.00	254.0		
3	80	18.50	469.9	9.25	235.0	18.50	469.9	12.00	304.8	12.00	304.8		
4	100	21.50	546.1	10.75	273.1	21.50	546.1	16.00	406.4	16.00	406.4		
5	125	26.50	673.1	13.25	336.6	26.50	673.1	19.00	482.6	19.00	482.6		
6	150	27.75	704.9	13.88	352.6	27.75	704.9	22.00	558.8	22.00	558.8		
8	200	32.75	831.9	16.38	416.1	32.75	831.9	28.00	711.2	28.00	711.2		
10	250	39.00	990.6	19.50	495.3	39.00	990.6	34.00	863.6	34.00	863.6		
12	300	44.50	1130.3	22.25	565.2	44.50	1130.3	39.00	990.6	39.00	990.6		
14	350	49.50	1257.3	24.75	628.7	49.50	1257.3	42.00	1066.8	42.00	1066.8		
16	400	-	-	-	-	54.50	1384.3	47.00	1193.8	47.00	1193.8		
18	450	-	-	-	-	60.50	1536.7	-	-	-	-		
20	500	-	-	-	-	65.50	1663.7	-	-	-	-		
22	550	-	-	-	-	-	-	-	-	-	-		
24	600	-	-	-	-	76.50	1943.1	-	-	-	-		

Table #40B - Steel Globe/Check Valve (1500)

Steel Globe and Check Valves/Dimensions Given in Inches and mm taken from Face to Face and End to End for Straightway Valves and Center to Face and Center to End for Angle Valves													
Nominal Pipe Size		Steel Globe and Check Valves / ANSI Class 2500											
		Long Pattern								Short Pattern			
		Raised Face and Beveled End / Class 2500								Beveled End / Class 2500			
		Straightway Globe		Angle Globe		Swing Check				Straightway Globe		Swing Check	
Inch	mm	Inch	mm	Inch	mm	Inch	mm			Inch	mm	Inch	mm
1/2	15	10.38	263.7	5.19	131.8	10.38	263.7			-	-	-	-
3/4	20	10.75	273.1	5.38	136.7	10.75	273.1			-	-	-	-
1	25	12.12	307.9	6.06	153.9	12.12	307.9			-	-	-	-
1 1/4	32	13.75	349.3	6.88	174.8	13.75	349.3			-	-	-	-
1 1/2	40	15.12	384.0	7.56	192.0	15.12	384.0			-	-	-	-
2	50	17.75	450.9	8.88	225.6	17.75	450.9			11.00	279.4	11.00	279.4
2 1/2	65	20.00	508.0	10.00	254.0	20.00	508.0			13.00	330.2	13.00	330.2
3	80	22.75	577.9	11.38	289.1	22.75	577.9			14.50	368.3	14.50	368.3
4	100	26.50	673.1	13.25	336.6	26.50	673.1			18.00	457.2	18.00	457.2
5	125	31.25	793.8	15.62	396.8	31.25	793.8			21.00	533.4	21.00	533.4
6	150	36.00	914.4	18.00	457.2	36.00	914.4			24.00	609.6	24.00	609.6
8	200	40.25	1022.4	20.12	511.1	40.25	1022.4			30.00	762.0	30.00	762.0
10	250	50.00	1270.0	25.00	635.0	50.00	1270.0			36.00	914.0	36.00	914.0
12	300	56.00	1422.4	28.00	711.2	56.00	1422.4			41.00	1041.4	41.00	1041.4
14	350	-	-	-	-	-	-			-	-	-	-
16	400	-	-	-	-	-	-			-	-	-	-
18	450	-	-	-	-	-	-			-	-	-	-

NOTE:
1. (-) indicates valve size is not shown in ASME/ANSI standards, but may be commercially available.
2. Millimetre dimensions are rounded off to the first decimal point.
3. Dimensions based on ASME/ANSI B16.10 standards.
4. Standard straightway lift check valve dimensions are interchangeable with straightway globe valve dimensions given above.
5. Angle lift check valve dimensions are interchangeable with angle globe valve dimensions given in tables.
6. Short pattern dimensions apply to pressure seal or flangeless bonnet valves (option on bolted bonnets).

Table #41 - Steel Globe/Check Valve (2500)

Added Dimensions Used to Establish End to End Dimensions of Ring Joint Flanges															
Pipe Size		Class 150		Class 300		Class 400		Class 600		Class 900		Class 1500		Class 2500	
Inch	mm	Inch	mm	Inch	mm	Inch	mm	Inch	mm	Inch	mm	Inch	mm	Inch	mm
1/2	15	-	-	0.44	11.2	-0.06	-1.5	-0.06	-1.5	0	0	0	0	0	0
3/4	20	-	-	0.50	12.7	0	0	0	0	0	0	0	0	0	0
1	25	0.50	12.7	0.50	12.7	0	0	0	0	0	0	0	0	0	0
1 1/4	32	0.50	12.7	0.50	12.7	0	0	0	0	0	0	0	0	0.12	3.0
1 1/2	40	0.50	12.7	0.50	12.7	0	0	0	0	0	0	0	0	0.12	3.0
2	50	0.50	12.7	0.62	15.7	0.12	3.0	0.12	3.0	0.12	3.0	0.12	3.0	0.12	3.0
2 1/2	65	0.50	12.7	0.62	15.7	0.12	3.0	0.12	3.0	0.12	3.0	0.12	3.0	0.25	6.4
3	80	0.50	12.7	0.62	15.7	0.12	3.0	0.12	3.0	0.12	3.0	0.12	3.0	0.25	6.4
4	100	0.50	12.7	0.62	15.7	0.12	3.0	0.12	3.0	0.12	3.0	0.12	3.0	0.38	9.7
5	125	0.50	12.7	0.62	15.7	0.12	3.0	0.12	3.0	0.12	3.0	0.12	3.0	0.50	12.7
6	150	0.50	12.7	0.62	15.7	0.12	3.0	0.12	3.0	0.12	3.0	0.12	6.4	0.50	12.7
8	200	0.50	12.7	0.62	15.7	0.12	3.0	0.12	3.0	0.12	3.0	0.12	9.7	0.62	15.7
10	250	0.50	12.7	0.62	15.7	0.12	3.0	0.12	3.0	0.12	3.0	0.12	9.7	0.88	22.4
12	300	0.50	12.7	0.62	15.7	0.12	3.0	0.12	3.0	0.12	3.0	0.12	15.8	0.88	22.4
14	350	0.50	12.7	0.62	15.7	0.12	3.0	0.12	3.0	0.38	9.7	0.75	19.1	-	-
16	400	0.50	12.7	0.62	15.7	0.12	3.0	0.12	3.0	0.38	9.7	0.88	22.4	-	-
18	450	0.50	12.7	0.62	15.7	0.12	3.0	0.12	3.0	0.50	12.7	0.88	22.4	-	-
20	500	0.50	12.7	0.75	19.1	0.25	6.4	0.25	6.4	0.50	12.7	0.88	22.4	-	-
22	550	0.50	12.7	0.88	22.4	0.38	9.7	0.38	9.7	-	-	-	-	-	-
24	600	0.50	12.7	0.88	22.4	0.38	9.7	0.38	9.7	0.75	19.1	1.12	28.4	-	-
26	650	-	-	1.00	25.4	0.50	12.7	0.50	12.7	-	-	-	-	-	-
28	700	-	-	1.00	25.4	0.50	12.7	0.50	12.7	-	-	-	-	-	-
30	750	-	-	1.00	25.4	0.50	12.7	0.50	12.7	-	-	-	-	-	-
32	800	-	-	1.12	28.4	0.62	15.7	0.62	15.7	-	-	-	-	-	-
34	850	-	-	1.12	28.4	0.62	15.7	0.62	15.7	-	-	-	-	-	-
36	900	-	-	1.12	28.4	0.62	15.7	0.62	15.7	-	-	-	-	-	-

NOTE:
1. To establish a ring joint flange dimension, add the appropriate dimension from this table to the raised face steel flange dimension given in the preceding tables.
2. To establish a dimension for an angle globe or angle lift check valve, use one half of the dimensions given in this table.
3. Millimetre dimensions are rounded off to the first decimal point.

Table #42 - Ring Joint Flanges

SECTION FOUR

FITTINGS

Pipe Fitting Types

Pipe fittings are the joining components that make possible the assembly of equipment, valves, and pipe into functioning piping systems. Fittings are manufactured to perform one or more of the following functions:

- Change direction of piping in system.
- Connect or join piping and/or equipment.
- Provide for branches, access, take-offs or auxiliary connections.
- Block or regulate flow within piping or equipment.

Fittings are specified or identified by:

Nominal pipe size or tube size that the fitting is manufactured to fit.

- Type or description of the fitting. For example, tees, wyes, elbows, crosses, couplings, etc.
- Joining or connecting method of the fitting. For example, threaded, soldered, welded, etc.
- Material that the fitting is manufactured from. For example, copper, cast iron, steel, plastic, etc.
- Pressure temperature rating or class designation.

Example: 2" class 3000 carbon steel (ASTM 105) threaded straight tee.

Elbows

Fittings that change direction in a piping system are generally referred to as elbows. Elbows are designated or described by the amount of directional change they make in a piping system. ***This directional change is given in degrees or fractions of a circle. Most elbows use degree designations, such as: 22½, 45, 60, or 90 degrees for classifying their change of direction.***

Elbows

Cast iron soil fittings on the other hand, are referred to in fractions of a circle, such as: $^1/_4$ bend, $^1/_8$ bend and $^1/_{16}$ bend etc. See illustration #67 for elbow directional change classifications.

Elbows are designated by the angle or degree of change they make relative to a circle. Angles for elbows range from $11^1/_4$ through 180 degrees.

Cast iron bend designations are determined by dividing the elbow fitting angle by 360° (degrees in a circle). Cast iron designations are expressed as a fraction.

Example for cast iron:

$$\text{Bend} = \frac{\text{Elbow Fitting Angle}}{360^\circ}$$

$$\text{Bend} = \frac{45^\circ}{360^\circ} = {}^1/_8$$

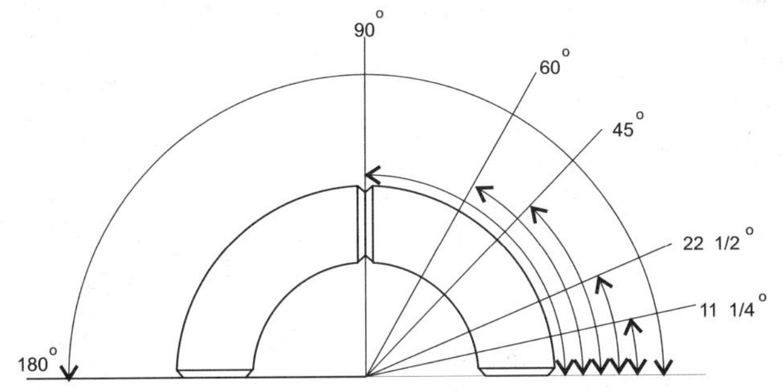

Illustration #67 - Elbow Fitting Angles

Elbows

The distance from the end to center for various types of welded elbows can be calculated quickly using figures given in illustration #68.

Easy Calculation Methods For Determining Elbow End To Center Measurements (dimension A):

- 90° long radius elbow = $1^1/_2$ x NPS
- Reducing long radius elbow = $1^1/_2$ x largest NPS
- Long radius return bend = 3 x NPS
- 90° short radius elbow = 1 x NPS
- 45° long radius elbow = $^5/_8$ x NPS
- 45° long radius elbow alternative = $^1/_2$ the NPS 3 times, add the first & last answers

Example: 4 inch (100 mm) 45° elbow

a. 4 in. (100 mm) b. 2 in. (50 mm)

c. 1 in. (25 mm) d. $^1/_2$ in. (15 mm)

= 2 in. + $^1/_2$ in. = $2^1/_2$ in. end to center

2.5 in. x 25.4 mm = 63.5 mm

Note: Calculations for 45° elbows are accurate between sizes 4"- 20" (100 - 500 mm).

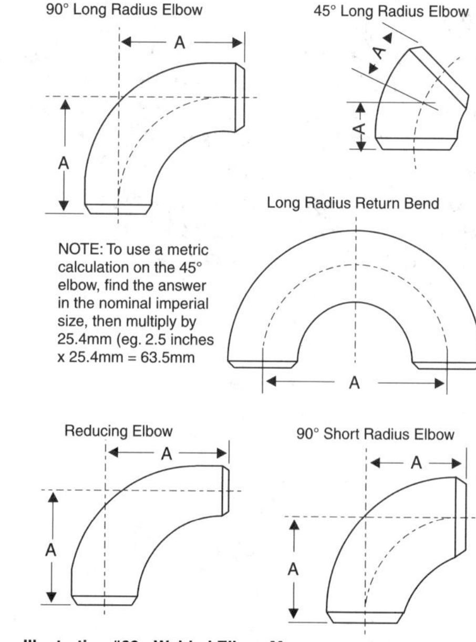

Illustration #68 - Welded Elbow Measurements

Unions

Couplings and unions are used to connect or join together pipe and equipment in piping systems. Even though joining by welding is used extensively in the petrochemical and power generation field, couplings and unions are also common. Unions are used to make joining or dismantling of piping and/ or equipment easier and quicker.

Unions are available in many styles but generally are classified as either ground joint or gasket types. Both types of unions are shown in illustrations #69A and #69B. Gasket unions require a gasket or washer to seal between the two union parts. Gasket material for the union is available in various materials depending on intended service.

Ground joint unions rely on a ground metal joint to seal between the two union parts. Common seats on the ground joint unions include:

Steel to steel, bronze to steel, stainless steel to steel, iron to iron, and copper or copper alloy to iron.

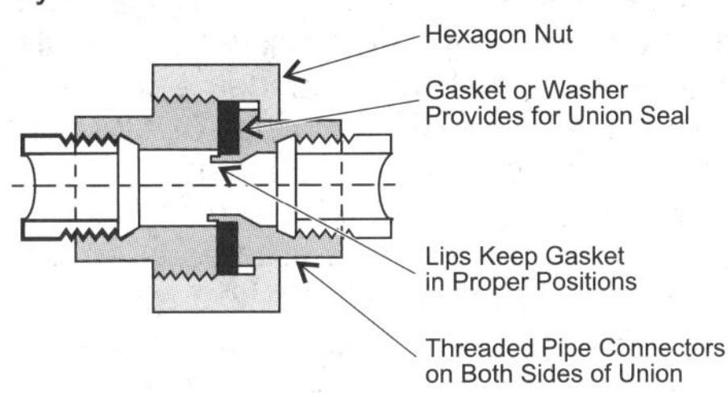

Illustration #69A - Gasket Type Threaded Union

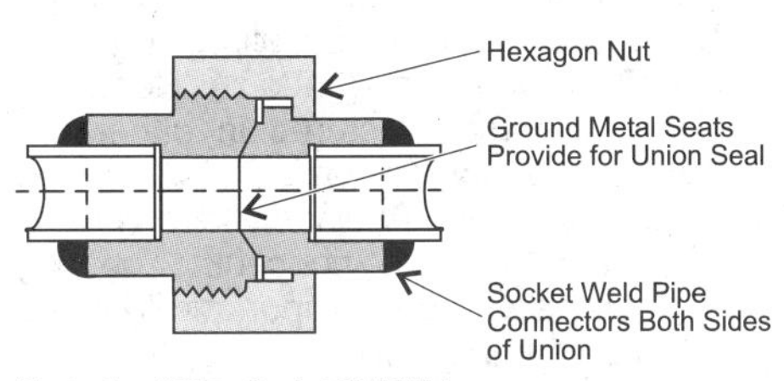

Illustration #69B - Socket Weld Union

Couplings

Couplings are used to join two pieces of pipe (the same size or different sizes) together in a straight line. Threaded couplings, as a rule, are supplied with right handed threads on both ends. However, couplings can be supplied with special right/left hand threads. Right/left couplings can be used instead of unions, but remember that threads on one end of the pipe must be cut with a special left hand die.

Threaded and non-threaded couplings are supplied in both straight and reducing styles. Straight couplings are used to join pipe of the same size, and reducing couplings are used to join two lines of unequal size. Reducing couplings are manufactured in either the concentric or eccentric configuration. These two coupling styles are displayed in illustration #70A and #70B. The concentric coupling is used in lines where it is important to keep a constant center line.

Eccentric couplings are used where either the top or bottom of a line must remain level.

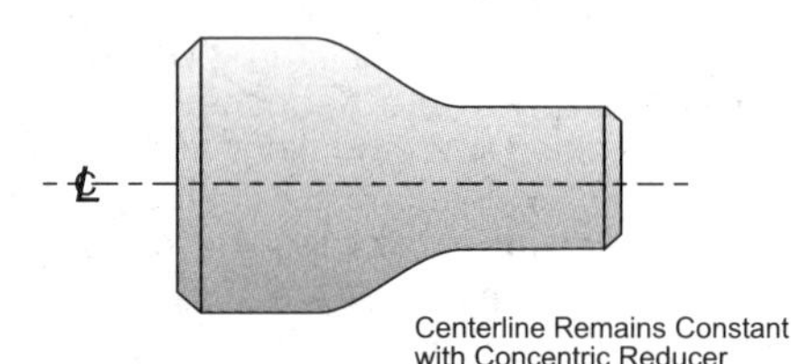

Illustration #70A - Concentric Reducer

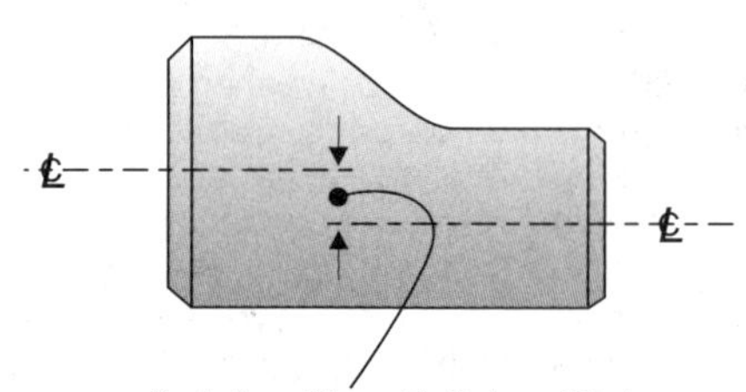

Illustration #70B - Eccentric Reducer

Tees, Wyes, Crosses

These fittings provide for branches, take-offs and/or auxiliary connections within a piping system.

When ordering or identifying tees, wyes and crosses, the size of the run is referred to first (largest opening first), followed by the branch outlet(s). If the fitting is a cross, the largest branch outlet is given after the run size, followed by the smaller cross branch outlet. The method of size designation for various fittings is given in illustration #71A - #71D.

NOTES:

- This method of designating or naming straight and reducing fittings applies to all fittings including: threaded, welded, soldered & flanged.
- Straight fittings (with no reduction) are designated or named: size x description e.g. 4" (200 mm) Tee, 4" (200 mm) 90° Elbow, 4" (200 mm) Cross etc.
- On a side outlet fitting the side outlet is always designated last.

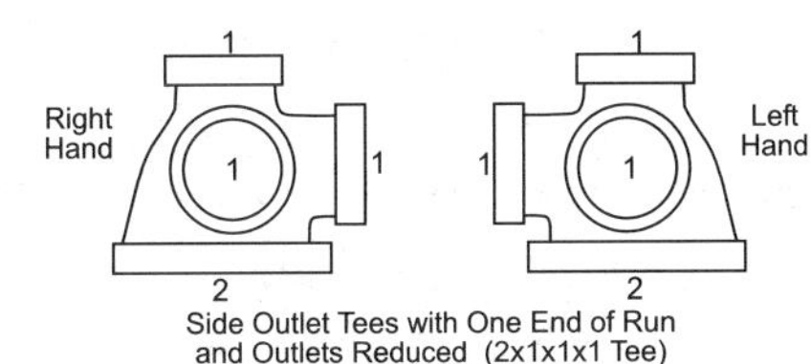

Side Outlet Tees with One End of Run and Outlets Reduced (2x1x1x1 Tee)

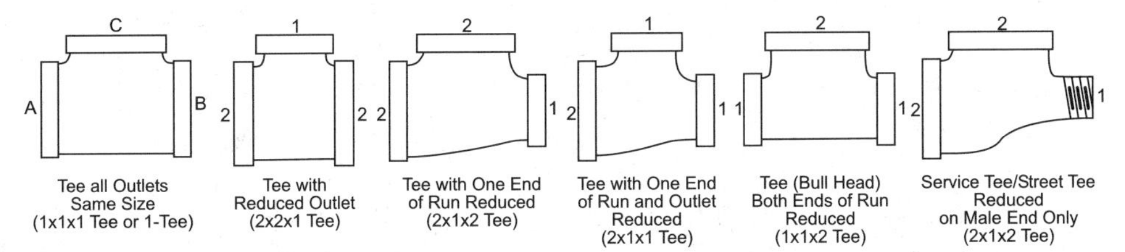

Illustration #71A - Designation & Naming of Tees

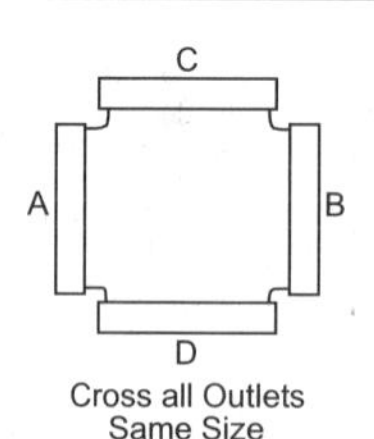

Cross all Outlets Same Size (1x1x1x1 Cross or 1-Cross)

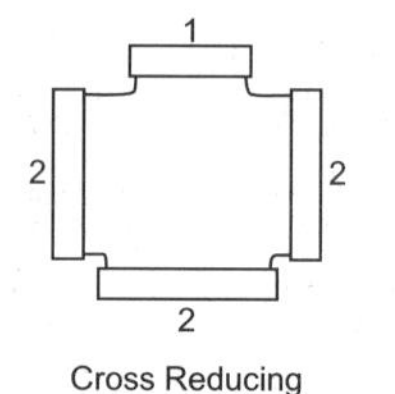

Cross Reducing on One Outlet Only (2x2x2x1 Cross)

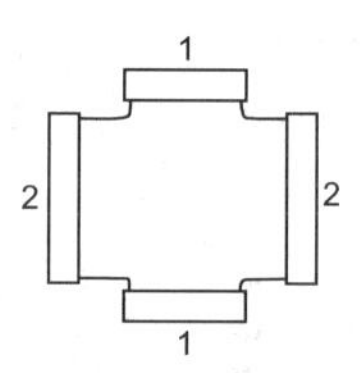

Cross Reducing on Both Outlets (2x2x1x1 Cross)

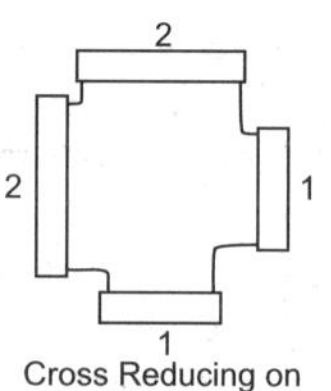

Cross Reducing on One End of Run and on One Outlet (2x1x2x1 Cross)

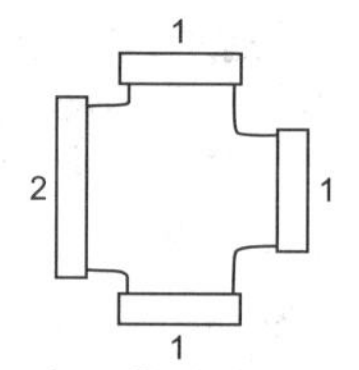

Cross Reducing on One End of Run and Both Outlets (2x1x1x1 Cross)

Illustration #71B - Designation & Naming of Crosses

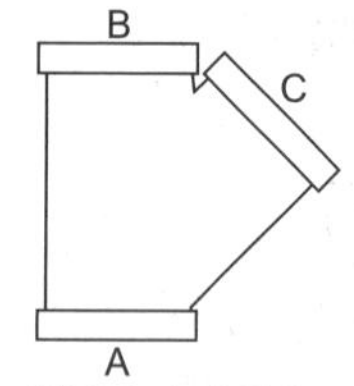

Y - Branch all Outlets the Same Size (1x1x1 Wye or 1-Wye)

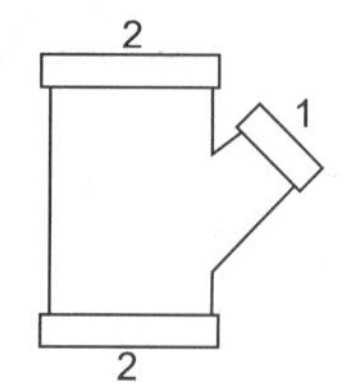

45° Y - Branch (Lateral) Reducing on Outlet Only (2x2x1 Wye)

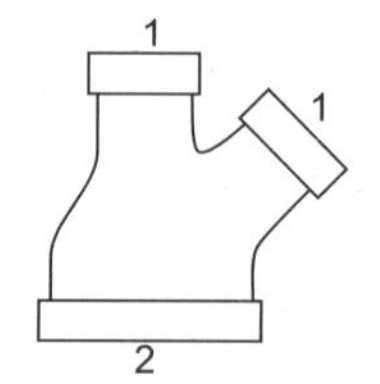

45° Y - Branch (Lateral) Reducing on Run and Outlet (2x1x1 Wye)

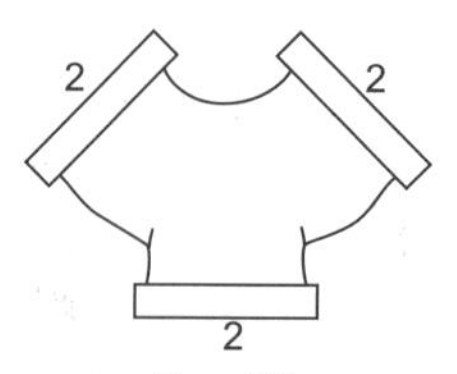

True "Y" (2x2x2 Wye)

Illustration #71C - Designation & Naming of Wyes

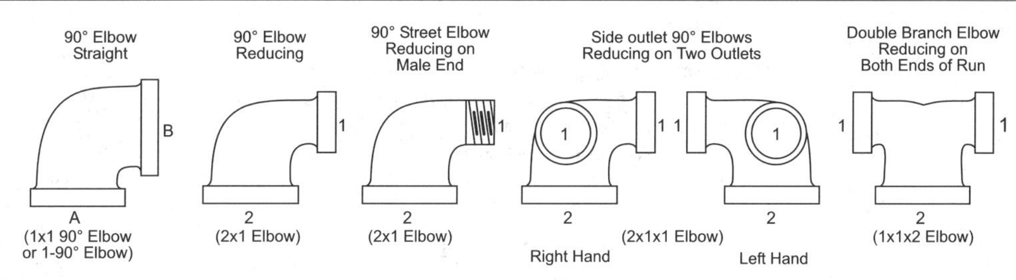

Illustration #71D - Designation & Naming of Elbows

Copper Solder Joint Fittings

Both wrought copper and cast copper alloy solder joint pressure fittings are produced for use with copper water tube. Pressure and temperature ratings for the fittings are equal to that of type L copper tube. However, in most cases the solder used to join the tube and fitting will determine the safe working pressure and temperature of the system. Recommended solder joint pressure and temperature ratings are given in table #43.

Dimensions for common cast copper alloy solder joints are given in table #44. The term “laying length” in reference to copper solder fittings refers to the distance from the center of the fitting to the shoulder or stop at the bottom end of the socket. American National Standards Institute (ANSI B16.221980) has established laying length and sizing designations for cast fittings only. It should be noted that because of the various forming methods, no standardized dimensions are established for wrought copper solder fittings.

RATED PRESSURE / TEMPERATURE FOR SOLDER AND BRAZED JOINTS														
Solder or Brazing Alloy	Working Temp.		Copper Tube Nominal Sizes, Inches and Millimetres										Saturated Steam	
			1/8" thru 1"		1 1/4" thru 2"		2 1/2" thru 4"		5" thru 8"		10" thru 12"			
			6mm - 25mm		32mm - 50mm		65mm - 100mm		125 - 200mm		250 - 300mm			
	°F	°C	PSI	kPa	PSI	kPa	PSI	kPa	PSI	kPa	PSI	kPa	PSI	kPa
50 - 50 Tin-Lead Solder	100	38	200	1379.0	175	1206.6	150	1034.2	135	930.8	100	689.5	-	-
	150	66	150	1034.2	125	861.8	100	689.5	90	620.5	70	482.6	-	-
	200	93	100	689.5	90	620.5	75	517.1	70	482.6	50	344.7	-	-
	250	120	85	586.0	75	517.1	50	344.7	45	310.2	40	275.8	15	103.4
95 - 5 Tin-Antimony Solder	100	38	500	3447.5	400	2758.0	300	2068.5	270	1861.6	150	1034.2	-	-
	150	66	400	2758.0	350	2413.2	275	1896.1	250	1723.7	150	1034.2	-	-
	200	93	300	2068.5	250	1723.7	200	1379.0	180	1241.1	140	965.3	-	-
	250	120	200	1379.0	175	1206.6	150	1034.2	135	930.8	110	758.4	15	103.4
Brazing Alloys Melting at or above 1000°F/ 540°C	250	120	300	2068.5	210	1447.9	170	1172.1	150	1034.2	150	1034.2	-	-
	350	176	270	1861.6	190	1310.0	150	1034.2	150	1034.2	150	1034.2	120	827.4

NOTE:
1. The pressure unit of 1 bar is equal to 14.5 PSI or 100,000 Newtons per square meter.
2. Saturated steam at 15 PSI (103.4 kPa) is produced at 250°F (121.1°C).
3. Saturated steam at 120 PSI (827.4 kPa) is produced at 350°F (176.7°C).
4. Brazing Alloys are recommended for low temperature service between 0°F to -100°F (-18°C to -73°C).

Table #43 - Pressure and Temperature for Solder and Brazed Joints

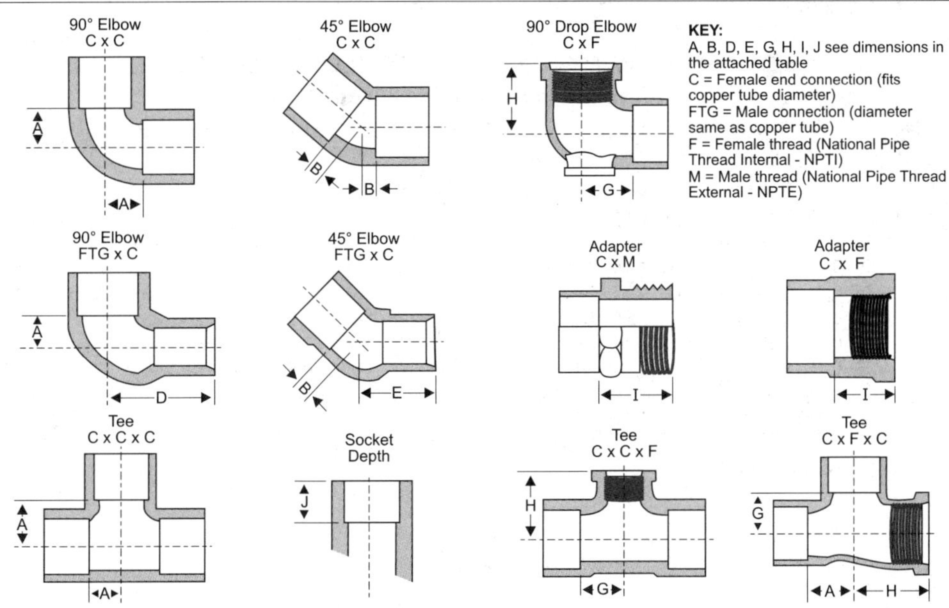

KEY:
A, B, D, E, G, H, I, J see dimensions in the attached table
C = Female end connection (fits copper tube diameter)
FTG = Male connection (diameter same as copper tube)
F = Female thread (National Pipe Thread Internal - NPTI)
M = Male thread (National Pipe Thread External - NPTE)

Table #44 Diagrams - Dimensions for Copper Solder Fittings

DIMENSIONS FOR COMMON CAST COPPER ALLOY SOLDER FITTINGS
(Given in Inches and Millimetres)

Tube Size	Dimensions - Inches								Tube Size	Dimensions - Millimetres							
	A	B	D	E	G	H	I	J		A	B	D	E	G	H	I	J
1/4	0.25	-	0.75	-	0.38	0.56	0.62	0.31	8	6.5	-	19.0	-	9.5	14.5	16.0	7.87
3/8	0.31	0.19	0.88	0.75	0.44	0.69	0.62	0.38	10	8.0	5.0	22.0	19.0	11.0	17.5	16.0	9.65
1/2	0.44	0.19	1.12	0.88	0.56	0.88	0.75	0.50	15	11.0	5.0	28.5	22.0	14.5	22.0	18.0	12.70
3/4	0.56	0.25	1.50	1.19	0.69	1.00	0.88	0.75	20	14.5	6.5	38.0	30.0	17.5	25.5	22.0	19.05
1	0.75	0.31	1.84	1.31	0.88	1.25	1.00	0.91	25	19.0	8.0	47.0	33.5	22.0	32.0	25.5	23.11
1 1/4	0.88	0.44	2.03	1.56	1.00	1.50	1.06	0.97	32	22.0	11.0	51.5	39.5	25.5	38.0	27.0	24.64
1 1/2	1.00	0.50	2.28	1.72	1.12	1.62	1.06	1.09	40	25.5	12.5	58.0	44.5	28.5	41.5	27.0	27.69
2	1.25	0.56	2.78	2.12	1.38	1.94	1.12	1.34	50	32.0	14.5	70.5	54.0	35.0	49.0	28.5	34.04

NOTE:
1. Dimensions apply to cast fittings only. Dimensions for wrought fittings have not been standardized.
2. Dimension for table-based on fittings manufactured to ANSI standards.
3. Tube size dimensions are nominal sizes.

Table #44 - Cast Copper Alloy Solder Fittings

Threaded Fittings

Four common styles of tapered threaded fittings are:

- Malleable Iron Threaded Fittings Classes 150 and 300.
- Forged Steel Threaded Fittings Class 2000, 3000 and Class 6000.
- Cast Iron Threaded Fittings Classes 125 and 250.
- Cast Iron Threaded Drainage Fittings.

All of these styles of threaded fittings use internal or external threads (female or male threads) that comply with American National Standard for taper pipe thread (NPT). Tables #46A & 46B give common thread information and fitting make-up engagement for all fittings that are threaded to American National Standard (ANSI/ASME B1.20.1).

The amount of torque required to make up leak-proof joints will vary depending on the material, size, and thread quality used in the joint to be made up.

Normal pressure tight screwed joints require wrench tight make up with the use of threading compound or TFE (Teflon) tape applied to the male threads only.

Note: Special dry seal threads may be used without threading compound. The recommended pipe wrench size used to make up various threaded fittings and pipe sizes are given in Table #45.

RECOMMENDED PIPE WRENCH SIZE - INCHES	
Nominal Pipe & Fitting Size	**Pipe Wrench Size**
1/8" - 1/2"	6
1/4" - 3/4"	8
1/4" - 1"	10
1/2" - 1 1/2"	12
1/2" - 1 1/2"	14
1" - 2"	18
1 1/2" - 2 1/2"	24
2" - 3 1/2"	36
3" - 5"	48
3" - 8"	60

Table #45 - Recommended Pipe Wrench Size - Inches

THREADING ENGAGEMENT BASED ON AMERICAN NATIONAL STANDARD PIPE THREADS GENERAL PURPOSE INCH									
Nominal Pipe Size (NPS)	Outside Pipe Diameter (OD)	Threads Per Inch (TPI)	Thread Pitch	Total Length of External Thread	Hand Tight Number of Threads	Hand Tight Engagement Length	Wrench Make-up Number of Threads	Wrench Make-up Engagement Length	Practical Make-up Hand & Wrench Engagement Length
1/16	0.3125	27	0.03704	3/8	4 1/4	0.1600	3	0.1111	1/4
1/8	0.405	27	0.03704	3/8	4 1/4	0.1615	3	0.1111	1/4
1/4	0.540	18	0.05556	5/8	4	0.2278	3	0.1667	5/16
3/8	0.675	18	0.05556	5/8	4 1/4	0.2400	3	0.1667	7/16
1/2	0.840	14	0.07143	13/16	4 1/2	0.3200	3	0.2143	1/2
3/4	1.050	14	0.07143	13/16	4 3/4	0.3390	3	0.2143	9/16
1	1.315	11.5	0.08696	1	4 1/2	0.4000	3	0.2609	11/16
1 1/4	1.660	11.5	0.08696	1	4 3/4	0.4200	3	0.2609	11/16
1 1/2	1.900	11.5	0.08696	1	4 3/4	0.4200	3	0.2609	11/16
2	2.375	11.5	0.08696	1 1/16	5	0.4360	3	0.2609	3/4
2 1/2	2.875	8	0.12500	1 9/16	5 1/2	0.6820	2	0.2500	3/4
3	3.500	8	0.12500	1 5/8	6 1/4	0.7660	2	0.2500	1
3 1/2	4.00	8	0.12500	1 11/16	6 1/2	0.8210	2	0.2500	1 1/16
4	4.500	8	0.12500	1 3/4	6 3/4	0.8440	2	0.2500	1 1/8
5	5.563	8	0.12500	1 13/16	7 1/2	0.9370	2	0.2500	1 1/4
6	6.625	8	0.12500	1 7/8	7 3/4	0.9580	2	0.2500	1 3/16
8	8.625	8	0.12500	2 1/8	8 1/2	1.0630	2	0.2500	1 5/16

NOTE:
1. Engagement tolerances of plus or minus one turn.
2. Fractions rounded to the nearest 1/16 inch.
3. Decimal places are gage dimensions and are used by ANSI/ASME for computations.

Table #46A - Threading Engagement Dimensions - Inches

THREADING ENGAGEMENT BASED ON AMERICAN NATIONAL STANDARD PIPE THREADS GENERAL PURPOSE MILLIMETRES									
Nominal Pipe Size (NPS)	Outside Pipe Diameter (OD)	Threads Per Inch (TPI)	Thread Pitch	Total Length of External Thread	Hand Tight Number of Threads	Hand Tight Engagement Length	Wrench Make-up Number of Threads	Wrench Make-up Engagement Length	Practical Make-up Hand & Wrench Engagement Length
3	7.94	27	0.94	9.90	4 1/4	4.06	3	2.82	6.35
6	10.29	27	0.94	9.97	4 1/4	4.10	3	2.82	6.35
8	13.72	18	1.41	15.10	4	5.79	3	4.23	9.65
10	17.15	18	1.41	15.26	4 1/4	6.10	3	4.23	9.65
15	21.34	14	1.81	19.85	4 1/2	8.13	3	5.44	12.70
20	26.67	14	1.81	20.15	4 3/4	8.61	3	5.44	14.22
25	33.40	11.5	2.21	25.01	4 1/2	10.16	3	6.63	17.53
32	42.16	11.5	2.21	25.62	4 3/4	10.67	3	6.63	17.53
40	48.26	11.5	2.21	26.04	4 3/4	10.67	3	6.63	17.53
50	60.33	11.5	2.21	26.88	5	11.07	3	6.63	19.05
65	73.03	8	3.18	39.91	5 1/2	17.32	2	6.35	23.88
80	88.90	8	3.18	41.50	6 1/4	19.46	2	6.35	25.4
90	101.60	8	3.18	42.77	6 1/2	20.85	2	6.35	26.92
100	114.30	8	3.18	44.04	6 3/4	21.44	2	6.35	28.45
125	141.30	8	3.18	46.74	7 1/2	23.80	2	6.35	31.75
150	168.28	8	3.18	49.43	7 3/4	24.33	2	6.35	33.27
200	219.08	8	3.18	54.51	8 1/2	27.00	2	6.35	36.58

NOTE:
1. Engagement tolerances of plus or minus one turn.
2. Millimetres rounded to the nearest second decimal place.

Table #46B - Threading Engagement Dimensions - Millimetres

Malleable Iron Threaded Fittings

Malleable iron fittings are manufactured in two general designations or classes: Class 150 and Class 300. Both classes of fittings can be supplied in either black or galvanized finishes. Pressure and temperature ratings for each class is given in table #47.

Malleable Couplings

Ribs or bars running along the length of malleable couplings are often used as thread type identification marks. A malleable coupling having 2 ribs or bars identifies the coupling as being a standard right-hand thread. Right/left-hand threaded couplings have 4 or more ribs unless the left-hand opening is clearly marked with an "L" for identification. Established ANSI/ASME B16.3 fitting dimensions for Class 150 are given in tables #48A and #48B.

PRESSURE/TEMPERATURE RATINGS FOR CLASS 150 & 300 MALLEABLE IRON FITTINGS
CLASS 150
- 150 PSI (1034.25 kPa) Saturated Steam - 300 PSI (2068.50 kPa) Liquid and Gas Non-Shock Service at max. 150°F (66°C)
CLASS 300
- 300 PSI (2068.5 kPa) Saturated Steam - 1/4 to 1 inch (8 to 25 mm) 2000 PSI (13790.0 kPa) - 1 1/4 to 2 inch (32 to 50 mm) 1500 PSI (10342.5 kPa) - 2 1/2 to 3 inch (65 to 80 mm) 1000 PSI (6895.0 kPa) - Liquid and Gas Non-Shock Service at max. 150°F (66°C)
NOTE: At high temperatures the allowable pressure rating is lower than stated.

Table #47 - Pressure and Temperature for 150/300 Malleable Iron Fittings

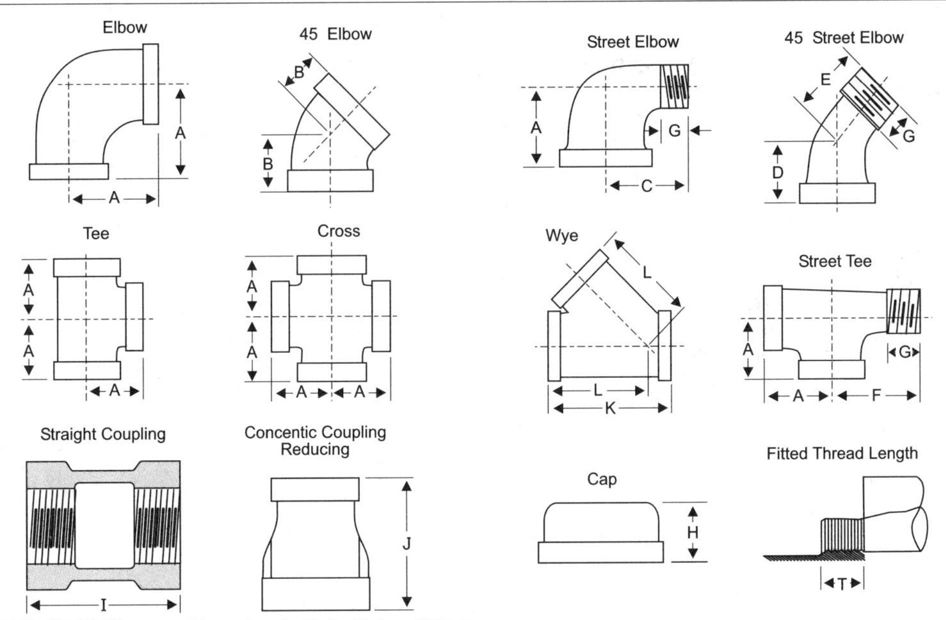

Table #48A,B Diagrams - Dimensions for Malleable Iron Fittings

DIMENSIONS FOR MALLEABLE IRON FITTINGS (CLASS 150)

Nominal Pipe Size Inches	Dimensions - Inches												
	A	B	C	D	E	F	G	H	I	J	K	L	T
1/8	0.69	-	1.00	-	-	-	0.26	0.53	0.96	-	-	-	0.25
1/4	0.81	0.73	1.19	0.73	0.94	1.19	0.40	0.63	1.06	1.00	-	-	0.32
3/8	0.95	0.80	1.44	0.80	1.03	1.44	0.41	0.74	1.16	1.13	1.93	1.43	0.36
1/2	1.12	0.88	1.63	0.88	1.15	1.63	0.53	0.87	1.34	1.25	2.32	1.71	0.43
3/4	1.31	0.98	1.89	0.98	1.29	1.89	0.55	0.97	1.52	1.44	2.77	2.05	0.50
1	1.50	1.12	2.14	1.12	1.47	2.14	0.68	1.16	1.67	1.69	3.28	2.43	0.58
1 1/4	1.75	1.29	2.45	1.29	1.71	2.45	0.71	1.28	1.93	2.06	3.94	2.92	0.67
1 1/2	1.94	1.43	2.69	1.43	1.88	2.69	0.72	1.33	2.15	2.31	4.38	3.28	0.70
2	2.25	1.68	3.26	1.68	2.22	3.26	0.76	1.45	2.53	2.81	5.17	3.93	0.75
2 1/2	2.70	1.95	3.86	1.95	2.57	-	1.14	1.70	2.88	3.25	6.25	4.73	0.92
3	3.08	2.17	4.51	2.17	3.00	-	1.20	1.80	3.18	3.69	7.26	5.55	0.98
3 1/2	3.42	2.39	-	-	-	-	-	1.90	-	-	-	-	1.03
4	3.79	2.61	5.69	2.61	3.70	5.69	1.30	2.08	3.69	4.38	8.98	6.97	1.08
5	4.50	3.05	6.86	-	-	-	1.41	2.32	-	-	-	-	1.18
6	5.13	3.46	8.03	-	-	-	1.51	2.55	-	-	-	-	1.28

NOTE: Dimensions for table are based on fittings manufactured to ANSI/ASME B16.3 Standard.

Table #48A - Malleable Iron Fittings (150) Imperial

DIMENSIONS FOR MALLEABLE IRON FITTINGS (CLASS 150)													
Nominal Pipe Size mm	Dimensions - Millimetres												
	A	B	C	D	E	F	G	H	I	J	K	L	T
6	17.5	-	25.4	-	-	25.4	6.70	13.5	24.4	-	-	-	6.4
8	20.6	18.5	30.2	18.5	23.9	30.2	10.20	16.0	26.9	25.4	-	-	8.1
10	24.1	20.3	36.6	20.3	26.2	36.6	10.36	18.8	29.5	28.7	49.0	36.3	9.1
15	28.5	22.4	41.2	22.4	29.2	41.2	13.56	22.1	34.0	31.8	58.9	43.4	10.9
20	33.3	24.9	48.0	24.9	32.8	48.0	13.86	24.6	38.6	36.6	70.4	52.1	12.7
25	38.1	28.5	54.4	28.5	37.3	54.4	17.34	29.5	42.4	42.9	83.3	61.7	14.7
32	44.5	32.8	62.2	32.8	43.4	62.2	17.94	32.5	49.0	52.3	100.1	74.2	17.0
40	49.3	36.3	68.3	36.3	47.8	68.3	18.38	33.8	54.6	58.7	111.3	83.3	17.8
50	57.2	42.7	82.8	42.7	56.4	82.8	19.22	36.8	64.3	71.4	131.3	99.8	19.1
65	68.6	49.5	98.0	49.5	65.3	98.0	28.96	43.2	73.2	82.6	158.8	120.1	23.4
80	78.2	55.1	114.6	55.1	76.2	114.6	30.48	45.7	80.8	93.7	184.4	141.0	24.9
90	86.9	60.7	-	-	-	-	-	48.3	-	-	-	-	26.2
100	96.3	66.3	144.5	66.3	94.0	144.5	33.02	52.8	93.7	111.3	228.1	177.0	27.4
125	114.3	77.5	174.2	-	-	174.2	35.72	58.9	-	-	-	-	30.0
150	130.3	87.9	204.0	-	-	204.0	38.42	64.8	-	-	-	-	32.5

NOTE: Dimensions for table are based on fittings manufactured to ANSI/ASME B16.3 Standard.

Table #48B - Malleable Iron Fittings (150) Metric

Forged Steel Threaded Fittings

Threaded steel fittings are produced in three pressure designations:

Class 2000

Class 3000

Class 6000

Pressure and temperature ratings for each of the pressure class fittings are taken to be equivalent to the following pipe wall thickness designations.

FITTING CLASS	SCHEDULE NO.	WEIGHT
2000	80	XS
3000	160	-
6000	-	XXS

Common forged steel threaded fitting dimensions for all classes are given in tables #49A and #49B.

Cast Iron Threaded Fittings

Cast iron threaded fittings are available in classes 125 and 250. Maximum pressure/temperature ratings for each class are given in table #50.

Cast iron fittings can be distinguished from malleable fittings by the larger surrounding bands placed at the end of fittings which inter-join to each other. Refer to illustration #72A and #72B for a comparison between both fittings.

Another unique feature of cast iron screwed fittings is that (when necessary) they can be removed by breaking the fitting with a sharp blow from a hammer.

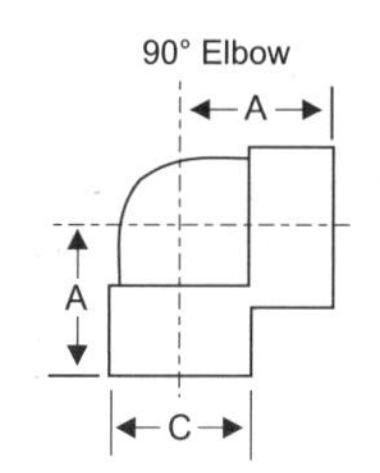

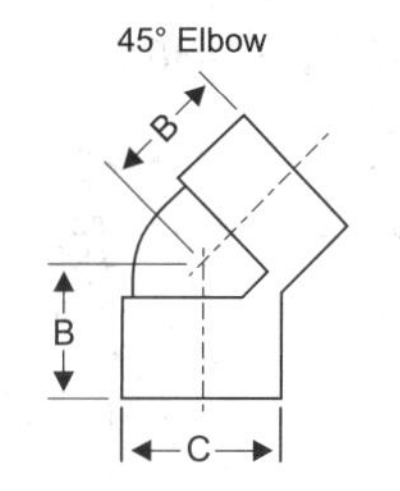

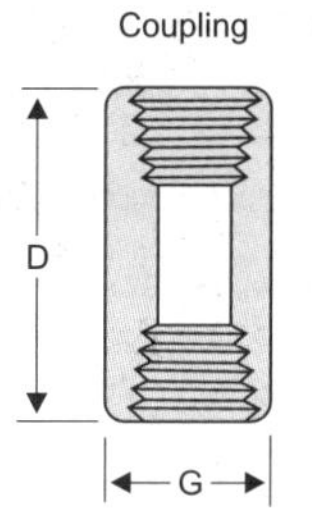

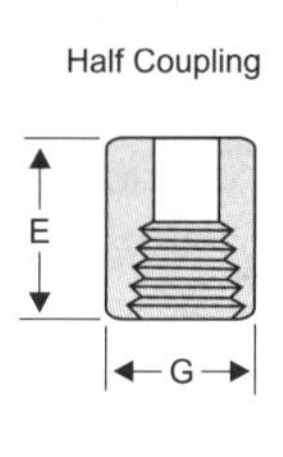

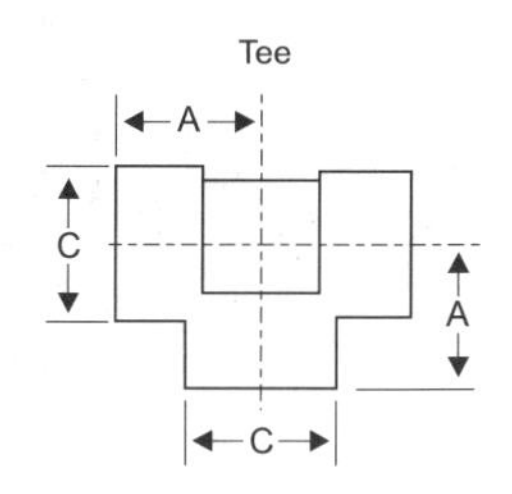

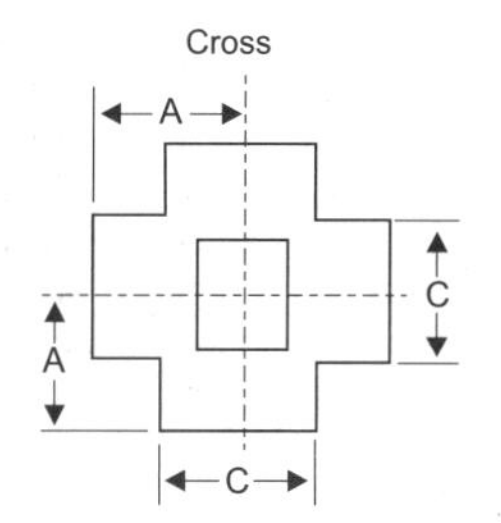

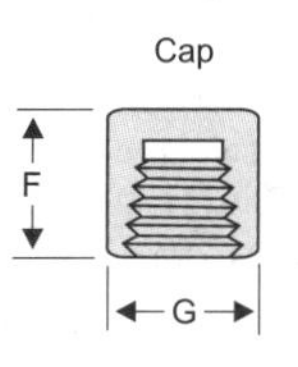

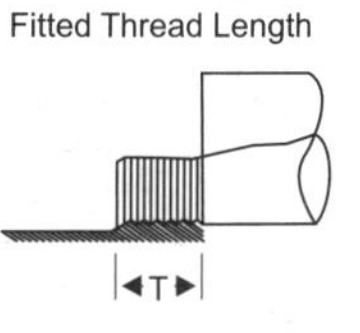

Table #49A,B Diagrams - Dimensions for Forged Steel Threaded Fittings

DIMENSIONS FOR FORGED STEEL THREADED FITTINGS (CLASS 2000, 3000 & 6000)																
Nominal Pipe Size Inches	Dimensions - Inches															
	A			B			C			D	E	F		G		T
	2000	3000	6000	2000	3000	6000	2000	3000	6000	3000/ 6000	3000/ 6000	3000	6000	3000	6000	ALL
1/8	0.81	0.81	0.97	0.69	0.69	0.75	0.88	0.88	1.00	1.25	0.63	0.75		0.62	0.88	0.26
1/4	0.81	0.97	1.12	0.69	0.75	0.88	0.88	1.00	1.31	1.38	0.69	1.00	1.06	0.75	1.00	0.40
3/8	0.97	1.12	1.31	0.75	0.88	1.00	1.00	1.31	1.50	1.50	0.75	1.00	1.06	0.88	1.25	0.41
1/2	1.12	1.31	1.50	0.88	1.00	1.12	1.31	1.50	1.81	1.88	0.94	1.25	1.31	1.12	1.50	0.53
3/4	1.31	1.50	1.75	1.00	1.12	1.31	1.50	1.81	2.19	2.00	1.00	1.44	1.50	1.38	1.75	0.55
1	1.50	1.75	2.00	1.12	1.31	1.38	1.81	2.19	2.44	2.38	1.19	1.62	1.69	1.75	2.25	0.68
1 1/4	1.75	2.00	2.38	1.31	1.38	1.69	2.19	2.44	2.97	2.62	1.31	1.75	1.81	2.25	2.50	0.71
1 1/2	2.00	2.38	2.50	1.38	1.69	1.72	2.44	2.97	3.31	3.12	1.56	1.75	1.88	2.50	3.00	0.72
2	2.38	2.50	3.25	1.69	1.72	2.06	2.97	3.31	4.00	3.38	1.69	1.88	2.00	3.00	3.62	0.76
2 1/2	3.00	3.25	3.75	2.06	2.06	2.50	3.62	4.00	4.75	3.62	1.81	2.38	2.50	3.62	4.25	1.14
3	3.38	3.75	4.19	2.50	2.50	3.12	4.31	4.75	5.75	4.25	2.13	2.56	2.69	4.25	5.00	1.20
4	4.19	4.50	4.50	3.12	3.12	3.12	5.75	6.00	6.00	4.75	2.38	2.69	2.94	5.50	6.25	1.30
NOTE: Dimensions for table are based on fittings manufactured to ANSI/ASME B16.3 Standard.																

Table #49A - Forged Steel Threaded Fittings (Imperial)

DIMENSIONS FOR FORGED STEEL THREADED FITTINGS (CLASS 2000, 3000 & 6000)

Nominal Pipe Size mm	Dimensions - Millimetres															
	← A →			← B →			← C →			← D →	← E →	← F →		← G →		← T →
	2000	3000	6000	2000	3000	6000	2000	3000	6000	3000/ 6000	3000/ 6000	3000	6000	3000	6000	ALL
6	21	21	25	17	17	19	22	22	25	32	16.0	19		16	22	6.5
8	21	25	29	17	19	22	22	25	33	35	17.5	25	27	19	25	10.0
10	25	29	33	19	22	25	25	33	38	38	19.0	25	27	22	32	10.5
15	29	33	38	22	25	29	33	38	46	48	24.0	32	33	29	38	13.5
20	33	38	44	25	29	33	38	46	56	51	25.5	37	38	35	44	14.0
25	38	44	51	29	33	35	46	56	62	60	30.0	41	43	44	57	17.5
32	44	51	60	33	35	43	56	62	75	67	33.5	44	46	57	64	18.0
40	51	60	64	35	43	44	62	75	84	79	39.5	44	48	64	76	18.5
50	60	64	83	43	45	52	75	84	102	86	43.0	48	51	76	92	19.0
65	76	83	95	52	52	64	92	102	121	92	46.0	60	64	92	108	29.0
80	86	95	106	64	64	79	110	121	146	108	54.0	65	68	108	127	30.5
100	106	114	114	79	79	79	146	152	152	121	60.5	68	75	140	159	33.0

NOTE: Dimensions for table are based on fittings manufactured to ANSI/ASME B16.3 Standard.

Table #49B - Forged Steel Threaded Fittings (Metric)

PRESSURE/TEMPERATURE RATINGS FOR CLASS 125 & 250 CAST IRON THREADED FITTINGS
CLASS 125
- 125 PSI (861.88 kPa) Saturated Steam - 175 PSI (1206.63 kPa) Liquid and Gas Non-Shock Service at max. 150°F (66°C)
CLASS 250
- 250 PSI (1723.75 kPa) Saturated Steam - 400 PSI (2758 kPa) Liquid and Gas Non-Shock Service at max. 150°F (66°C)
NOTE: At high temperatures the allowable pressure rating is lower than stated.

Table #50 - Pressure and Temperature for Cast Iron Threaded Fittings

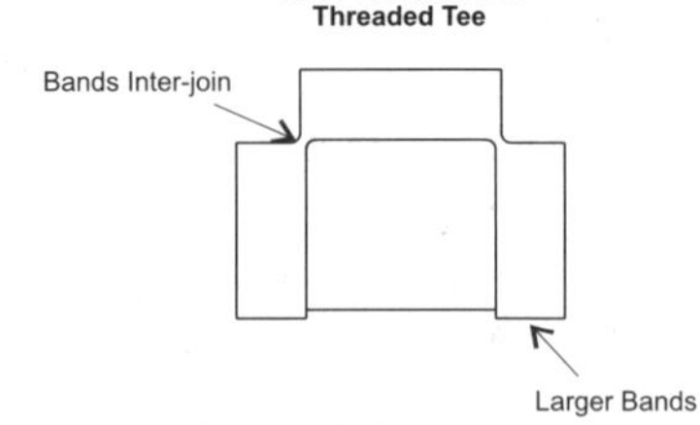

Illustration #72A - Cast Iron Fitting

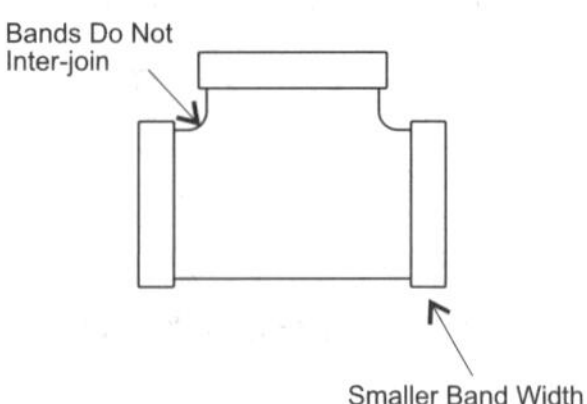

Illustration #72B - Malleable Iron Fitting

Cast Iron Threaded Drainage Fittings

Cast iron threaded drainage fittings are designed for use in gravity flow drainage systems and not intended for pressurized applications. The fitting has an inside shoulder making a smooth, flush connection when a pipe is screwed into the fitting. Illustration #73A and #73B show typical inside drainage fitting designs. This joining design prevents material in the drainage system from catching on the fitting or pipe and thus possibly blocking the drainage flow. Fittings with openings of 90 degrees from the vertical are tapped to provide 1/4 in. per ft. (21 mm/m) pitch for proper grade on drainage lines.

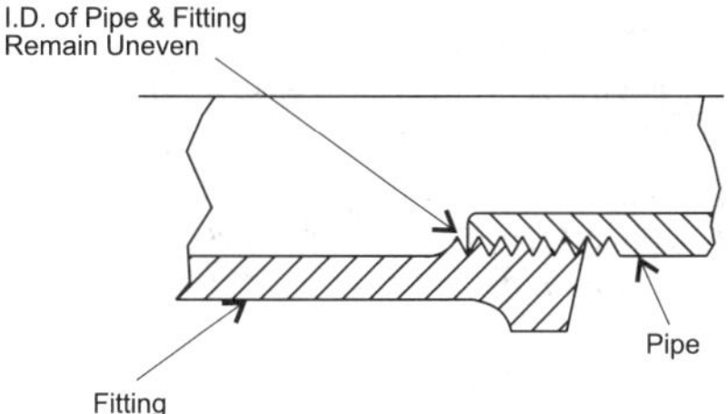

Illustration #73A - Standard Threaded Fitting

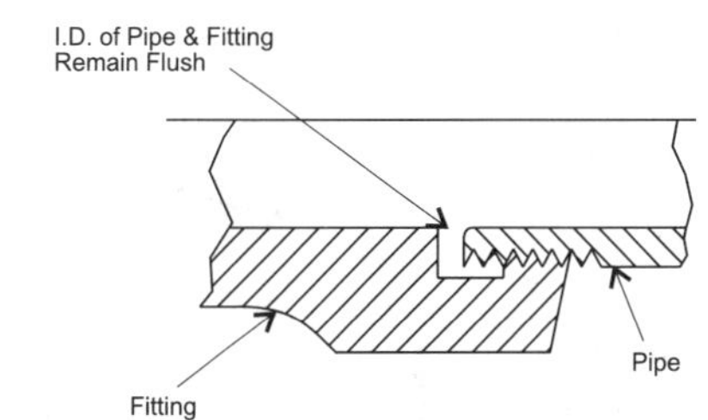

Illustration #73B - Cast Iron Drainage Fitting

Welded Fittings

Fittings are available for welded joints in either socket welded style or butt welded style. Typical cross sections of both socket and butt welded joints are displayed in illustration #74A and #74B.

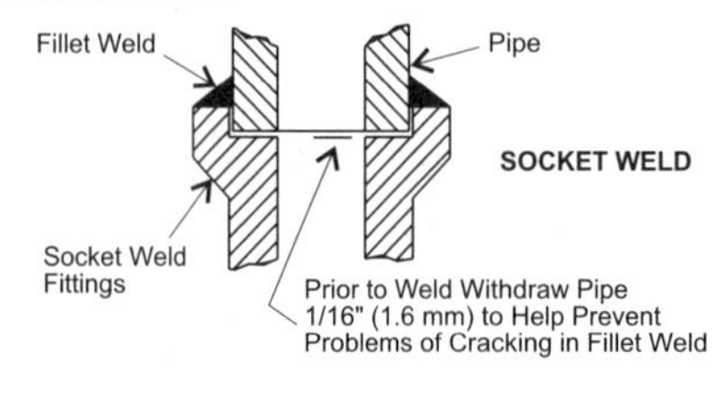

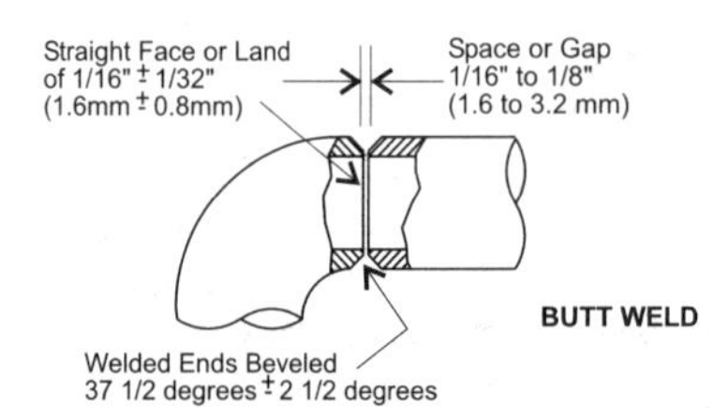

Illustration #74A, B - Typical Socket & Butt Weld Fittings

Socket Weld Fittings

Socket weld fittings are used to join smaller sizes of pipe (usually 2 in. (50 mm) and under) which require the strength and security of a welded joint. Pressure-temperature ratings for socket weld fittings are taken to be equivalent to the following pipe wall thickness designations:

FITTING CLASS	SCHEDULE NO.	WEIGHT
3000	80	XS
6000	160	-
9000	-	XXS

The socket welded joint is made by fitting pipe into the socket of the fitting and fillet welding around the pipe and the top of fitting. Dimensions for standard forged steel socket welded fittings, pressure classes 3000, 6000 and 9000 are given in tables #51 and #52.

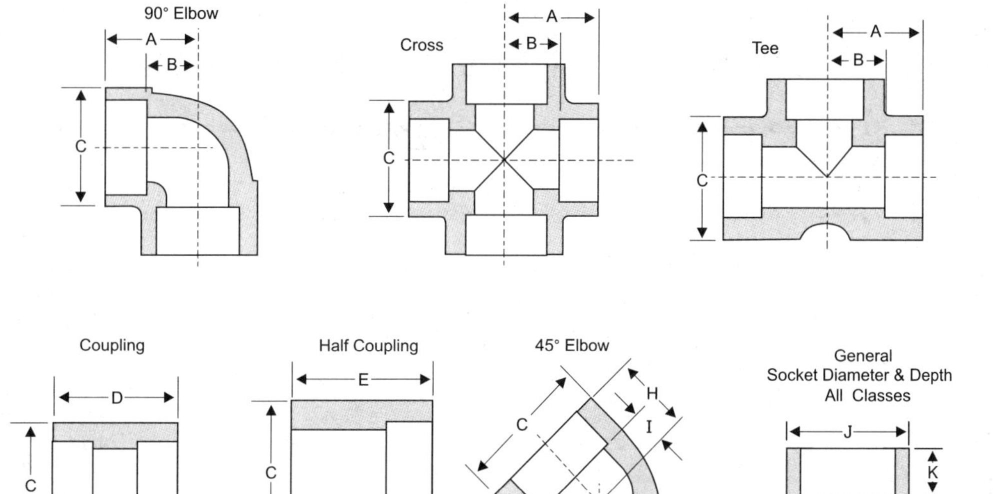

Table #51, 52 Diagrams - Dimensions for Forged Steel Socket Weld Fittings

STANDARD FORGED STEEL SOCKET WELDED FITTINGS

Nominal Pipe Size Inches	Dimensions - Inches																					
	A			B			C			D	E	F	G	H			I			J		K
	3000	6000	9000	3000	6000	9000	3000	6000	9000	ALL	ALL	ALL	ALL	3000	6000	9000	3000	6000	9000	max	min	min
1/8	0.82	0.82	-	0.44	0.44	-	0.68	0.74	-	1.01	1.00	0.25	0.62	0.69	0.69	-	0.31	0.31	-	0.430	0.420	0.38
1/4	0.82	0.91	-	0.44	0.53	-	0.86	0.92	-	1.01	1.00	0.25	0.62	0.69	0.69	-	0.31	0.31	-	0.565	0.555	0.38
1/8	0.91	1.00	-	0.53	0.62	-	1.01	1.09	-	1.01	1.07	0.25	0.69	0.69	0.82	-	0.31	0.44	-	0.700	0.690	0.38
1/2	1.00	1.13	1.38	0.62	0.75	1.00	1.23	1.33	1.60	1.14	1.26	0.38	0.88	0.82	0.88	1.00	0.44	0.50	0.62	0.865	0.855	0.38
3/4	1.25	1.38	1.62	0.75	0.88	1.12	1.46	1.62	1.84	1.38	1.44	0.38	0.94	1.00	1.06	1.25	0.50	0.56	0.75	1.075	1.065	0.50
1	1.33	1.56	1.75	0.88	1.06	1.25	1.78	1.96	2.23	1.50	1.62	0.50	1.12	1.06	1.19	1.31	0.56	0.69	0.81	1.340	1.330	0.50
1 1/4	1.56	1.75	1.88	1.06	1.25	1.38	2.16	2.30	2.64	1.50	1.69	0.50	1.19	1.19	1.31	1.38	0.69	0.81	0.88	1.685	1.675	0.50
1 1/2	1.75	2.00	2.00	1.25	1.50	1.50	2.42	2.62	2.92	1.50	1.75	0.50	1.25	1.31	1.50	1.50	0.81	1.00	1.00	1.925	1.915	0.50
2	2.12	2.74	2.74	1.50	1.62	2.12	2.96	3.27	3.50	1.99	2.24	0.75	1.62	1.62	1.74	1.74	1.00	1.12	1.12	2.416	2.406	0.62
2 1/2	2.24	-	-	1.62	-	-	3.60	-	-	1.99	2.31	0.75	1.69	1.72	-	-	1.12	-	-	2.721	2.906	0.62
3	2.87	-	-	2.25	-	-	4.29	-	-	1.99	2.37	0.75	1.75	1.87	-	-	1.25	-	-	3.550	3.535	0.62
4	3.37	-	-	2.62	-	-	5.39	-	-	2.25	2.63	0.75	1.88	2.37	-	-	1.62	-	-	4.560	4.545	0.75

NOTE: 1. Dimensions for table are based on fittings manufactured to ANSI/ASME B16.11 Standard.
2. Slight variations between inch and millimetre dimensions are due to rounding factors and permitted tolerances in standard.

Table #51 - Socket Weld Fittings (Imperial)

STANDARD FORGED STEEL SOCKET WELDED FITTINGS

Nom. Pipe Size mm	Dimensions - Millimetres																							
	A			B			C			D	E	F	G	H			I			J		K
	3000	6000	9000	3000	6000	9000	3000	6000	9000	ALL	ALL	ALL	ALL	3000	6000	9000	3000	6000	9000	max	min	min
6	20.8	20.8	-	11.2	11.2	-	17.3	18.8	-	25.7	25.4	6.4	15.8	18.0	18.0	-	8.0	8.0	-	10.9	10.7	10
8	20.8	23.1	-	11.2	13.5	-	21.8	23.4	-	25.7	25.4	6.4	15.8	18.0	18.0	-	8.0	8.0	-	14.4	14.1	10
10	23.1	25.4	-	13.5	15.8	-	25.7	27.7	-	25.7	27.2	6.4	17.5	18.0	21.5	-	8.0	11.5	-	17.8	17.6	10
15	25.4	28.7	35.1	15.8	19.1	25.4	31.2	33.8	40.6	29.0	32.0	9.7	22.4	21.5	22.5	25.5	11.5	12.5	15.5	22.0	21.7	10
20	31.8	35.1	41.2	19.1	22.4	28.5	37.1	41.2	46.7	35.1	36.6	9.7	23.9	25.0	27.5	32.0	12.0	14.5	19.0	27.3	27.1	13
25	33.8	39.6	44.5	22.4	26.9	31.8	45.2	49.8	56.6	38.1	41.2	12.7	28.5	27.0	30.0	34.0	14.0	17.0	21.0	34.1	33.8	13
32	39.6	44.5	47.8	26.9	31.8	35.1	54.9	58.4	67.1	38.1	42.9	12.7	30.2	30.0	34.0	35.0	17.0	21.0	22.0	42.8	42.6	13
40	44.5	50.8	50.8	31.8	38.1	38.1	61.5	66.6	74.2	38.1	44.5	12.7	31.8	34.0	38.0	38.5	21.0	25.0	25.5	48.9	48.7	13
50	53.9	56.9	69.6	38.1	41.2	53.9	75.2	83.1	88.9	50.6	56.9	19.1	41.2	41.0	45.0	44.5	25.0	29.0	28.5	61.4	61.1	16
65	56.9	-	-	41.2	-	-	91.4	-	-	50.6	58.7	19.1	42.9	45.0	-	-	29.0	-	-	74.2	73.8	16
80	72.9	-	-	57.2	-	-	109.0	-	-	50.6	60.2	19.1	44.5	47.5	-	-	31.5	-	-	90.2	89.9	16
100	85.6	-	-	66.6	-	-	136.9	-	-	57.2	66.8	19.1	47.8	60.5	-	-	41.5	-	-	115.8	115.5	19

NOTE: 1. Dimensions for table are based on fittings manufactured to ANSI/ASME B16.11 Standard.
2. Slight variations between inch and millimetre dimensions are due to rounding factors and permitted tolerances in standard.
3. All dimensions have been rounded to one decimal point.

Table #52 - Socket Weld Fittings (Metric)

Butt Weld Fittings

The most common style of welded fitting used in welded pipe systems (primarily in pipe sizes over 2 inches (50 mm)) is the butt weld fitting.

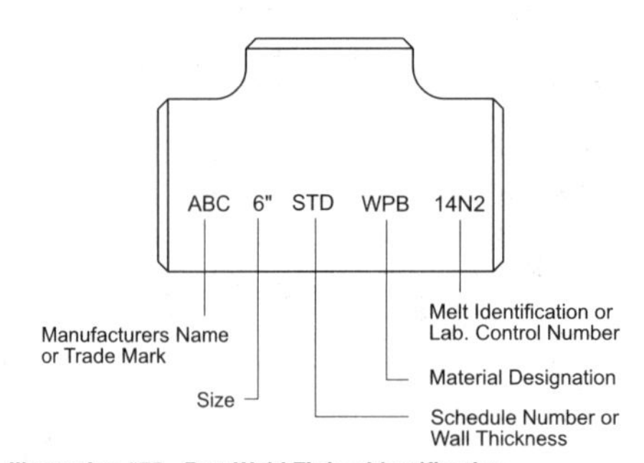

Illustration #75 - Butt Weld Fitting Identification

Illustration #75 shows a typical butt weld fitting and explains the identification markings required on fittings.

Pressure/temperature ratings for the fittings duplicate that of seamless pipe of the same material size, and wall thickness. Standard sizes for butt weld fittings are available in wall thickness and schedule numbers paralleling that of steel pipe.

Specific dimensions for butt welded fittings that are displayed in illustration #76 are provided in tables #53, #54, #55, #56.

90° Long Radius Elbow

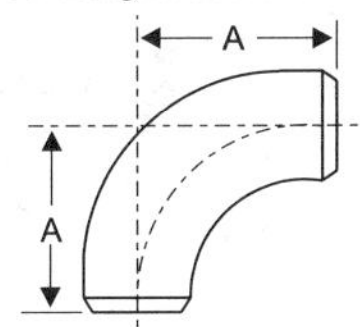

45° Elbow

Straight Tee

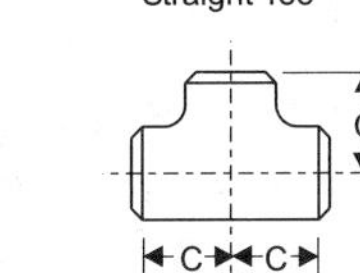

Straight Cross

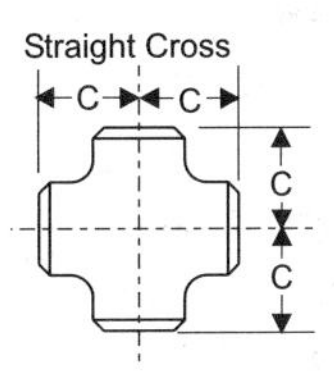

Concentric Reducer

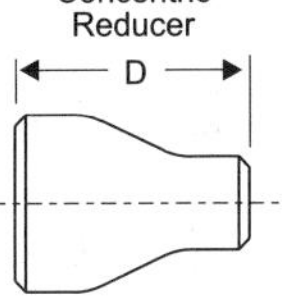

Long Radius Return Bend

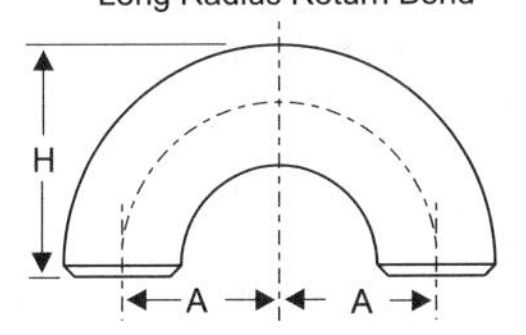

90° Long Radius Reducing Elbow

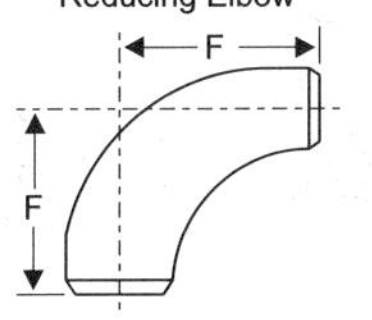

90° Short Radius Elbow

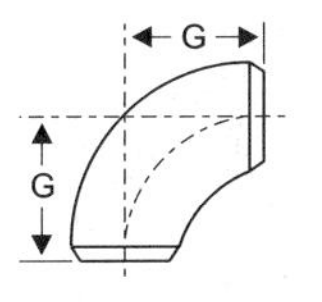

Eccentric Reducer

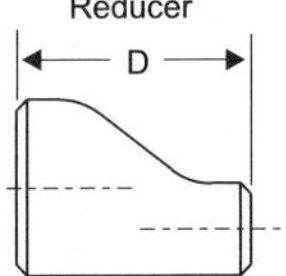

Short Radius Return Bend

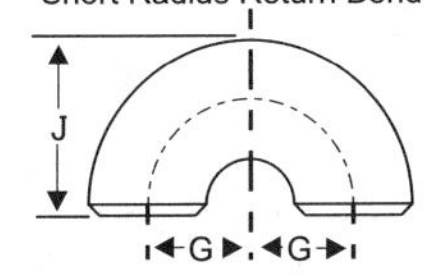

Reducing Outlet Tee

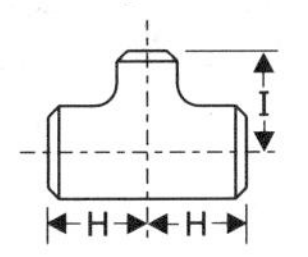

Reducing Outlet Cross

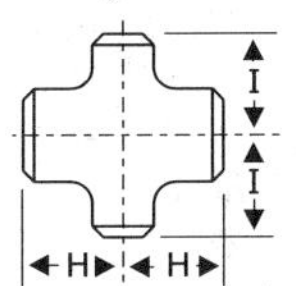

Cap

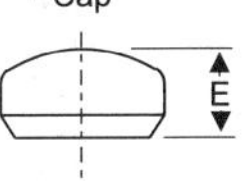

Illustration # 76 - Butt Weld Fittings

DIMENSIONS FOR BUTT WELD FITTING - INCHES

Nominal Pipe Size	A	B	C	D	E	F	G	H	J
1/2	1.50	0.62	1.00	-	1.00	-	-	1.88	-
3/4	1.12	0.44	1.12	1.50	1.00	-	-	1.69	-
1	1.50	0.88	1.50	2.00	1.50	-	1.00	2.19	1.62
1 1/4	1.88	1.00	1.88	2.00	1.50	-	1.25	2.75	2.06
1 1/2	2.25	1.12	2.25	2.50	1.50	-	1.50	3.25	2.44
2	3.00	1.38	2.50	3.00	1.50	3.00	2.00	4.19	3.19
2 1/2	3.75	1.75	3.00	3.50	1.50	3.75	2.50	5.19	3.94
3	4.50	2.00	3.38	3.50	2.00	4.50	3.00	6.25	4.75
3 1/2	5.25	2.25	3.75	4.00	2.50	5.25	3.50	7.25	5.50
4	6.00	2.50	4.12	4.00	2.50	6.00	4.00	8.25	6.25
5	7.50	3.12	4.88	5.00	3.00	7.50	5.00	10.31	7.75
6	9.00	3.75	5.62	5.50	3.50	9.00	6.00	12.31	9.31
8	12.00	5.00	7.00	6.00	4.00	12.00	8.00	16.31	12.31
10	15.00	6.25	8.50	7.00	5.00	15.00	10.00	20.38	15.38
12	18.00	7.50	10.00	8.00	6.00	18.00	12.00	24.38	18.38
14	21.00	8.75	11.00	13.00	6.50	21.00	14.00	28.00	21.00
16	24.00	10.00	12.00	14.00	7.00	24.00	16.00	32.00	24.00
18	27.00	11.25	13.50	15.00	8.00	27.00	18.00	36.00	27.00
20	30.00	12.50	15.00	20.00	9.00	30.00	20.00	40.00	30.00
22	33.00	13.50	16.50	20.00	10.00	-	22.00	44.00	-
24	36.00	15.00	17.00	20.00	10.50	36.00	24.00	48.00	36.00
26	39.00	16.00	19.50	24.00	10.50	-	-	-	-
28	42.00	17.25	20.50	24.00	10.50	-	-	-	-
30	45.00	18.50	22.00	24.00	10.50	-	-	-	-
32	48.00	19.75	23.50	24.00	10.50	-	-	-	-
34	51.00	21.00	25.00	24.00	10.50	-	-	-	-
36	54.00	22.25	26.50	24.00	10.50	-	-	-	-
38	57.00	23.62	28.00	24.00	12.00	-	-	-	-
40	60.00	24.88	29.50	24.00	12.00	-	-	-	-
42	63.00	26.00	30.00	24.00	12.00	-	-	-	-
44	66.00	27.38	32.00	24.00	13.50	-	-	-	-
46	69.00	28.62	33.50	28.00	13.50	-	-	-	-
48	72.00	29.88	35.00	28.00	13.50	-	-	-	-

LEGEND

A = 90° Long Radius Elbow

B = 45° Elbows

C = Tees and Crosses

D = Reducing Couplings Con/Ecc

E = Caps

F = Reducing 90° Elbows

G = Short Radius 90° Elbows

H = Long Radius Return Bends

J = Short Radius Return Bends

Table #53 - Butt Weld Fittings (Imperial)

DIMENSIONS FOR BUTT WELD FITTING - MILLIMETRES

Nominal Pipe Size	A	B	C	D	E	F	G	H	J
15	38	16	25	-	25	-	-	48	-
20	29	11	29	38	25	-	-	43	-
25	38	22	38	51	38	-	25	56	41
32	48	25	48	51	38	-	32	70	52
40	57	29	57	64	38	-	38	83	62
50	76	35	64	76	38	76	51	106	81
65	95	44	76	89	38	95	64	132	100
80	114	51	86	89	51	114	76	159	121
90	133	57	95	102	64	133	89	184	140
100	152	64	105	102	64	152	102	210	159
125	190	79	124	127	76	190	127	262	197
150	229	95	143	140	89	229	152	313	237
200	305	127	178	152	102	305	203	414	313
250	381	159	216	178	127	381	254	518	391
300	457	190	254	203	152	457	305	619	467
350	533	222	279	330	165	533	356	711	533
400	610	254	305	356	178	610	406	813	610
450	686	286	343	381	203	686	457	914	686
500	762	318	381	508	229	762	508	1016	762
550	838	343	419	508	254	-	559	1118	-
600	914	381	432	508	267	914	610	1119	914
650	991	406	495	610	267	-	-	-	-
700	1067	438	521	610	267	-	-	-	-
750	1143	470	559	610	267	-	-	-	-
800	1219	502	597	610	267	-	-	-	-
850	1295	533	635	610	267	-	-	-	-
900	1372	565	673	610	267	-	-	-	-
950	1448	600	711	610	305	-	-	-	-
1000	1524	632	749	610	305	-	-	-	-
1050	1600	660	762	610	305	-	-	-	-
1100	1676	695	813	610	343	-	-	-	-
1150	1723	727	851	711	343	-	-	-	-
1200	1829	759	889	711	343	-	-	-	-

LEGEND

- A = 90° Long Radius Elbow
- B = 45° Elbows
- C = Tees and Crosses
- D = Reducing Couplings Con/Ecc
- E = Caps
- F = Reducing 90° Elbows
- G = Short Radius 90° Elbows
- H = Long Radius Return Bends
- J = Short Radius Return Bends

Table #54 - Butt Weld Fittings (Metric)

DIMENSIONS FOR REDUCING OUTLET TEES & CROSSES - INCHES		
Nominal Pipe Size	RUN ← H →	OUTLET ← I →
1/2 x 1/2 x 3/8	1.00	1.00
1/2 x 1/2 x 1/4	1.00	1.00
3/4 x 3/4 x 1/2	1.12	1.12
3/4 x 3/4 x 3/8	1.12	1.12
1 x 1 x 3/4	1.50	1.50
1 x 1 x 1/2	1.50	1.50
11/4 x 11/4 x 1	1.88	1.88
11/4 x 11/4 x 3/4	1.88	1.88
11/4 x 11/4 x 1/2	1.88	1.88
11/2 x 11/2 x 11/4	2.25	2.25
11/2 x 11/2 x 1	2.25	2..25
11/2 x 11/2 x 3/4	2.25	2.25
11/2 x 11/2 x 1/2	2.25	2.25
2 x 2 x 11/2	2.50	2.38
2 x 2 x 11/4	2.50	2.25
2 x 2 x 1	2.50	2.00
2 x 2 x 3/4	2.50	1.75
21/2 x 21/2 x 2	3.00	2.75
21/2 x 21/2 x 11/2	3.00	2.62
21/2 x 21/2 x 11/4	3.00	2.50
21/2 x 21/2 x 1	3.00	2.25

DIMENSIONS FOR REDUCING OUTLET TEES & CROSSES - INCHES		
Nominal Pipe Size	RUN ← H →	OUTLET ← I →
3 x 3 x 21/2	3.38	3.25
3 x 3 x 2	3.38	3.00
3 x 3 x 11/2	3.38	2.88
3 x 3 x 11/4	3.38	2.75
31/2 x 31/2 x 3	3.75	3.62
31/2 x 31/2 x 21/2	3.75	3.50
31/2 x 31/2 x 2	3.75	3.25
31/2 x 31/2 x 11/2	3.75	3.12
4 x 4 x 31/2	4.12	4.00
4 x 4 x 3	4.12	3.88
4 x 4 x 21/2	4.12	3.75
4 x 4 x 2	4.12	3.50
4 x 4 x 11/2	4.12	3.38
5 x 5 x 4	4.88	4.62
5 x 5 x 31/2	4.88	4.50
5 x 5 x 3	4.88	4.38
5 x 5 x 21/2	4.88	4.25
5 x 5 x 2	4.88	4.12
6 x 6 x 5	5.62	5.38
6 x 6 x 4	5.62	5.12
6 x 6 x 31/2	5.62	5.00
6 x 6 x 3	5.62	4.88
6 x 6 x 21/2	5.62	4.75

DIMENSIONS FOR REDUCING OUTLET TEES & CROSSES - INCHES		
Nominal Pipe Size	RUN ← H →	OUTLET ← I →
8 x 8 x 6	7.00	6.62
8 x 8 x 5	7.00	6.38
8 x 8 x 4	7.00	6.12
8 x 8 x 31/2	7.00	6.00
10 x 10 x 8	8.50	8.00
10 x 10 x 6	8.50	7.62
10 x 10 x 5	8.50	7.50
10 x 10 x 4	8.50	7.25
12 x 12 x 10	10.00	9.50
12 x 12 x 8	10.00	9.00
12 x 12 x 6	10.00	8.62
12 x 12 x 5	10.00	8.50
14 x 14 x 12	11.00	10.62
14 x 14 x 10	11.00	10.12
14 x 14 x 8	11.00	9.75
14 x 14 x 6	11.00	9.38
16 x 16 x 14	12.00	12.00
16 x 16 x 12	12.00	11.62
16 x 16 x 10	12.00	11.12
16 x 16 x 8	12.00	10.75
16 x 16 x 6	12.00	10.38

Table #55A - Reducing Outlet Tees and Crosses (Imperial)

DIMENSIONS FOR REDUCING OUTLET TEES & CROSSES - INCHES		
Nominal Pipe Size	RUN ← H →	OUTLET ← I →
18 x 18 x 16	13.50	13.00
18 x 18 x 14	13.50	13.00
18 x 18 x 12	13.50	12.62
18 x 18 x 10	13.50	12.12
18 x 18 x 8	13.50	11.75
20 x 20 x 18	15.00	14.50
20 x 20 x 16	15.00	14.00
20 x 20 x 14	15.00	14.00
20 x 20 x 12	15.00	13.62
20 x 20 x 10	15.00	13.12
20 x 20 x 8	15.00	12.75
22 x 22 x 20	16.50	16.00
22 x 22 x 18	16.50	15.50
22 x 22 x 16	16.50	15.00
22 x 22 x 14	16.50	15.00
22 x 22 x 12	16.50	14.62
22 x 22 x 10	16.50	14.12
24 x 24 x 22	17.00	17.00
24 x 24 x 20	17.00	17.00
24 x 24 x 18	17.00	16.50
24 x 24 x 16	17.00	16.00
24 x 24 x 14	17.00	16.00
24 x 24 x 12	17.00	15.62
24 x 24 x 10	17.00	15.12

DIMENSIONS FOR REDUCING OUTLET TEES & CROSSES - INCHES		
Nominal Pipe Size	RUN ← H →	OUTLET ← I →
26 x 26 x 24	19.50	19.00
26 x 26 x 22	19.50	18.50
26 x 26 x 20	19.50	18.00
26 x 26 x 18	19.50	17.50
26 x 26 x 16	19.50	17.00
26 x 26 x 14	19.50	17.00
26 x 26 x 12	19.50	16.62
28 x 28 x 26	20.50	20.50
28 x 28 x 24	20.50	20.00
28 x 28 x 22	20.50	19.50
28 x 28 x 20	20.50	19.00
28 x 28 x 18	20.50	18.50
28 x 28 x 16	20.50	18.00
28 x 28 x 14	20.50	18.00
28 x 28 x 12	20.50	17.62
30 x 30 x 28	22.00	21.50
30 x 30 x 26	22.00	21.50
30 x 30 x 24	22.00	21.00
30 x 30 x 22	22.00	20.50
30 x 30 x 20	22.00	20.00
30 x 30 x 18	22.00	19.50
30 x 30 x 16	22.00	19.00
30 x 30 x 14	22.00	19.00
30 x 30 x 12	22.00	18.62
30 x 30 x 10	22.00	18.12

DIMENSIONS FOR REDUCING OUTLET TEES & CROSSES - INCHES		
Nominal Pipe Size	RUN ← H →	OUTLET ← I →
32 x 32 x 30	23.50	23.00
32 x 32 x 28	23.50	22.50
32 x 32 x 26	23.50	22.50
32 x 32 x 24	23.50	22.00
32 x 32 x 22	23.50	21.50
32 x 32 x 20	23.50	21.00
32 x 32 x 18	23.50	20.50
32 x 32 x 16	23.50	20.00
32 x 32 x 14	23.50	20.00
34 x 34 x 32	25.00	24.50
34 x 34 x 30	25.00	24.00
34 x 34 x 28	25.00	23.50
34 x 34 x 26	25.00	23.50
34 x 34 x 24	25.00	23.00
34 x 34 x 22	25.00	22.50
34 x 34 x 20	25.00	22.00
34 x 34 x 18	25.00	21.50
34 x 34 x 16	25.00	21.00
36 x 36 x 34	26.50	26.00
36 x 36 x 32	26.50	25.50
36 x 36 x 30	26.50	25.00
36 x 36 x 28	26.50	24.50
36 x 36 x 26	26.50	24.50
36 x 36 x 24	26.50	24.00
36 x 36 x 22	26.50	23.50

Table #55B - Reducing Outlet Tees and Crosses (Imperial)

DIMENSIONS FOR REDUCING OUTLET TEES & CROSSES - INCHES		
Nominal Pipe Size	RUN ←H→	OUTLET ←I→
36 x 36 x 20	26.50	23.00
36 x 36 x 18	26.50	22.50
36 x 36 x 16	26.50	22.00
38 x 38 x 36	28.00	28.00
38 x 38 x 34	28.00	27.50
38 x 38 x 32	28.00	27.00
38 x 38 x 30	28.00	26.50
38 x 38 x 28	28.00	25.50
38 x 38 x 26	28.00	25.50
38 x 38 x 24	28.00	25.00
38 x 38 x 22	28.00	24.50
38 x 38 x 20	28.00	24.00
38 x 38 x 18	28.00	23.50
40 x 40 x 48	29.50	29.50
40 x 40 x 36	29.50	29.00
40 x 40 x 34	29.50	28.50
40 x 40 x 32	29.50	28.00
40 x 40 x 30	29.50	27.50
40 x 40 x 28	29.50	26.50
40 x 40 x 26	29.50	26.50
40 x 40 x 24	29.50	26.00
40 x 40 x 22	29.50	25.50
40 x 40 x 20	29.50	25.00
40 x 40 x 18	29.50	24.50

DIMENSIONS FOR REDUCING OUTLET TEES & CROSSES - INCHES		
Nominal Pipe Size	RUN ←H→	OUTLET ←I→
42 x 42 x 40	30.00	28.00
42 x 42 x 38	30.00	28.00
42 x 42 x 36	30.00	28.00
42 x 42 x 34	30.00	28.00
42 x 42 x 32	30.00	28.00
42 x 42 x 30	30.00	28.00
42 x 42 x 28	30.00	27.50
42 x 42 x 26	30.00	27.50
42 x 42 x 24	30.00	26.00
42 x 42 x 22	30.00	26.00
42 x 42 x 20	30.00	26.00
42 x 42 x 18	30.00	25.50
42 x 42 x 16	30.00	25.00
44 x 44 x 42	32.00	30.00
44 x 44 x 40	32.00	29.50
44 x 44 x 38	32.00	29.00
44 x 44 x 36	32.00	28.50
44 x 44 x 34	32.00	28.50
44 x 44 x 32	32.00	28.00
44 x 44 x 30	32.00	28.00
44 x 44 x 28	32.00	27.50
44 x 44 x 26	32.00	27.50
44 x 44 x 24	32.00	27.50
44 x 44 x 22	32.00	27.00
44 x 44 x 20	32.00	27.00

DIMENSIONS FOR REDUCING OUTLET TEES & CROSSES - INCHES		
Nominal Pipe Size	RUN ←H→	OUTLET ←I→
46 x 46 x 44	33.50	31.50
46 x 46 x 42	33.50	31.00
46 x 46 x 40	33.50	30.50
46 x 46 x 38	33.50	30.00
46 x 46 x 36	33.50	30.00
46 x 46 x 34	33.50	29.50
46 x 46 x 32	33.50	29.50
46 x 46 x 30	33.50	29.00
46 x 46 x 28	33.50	29.00
46 x 46 x 26	33.50	29.00
46 x 46 x 24	33.50	28.50
46 x 46 x 22	33.50	28.50
48 x 48 x 46	35.00	33.00
48 x 48 x 44	35.00	33.00
48 x 48 x 42	35.00	32.00
48 x 48 x 40	35.00	32.00
48 x 48 x 38	35.00	32.00
48 x 48 x 36	35.00	31.00
48 x 48 x 34	35.00	31.00
48 x 48 x 32	35.00	31.00
48 x 48 x 30	35.00	30.00
48 x 48 x 28	35.00	30.00
48 x 48 x 26	35.00	30.00
48 x 48 x 24	35.00	29.00
48 x 48 x 22	35.00	29.00

Table #55C - Reducing Outlet Tees and Crosses (Imperial)

DIMENSIONS FOR REDUCING OUTLET TEES & CROSSES - mm		
Nominal Pipe Size	RUN ←H→	OUTLET ← I →
15 x 15 x 10	25	25
15 x 15 x 8	25	25
20 x 20 x 15	29	29
20 x 20 x 10	29	29
25 x 25 x 20	38	38
25 x 25 x 15	38	38
32 x 32 x 25	48	48
32 x 22 x 20	48	48
32 x 32 x 15	48	48
40 x 40 x 32	57	57
40 x 40 x 25	57	57
40 x 40 x 20	57	57
40 x 40 x 15	57	57
50 x 50 x 40	64	60
50 x 50 x 32	64	57
50 x 50 x 25	64	51
50 x 50 x 20	64	44
65 x 65 x 50	76	70
65 x 65 x 40	76	67
65 x 65 x 32	76	64
65 x 65 x 25	76	57

DIMENSIONS FOR REDUCING OUTLET TEES & CROSSES - mm		
Nominal Pipe Size	RUN ←H→	OUTLET ← I →
80 x 80 x 65	86	83
80 x 80 x 50	86	76
80 x 80 x 40	86	73
80 x 80 x 32	86	70
90 x 90 x 80	95	92
90 x 90 x 65	95	89
90 x 90 x 50	95	83
90 x 90 x 40	95	79
100 x 100 x 90	105	102
100 x 100 x 80	105	98
100 x 100 x 65	105	95
100 x 100 x 50	105	89
100 x 100 x 40	105	86
125 x 125 x 100	124	117
125 x 125 x 90	124	114
125 x 125 x 80	124	111
125 x 125 x 65	124	108
125 x 125 x 50	124	105
150 x 150 x 125	143	137
150 x 150 x 100	143	130
150 x 150 x 90	143	127
150 x 150 x 80	143	124
150 x 150 x 65	143	121

DIMENSIONS FOR REDUCING OUTLET TEES & CROSSES - mm		
Nominal Pipe Size	RUN ←H→	OUTLET ← I →
200 x 200 x 150	178	168
200 x 200 x 125	178	162
200 x 200 x 100	178	156
200 x 200 x 90	178	152
250 x 250 x 200	216	203
250 x 250 x 150	216	194
250 x 250 x 125	216	191
250 x 250 x 100	216	184
300 x 300 x 250	254	241
300 x 300 x 200	254	229
300 x 300 x 150	254	219
300 x 300 x 125	254	216
350 x 350 x 300	279	270
350 x 350 x 250	279	257
350 x 350 x 200	279	248
350 x 350 x 150	279	238
400 x 400 x 350	305	305
400 x 400 x 300	305	295
400 x 400 x 250	305	283
400 x 400 x 200	305	273
400 x 400 x 150	305	264
450 x 450 x 400	343	330

Table #56A - Reducing Outlet Tees and Crosses (Metric)

DIMENSIONS FOR REDUCING OUTLET TEES & CROSSES - mm		
Nominal Pipe Size	RUN ←H→	OUTLET ←I→
450 x 450 x 350	343	330
450 x 450 x 300	343	321
450 x 450 x 250	343	308
450 x 450 x 200	343	298
500 x 500 x 450	381	368
500 x 500 x 400	381	356
500 x 500 x 350	381	356
500 x 500 x 300	381	346
500 x 500 x 250	381	333
500 x 500 x 200	381	324
550 x 550 x 500	419	406
550 x 550 x 450	419	394
550 x 550 x 400	419	381
550 x 550 x 350	419	381
550 x 550 x 300	419	371
550 x 550 x 250	419	359
600 x 600 x 550	432	432
600 x 600 x 500	432	432
600 x 600 x 450	432	419
600 x 600 x 400	432	406
600 x 600 x 350	432	406
600 x 600 x 300	432	397
600 x 600 x 250	432	384

DIMENSIONS FOR REDUCING OUTLET TEES & CROSSES - mm		
Nominal Pipe Size	RUN ←H→	OUTLET ←I→
650 x 650 x 600	495	483
650 x 650 x 550	495	470
650 x 650 x 500	495	457
650 x 650 x 450	495	444
650 x 650 x 400	495	432
650 x 650 x 350	495	432
650 x 650 x 300	495	422
700 x 700 x 650	521	521
700 x 700 x 600	521	508
700 x 700 x 550	521	495
700 x 700 x 500	521	483
700 x 700 x 450	521	470
700 x 700 x 400	521	457
700 x 700 x 350	521	457
700 x 700 x 300	521	448
750 x 750 x 700	559	546
750 x 750 x 650	559	546
750 x 750 x 600	559	533
750 x 750 x 550	559	521
750 x 750 x 500	559	508
750 x 750 x 450	559	495
750 x 750 x 400	559	483
750 x 750 x 350	559	483
750 x 750 x 300	559	473
750 x 750 x 250	559	460

DIMENSIONS FOR REDUCING OUTLET TEES & CROSSES - mm		
Nominal Pipe Size	RUN ←H→	OUTLET ←I→
800 x 800 x 750	597	584
800 x 800 x 700	597	572
800 x 800 x 650	597	572
800 x 800 x 600	597	559
800 x 800 x 550	597	546
800 x 800 x 500	597	533
800 x 800 x 450	597	521
800 x 800 x 400	597	508
800 x 800 x 350	597	508
850 x 850 x 800	635	622
850 x 850 x 750	635	610
850 x 850 x 700	635	597
850 x 850 x 650	635	597
850 x 850 x 600	635	584
850 x 850 x 550	635	572
850 x 850 x 500	635	559
850 x 850 x 450	635	546
850 x 850 x 400	635	433
900 x 900 x 850	673	660
900 x 900 x 800	673	648
900 x 900 x 750	673	635
900 x 900 x 700	673	622
900 x 900 x 650	673	622
900 x 900 x 600	673	610
900 x 900 x 550	673	597

Table #56B - Reducing Outlet Tees and Crosses (Metric)

DIMENSIONS FOR REDUCING OUTLET TEES & CROSSES - mm		
Nominal Pipe Size	RUN ←H→	OUTLET ←I→
900 x 900 x 500	673	584
900 x 900 x 500	673	572
900 x 900 x 500	673	559
950 x 950 x 900	711	711
950 x 950 x 850	711	698
950 x 950 x 800	711	686
950 x 950 x 750	711	673
950 x 950 x 700	711	648
950 x 950 x 650	711	648
950 x 950 x 600	711	635
950 x 950 x 550	711	622
950 x 950 x 500	711	610
950 x 950 x 450	711	597
1000 x 1000 x 950	749	749
1000 x 1000 x 900	749	737
1000 x 1000 x 850	749	724
1000 x 1000 x 800	749	711
1000 x 1000 x 750	749	698
1000 x 1000 x 700	749	673
1000 x 1000 x 650	749	673
1000 x 1000 x 600	749	660
1000 x 1000 x 550	749	648
1000 x 1000 x 500	749	635
1000 x 1000 x 450	749	622

DIMENSIONS FOR REDUCING OUTLET TEES & CROSSES - mm		
Nominal Pipe Size	RUN ←H→	OUTLET ←I→
1050 x 1050 x 1000	762	711
1050 x 1050 x 950	762	711
1050 x 1050 x 900	762	711
1050 x 1050 x 850	762	711
1050 x 1050 x 800	762	711
1050 x 1050 x 750	762	711
1050 x 1050 x 700	762	698
1050 x 1050 x 650	762	698
1050 x 1050 x 600	762	660
1050 x 1050 x 550	762	660
1050 x 1050 x 500	762	660
1050 x 1050 x 450	762	648
1050 x 1050 x 400	762	635
1100 x 1100 x 1050	813	762
1100 x 1100 x 1000	813	749
1100 x 1100 x 950	813	737
1100 x 1100 x 900	813	724
1100 x 1100 x 850	813	724
1100 x 1100 x 800	813	711
1100 x 1100 x 750	813	711
1100 x 1100 x 700	813	698
1100 x 1100 x 650	813	698
1100 x 1100 x 600	813	698
1100 x 1100 x 550	813	686
1100 x 1100 x 500	813	686

DIMENSIONS FOR REDUCING OUTLET TEES & CROSSES - mm		
Nominal Pipe Size	RUN ←H→	OUTLET ←I→
1150 x 1150 x 1100	851	800
1150 x 1150 x 1050	851	787
1150 x 1150 x 1000	851	775
1150 x 1150 x 950	851	762
1150 x 1150 x 900	851	762
1150 x 1150 x 850	851	749
1150 x 1150 x 800	851	749
1150 x 1150 x 750	851	737
1150 x 1150 x 700	851	737
1150 x 1150 x 650	851	737
1150 x 1150 x 600	851	724
1150 x 1150 x 550	851	724
1200 x 1200 x 1150	889	838
1200 x 1200 x 1100	889	838
1200 x 1200 x 1050	889	813
1200 x 1200 x 1000	889	813
1200 x 1200 x 950	889	813
1200 x 1200 x 900	889	787
1200 x 1200 x 850	889	787
1200 x 1200 x 800	889	787
1200 x 1200 x 750	889	762
1200 x 1200 x 700	889	762
1200 x 1200 x 650	889	762
1200 x 1200 x 600	889	737
1200 x 1200 x 550	889	737

Table #56C - Reducing Outlet Tees and Crosses (Metric)

Weld Outlets

When outlets or branch connections are required on a run of pipe, weld outlet fittings are often used. Weld outlet fittings provide branch connections that are considerably stronger than welding a pipe directly into the run (stub in connection). The outlets are made for threaded, socket weld, and butt weld branch connections. These connections are available in a wide variety of styles including: elbow outlets, lateral outlets, flat surface outlets, nipple outlets and standard beveled outlets. Nominal dimensions for weld outlet fittings are provided in tables #57 and #58.

Grooved Fitting Joints

Joining of pipe and/or grooved end fittings is accomplished by the use of special grooved pipe couplings. These couplings use an elastomer gasket seal and a bolted split metallic collar held in place by grooves made into the pipe.

Grooves are either cut or rolled into the pipe, with matching surfaces manufactured into the design of the grooved fittings. A cross sectional view of a grooved joint is displayed in illustration #77. It is important when installing grooved fittings that the elastomer gaskets supplied are checked to be certain they are suitable for the services intended.

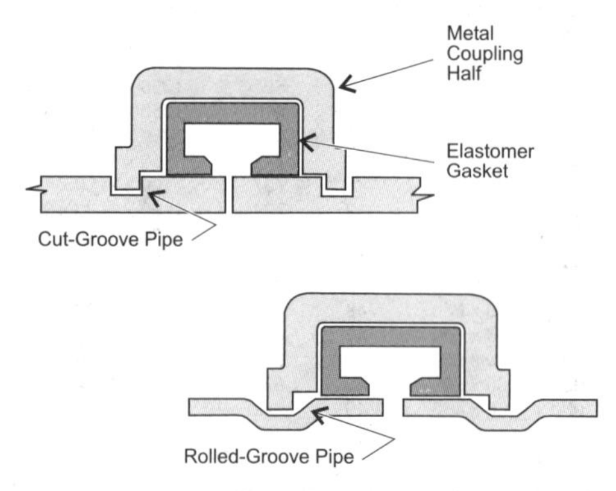

Illustration #77 - Cut & Rolled Grooved Joints

WELD OUTLET FITTINGS - INCHES						
Nominal Pipe Size	Buttweld Outlet		Threaded Outlet		Socket Weld Outlet	
	Standard	Extra Heavy	3000 lbs.	6000 lbs.	3000 lbs.	6000 lbs.
1/2	3/4	3/4	1	1 1/4	1	1
3/4	7/8	7/8	1 1/16	1 7/16	1 1/16	1 1/16
1	1 1/16	1 1/16	1 5/16	1 9/16	1 5/16	1 5/16
1 1/4	1 1/4	1 1/4	1 5/16	1 5/8	1 5/16	1 5/16
1 1/2	1 5/16	1 5/16	1 3/8	1 11/16	1 3/8	1 3/8
2	1 1/2	1 1/2	1 1/2	2 1/16	1 1/2	1 1/2
2 1/2	1 5/8	1 5/8	1 13/16		1 13/16	
3	1 3/4	1 3/4	2		2	
3 1/2	2 *	2 *	2 1/8		2 7/8	
4	2	2	2 1/4		2 1/4	
5	2 1/8 *	2 1/16 *				
6	2 3/8	3 1/16				
8	2 3/4	3 7/8				
10	3 1/16	3 1/2 *				
12	3 3/8	3 15/16 *				
14	3 1/2	4 1/8 *				
16	3 11/16	4 7/16				
18	4 1/16 *	4 11/16 *				
20	4 5/8 *	5 *				
24	5 3/8 *	5 1/2				

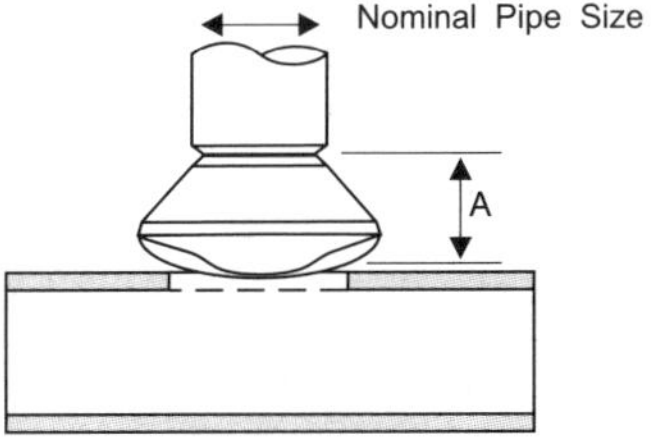

Nominal Pipe Size

A

NOTE: * Indicates dimensions may vary with branch size.

Table #57 - Weld Outlet Fittings (Imperial)

WELD OUTLET FITTINGS - MILLIMETRES						
Nominal Pipe Size	Buttweld Outlet		Threaded Outlet		Socket Weld Outlet	
	Standard	Extra Heavy	3000 lbs.	6000 lbs.	3000 lbs.	6000 lbs.
15	19	19	25	32	25	25
20	22	22	27	37	27	27
25	27	27	33	41	33	33
32	32	32	33	43	33	33
40	33	33	35	52	35	35
50	38	38	38		38	38
65	41	41	46		46	46
80	44	44	50		50	
90	51 *	51 *	54		54	
100	51	51	57		57	
125	54 *	52 *				
150	60	78				
200	70	98				
250	78	89 *				
300	86	100 *				
350	90	105 *				
400	94	113 *				
450	103 *	119 *				
500	117 *	127 *				
600	137 *	140				

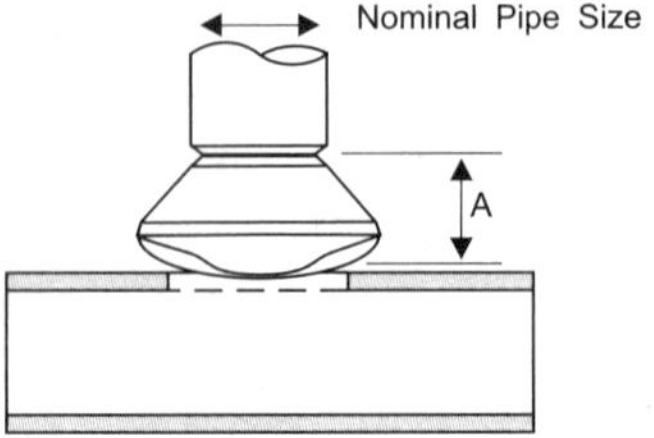

NOTE: * Indicates dimensions may vary with branch size.

Table #58 - Weld Outlet Fittings (Metric)

General Assembly and Disassembly Procedure for Grooved Fittings and Pipe

1. Check gasket for service suitability and apply lubricant to gasket if required by manufacturer.
2. Inspect pipe end for indents, dirt, rust and generally for anything that may interfere with a leak tight gasket seal.
3. Inspect groove for proper dimensions (most manufacturers will supply an inspection gage or groove specification chart to check dimensions).
4. Pull elastomer gasket completely over pipe or fitting, leaving the end of the pipe or fitting extending out slightly.
5. Align the pipe or fitting to be joined together and position the gasket evenly between the two pieces.
6. Place both ends of the split coupling over the gasket keeping the coupling in the same plane as any others in the assembly.
7. Install coupling bolts and nuts finger tight, making sure metal coupling lips engage fully into the groove.
8. Tighten nuts uniformly until coupling halves (metal to metal) firmly touch. To disassemble, follow procedure in reverse.

Note: Before disassembly, all pressure must be relieved from system.

Illustration #78 displays the procedures in assembling grooved joints.

Grooved Fitting Assembly

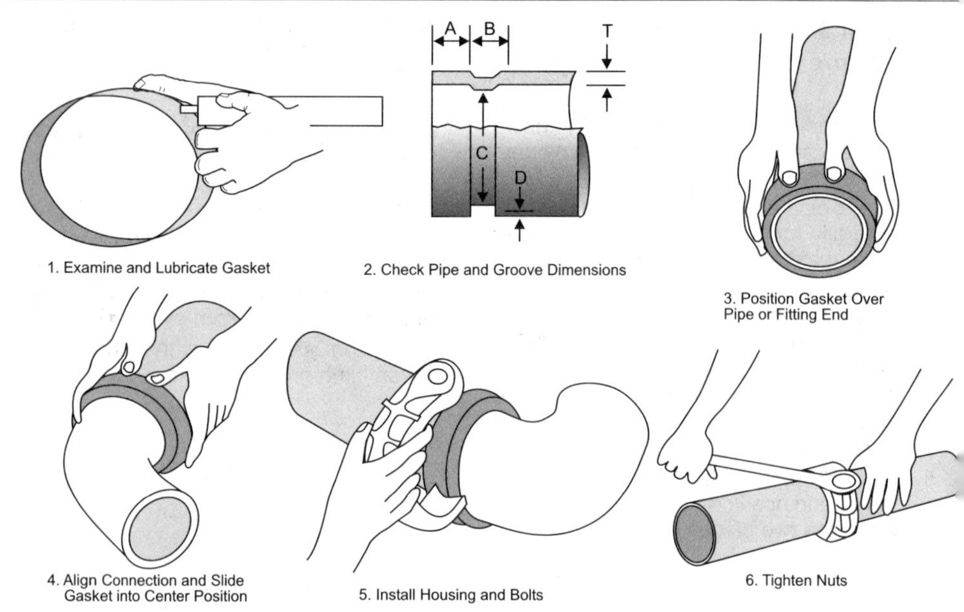

1. Examine and Lubricate Gasket

2. Check Pipe and Groove Dimensions

3. Position Gasket Over Pipe or Fitting End

4. Align Connection and Slide Gasket into Center Position

5. Install Housing and Bolts

6. Tighten Nuts

Illustration #78 - Assembling Grooved Joints

Tube Fittings

The flare-less mechanical grip fitting is the type of tube fitting used most often in modern day industry. These fittings use either a single ferrule or double ferrule arrangement to connect and seal together the tubing and the fitting. Illustration #79 shows a double ferrule flare-less mechanical grip tube fitting and its essential parts placed in proper assembly sequence.

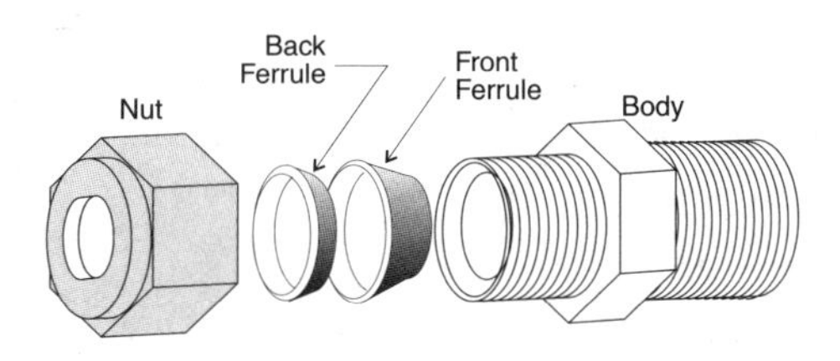

Illustration #79 - Double Ferrule Flare-less Tube Fitting

It is customary when specifying a type of tube fitting to designate the fitting in the following order:

1. Fitting material
2. Tube size in Outside Diameter (OD)
3. Tube thread size (NPT)
4. Fitting standard name

Example:

The stainless steel elbow shown in illustration #80 would be specified as:

Stainless Steel 1/4 inch x 3/8 inch NPT Male Elbow

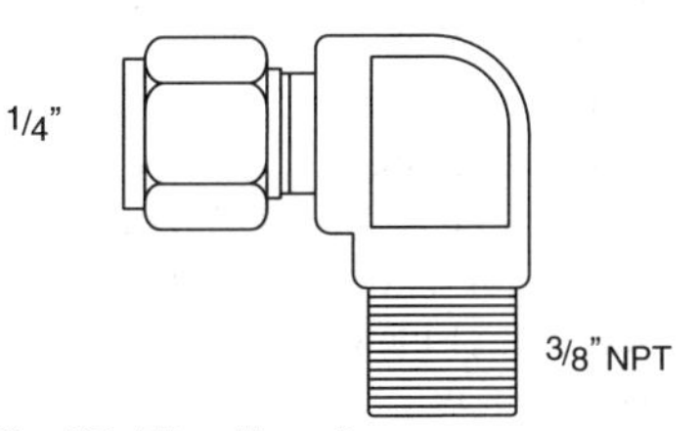

Illustration #80 - Elbow Example

Swagelok Numbering System

Tube fitting manufacturers often use order or catalogue numbers to simplify ordering and fitting identification. For example, Swagelok uses a sequential number code system to identify material, size, series, component type, and end designation for their tube fittings. Table #59 listing of the numbers and letters used in Swagelok's identification code. As an example, the stainless steel 1/4 inch x 3/8 inch NPT male elbow shown in illustration #80 would be specified as follows:

SS - 400 - 2 - 6

SS = Stainless Steel

4 = 4/16" or 1/4" Tube OD Size

0 = Fractional Size 1/16" to 3/8"

0 = Complete Fitting

2 = Male Elbow

6 = 6/16" or 3/8" NPT Male Pipe Size

All numbers are prefaced by a ***Material*** Designator code.

Code	Material
A-	Aluminum
B-	Brass
C20-	Alloy 20
HC-	Alloy C-276
INC-	Alloy 600
M-	Alloy 400/R-405
NY-	Nylon
S-	Steel
SS-	316 Stainless Steel
T-	TFE
TI-	Titanium

Fitting Series Designator identifies the ***Design Size***.

Code	Design Sizes
0	Fractional 1/16" to 3/8" 1 1/4" to 2"
1	Fractional 1/2" to 1"
M	millimetre Tube Sizes
F	Female Swagelok
TI-	Titanium

Size Designator indicates the tubing O.D. in sixteenths of an inch or millimetres.

Code	Tube O.D. (inches)	Code	Tube O.D. (mm)
		-2	2
-1	1/16	-3	3
-2	1/8	-4	4
-3	3/16	-6	5
-4	1/4	-8	8
-5	5/16	-10	10
-6	3/8	-12	12
-8	1/2	-14	14
-10	5/8	-15	15
-12	3/4	-16	16
-14	7/8	-18	18
-16	1	-20	20
		-22	22
		-25	25

Table #59 - Swagelok Fitting Designation

Component Designator identifies the Type of Component(s).	
Code	**Component(s)**
0	Complete Fitting Assembly
1	Body only (such as Port Connector)
2	Nut
3	Front Ferrule
4	Back Ferrule
5	Insert

Illustration #81A,B,C shows common tube fittings, their names, and grouping by the function they are designed to perform in a tubing system.

The ***Type of Fitting*** Designator is then indicated, followed by a dash if a suffix is required.	
Code	**Type of Fitting**
-1-	Male Connector
-2-	Male Elbow - 90°
-3	Tee, Union
-3TTF	Tee, Female Branch
-3TFT	Tee, Female Run
-3TTM	Tee, Male Branch
-3TMT	Tee, Male Run
-3TST	Tee, Positionable Run
-3TTS	Tee, Positionable Branch
-4	Cross, Union
-5-	Male Elbow - 45°
-6	Union
-6-	Reducing Union
-7-	Female Connector
-8-	Female Elbow
-9	Elbow, Union
-11-	Bulkhead Male Connector
-61	Bulkhead Union
-71-	Bulkhead Female Connector
-A-	Adapter
-C-	Cap
-P-_	Plug
-PC	Port Connector
-R	Reducer
-RI	Bulkhead Reducer

A suffix denotes Reduced Size or Type of ***Other End Connection***.	
Code Suffix	**Typical End Connection**
AN	37° Male AN Flare
ANIF	37° Female AN Flare
BT	Bored-Through Fitting
GC	Gas Chromatograph Fitting
F	Female Thread
K	Knurled Nut
KN	Knurled Nut, Nylon Ferrules
KT	Knurled Nut, TFE Ferrules
LV	Low Volume Chromatograph Fitting
M	Metric Tube Ending
OR	O-Seal Connection
RIP	ISO Parallel Pipe Thread
RG	ISO Parallel Pipe Thread (Gage)
RS	ISO Parallel Pipe Thread
RT	ISO Tapered Pipe Thread
ST	StraightThread with O-Ring (for SAE/MS Ports)
W	Weld
ZV	Zero Volume Chromatograph Fitting

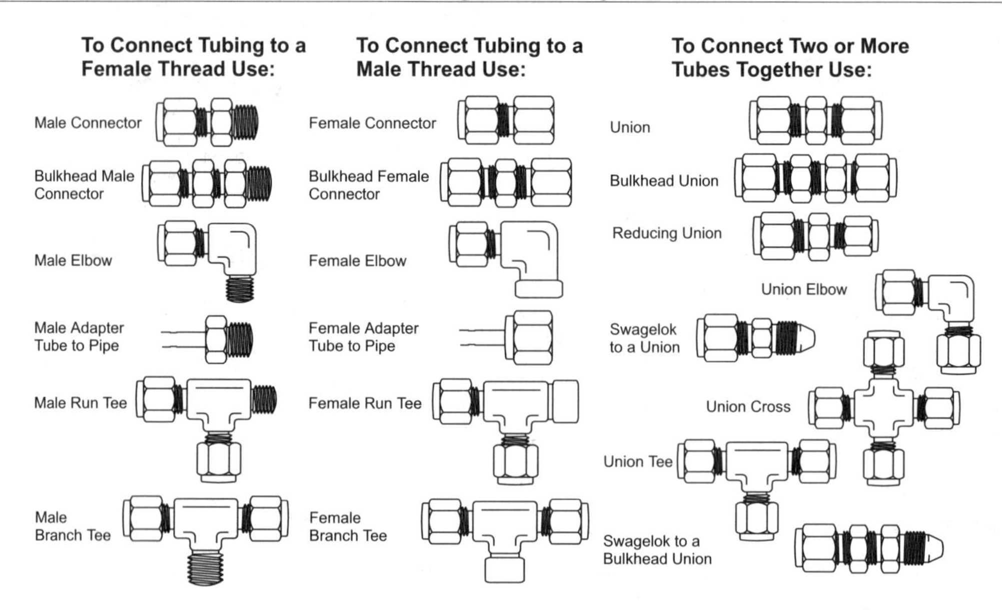

Illustration #81A - Swagelok Fittings

To Connect Two or More Tube Fittings Together Use:

Reducer

Bulkhead Reducer

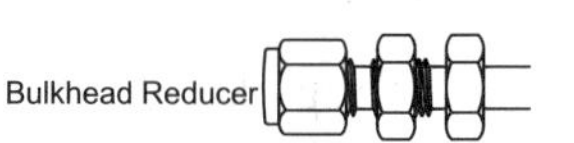

Swagelok to AN Adapter

Port Connector

To Cap a Tube or Plug a Fitting Use:

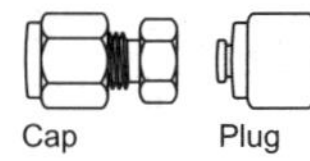

Cap Plug

To Connect Tubing to an All Welded System Use:

Swagelok to Male Pipe Weld Connector

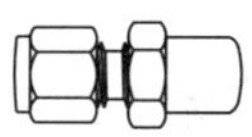

Swagelok to Male Pipe Weld Elbow

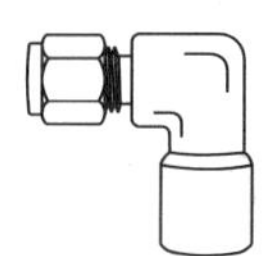

Swagelok to Tube Socket Weld Union

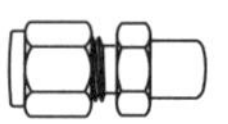

Swagelok to Tube Socket Weld Elbow

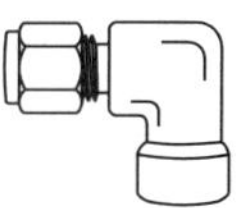

To Connect Tubing to Pipe or Straight Thread Port Using an O-Ring Use:

O-Seal Male Connector to Male Straight Thread

O-Seal Male Connector to Short Male NPT Thread

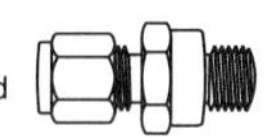

As Spare Parts Use:

Nut

Knurled Nut

Back Ferrule

Front Ferrule

Insert

Ferrule-Pak® Package

Illustration #81B - Swagelok Fittings

To Connect Tubing to SAE/MS Straight Thread Ports Use:

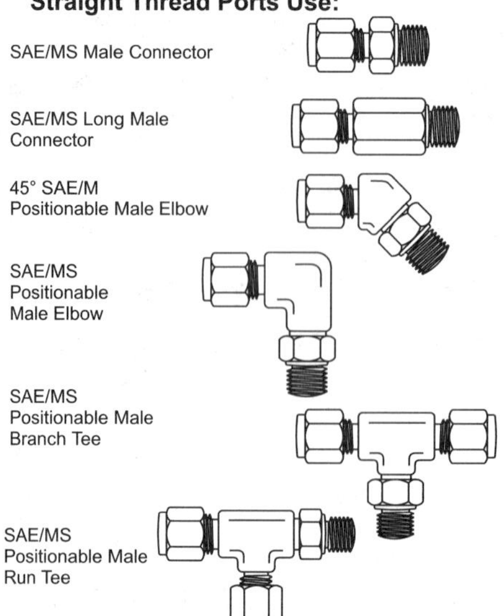

For Special Connections such as Chromatographs, Heat Exchangers or Thermocouples, Use:

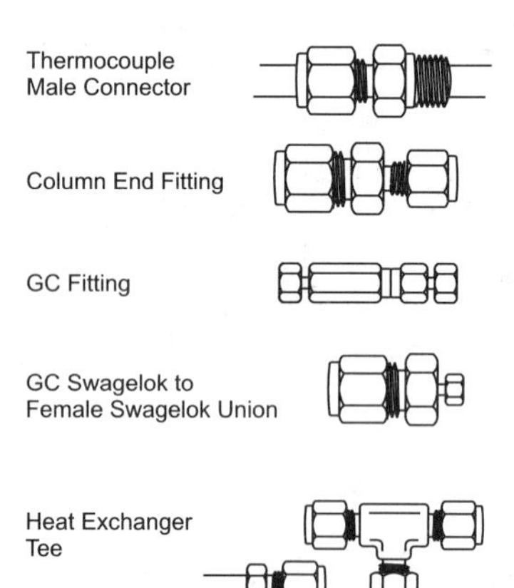

Illustration #81C - Swagelok Fittings

Swagelok Tube Fittings Assembly

Swagelok fittings come completely assembled, finger-tight, and are ready for use. Disassembly before use is unnecessary and could result in dirt getting into the fittings, resulting in leaks. See illustration #82.

Step 1: Insert the tubing into the fitting. Ensure that the tubing rests on the fittings shoulder and the nut is finger tight.

Step 2: Before tightening the nut, scribe the nut at the 6:00 o'clock position.

Step 3: While holding the fitting body with a backup wrench, tighten the nut. For sizes $^{1}/_{16}$, $^{1}/_{8}$, $^{3}/_{16}$ inch (2, 3, 4mm) tighten $^{3}/_{4}$ of a turn from finger tight. For $^{1}/_{4}$ inch and larger sizes, tighten 1 $^{1}/_{4}$ turns.

Note: Swagelok installation and related information reproduced courtesy Swagelok Co.

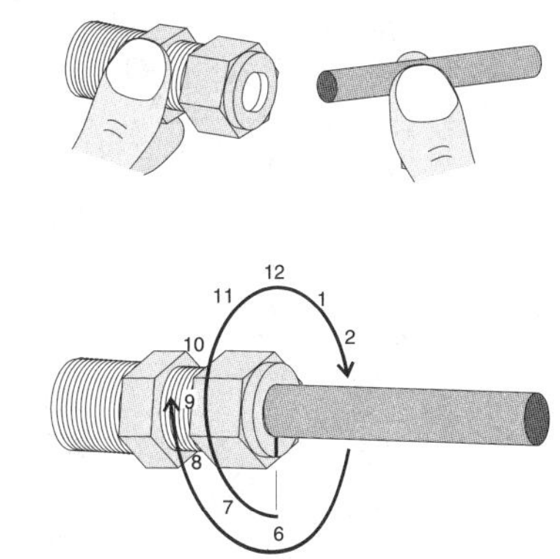

Illustration #82 - Swagelok Installation

ISA Tube Fittings

The following tube fitting abbreviations and symbols are the Instrumentation, Systems, and Automation Society (ISA) recommended practices for the representation of mechanical flared and flare-less tube fittings. See illustration #83.

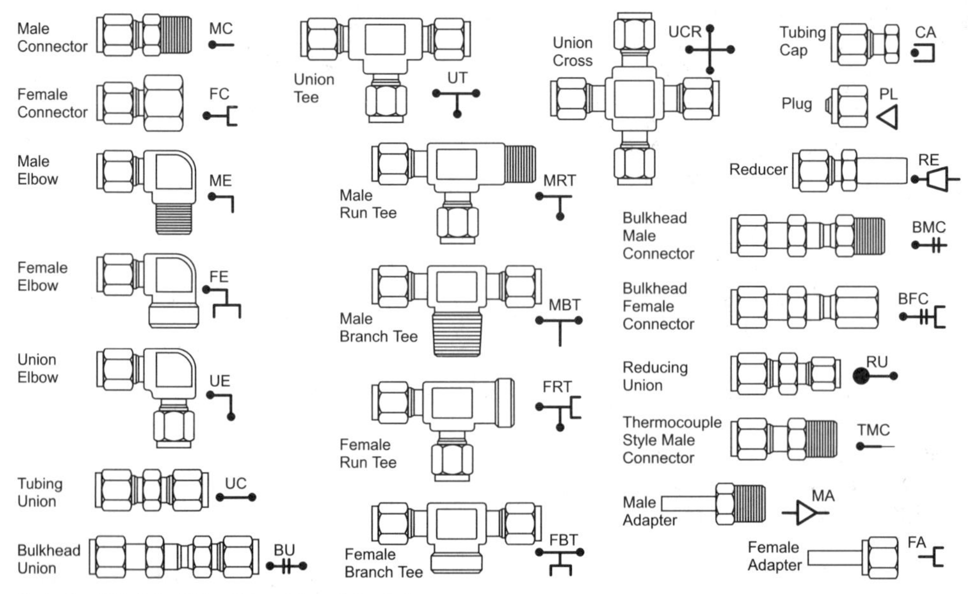

Illustration #83 - Tube Fitting Abbreviations & Symbols

SECTION FIVE

GASKETED JOINTS

Flanges

Flanges connect piping and components together in a system by the use of bolted connections. This type of connection eases the disassembling and separation of piping, and equipment for repair and regular maintenance.

Illustrations #84 through #87 display the different types of flange styles that are available. The following gives a brief description of each.

Welding-neck Flange

This flange type is designed to be connected by butt-welding the protruding neck of the flange to either a fitting, pipe, or equipment requiring a flanged joint. Welding-neck flanges provide good service under a wide range of temperature and pressure conditions in both static or intermittent flows (Illustration #84A).

Slip-on Flange

The slip-on flange is designed to slip over the end of the pipe. It allows easy positioning before welding.

Both the inside and outside of the flange is fillet welded to the pipe. The inside weld is accomplished by pulling the pipe back (approximately the wall thickness of the pipe) from the end of the flange and welding the end of the pipe to the inside of the flange (illustration #84B).

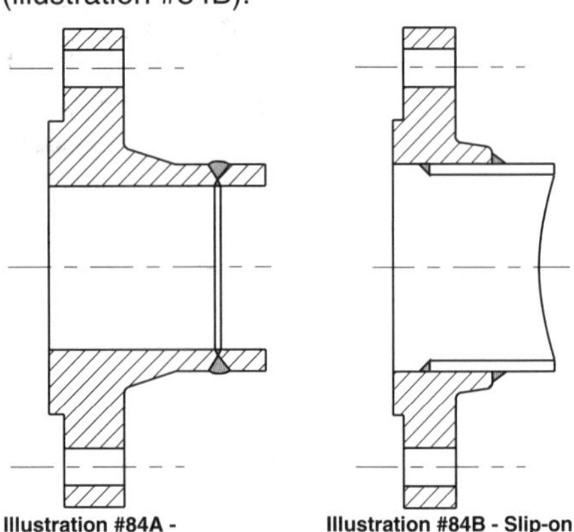

Illustration #84A - Weld-neck Flange

Illustration #84B - Slip-on Flange

Lap-joint (Van Stone Flange)

This flange arrangement consists of both a stub end and a flange. The flange itself is not welded but slips over the stub end which is butt welded to the fitting, pipe or equipment. This arrangement assists flange alignment in conditions where non-alignment may cause problems. Because the flange is not in contact with line fluids, it may be made from less costly carbon steel. The stub end is made from the pipe material (illustration #85A).

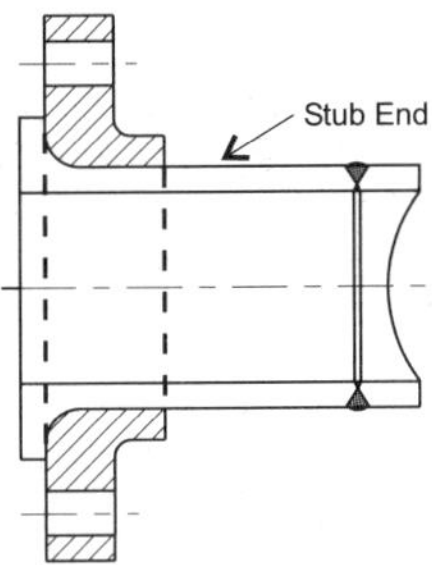

Illustration #85A - Lap-joint (Van Stone Flange)

Reducing Flange

This flange changes the line size without adding an extra fitting. The reducing flange changes the line size abruptly. It is not recommended where flow disturbance or turbulent conditions will cause problems (illustration #85B).

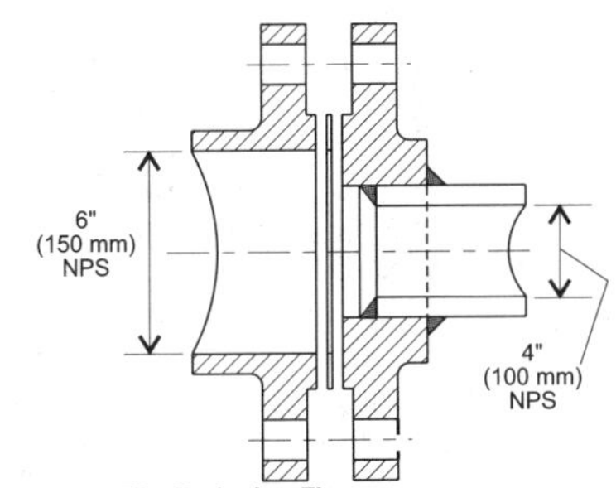

Illustration #85B - Reducing Flange

Socket Welding Flange

This flange is joined to pipe the same as socket welded fittings. It is used primarily on small piping and low pressure applications (illustration #86A).

Blind Flange

The blind flange is a solid flange plate used to close-off the end of pipe, fittings, valves, and/or equipment (illustration #86B).

Spectacle Blinds

Spectacle and line blinds are similar to a blind flange, but differ in that they fit between two flange connections. Spectacle blinds get their name from their similarity to a pair of eyeglasses or spectacles. One side of the spectacle blind is fully closed for 100% flow shut-off, while the other side of the spectacle is open for full flow. ***With this construction, the blind can be rotated without leaving a space when it is taken out of the line. Another advantage of the spectacle construction is that it can be seen at a glance if the line is open or closed off (illustration #86C).***

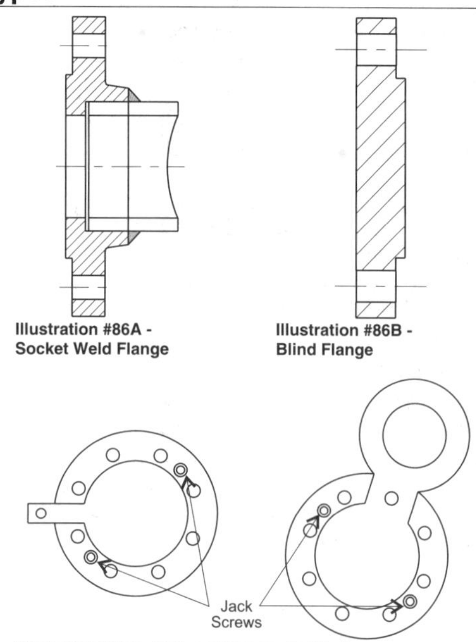

Illustration #86A - Socket Weld Flange

Illustration #86B - Blind Flange

Illustration #86C - Line and Spectacle Blinds

Screwed Flange

The screwed or threaded flange is often used in flanging applications where welding is not practical or desired. It is mostly used in commercial applications on low pressure and small piping (illustration #87A).

Orifice Flanges

Orifice flanges (always used in pairs) are used in conjunction with an orifice plate for measuring flow of liquids and gases within a piping system. They differ from other flanges in that they are pre-drilled with tapped holes made in the flange rims to accommodate metering piping. The flanges that make up the orifice flange arrangement are usually of the welding-neck end connection type. Slip-on and threaded types of end connections are sometimes used, but the pipe must be drilled to accommodate the tapped holes through which the pressure is sensed (illustration #87B).

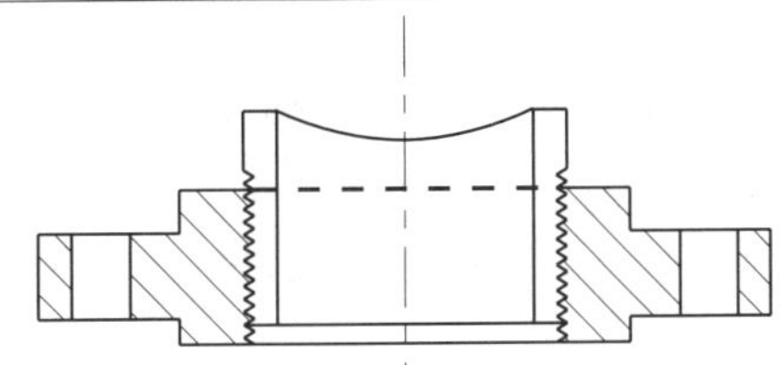

Illustration #87A - Screwed Flange

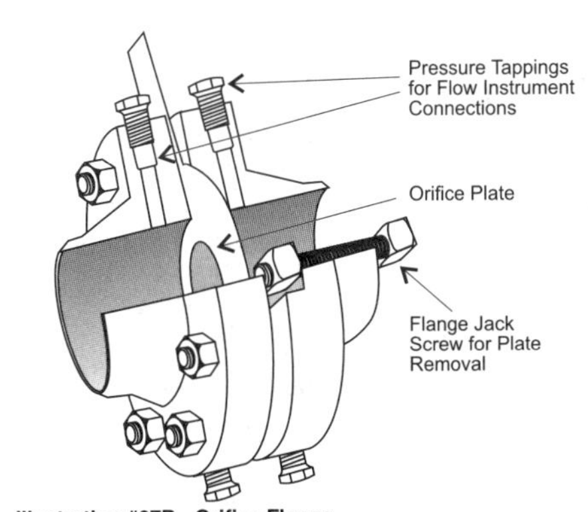

Illustration #87B - Orifice Flange

Flange Faces

The various types of flanges are manufactured with a variety of face types.These various face types are shown in illustrations #88 through #96. Even though there are many face types available, the most commonly used are:

- Raised face
- Flat face
- Lap joint (Van Stone flange)
- Ring joint

The flange face type should not be confused with flange "finish", which indicates the surface contact finish applied to the actual face of the flange. The major types of flange face finishes available are:

1. Smooth finish
2. Serrated finish
   a) serrated concentric grooves
   b) serrated spiral grooves

Note: Both serrated finishes have grooves 1/16 (1.6 mm) deep with 24 to 40 grooves per inch (25.4 mm). Flange face finishes other than these may be furnished as determined by end user.

Raised face:

Used in the majority of flange applications for pressures up to 900 psi (6200 Kpa). Face heights of 1/16" (1.6 mm) are used for flange classes 150, 250 and 300. Higher number flange classes use the 1/4" (6.35 mm) raised face (illustration #88).

Large Male and Female:

This face design provides ample gasket sealing area while still giving good gasket support (illustration #89).

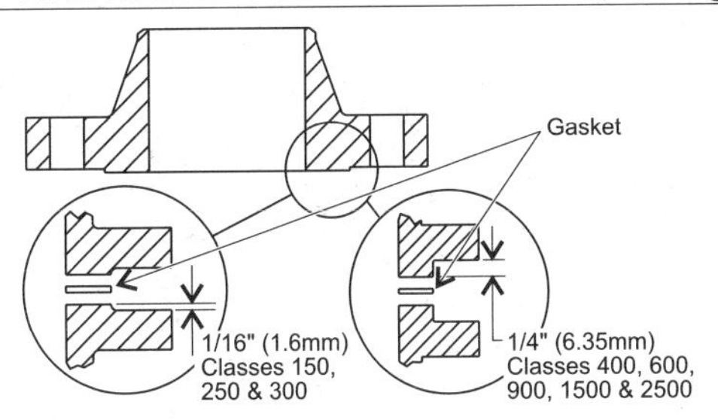

Illustration #88 - Raised Face Flange

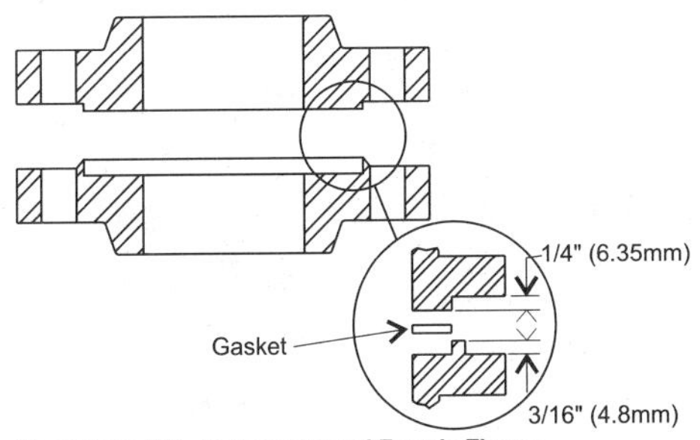

Illustration #89 - Large Male and Female Flange

Large Tongue and Groove:

The small contact sealing area in this design contributes to good gasket compression under low bolt loading. The grooved slot also provides excellent gasket containment under high pressure service (illustration #90).

Flat Face:

Used commonly on cast iron flanges and as mating flanges to pumps and valves in low pressure applications (illustration #91).

Ring Joint:

This flange face design consists of a grooved slot in which a metal ring gasket is used for sealing. The metal ring sealing face makes it an ideal flange face for corrosive and high pressure/temperature applications (illustration #92).

Lap Joint:

Similar face area design to that of the raised face flange, but differs in that the stub end makes up the sealing face (illustration #93).

Small Tongue and Groove:

Same usage as the large version, but the smaller sealing face gives better gasket compression under lower bolt loading (illustration #94).

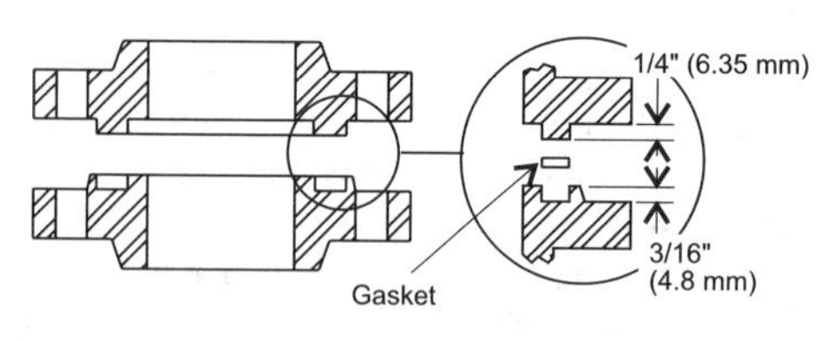

Illustration #90 - Large Tongue and Groove Flange

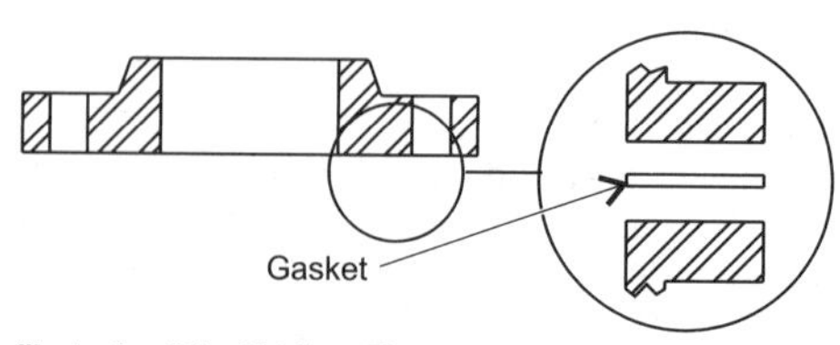

Illustration #91 - Flat Face Flange

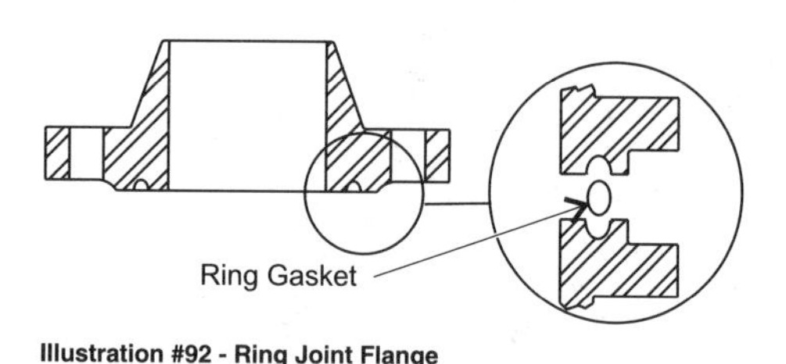

Illustration #92 - Ring Joint Flange

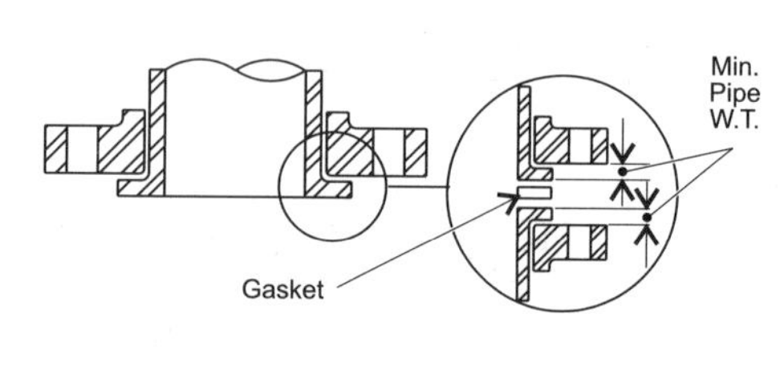

Illustration #93 - Lap-joint (Van Stone) Flange

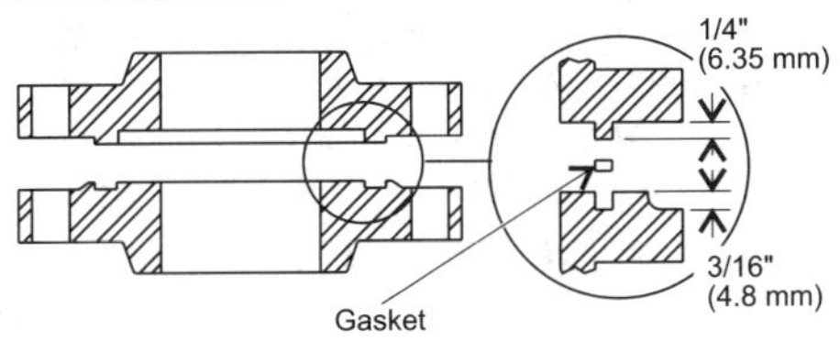

Illustration #94 - Small Tongue and Groove Flange

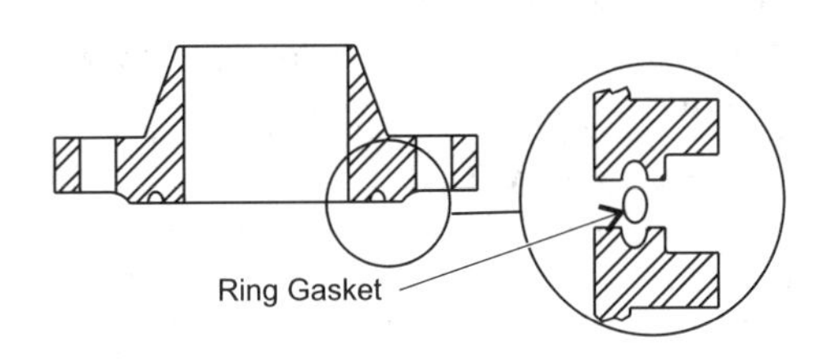

Small Male and Female:

Same general usage as the large version, but gasket is contained in a smaller groove area giving better retention and gasket compression (illustration #95).

Illustration #95 - Small Male and Female Flange

Threaded Pipe End:

The sealing area of the flange face is made by projecting one end of threaded pipe from one flange face into the recessed end of another threaded flange. Care should be taken to ensure that pipe thickness is sufficient to prevent crushing gaskets (Illustration #96).

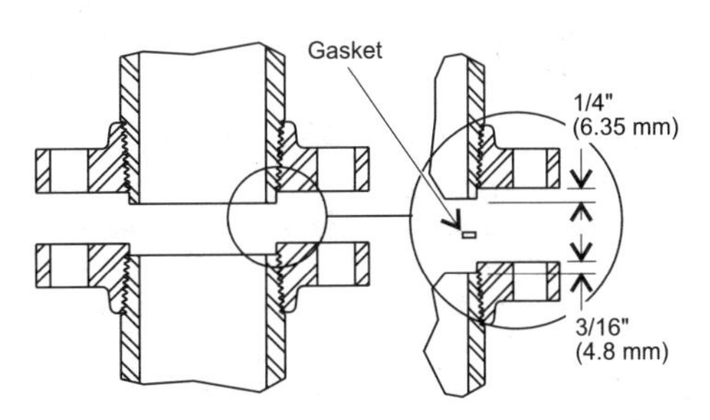

Illustration #96 - Small Male and Female (Threaded)

ASME/ANSI Flanges

Most standard steel and alloy flanges are covered under ASME/ANSI specification B16.5 for flange pressure classes of: 150, 300, 400, 600, 900,1500, and 2500.

Cast iron flange classes and specifications are described under ASME/ANSI B16.1 standard. These flanges (cast iron) are available in classes: 25, 125, 250, and 800. Cast iron pipe flanges are usually threaded connections; with class 25 and 125 having flat faces, and class 250 and 800 having raised faces of 1/16 inch and 1/4 inch (1.6 mm and 6.35 mm) respectively.

The American Petroleum Institute (API) also designates flange standards. Flanges manufactured to API standard are used primarily for the oil industry. They are rated for higher pressure applications and are usually used with high strength API tubular products.

Even though the dimensions of ASME/ANSI and API flanges are similar, they should not be interconnected because of the alternate pressure ratings.

Flange Markings

ASME/ANSI standards require flanges to be stamped or marked with the following information:

- Manufacturer's name or trademark.
- Nominal pipe size.
- Rating designation.
- Material designation.
- Melt code identification.
- Ring joint groove number (when applicable).

A typical example of flange marking is given in illustration #97.

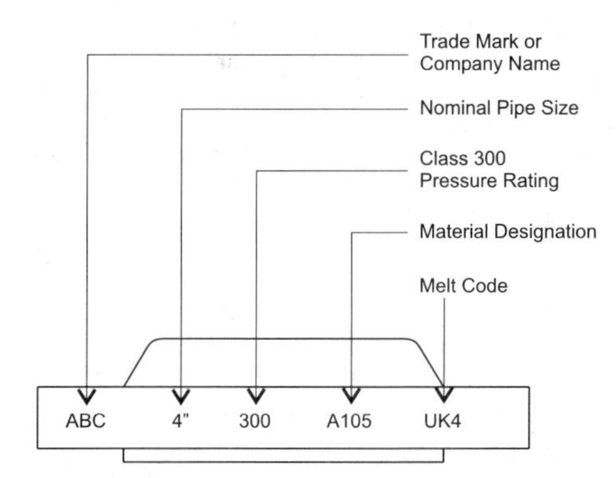

Illustration #97 - Flange Identification and Marking Example

Specific pipe flange and bolt dimensions for cast iron and steel flanges corresponding to ASME/ANSI standards are given in table #59A-G.

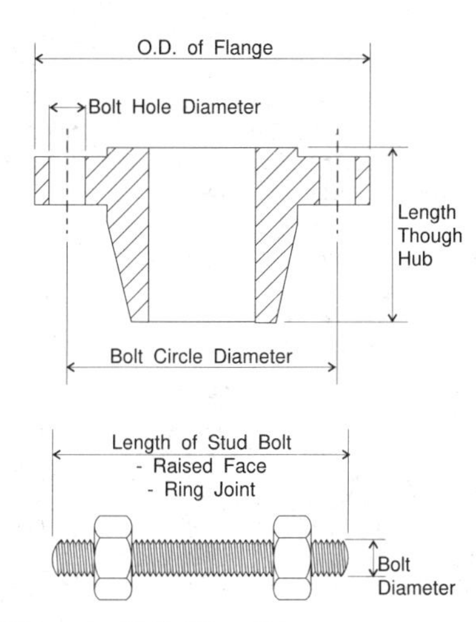

NOTE: see page 234 for Flange Notes

Table #59A - #59G

FLANGE AND BOLT DIMENSIONS - CLASS 150 STEEL AND CLASS 125 CAST IRON																				
Nominal Pipe Size	in	1/2	3/4	1	1 1/4	1 1/2	2	2 1/2	3	4	5	6	8	10	12	14	16	18	20	24
	mm	15	20	25	32	40	50	65	80	100	125	150	200	250	300	350	400	450	500	600
O.D. Flange	in	3.50	3.88	4.25	4.62	5.00	6.00	7.00	7.50	9.00	10.00	11.00	13.50	16.00	19.00	21.00	23.50	25.00	27.50	32.00
	mm	88.9	98.6	108.	117.	127.	152.	178.	191.	228.6	254.0	279.4	342.9	406.4	482.6	533.4	596.9	635.0	698.5	812.8
Bolt Circle Diameter	in	2.38	2.75	3.12	3.50	3.88	4.75	5.50	6.00	7.50	8.50	9.50	11.75	14.25	17.00	18.75	21.25	22.75	25.00	29.50
	mm	60.5	69.9	79.3	88.9	98.6	120.	139.	152.	190.5	215.9	241.3	298.5	362.0	431.8	476.3	539.8	577.9	635.0	749.3
# of Bolts		4	4	4	4	4	4	4	4	8	8	8	8	12	12	12	16	16	20	20
Bolt Diameter	in	1/2	1/2	1/2	1/2	1/2	5/8	5/8	5/8	5/8	3/4	3/4	3/4	7/8	7/8	1	1	1 1/8	1 1/8	1 1/4
	mm	12.7	12.7	12.7	12.7	12.7	16.0	16.0	16.0	16.0	20.0	20.0	20.0	23.0	23.0	25.0	25.0	29.0	29.0	32.0
Bolt Length Raised Face	in	2.25	2.50	2.50	2.75	2.75	3.25	3.50	3.50	3.50	3.75	4.00	4.25	4.50	4.75	5.25	5.25	5.75	6.25	6.75
	mm	57.2	63.5	63.5	69.9	69.9	82.6	88.9	88.9	88.9	95.3	101.6	108.0	114.3	120.7	133.4	133.4	146.0	158.8	171.5
Bolt Length Ring Joint	in	n/a	n/a	3.00	3.25	3.25	3.75	4.00	4.00	4.00	4.25	4.50	4.75	5.00	5.25	5.75	5.75	6.25	6.75	7.25
	mm	n/a	n/a	76.2	82.6	82.6	95.3	101.	101.	101.	108.0	114.3	120.7	127.0	133.4	146.1	146.1	158.8	171.5	184.2
Bolt Hole Diameter	in	0.62	0.62	0.62	0.62	0.62	0.75	0.75	0.75	0.75	0.88	0.88	0.88	1.00	1.00	1.12	1.12	1.25	1.25	1.38
	mm	16.0	16.0	16.0	16.0	16.0	19.0	19.0	19.0	19.0	22.0	22.0	22.0	26.0	26.0	29.0	29.0	32.0	32.0	35.0
Welding Neck Length	in	1.88	2.06	2.19	2.25	2.44	2.50	2.75	2.75	3.00	3.50	3.50	4.00	4.00	4.50	5.00	5.00	5.50	5.69	6.00
	mm	47.8	52.3	55.6	57.2	62.0	63.5	69.9	69.9	76.2	88.9	88.9	101.6	101.6	114.3	127.0	127.0	139.7	144.5	152.4
Slip-on & Socket Length	in	0.62	0.62	0.69	0.81	0.88	1.00	1.12	1.19	1.31	1.44	1.56	1.75	1.94	2.19	2.25	2.50	2.69	2.88	3.25
	mm	15.8	15.8	17.5	20.6	22.4	25.4	28.5	30.2	33.3	36.6	39.6	44.5	49.3	55.6	57.2	63.5	68.3	73.2	82.6

NOTE: Shaded boxes do not show the first decimal place because of space constraints.

Table #59A - Flange and Bolt Dimensions (Class 150/125)

FLANGE AND BOLT DIMENSIONS - CLASS 300 STEEL OR 250 CAST IRON FLANGES																				
Nominal Pipe Size	in	1/2	3/4	1	1 1/4	1 1/2	2	2 1/2	3	4	5	6	8	10	12	14	16	18	20	24
	mm	15	20	25	32	40	50	65	80	100	125	150	200	250	300	350	400	450	500	600
O.D. Flange	in	3.75	4.62	4.88	5.25	6.12	6.50	7.50	8.25	10.00	11.00	12.50	15.00	17.50	20.50	23.00	25.50	28.00	30.50	36.00
	mm	95.3	117.	124.	133.	155.	165.	190.	109.	254.0	279.4	317.5	381.0	444.5	520.7	584.2	647.7	711.2	774.7	914.4
Bolt Circle Diameter	in	2.62	3.25	3.50	3.88	4.50	5.00	5.88	6.62	7.88	9.25	10.62	13.00	15.25	17.75	20.25	22.50	24.75	27.00	32.00
	mm	66.5	82.6	88.9	98.6	114.	127.	149.	168.	200.2	235.0	269.7	330.2	387.4	450.9	514.4	571.5	628.7	685.8	812.8
# of Bolts		4	4	4	4	4	8	8	8	8	8	12	12	16	16	20	20	24	24	24
Bolt Diameter	in	1/2	5/8	5/8	5/8	3/4	5/8	3/4	3/4	3/4	3/4	3/4	7/8	1	1 1/8	1 1/8	1 1/4	1 1/4	1 1/4	1 1/2
	mm	12.7	15.9	15.9	15.9	19.1	16.0	20.0	20.0	20.0	20.0	20.0	23.0	25.0	29.0	29.0	32.0	32.0	32.0	38.0
Bolt Length Raised Face	in	2.50	3.00	3.00	3.25	3.50	3.50	4.00	4.25	4.50	4.75	4.75	5.50	6.25	6.75	7.00	7.50	7.75	8.00	9.00
	mm	63.5	76.2	76.2	82.6	88.9	90.0	101.	108.	114.3	120.7	120.7	140.0	158.8	171.5	177.8	190.5	196.9	203.2	228.6
Bolt Length Ring Joint	in	3.00	3.50	3.50	3.75	4.00	4.00	4.50	4.75	5.00	5.25	5.50	6.00	6.75	7.25	7.50	8.00	8.25	8.75	10.00
	mm	76.2	88.9	88.9	95.3	101.	101.	114.	120.	127.0	133.4	140.0	152.4	171.5	184.2	190.5	203.2	209.6	222.3	254.0
Bolt Hole Diameter	in	0.62	0.75	0.75	0.75	0.88	0.75	0.88	0.88	0.88	0.88	0.88	1.00	1.12	1.25	1.25	1.38	1.38	1.38	1.62
	mm	16.0	19.0	19.0	19.0	22.0	19.0	22.0	22.0	22.0	22.0	22.0	26.0	29.0	32.0	32.0	35.0	35.0	35.0	42.0
Welding Neck Length	in	2.06	2.25	2.44	2.56	2.69	2.75	3.00	3.12	3.38	3.88	3.88	4.38	4.62	5.12	5.62	5.75	6.25	6.38	6.62
	mm	52.3	57.2	62.0	65.0	68.3	69.9	76.2	79.3	85.9	98.6	98.6	111.3	117.4	130.1	142.8	146.1	158.8	162.1	168.2
Slip-on & Socket Length	in	0.88	1.00	1.06	1.06	1.19	1.31	1.50	1.69	1.88	2.00	2.06	2.44	2.62	2.88	3.00	3.25	3.50	3.75	4.19
	mm	22.4	25.4	27.0	27.0	30.2	33.3	38.1	43.0	47.8	50.8	52.4	62.0	66.6	73.2	76.2	82.6	88.9	95.3	106.4
NOTE: Shaded boxes do not show the first decimal place because of space constraints.																				

Table #59B - Flange and Bolt Dimensions (Class 300/250)

FLANGE AND BOLT DIMENSIONS - CLASS 400 STEEL AND ALLOY FLANGES																				
Nominal Pipe Size	in	1/2	3/4	1	1 1/4	1 1/2	2	2 1/2	3	4	5	6	8	10	12	14	16	18	20	24
	mm	15	20	25	32	40	50	65	80	100	125	150	200	250	300	350	400	450	500	600
O.D. Flange	in	3.75	4.62	4.88	525	6.12	6.50	7.50	8.25	10.0	11.0	12.5	15.00	17.50	20.50	23.00	25.50	28.00	30.50	36.00
	mm	95.3	117.	124.	133.	155.	165.	190.	209.	254.0	279.4	317.5	381.0	444.5	520.7	584.2	647.7	711.2	774.7	914.4
Bolt Circle Diameter	in	2.62	3.25	3.50	3.88	4.50	5.50	5.88	6.62	7.88	9.25	10.62	13.00	15.25	17.75	20.25	22.50	24.75	27.00	32.00
	mm	66.5	82.6	88.9	98.6	114.	127.	149.	168.	200.2	235.0	269.7	330.2	387.4	450.9	514.4	571.5	628.7	685.8	812.8
# of Bolts		4	4	4	4	4	8	8	8	8	8	12	12	16	16	20	20	24	24	24
Bolt Diameter	in	1/2	5/8	5/8	5/8	3/4	5/8	3/4	3/4	7/8	7/8	7/8	1	1 1/8	1 1/4	1 1/4	1 3/8	1 3/8	1 1/2	1 3/4
	mm	12.7	15.9	15.9	15.9	19.1	16.0	20.0	20.0	23.0	23.0	23.0	25.0	29.0	32.0	32.0	35.0	35.0	38.0	45.0
Bolt Length Raised Face	in	3.00	3.50	3.50	3.75	4.25	4.25	4.75	5.00	5.50	5.75	6.00	6.75	7.50	8.00	8.25	8.75	9.00	9.50	10.50
	mm	76.2	88.9	88.9	95.3	108.	108.	120.	127.	139.7	146.0	152.4	171.5	190.5	203.2	209.6	222.3	228.6	241.3	266.7
Bolt Length Ring Joint	in	3.00	3.50	3.50	3.75	4.25	4.25	4.75	5.00	5.50	5.75	6.00	6.75	7.50	8.00	8.25	8.75	9.00	9.75	11.00
	mm	76.2	88.9	88.9	95.3	108.	108.	120.	127.	139.7	146.0	152.4	171.5	190.5	203.2	209.6	222.3	228.6	247.7	279.4
Bolt Hole Diameter	in	0.62	0.75	0.75	0.75	0.88	0.75	0.88	0.88	1.00	1.00	1.00	1.12	1.25	1.38	1.38	1.50	1.50	1.62	1.88
	mm	16.0	19.0	19.0	19.0	22.0	19.0	22.0	22.0	26.0	26.0	26.0	29.0	32.0	35.0	35.0	38.0	38.0	42.0	47.8
Welding Neck Length	in	2.31	2.50	2.69	2.87	3.00	3.13	3.37	3.50	3.75	4.25	4.31	4.87	5.13	5.63	6.13	6.25	6.75	6.87	7.13
	mm	58.7	63.5	68.3	72.9	76.2	79.5	85.6	88.9	95.3	108.0	109.5	123.7	130.3	143.0	155.7	158.8	171.5	174.5	181.1
Slip-on & Socket Length	in	1.13	1.25	1.31	1.37	1.50	1.69	1.87	2.06	2.25	2.37	2.50	2.94	3.13	3.37	3.56	3.94	4.13	4.25	4.75
	mm	28.7	31.8	33.3	34.8	38.1	42.9	47.5	52.3	57.2	60.2	63.5	74.7	79.5	85.6	90.4	100.1	104.9	108.0	120.7
NOTE: Shaded boxes do not show the first decimal place because of space constraints.																				

Table #59C - Flange and Bolt Dimensions (Class 400)

FLANGE AND BOLT DIMENSIONS - CLASS 600 STEEL AND ALLOY FLANGES																				
Nominal Pipe Size	in	1/2	3/4	1	1 1/4	1 1/2	2	2 1/2	3	4	5	6	8	10	12	14	16	18	20	24
	mm	15	20	25	32	40	50	65	80	100	125	150	200	250	300	350	400	450	500	600
O.D. Flange	in	3.75	4.62	4.88	5.25	6.12	6.50	7.50	8.25	10.75	13.00	14.00	16.50	20.00	22.00	23.75	27.00	29.25	32.00	37.00
	mm	95.3	117.	124.	133.	155.	165.	190.	209.	273.0	330.2	355.6	419.0	508.0	558.8	603.3	685.8	743.0	812.8	939.8
Bolt Circle Diameter	in	2.62	3.25	3.50	3.88	4.50	5.00	5.88	6.62	8.50	10.50	11.50	13.75	17.00	19.25	20.75	23.75	25.75	28.50	33.00
	mm	66.5	82.6	88.9	98.6	114.	127.	149.	168.	215.9	266.7	292.1	349.3	431.8	489.0	527.1	603.3	654.1	724.0	838.2
# of Bolts		4	4	4	4	4	8	8	8	8	8	12	12	16	20	20	20	20	24	24
Bolt Diameter	in	1/2	5/8	5/8	5/8	3/4	5/8	3/4	3/4	7/8	1	1	1 1/8	1 1/4	1 1/4	1 3/8	1 1/2	1 5/8	1 5/8	1 7/8
	mm	12.7	15.9	15.9	15.9	19.1	16.0	20.0	20.0	22.0	25.0	25.0	29.0	32.0	32.0	35.0	38.0	42.0	42.0	48.0
Bolt Length Raised Face	in	3.00	3.50	3.50	3.75	4.25	4.25	4.75	5.00	5.75	6.50	6.75	7.50	8.50	8.75	9.25	10.00	10.75	11.25	13.00
	mm	76.2	88.9	88.9	95.3	108.	108.	120.	127.	146.0	165.0	171.5	190.5	216.0	222.3	235.0	254.0	273.0	285.8	330.2
Bolt Length Ring Joint	in	3.00	3.50	3.50	3.75	4.25	4.25	4.75	5.00	5.75	6.50	6.75	7.75	8.50	8.75	9.25	10.00	10.75	11.50	13.25
	mm	76.2	88.9	88.9	95.3	108.	108.	120.	127.	146.0	165.0	171.5	196.9	216.0	222.3	235.0	254.0	273.0	292.0	336.6
Bolt Hole Diameter	in	0.62	0.75	0.75	0.75	0.88	0.75	0.88	0.88	1.00	1.12	1.12	1.25	1.38	1.38	1.50	1.62	1.75	1.75	2.00
	mm	16.0	19.0	19.0	19.0	22.0	19.0	22.0	22.0	26.0	29.0	29.0	32.0	35.0	35.0	38.0	42.0	45.0	45.0	51.0
Welding Neck Length	in	2.31	2.50	2.69	2.87	3.00	3.13	3.37	3.50	4.25	4.75	4.87	5.50	6.25	6.37	6.75	7.25	7.50	7.75	8.25
	mm	58.7	63.5	68.3	72.9	76.2	79.5	85.6	88.9	108.0	120.7	123.7	139.7	158.8	161.8	171.5	184.2	190.5	196.9	209.6
Slip-on & Socket Length	in	1.13	1.25	1.31	1.37	1.50	1.69	1.87	2.06	2.37	2.63	2.87	3.25	3.63	3.87	3.94	4.44	4.87	5.25	5.75
	mm	28.7	31.8	33.3	34.8	38.1	42.9	47.5	52.3	60.2	66.8	72.9	82.5	92.2	98.3	100.1	112.8	123.7	133.4	146.1
NOTE: Shaded boxes do not show the first decimal place because of space constraints.																				

Table #59D - Flange and Bolt Dimensions (Class 600)

FLANGE AND BOLT DIMENSIONS - CLASS 900 STEEL AND ALLOY FLANGES																				
Nominal Pipe Size	in	$\frac{1}{2}$	$\frac{3}{4}$	1	$1\frac{1}{4}$	$1\frac{1}{2}$	2	$2\frac{1}{2}$	3	4	5	6	8	10	12	14	16	18	20	24
	mm	15	20	25	32	40	50	65	80	100	125	150	200	250	300	350	400	450	500	600
O.D. Flange	in	4.75	5.12	5.88	6.25	7.00	8.50	9.62	9.50	11.50	13.75	15.00	18.50	21.50	24.00	25.25	27.75	31.00	33.75	41.00
	mm	120.	130.	149.	158.	177.	215.	244.	241.	292.1	349.3	381.0	469.9	546.1	609.6	641.4	704.9	787.4	857.3	1041
Bolt Circle Diameter	in	3.25	3.50	4.00	4.38	4.88	6.50	7.50	7.50	9.25	11.00	12.50	15.50	18.50	21.00	22.00	24.25	27.00	29.50	35.50
	mm	82.6	88.9	101.	111.	124.	165.	190.	190.	235.0	279.4	317.5	393.7	469.9	533.4	558.8	616.0	685.8	749.3	901.7
# of Bolts		4	4	4	4	4	8	8	8	8	8	12	12	16	20	20	20	20	20	20
Bolt Diameter	in	$\frac{3}{4}$	$\frac{3}{4}$	$\frac{7}{8}$	$\frac{7}{8}$	1	$\frac{7}{8}$	1	$\frac{7}{8}$	$1\frac{1}{8}$	$1\frac{1}{4}$	$1\frac{1}{8}$	$1\frac{3}{8}$	$1\frac{3}{8}$	$1\frac{3}{8}$	$1\frac{1}{2}$	$1\frac{5}{8}$	$1\frac{7}{8}$	2	$2\frac{1}{2}$
	mm	19.1	19.1	22.2	22.2	25.4	23.0	25.0	23.0	29.0	32.0	29.0	35.0	35.0	35.0	38.0	42.0	48.0	50.0	64.0
Bolt Length Raised Face	in	4.25	4.50	5.00	5.00	5.50	5.75	6.25	5.75	6.75	7.50	7.50	8.75	9.25	10.00	10.75	11.25	12.75	13.75	17.25
	mm	108.	114.	127.	127.	139.	146.	158.	146.	171.5	190.5	190.5	222.3	235.0	254.0	273.0	285.8	323.9	349.3	438.2
Bolt Length Ring Joint	in	4.25	4.50	5.00	5.00	5.50	5.75	6.25	5.75	6.75	7.50	7.75	8.75	9.25	10.00	11.00	11.50	13.25	14.25	18.00
	mm	108.	114.	127.	127.	139.	146.	158.	146.	171.5	190.5	196.9	222.3	235.0	254.0	279.4	292.0	336.6	362.0	457.2
Bolt Hole Diameter	in	0.88	0.88	1.00	1.00	1.12	1.00	1.12	1.00	1.25	1.38	1.25	1.50	1.50	1.50	162	1.75	2.00	2.12	2.62
	mm	22.0	22.0	26.0	26.0	29.0	26.0	29.0	26.0	32.0	35.0	32.0	38.0	38.0	38.0	42.0	45.0	52.0	54.0	67.0
Welding Neck Length	in	2.63	3.00	3.13	3.13	3.50	4.25	4.37	4.25	4.75	5.25	5.75	6.63	7.50	8.13	8.63	8.75	9.25	10.00	11.75
	mm	66.8	76.2	79.5	79.5	88.9	108.	111.	108.	120.7	133.4	146.1	168.4	190.5	206.5	219.2	222.3	235.0	254.0	298.5
Slip-on & Socket Length	in	1.50	1.63	1.87	1.87	2.00	2.50	2.75	2.37	3.00	3.37	3.63	4.25	4.50	4.87	5.37	5.50	6.25	6.50	8.25
	mm	38.1	41.4	47.5	47.5	50.8	63.5	69.9	60.2	76.2	85.6	92.2	108.0	114.3	123.7	136.4	139.7	158.8	165.1	209.6

NOTE: Shaded boxes do not show the first decimal place because of space constraints.

Table #59E - Flange and Bolt Dimensions (Class 900)

FLANGE AND BOLT DIMENSIONS - CLASS 1500 STEEL AND ALLOY FLANGES																				
Nominal Pipe Size	in	1/2	3/4	1	1 1/4	1 1/2	2	2 1/2	3	4	5	6	8	10	12	14	16	18	20	24
	mm	15	20	25	32	40	50	65	80	100	125	150	200	250	300	350	400	450	500	600
O.D. Flange	in	4.75	5.12	5.88	6.25	7.00	8.50	9.62	10.5	12.25	14.75	15.50	19.00	23.00	26.50	29.50	32.50	36.00	38.75	46.00
	mm	120.	130.	149.	158.	177.	215.	244.	266.	311.2	374.7	393.7	482.6	584.2	673.1	749.3	825.5	914.4	984.3	1168
Bolt Circle Diameter	in	3.25	3.50	4.00	4.38	4.88	6.50	7.50	8.00	9.50	11.50	12.50	15.50	19.00	22.50	25.00	27.75	30.50	32.75	39.00
	mm	82.6	88.9	101.	111.	124.	165.	190.	203.	241.3	292.1	317.5	393.7	482.6	571.5	635.0	704.9	774.7	831.9	990.6
# of Bolts		4	4	4	4	4	8	8	8	8	8	12	12	12	16	16	16	16	16	16
Bolt Diameter	in	3/4	3/4	7/8	7/8	1	7/8	1	1 1/8	1 1/4	1 1/2	1 3/8	1 5/8	1 7/8	2	2 1/4	2 1/2	2 3/4	3	3 1/2
	mm	19.1	19.1	22.2	22.2	25.4	23.0	25.0	29.0	32.0	38.0	35.0	42.0	48.0	50.0	54.0	64.0	70.0	76.0	89.0
Bolt Length Raised Face	in	4.25	4.50	5.00	5.00	5.50	5.75	6.25	7.00	7.75	9.75	10.25	11.50	13.25	14.75	16.00	17.50	19.50	21.25	24.25
	mm	108.	114.	127.	127.	139.	146.	158.	177.	196.9	247.7	260.4	292.0	336.6	374.7	406.4	444.5	495.3	539.8	616.0
Bolt Length Ring Joint	in	4.25	4.50	5.00	5.00	5.50	5.75	6.25	7.00	7.75	9.75	10.50	12.75	13.50	15.25	16.75	18.50	20.75	22.25	25.50
	mm	108.	114.	127.	127.	139.	146.	158.	177.	196.9	247.7	266.7	323.9	342.9	387.4	425.5	469.9	527.1	565.2	647.7
Bolt Hole Diameter	in	0.88	0.88	1.00	1.00	1.12	1.00	1.12	1.25	1.38	1.62	1.50	1.75	2.00	2.12	2.38	2.62	2.88	3.12	3.62
	mm	22.0	22.0	26.0	26.0	29.0	26.0	29.0	32.0	35.0	42.0	38.0	45.0	52.0	54.0	61.0	67.0	74.0	80.0	92.0
Welding Neck Length	in	2.63	3.00	3.13	3.13	3.50	4.25	4.37	4.87	5.13	6.37	7.00	8.63	10.25	11.37	12.00	12.50	13.13	14.25	16.25
	mm	66.8	76.2	79.5	79.5	88.9	108.	111.	123.	130.3	161.8	177.8	219.2	260.4	288.8	304.8	317.5	333.5	362.0	412.8
Slip-on & Socket Length	in	1.50	1.63	1.87	1.87	2.00	2.50	2.75	3.13	3.81	4.37	4.94	5.87	6.50	7.37	n/a	n/a	n/a	n/a	n/a
	mm	38.1	41.4	47.5	47.5	50.8	63.5	69.9	79.5	96.8	111.0	125.5	149.1	165.1	187.2	n/a	n/a	n/a	n/a	n/a

NOTE: Shaded boxes do not show the first decimal place because of space constraints.

Table #59F - Flange and Bolt Dimensions (Class 1500)

FLANGE AND BOLT DIMENSIONS - CLASS 2500 STEEL AND ALLOY FLANGES

Nominal Pipe Size	in	1/2	3/4	1	1 1/4	1 1/2	2	2 1/2	3	4	5	6	8	10	12
	mm	15	20	25	32	40	50	65	80	100	125	150	200	250	300
O.D. Flange	in	5.25	5.50	6.25	7.25	8.00	9.25	10.50	12.00	14.00	16.50	19.00	21.75	26.50	30.00
	mm	133.4	139.7	158.8	184.2	203.2	235.0	266.7	304.8	355.6	419.1	482.6	552.5	673.1	762.0
Bolt Circle Diameter	in	3.50	3.75	4.25	5.12	5.75	6.75	7.75	9.00	10.75	12.75	14.50	17.25	21.25	24.38
	mm	88.9	95.3	108.	130.0	146.1	171.5	196.9	228.6	273.1	323.9	368.3	438.2	539.8	619.3
# of Bolts		4	4	4	4	4	8	8	8	8	8	8	12	12	12
Bolt Diameter	in	3/4	3/4	7/8	1	1 1/8	1	1 1/8	1 1/4	1 1/2	1 3/4	2	2	2 1/2	2 3/4
	mm	19.1	19.1	22.2	25.4	28.6	25.0	29.0	32.0	38.0	45.0	50.0	50.0	64.0	70.0
Bolt Length Raised Face	in	4.75	5.00	5.50	6.00	6.75	7.00	7.75	8.75	10.00	11.75	13.50	15.00	19.25	21.25
	mm	120.7	127.0	139.7	152.4	171.5	177.8	196.9	222.3	254.0	298.5	342.9	381.0	489.0	539.8
Bolt Length Ring Joint	in	4.75	5.00	5.50	6.00	6.75	7.00	8.00	9.00	10.25	12.25	14.00	15.50	20.00	22.00
	mm	120.7	127.0	139.7	152.4	171.5	177.8	203.2	228.6	260.4	311.2	355.6	393.7	508.0	558.8
Bolt Hole Diameter	in	0.88	0.88	1.00	1.12	1.25	1.12	1.25	1.38	1.62	1.88	2.12	2.12	2.62	2.88
	mm	22.0	22.0	26.0	29.0	32.0	29.0	32.0	35.0	42.0	48.0	54.0	54.0	67.0	74.0
Welding Neck Length	in	3.13	3.37	3.75	4.00	4.63	5.25	5.87	6.87	7.75	9.25	11.00	12.75	16.75	18.50
	mm	79.5	85.6	95.3	101.6	117.6	133.4	149.1	174.5	196.9	235.0	279.4	323.9	425.5	469.9
Slip-on & Socket Length	in	1.81	1.94	2.13	2.31	2.63	3.00	3.37	3.87	4.50	5.37	6.25	7.25	9.25	10.25
	mm	46.0	49.3	54.1	57.9	66.8	76.2	85.6	98.3	114.3	136.4	158.8	184.2	235.0	260.4

NOTE: Shaded boxes do not show the first decimal place because of space constraints.

Flange Dimension Notes:

1. Dimensions based on ASME/ ANSI pipe flanges.

2. Dimensions for flange length thru hub include 0.06 in. (1.5mm) raised face for Classes 150 and 300.

3. Dimensions for flange length thru hub include 0.25 in. (6.4mm) raised face for Classes 400 and higher.

4. Threaded, Slip-on, or Socket flange styles may not be commercially available in all sizes stated.

Table #59G - Flange and Bolt Dimensions (Class 2500)

Flange Gaskets

Pipe flange gaskets are used to provide a leak-tight seal between two flange faces. ***In order to provide this seal, the gasket must be able to flow or form under bolt compression to fill all irregularities in the flange sealing face. The gaskets must also withstand possible high service temperatures and corrosive products while also being strong enough to prevent the system pressure from blowing it out.*** Gaskets are generally available in the following three flange face types:

- Full face gasket - covers the full face or area of the flange.
- Flat ring gasket - covers only the sealing area or raised face of the flange.
- Metallic ring gasket - designed to fit between the U-shaped machined grooves of ring type flanges.

An example of each of these three flange styles are shown in illustration #98.

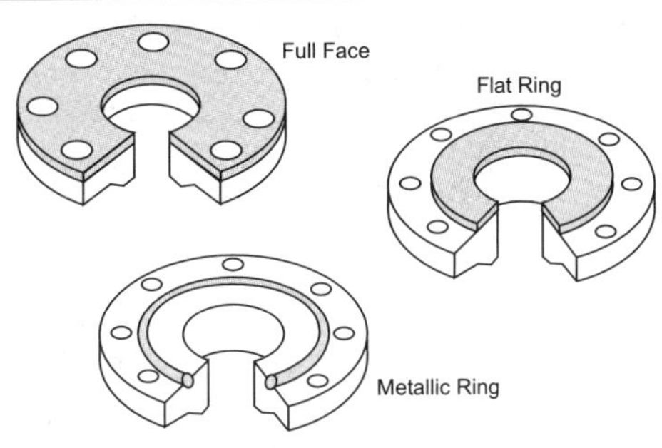

Illustration #98 - Gasket Face Types

Metallic Ring Gaskets

The metallic ring gasket is generally made from a soft iron or a softer grade of the same metal type as the joining flange. The cross sectional shapes of both the oval and octagonal rings are displayed in illustration #99 along with ANSI material identification markings.

Metallic Ring Gasket

The ring gasket is considered one of the best gasket types for corrosive process, and high pressure and temperature service.

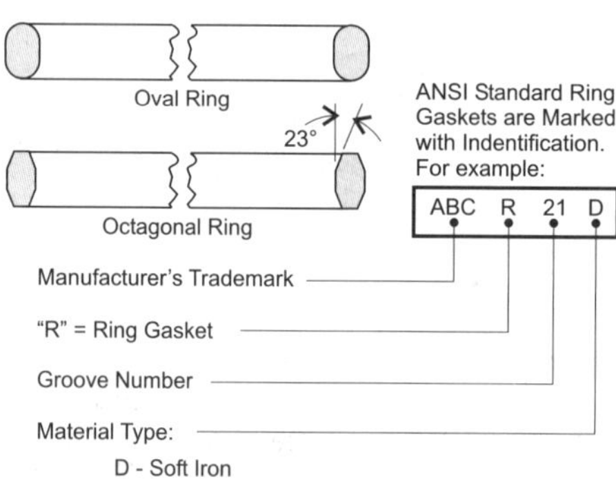

Illustration #99 - Oval and Octagonal Gaskets with ANSI Markings

When disassembling ring face flanges, it should be remembered they can only be separated in a straight line pull directly opposite to each other as shown in illustration #100. This is because of the way the flange groove fits over the U-shaped seal of the gasket. It does not allow lateral movement.

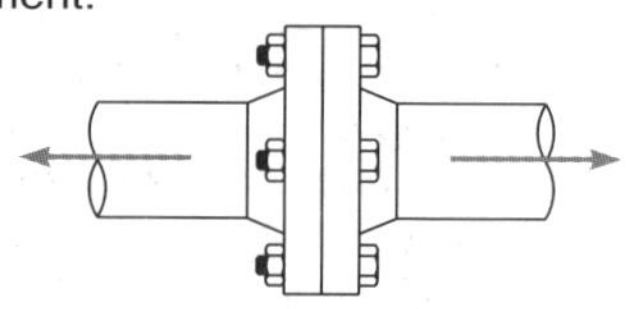

Ring Joint Facing can only be Disassembled in the Axial Direction as shown by the Arrows

Illustration #100 - Metal Ring Joint Flange Disassembly

Flange Disassembly Procedures

It is very important before disassembling any flange joint that the line is checked to ensure that it has been de-pressurized. Make sure seepage, accidental valve opening, liquid vaporization and/or melting hydrates have not allowed the line to pressurize again.

When disassembling the flange, always position yourself out of the line of fire and if no procedure is in place for disassembling, follow the reverse order of tightening the flange:

- Un-tighten bolts in a crisscross pattern.
- Use several passes to loosen the nuts and bolts, slightly loosen each fastener before loosening the next.
- Do not remove the nuts.
- Verify the seal of the flange joint has been broken before removing the nuts.
- Be prepared if the piping springs or equipment shifts once the fasteners are removed.
- Take care in handling the flange faces and fasteners to avoid damage.
- If necessary, scrape out remnants of the gasket using a brush or a drift that is softer than the flange material.

Full Face and Flat Ring Gaskets

Both full face and flat ring gaskets are designed for use with flat face and raised face flanges respectively.

A variety of material can be used for each type of gasket depending on intended service. Common gasket materials include: synthetic rubber, asbestos composition, Teflon, metals and various metal non-metal combinations.

Generally, synthetic rubber gaskets are used for low pressure and temperature water service [maximum 250°F (120°C)].

Asbestos compositions and Teflon gaskets are utilized for intermediate pressure and temperature services [Asbestos to a maximum 750°F (500°C) and Teflon to 400°F (260°C)]. High pressure and temperature [1100°F (593°C)] service usually require metal-asbestos spiral wound gaskets.

Note: See asbestos note on table #60B.

Refer to table #60 for specific information on maximum temperatures and service properties of various gasket materials.

GASKET MATERIAL SELECTION			
	MAX TEMP.		
MATERIAL	°F	°C	SERVICE PROPERTIES
Acrylic	450	(232)	Moderate heat resistance, but poor cold handling capability. Good resistance to oils, aliphatic and aromatic hydrocarbons. Poor resistance to water, alkalies and some acids.
Asbestos: Compressed asbestos and composites	750	(398)	Best general gasket material, but because of its low strength and high porosity, pure asbestos is seldom used. It is usually mixed with plastic or rubber to form compressed asbestos fibre (CAF). Large number of combinations available; properties vary widely depending on materials used with asbestos. CAUTION: See note at end of table #60B.
Asbestos - TFE (Teflon)	500	(260)	Combines heat resistance and sealing properties of asbestos with the chemical inertness of Teflon.
Butyl	300	(148)	Good resistance to water, alkalies and dilute acids. Poor resistance to oils, gasoline and most solvents.
Cellulose fibre	300	(148)	Moisture and humidity changes the physical dimensions and hardness and softness of the gasket. Generally good resistance to chemicals, except strong acids.
Cork compositions surfaces	250	(121)	Conforms well to irregular flange. High resistance to oils, water, and many chemicals. Should not be used with inorganic acids, alkalies, oxidizing solutions, and live steam.
Cork Rubber	300	(148)	Good conformability and resistance to fatigue under bolting. Chemical inertness of gasket depends on rubber type used.
Neoprene	250	(121)	One of the most common gasket materials with excellent mechanical properties. It has good resistance to water, alkalies, most oil and solvents (except aromatic, chlorinated, or ketone types).
Nitrile	300	(148)	Good resistance to water, oils, gasoline, and dilute acids.
Metal: Aluminum	800	(426)	High corrosion resistance, but may be slightly attacked by strong acids and alkalies.
Copper	600	(315)	Both copper and brass offer good corrosion resistance at moderate temperatures.
Brass	500	(260)	
Inconel	2000	(1093)	Excellent heat and oxidation resistance.

Table #60A - Gasket Material Selection

GASKET MATERIAL SELECTION			
	MAX TEMP.		
MATERIAL	°F	°C	SERVICE PROPERTIES
Lead	500	(260)	Good chemical resistance and the best conformability of all metal gaskets.
Monel	1500	(815)	High corrosion resistance against most acids and alkalies. Attacked by strong hydrochloric and strong oxidizing acids.
Nickel	1400	(760)	Generally high temperature and corrosion resistance.
Stainless steel:			
302	1150	(621)	Stainless steels in general are highly corrosion resistant to most chemicals. Unsatisfactory for wet chlorine gas and liquid, chlorides and some acids.
304	1000	(537)	
316	1000	(537)	
410	1200	(648)	
430	1400	(760)	
(Metal Composites)	-	-	Many metal and non-metal combinations are available; properties vary widely depending on materials used. Non-metallic material inserted with metal gaskets or metal combinations may affect the gasket's temperature limit.
Rubber natural	225	(107)	Good mechanical properties, but poor weathering and aging properties. Impervious to air and water. Fair to good resistance to acids and alkalies. Unsuitable for oils and gasoline.
Silicon	600	(315)	Good heat resistance properties, but poor resistance to high pressure steam. Fair to average resistance to water, acids and alkalies. Poor resistance to oil and solvents.
Styrene-butadiene	250	(121)	Similar to natural rubber, but better water resistance properties. Fair to good resistance to acids and alkalies. Unsuitable for oils, gasoline and solvents.
Teflon (TEF)	500	(260)	Is resistant to almost all chemicals and solvents. Good heat resistance and sealing properties at low temperatures.
Viton and Fluorel	450	(232)	Resistant to fuels, oils, lubricants and hydraulic fluids. Good resistance to ozone and weathering combined with good mechanical properties.
NOTE: Asbestos fibres are considered to be a health hazard and care must be taken when cutting or grinding asbestos gasket material. Do not cut or grind any gasket material before knowing the composition of the material. If uncertain about the gasket material to be fabricated, simply don't until all safety concerns can be accurately answered.			

Table #60B - Gasket Material Selection

Gasket Design Configurations

The basic design of a flange gasket must conform to the round physical design of the flange face to which it must seal. However, this basic geometric circle design can have various cross sectional and structural configurations. Illustrations #101 through #107 display common gasket configurations. The following gives a brief description of each gasket.

Flat Gaskets

The flat gasket design uses both metallic and non-metallic gasket material in its construction. Thickness varies from 1/64 to 1/8 of an inch (0.4 to 3.18 mm) depending on material and intended service. Metallic material includes aluminum, soft copper, iron, and stainless steel. Non-metallic material includes compressed asbestos, elastomers, rubber, Teflon, and paper. Combinations of metallic and non-metallic materials are also common (Illustration #101).

Reinforced and Multi-Ply Gaskets

Wire or fabric insertions are used in reinforced gaskets to give better torque retention and blow-out resistance to non-metallic flat gaskets. Layers of reinforced material may also be bonded together to form thicker double and triple ply gaskets (Illustration #102).

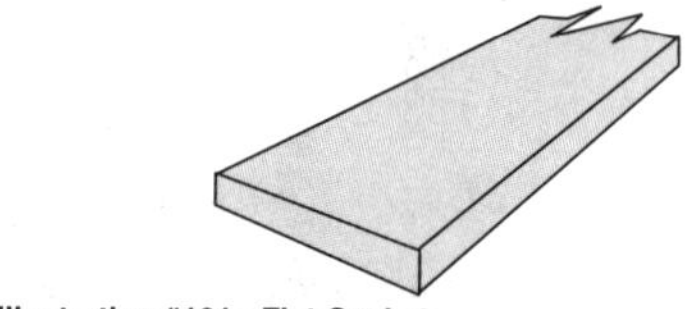

Illustration #101 - Flat Gasket

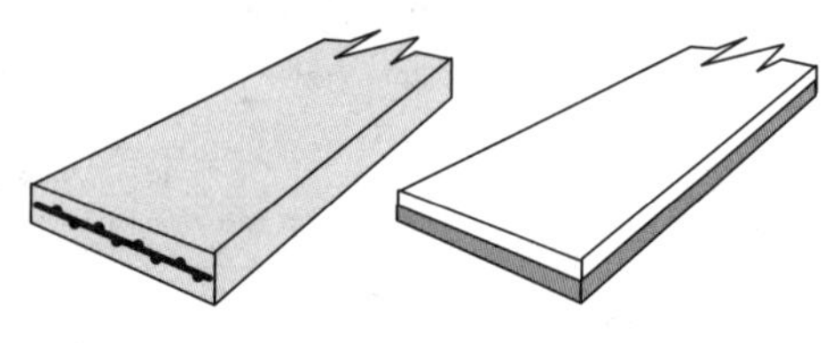

Illustration #102 - Reinforced and Multi-ply Gaskets

Gasket Design Configurations

Corrugated Metal Gaskets

Corrugated gaskets are used for moderate temperature and pressure service [maximum 1150°F (621 °C); 600 p.s.i. (4137 kPa)]. The gasket may be all metal construction or metal with asbestos filler between the grooves or an asbestos jacket over the metal grooves (Illustration #103).

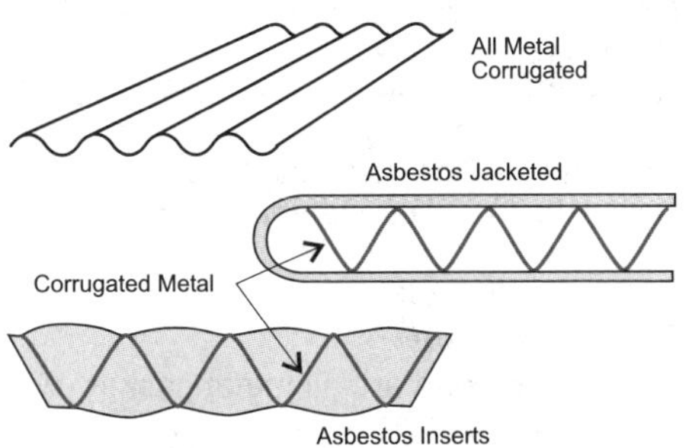

Illustration #103 - Corrugated Metal Gaskets

Spiral-Wound Gasket (Flexatallic)

This gasket consists of spiral wound metal strips (stainless steel, carbon steel or monel, etc.) with an asbestos or other non-metal filler between the metal spirals. Metal inner and outer rings are often used for centering the gasket and for compression control. They are excellent for high temperature service on raised face and male and female flanges (Illustration #104).

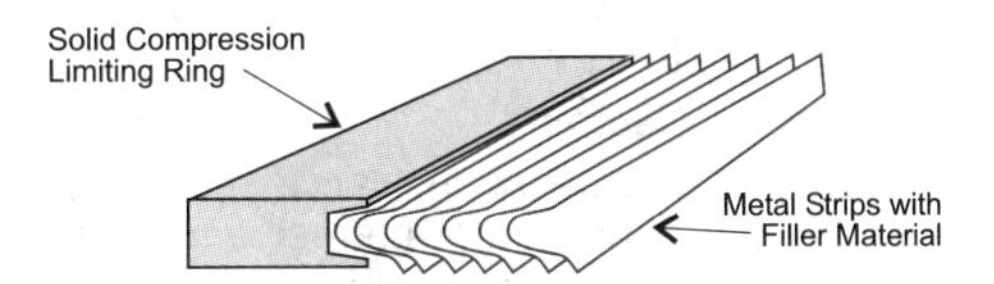

Illustration #104 - Spiral Wound Gaskets

Gasket Design Configurations

Metal Jacketed Gasket

In this gasket design, a filler material of asbestos or other non-metal is wrapped in a metal jacket. The metal jacket provides protection to the filler material, and in higher temperature service, non-metal filler can be replaced with metal filler (Illustration #105).

Serrated Gasket

Serrated gaskets are all metal gaskets with V-grooves machined into the face area to provide for multiple sealing surfaces with low bolt loading. They are available in raised face, male/female and tongue and groove, but are preferably used with smooth finished flanges (Illustration #106).

Metal Ring Gasket

This gasket is used extensively in the oil and petrochemical industry in corrosive services where other gasket materials are not suitable. It is constructed from all metal in both oval and octagonal cross sections (Illustration #107).

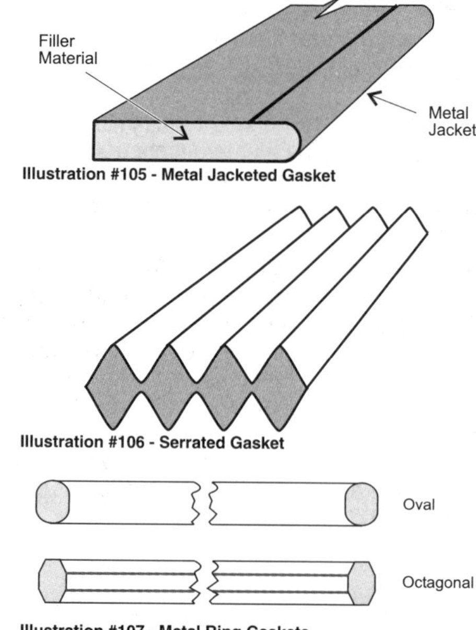

Illustration #105 - Metal Jacketed Gasket

Illustration #106 - Serrated Gasket

Illustration #107 - Metal Ring Gaskets

Cutting Non-Metallic Gaskets

Gaskets can be purchased either ready made or in large material sheets that need to be cut to suit particular flange faces. Fabrication of gaskets from sheet material is usually accomplished by either:

- trace and tap method
- layout and cut method

Trace and Tap Method

When using the trace and tap method of gasket fabrication, the sheet of gasket material is placed over the flange requiring the gasket. The flange is used as the tracing pattern for the gasket. ***A soft face hammer is then used to lightly tap the gasket material around the O.D. and I.D. edges of the flange. Bolt holes are formed in the gasket material by using the peen end of a ball peen hammer in the same manner as the soft face hammer was used.***

Note: Care should be taken to ensure hammering does not damage the flange face.

This method of fabrication is displayed in illustration #108.

If the gasket material is too thick or resilient, the hammer tap method may be used to trace the needed pattern onto the gasket material. The pattern (depending on material thickness and type) is then cut out using scissors, snips or a utility knife.

Gasket hole punches can be used to cut smooth round bolt holes in the traced gasket material.

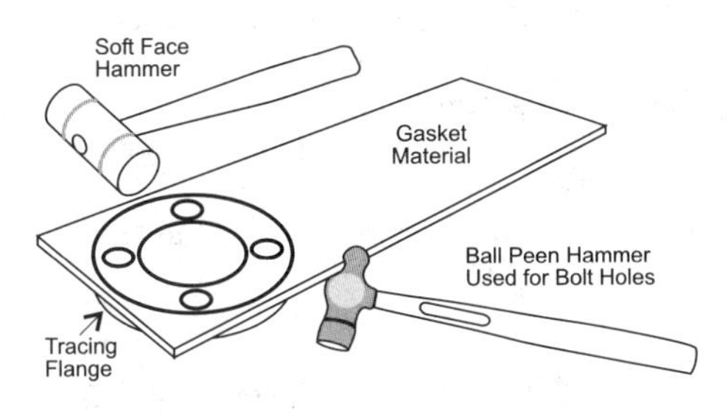

Illustration #108 - Gasket Fabrication (Trace and Tap)

Layout and Cut Method

The needed pattern for a gasket can be laid out directly on the material sheets by the use of a compass, divider or gasket cutter. These methods are presented in illustration #109. ***In order to layout the gasket, the I.D. and O.D. of the needed gasket must be determined. The I.D. of gasket should equal the outside diameter of the pipe for full and raised face flanges. The O.D. for a full face gasket equals the outside diameter of the flange. The O.D, for raised face flanges should equal the diameter extending across the inside edge of the bolt holes.***

I.D and O.D. Gasket Diameters

a. I.D. of full and flat ring gaskets = outside diameter of pipe.

b. I.D. of large male and female gaskets = outside diameter of pipe.

c. O.D. of full face gaskets = outside diameter of flange.

d. O.D. of flat ring gaskets = diameter of inside edge of flange bolt holes.

e. O.D. of large male and female gaskets = outside diameter of male flange face.

Note: I.D. and O.D. of gaskets for other flange faces should equal the flange's seal surface.

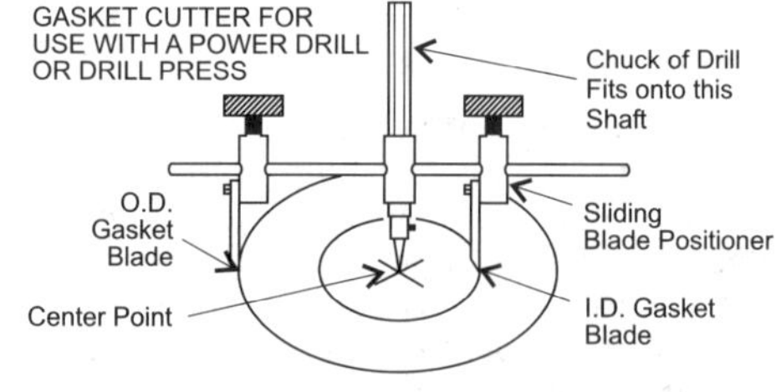

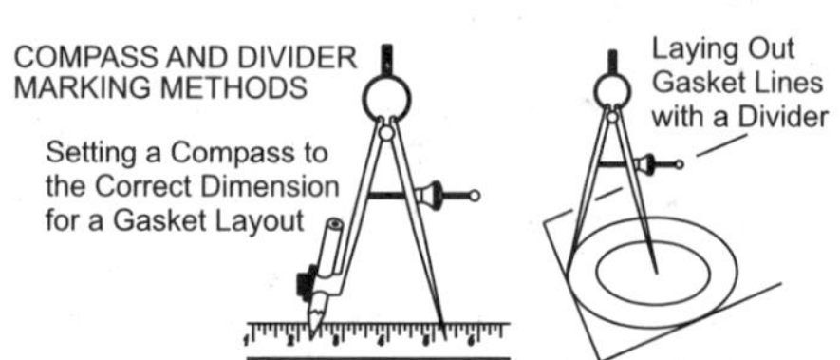

Illustration #109 - Gasket Fabrication (Layout Method)

Bolt Hole Layout

When using sheet gasket material to fabricate full face gaskets the bolt holes must also be laid out and cut. Bolt holes are laid out by first drawing the center circumference of the bolt holes around the gasket. The location for the bolt holes along the circumference line is determined by either the angle or chord method.

Angle or Degree Method of Bolt Hole Layout

The angle method uses degrees to divide the bolt hole circle into equal parts. Angles are measured in degrees, and for more accurate layout and measurement, can further be divided into minutes and seconds. The relationship between degrees, minutes and seconds are denoted in table #61. Bolt hole angles or degrees can be found by dividing 360 degrees (one complete circle) by the number of bolts in the flange.

DEGREES, MINUTES AND SECONDS	
ANGLES AND CIRCLES	
One complete circle = 360° (degrees)	
1 degree = 60' (minutes)	1 degree = 1/360 circle
1 minute = 60" (seconds)	90° = 1/4 circle
1 second = 1/60 (minute)	180° = 1/2 circle
1 minute = 1/60 (degree)	270° = 3/4 circle
	360° = full circle

Table #61 - Degrees, Minutes, Seconds

Bolt Angle = 360 degrees divided by number of bolts.

As an example, a flange with 8 bolts will have an angle between the bolts of 45 degrees. (360 degrees divided by 8 bolts = 45 degrees).

Commonly used flanges and the degrees between bolt holes are provided in table #62.

COMMONLY USED ANGLES FOR FLANGE BOLT HOLES	
No. of Holes	**Degree of Spacing**
4	90°
8	45°
10	36°
12	30°
16	22 1/2°
20	18°
24	15°
36	10°

Table #62 - Common Flange Bolt Hole Angles

The protractor shown in illustration #110 is one instrument that can be used in laying out angles. It is used by placing the base line and 90 degree marks on the horizontal and vertical gasket center lines respectively. Angles can be measured clockwise (outer degree numbers) or counter clockwise (using the inner degree numbers).

After the top half of the bolts are marked, the gasket can be rotated to mark the remaining half.

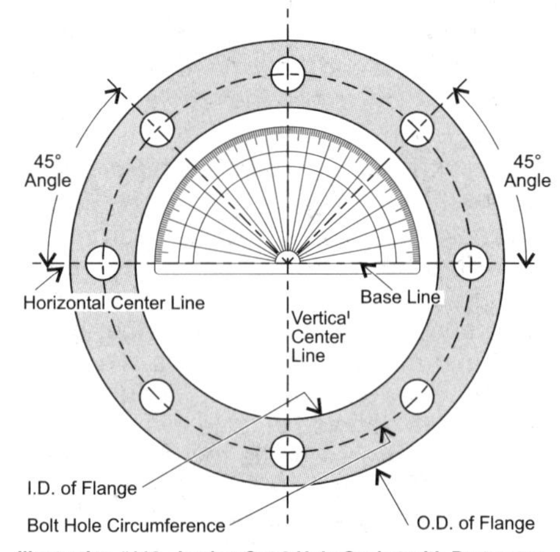

Illustration #110 - Laying Out 8 Hole Gasket with Protractor

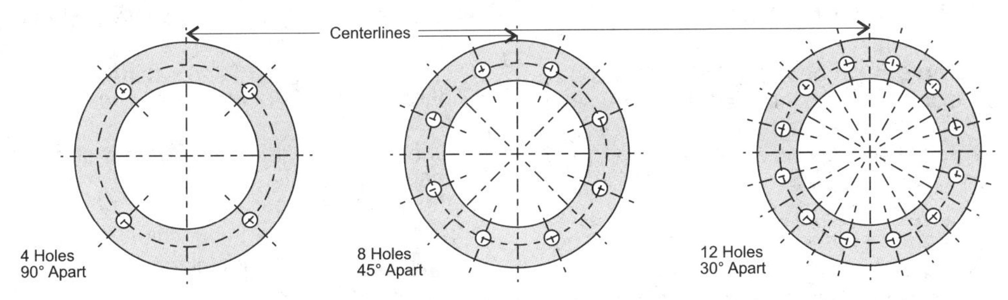

Illustration #111 - Flange Holes Straddling Centerlines

Note: ASME/ANSI flange bolt holes increase by multiples of four and are equally spaced in pairs between the center lines of the flange. See illustration #111.

Chord Chart Method for Bolt Hole Layout

The chord chart method for layout of flange bolt holes uses the chord of a circle to determine the location of bolt holes.

Illustration #112 displays the basic parts that make up a circle.

To find the desired chord length, the bolt hole diameter for a selected flange (the measurement extending from one bolt hole center through the center of the flange to the opposite bolt hole center) is multiplied by the chord factor (length) taken from the chord chart table #63.

Chord length = bolt hole diameter x chord factor

Example: A flange with 8 bolts and a bolt circle diameter of 7.5 inches (228.6 mm) would have a chord distance of 2.87 inches (72.9 mm) between bolt hole centers.

Bolt hole diameter = 7.5 inches (228.6 mm) x chord factor 0.382683 = 2.87 inches (72.9 mm) chord distance between bolt hole centers.

This calculated chord distance is the straight line measurement between each successive bolt hole. The measurement can be transferred to a compass or divider which is then used to intersect each consecutive bolt hole center.

Circle Components

- Circumference: The boundary line or distance around a circle.
- Radius: Any straight line extending from the center to the circumference.
- Segment: A portion of a circle bounded by an arc and a chord.
- Diameter: A straight line extending from one side to the other through the center of the circle.
- Chord: A straight line of length shorter than the diameter, from point to point on the circumference.
- Arc: Any portion of the circumference.
- Sector: A portion of a circle bounded by an arc and two radii.

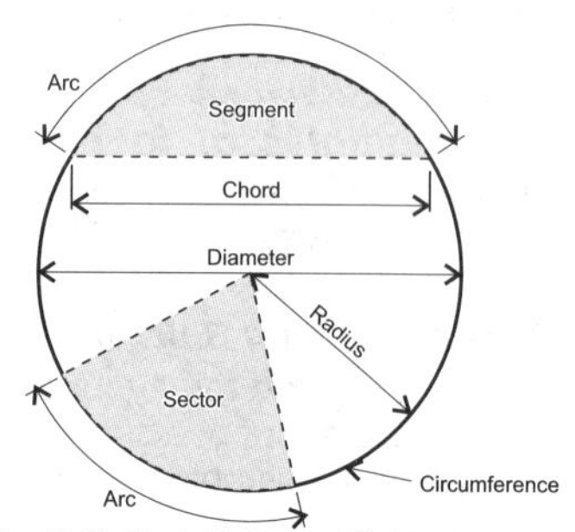

Illustration #112 - Basic Parts of a Circle

CHORD CHART					
No. of Spaces	Chord Factor	No. of Spaces	Chord Factor	No. of Spaces	Chord Factor
3	0.866025	44	0.071339	108	0.029084
4	0.707106	48	0.065403	112	0.028046
5	0.587785	52	0.060378	116	0.027079
6	0.500000	56	0.056070	120	0.026176
8	0.382683	60	0.052336	124	0.025332
10	0.309017	64	0.049067	128	0.024541
12	0.258819	68	0.046183	132	0.023797
14	0.222520	72	0.043619	136	0.023097
16	0.195090	76	0.041324	140	0.022438
18	0.173648	80	0.039259	144	0.021814
20	0.156434	84	0.037391	148	0.021225
24	0.130526	88	0.035692	152	0.020666
28	0.111964	92	0.034141	156	0.020137
32	0.098017	96	0.032719	160	0.019663
36	0.087155	100	0.031410	164	0.019154
40	0.078459	104	0.030202	168	0.018698

Table #63 - Chord Chart for Hole to Hole Spacing

Flange Bolting

There are two basic types of bolts used in making flange connections: a machine bolt that uses one nut, and the more commonly used stud bolt which uses two nuts. Both types are displayed in illustration #113. The most common thread classification used for flange bolts and nuts is the Unified Screw Thread Standard.

The Unified Screw Thread Standard is used in Canada, the United States, and Great Britain to classify bolts and nuts. ***Flange bolts and nuts are classified under this standard as: UNC (Unified Coarse) class 2, (medium fit) A for bolts and B for nuts.***

ASTM Material Bolt Specifications

ASME/ANSI standard B16.5 categorize ASTM flange bolting materials into the following groups:

- High Strength Bolting
- Intermediate Strength Bolting
- Low Strength Bolting
- Nickel and Special Alloy Bolting

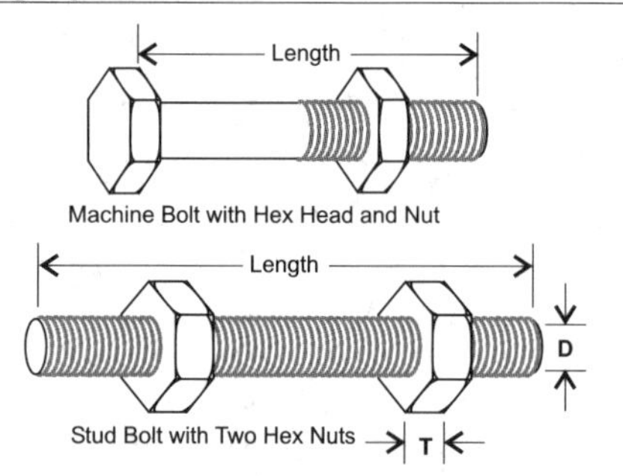

Illustration #113 - Machine Bolt and Stud Bolt

High Strength Bolting materials can be for use in any flange jointing application with all listed materials and gaskets within the standard.

ASTM Material Bolt Specifications

Intermediate Strength Bolting material is also used for any flange jointing applications, but the end user must verify the soundness of the flange joint under operating temperature and pressure.

Low Strength Bolting material is limited to flange classes 150 and 300.

Nickel and Special Alloy Bolting is used with comparable nickel and special alloy flange material.

The following listing gives the ASTM standards which are applicable under each strength grouping.

High Strength Bolting

- A 193/A 193M: Grades B7 and B16. Alloy Steel and Stainless Steel Bolting Material for High Temperature Service.
- A 320/A 320M: Grades L7, L7A, L7B, L7C, and L43. Alloy Steel Bolting Materials for Low-Temperature Service. (Use of A 194 Grade 4 or Grade 7 nuts is recommended).
- A 354: Grades BC and BD. Quenched and Tempered Alloy Steel Bolts, Studs, and Other Externally Threaded Fasteners.
- A 540: Grades B21, B22, B23, and B24. Alloy Steel Bolting Materials for Special Applications.

Intermediate Strength Bolts

- A 193/A 193M: Grades B5, B6, B6X, B7M, B8 C1.2, B8C C1.2, B8M C1.2, and B8T C1.2. Alloy Steel and Stainless Steel Bolting Material for High Temperature Service. (Use of A 194 nuts of corresponding material recommended).
- A 320/A 320M: Grades B8 C1.2, B8C C1.2, B8F C1.2, B8M C1.2, and B8T C1.2. Alloy Steel Bolting Materials for Low Temperature Service. (Use of A 194 nuts of corresponding material recommended).
- A 449: Quenched and Tempered Steel Bolts and Studs. (Use of A 194 nuts Grades 2 and 2H recommended).

Intermediate Strength Bolts

- A 453/A 453M, Grades: 651 and 660. Bolting Materials, High-Temperature, 50 to 120 psi (345 to 827 MPa) Yield Strength, with Expansion Coefficients Comparable to Austenitic Steels.

Low Strength Bolting

- A 193/A 193M, Grades: B8 C1.1, B8C C1.1, B8M C1.1, B8T C1.1, B8A, B8CA, B8MA, and B8TA. Alloy Steel and Stainless Steel Bolting Material for High Temperature Service. (Use of A 194 nuts of corresponding material recommended).
- A 307, Grade B. Carbon Steel Bolts and Studs, 60 000 psi Tensile Strength.
- A 320/A 320, Grades: B8 C1.1, B8C C1.1, B8M C1.1, and B8T C1.1. Alloy Steel Bolting Materials for Low Temperature Service.

Nickel and Special Alloy Bolting

- B 164 Nickel-Copper Alloy. (Nuts may conform to ASTM A 194), (Max. Temp. 500 degrees F/260 degrees C).
- B 166 Nickel-Chromium-Iron Alloys. (Nuts may conform to ASTM A 194), (Max. temp. 500 degrees F/260 degrees C).
- B 355, Grade N10665, Nickel-Molybdenum Alloy. (Nuts may conform to ASTM A 194).
- B 408 Nickel-Iron-Chromium Alloy. (Nuts may conform to ASTM A 194), (Max. temp. 500 degrees F/260 degrees C).
- B 473, Grades: UNS N08020, UNS N08026, and UNS N08024. Nickel Alloy. (Nuts may conform to ASTM A 194).
- B 574, Grade N10276. Nickel-Molybdenum-Chromium Alloy. (Nut may conform to ASTM A 194).

Bolting of Cast Iron and Steel Flanges

When it is necessary to bolt together a cast iron flange to a steel flange, the information in table #64 will help to prevent any possible damage or breakage to the cast iron flange.

BOLTING TOGETHER OF CAST IRON AND STEEL FLANGES			
FLANGE COMBINATION	FACE TYPE	GASKET TYPE	BOLTING MATERIAL STRENGTH
Class 125 Cast Iron to Class 150 Steel	1/16" (1.6mm) raised face on steel flange must be removed.	Ring Type Full Face	Low Strength Low, Intermediate or High
Class 250 Cast Iron to Class 300 Steel	No change in raised face is necessary. If faces are removed and a full face gasket is used Inter. or High strength bolting may be used.	Ring Type Full Face	Low Strength Low, Intermediate or High

Table #64 - Bolting Steel and Cast Iron

Size Diameter and Length

It is standard practice to furnish machine and stud bolts with heavy hexagon nuts. The heavy hex nut can be used as an easy way to determine the size or diameter of the stud or machine bolt being used. The thickness of the heavy hex nut equals the bolt diameter. For example, a 3/4 inch (19 mm) diameter machine bolt will have a heavy hex nut that is 3/4 of an inch (19 mm) thick.

Stud bolt length is measured from end to end, not including the point or crowned ends. Machine bolts are measured from the base of the head to end of the point or crown. Both machine and stud bolt lengths are measured to the nearest 1/4 inch (6.35 mm). Illustration #113 demonstrates the methods of sizing diameters and lengths for flange bolts and nuts.

Flange Bolt Tightening

When tightening flange bolts, the flange faces must first align properly to provide even contact on the gasket surface. Before installing stud bolts, apply thread lubricant to each bolt. The lubricant makes tightening and future dismantling easier. Bolts should be installed hand tight and then evenly tightened in a crisscross pattern.

Note: See illustration #65 in Section Three, Valves, for bolt tightening sequence of flanges.

The proper wrench size for Imperial heavy hex nuts can be calculated by:

Wrench size = (1 1/2 x Bolt Diameter) + 1/8" (Imperial sizes only).

Example: Find the wrench size for a heavy hex nut used with a stud bolt diameter of 1 1/4 inches.

Wrench size = (1 1/2 x 1 1/4) + 1/8
= (3/2 x 5/4) + 1/8
= 1 7/8 + 1/8
= 2 inches

Table #65 provides standard wrench sizes for heavy hex nuts used with stud bolts in both metric and inch sizes.

WRENCH SIZES FOR HEAVY HEX NUTS USED ON FLANGE STUD BOLTS				
Stud Size (inches)	Wrench Size (inches)		Stud Size (mm)	Wrench Size (mm)
1/2	7/8		M12	21
5/8	1 1/16		M14	24
3/4	1 1/4		M16	27
7/8	1 7/16		M20	34
1	1 5/8		M22	36
1 1/8	1 13/16		M24	41
1 1/4	2		M27	46
1 1/2	2 3/8		M30	50
1 5/8	2 9/16		M36	60
1 3/4	2 3/4		M42	70
1 7/8	2 15/16		M48	80
2	3 1/8		M56	90
2 1/4	3 1/2		M64	100
2 1/2	3 7/8		M72	110
2 3/4	4 1/4		M80	120
3	4 5/8		M90	135
3 1/2	5 3/8		M100	150

Table #65 - Wrench Sizes on Stud Bolts

SECTION SIX

PIPE OFFSETS

Introduction

In most piping systems, it is impossible to run pipe in a straight line without making offsets. Offsets can be made easily with 90 degree elbows, but because of high friction loss and installation costs, most offsets are made with lesser degree elbows. However, these lesser degree elbow offsets (e.g. $11^1/_4$, $22^1/_2$ and 45 degree) require a more complex method of calculating pipe length for the offsets.

Illustration #114 displays a typical 45 degree offset and the piping terms used to describe the various parts of the offset.

Offset: Sometimes referred to as set, is the perpendicular distance between the centerline of two parallel running pipes.

Rise: The vertical distance between the centers of bend of two parallel vertical running pipes.

Run: The horizontal distance between the centers of bend of two parallel horizontal running pipes.

Travel: The diagonal distance between the centerlines of two parallel running pipes.

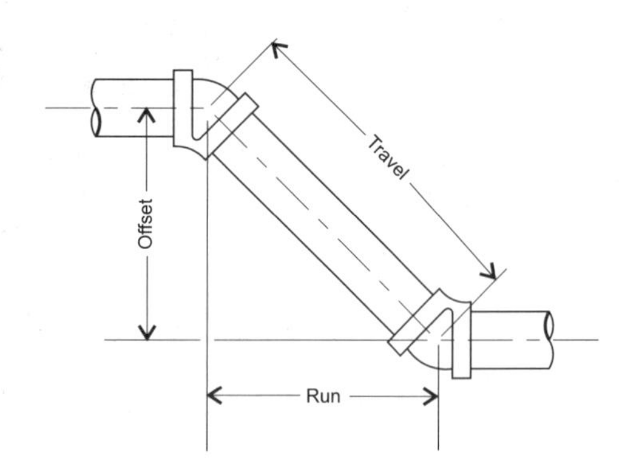

Illustration #114 - Typical Offset Terminology

Offset Calculation Methods

The various pipe lengths and distances that make up offsets can be calculated by one of the following methods:

- Pythagorean Theorem
- Trigonometry
- Constant Multiplier

Additional information on Pythagorean Theorem and Trigonometry used in solving offset problems can be found in Section Seven on Trigonometry Functions.

In most offsets, the necessary offset lengths can be calculated simply by multiplying a known side in the offset by a ratio number called a constant. The constants and formulas for various fitting angles used in common offsets are shown in table #66.

Table #66 can be used for any size or material of pipe as long as the offset matches the elbow fitting angle in the table.

Table #66 Use (Calculating Offsets)

In the formula column of the table, find the particular offset part whose length is unknown and must be determined. Match the required part with the known component of the offset in the same column. Across the top of the table, find the elbow fitting angle used in the offset. Multiply the constant number found under the elbow fitting angle with the formula that intersects it.

Example:

Find the travel length for a 22 1/2 degree offset given a run distance of 12 in. (304.8 mm).

Solution Steps:

1. Under the formula column, "Travel = Run x" "constant in table" is selected as the needed formula.
2. In the formula column, continue across the table horizontally to the constant number under the 22 1/2 degree elbow column.
3. Multiply the known run length of 12 in. (304,8 mm) by the constant number found under the 22 1/2 elbow fitting angle.

Imperial Calculations:
Travel = 12 inches x 1.082
= 12.984 inches (1 ft. x 1 in.)

Metric Calculations:
Travel = 304.8 mm x 1.082
= 329.8 mm

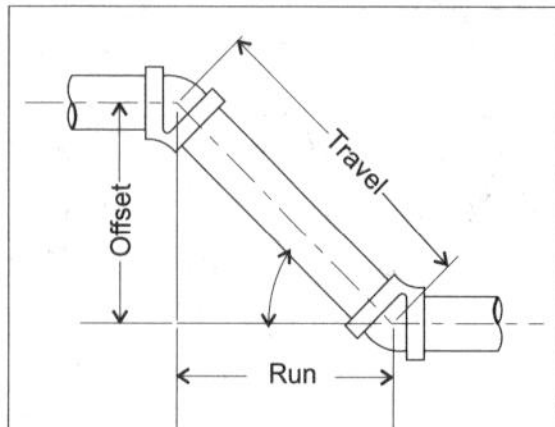

CONSTANTS AND FORMULAS FOR CALCULATING COMMON OFFSETS

Formula	Elbow Fitting Angles						
	72° Elbow	60° Elbow	45° Elbow	30° Elbow	22 1/2° Elbow	11 1/4° Elbow	5 5/8° Elbow
Travel = Offset x	1.052	1.155	1.414	2.000	2.613	5.126	10.187
Travel = Run or Rise x	3.236	2.000	1.414	1.155	1.082	1.019	1.004
Run or Rise = Offset x	0.325	0.577	1.000	1.732	2.414	5.027	10.158
Run or Rise = Travel x	0.309	0.500	0.707	0.866	0.924	0.980	0.995
Offset = Travel x	0.951	0.866	0.707	0.500	0.383	0.195	0.095
Offset = Run or Rise x	3.078	1.732	1.000	0.577	0.414	0.198	0.098

Table #66 - Common Offset Constants and Formulas

Finding Specific Constants for Offsets Using the Trigonometry Table

If the angle and constant for a particular bend is not found in table #66, the constant or ratio for the offset formula can be found in the Trigonometry Table, Section Seven. In the Trigonometry Table, the angle of the bend used in the offset is located. The known side of the offset is then multiplied by the necessary trigonometry ratio found alongside the angle. The following gives the applicable formula and trigonometry ratio to use for finding the length of the unknown side:

Travel = Offset x Cosecant
Travel = Run or Rise x Secant
Run or Rise = Offset x Cotangent
Run or Rise = Travel x Cosine
Offset = Travel x Sine
Offset = Run or Rise x Tangent

Note: Use of the trigonometry method is explained in greater detail in section seven.

Fitting Allowance

The calculated travel length for an offset is given as a center to center measurement. To determine the actual length of the pipe needed for the travel, subtract the allowance for each fitting used in the offset from the center to center measurement calculated:

Actual Pipe Length = Center to Center Offset - (minus) 2 Fitting Allowances

The methods of determining fitting allowance for typical welded and threaded fittings are shown in illustration #115.

Specific fitting allowance dimensions can be found in Section Four, Fittings of this book.

Calculating 45 Degree Offsets

The travel length for 45 degree offsets can be found easily by multiplying the offset needed by the constant 1.414. This constant is used extensively in pipefitting calculations and should be noted and memorized.

Examples for both vertical and horizontal 45 degree offsets are displayed in illustration #116.

This example should be used as a reference for calculating other fitting angles and their constants provided in table #66.

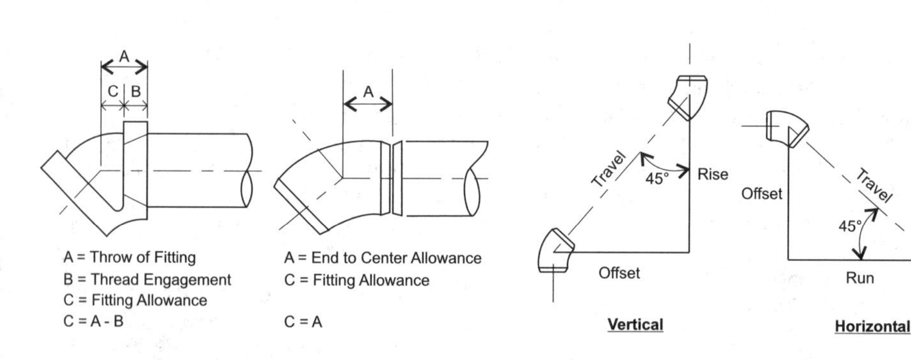

Illustration #115 - Threaded and Welding Fitting Allowance

Illustration #116 - 45° Offsets

Vertical Layout Measurements

Travel (45 degrees)

Travel = 1.414 x Rise

Travel = 1.414 x Offset

Rise

Rise = 0.707 x Travel

Rise = Offset

Offset

Offset = 0.707 x Travel

Offset = Rise

Horizontal Layout Measurements

Travel (45 degrees)

Travel = 1.414 x Rise

Travel = 1.414 x Offset

Run

Run = 0.707 x Travel

Run = Offset

Offset

Offset = 0.707 x Travel

Offset = Run

Example:

Find the travel length for the horizontal 45 degree offset shown in illustration #117 which has an offset of 15 inches (381 mm).

Solution Steps:

Imperial Calculations:

1. Offset = 15 in.
2. Travel = 1.414 x Offset
   = 1.414 x 15 in.
   = 21.21 inches
3. The decimal of an inch can be changed to a fraction of an inch by:

Nearest sixteenth: .21 in. x 16
   = 3.36 = 3.36/16
   = 3/16 inches (rounded off).

4. Travel = 21 3/16 inches

Note: Decimals of an inch and fractional equivalents can be found in the Equivalent Chart in the Appendices.

Find Travel Length

Metric Calculation:

1. Offset = 381 mm
2. Travel = 1.414 x Offset
   = 1.414 x 381 mm
   = 538.7 mm

Illustration #117 - Travel Length Example

Equal Spread Two Pipe Offsets

When two or more pipes running parallel are required to change route, it is normally accomplished by using equal spread offsets. Equal spread offsets provide for orderly and space saving turns. Illustration #118 displays a typical equal spread offset.

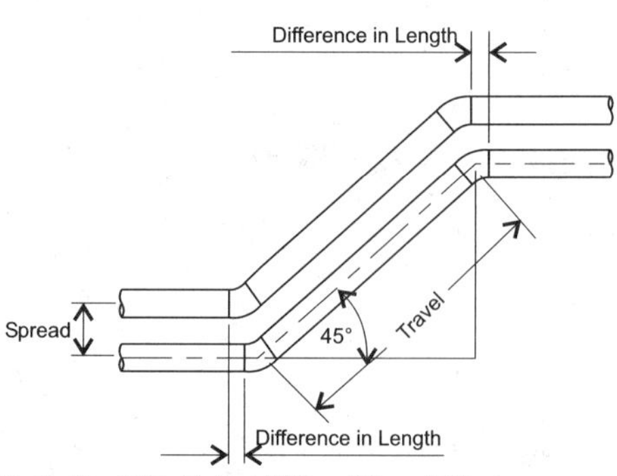

Illustration #118 - Typical 45° Equal Spread Offset

Equal Spread Offsets

Travel lengths for parallel offsets are equal and are calculated the same as any single offset. The major difference between an equal spread offset and a single offset is the distance where the offset starts and ends on each line. This difference in length is calculated by using the spread or distance between the parallel lines.

The spread length between the lines is multiplied by a constant number for the particular angle of the elbow or bend used in the offset:

Difference in Length = Spread x Constant

Table #67 gives the constant numbers needed for various equal spread offset angles.

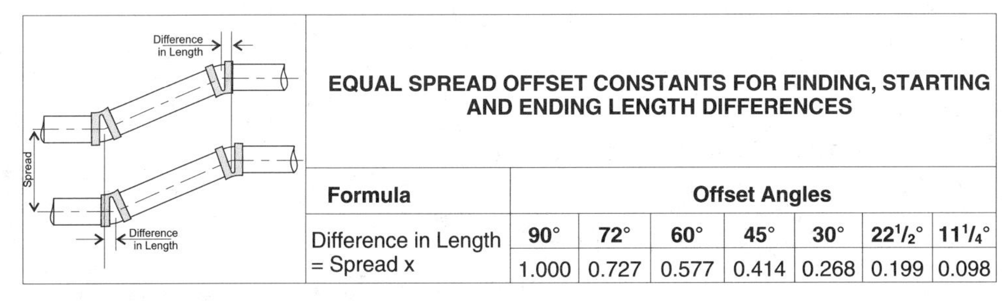

EQUAL SPREAD OFFSET CONSTANTS FOR FINDING, STARTING AND ENDING LENGTH DIFFERENCES

Formula	Offset Angles						
	90°	**72°**	**60°**	**45°**	**30°**	**22½°**	**11¼°**
Difference in Length = Spread x	1.000	0.727	0.577	0.414	0.268	0.199	0.098

Table #67 - Constants for Finding Length Differences

Calculating Common Equal Spread Offset Lengths (Refer to Table #66 and #67 for Constants) (Illustration #119)

Formulas Used in Calculating 22 1/2° Equal Spread Offset Lengths:

Difference in Length = Spread x Constant
Difference in Length = Spread x 0.199

Run = Offset x Constant
Run = Offset x 2.414

Travel 1 & 2 = Offset x Constant
Travel 1 & 2 = Offset x 2.613

Offset = Run x Constant
Offset = Run x 0.414

Note: Refer to table #66 for travel constants and table #67 for length difference constants.

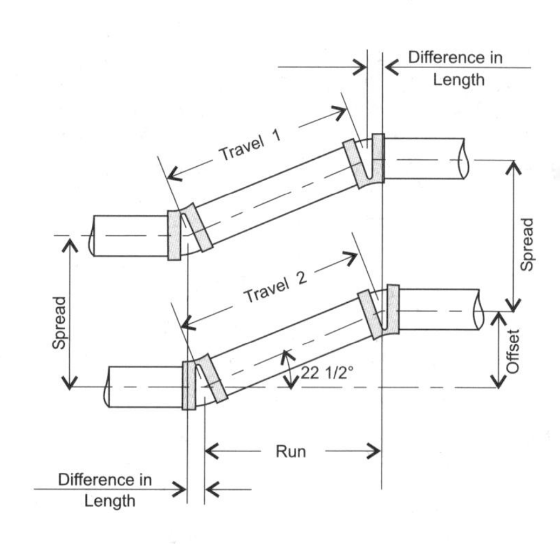

Illustration #119 - 22 1/2° Two Pipe Equal Spread Offset

Formulas Used in Calculating 45 Degree Equal Spread Offset Lengths: (Illustration #120)

Difference in Length = Spread x Constant
Difference in Length = Spread x 0.414

Run = Offset x Constant
Run = Offset x 1

Travel 1 & 2 = Offset x Constant
Travel 1 & 2 = Offset x 1.414

Offset = Run x Constant
Offset = Run x 1

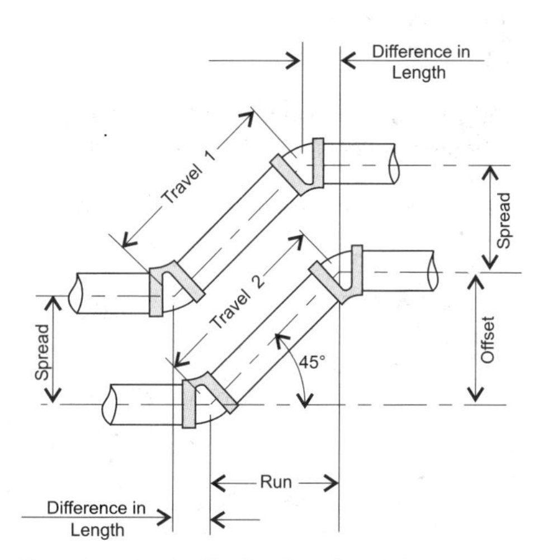

Illustration #120 - 45° Two Pipe Equal Spread Offset

Formulas Used in Calculating 60 Degree Equal Spread Offset Lengths: (Illustration #121)

Difference in Length = Spread x Constant
Difference in Length = Spread x 0.577

Run = Offset x Constant
Run = Offset x 0.577

Travel 1 & 2 = Offset x Constant
Travel 1 & 2 = Offset x 1.155

Offset = Run x Constant
Offset = Run x 1.732

Formulas Used in Calculating 60 Degree Equal Spread Offset Lengths: (Illustration #121)

Example:

Find the difference in length for the two pipe 60 degree equal spread offset shown in illustration #121.

The spread distance in the offset is 10 inches (254 mm) center to center between the two pipes.

Solution Steps:

Imperial Calculations:

Difference in Length = Spread x 0.577
= 10 in. x 0.577
= 5.77 (5 3/4) inches

Metric Calculations:

Difference in Length = Spread x 0.577
= 254 mm x 0.577
= 146.558 mm

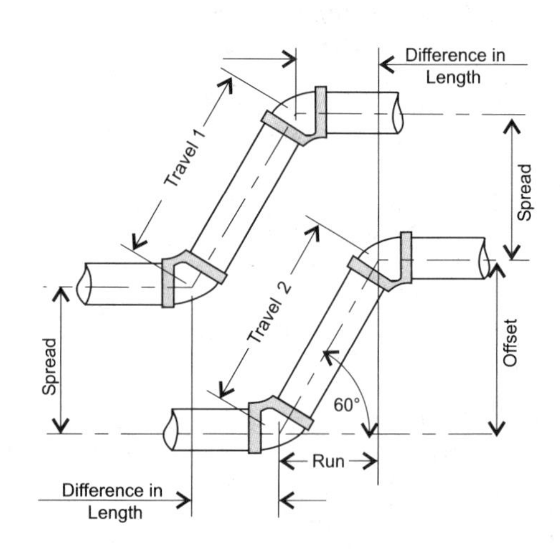

Illustration #121 = 60° Two Pipe Equal Spread Offset

Finding the Difference in Length For Starting Equal Spread Offsets Using Tangents

If the angle and constant for equal spread offsets are not found in table #67, the constant or ratio for the offset formula can be found in the Trigonometry Table, Section Seven.

Multiply the tangent of half the fitting angle times the spread length to get the difference in the length of the next pipe.

Difference in Length = Spread Length x Tangent of Angle A

Angle A = $1/2$ Offset Angle

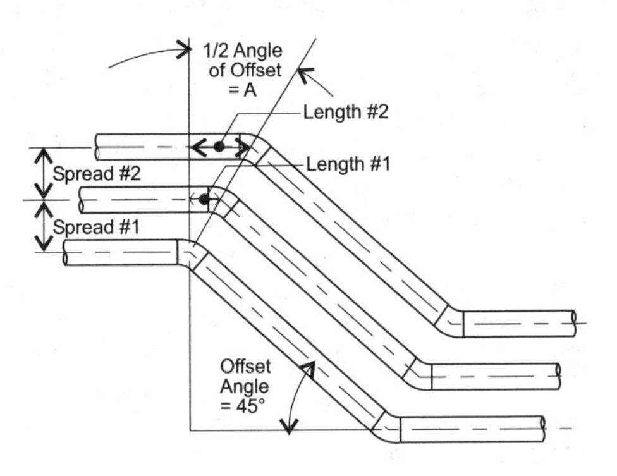

Illustration #122 - Using Tangent to Determine Offset Starting Point

90 Degree Turn Using Equal Spread 45 Degree Offsets

Equal spread offsets using 45 degree elbows are often used to make 90 degree directional changes in parallel piping runs. A typical 90 degree turn using 45 degree offsets is displayed in illustration #123.

In this type of piping offset, the piping run nearest to the turning point has the shortest travel length, while the run farthest from the turning point has the longest travel. The distance where the offset starts and ends on each line is calculated as other 45 degree equal spread offsets:

Difference in Length = Spread x 0.414

The travel distance calculation for each pipe in the offset differs from other offsets. The method of calculating each travel distance is as follows:

1. Shortest travel = 1.414 x Offset
2. Each following travel length = previous shortest travel + (2 x difference in length)

Example:

Find the difference in length and the travel lengths for a two pipe equal spread 45 degree offset which has a spread of 8 inches (203.2 mm) and an offset of 10 inches (254 mm).

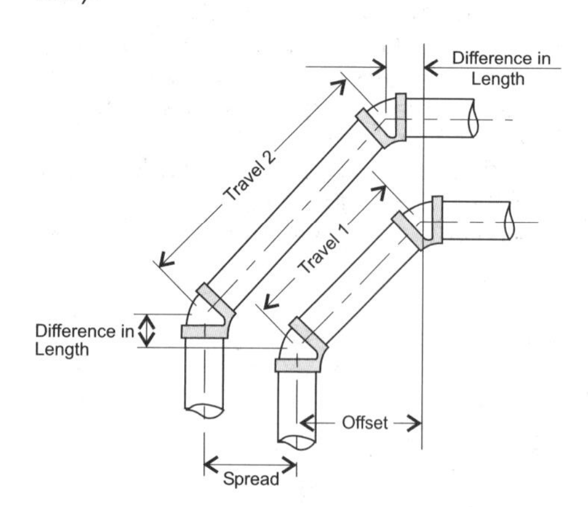

Illustration #123 - Offset Forming 90° Turn

Equal Spread 90 Degree Turn

Solution Steps:

Imperial Calculations:

1. Difference in length = spread x 0.414
   = 8 in. x 0.414
   = 3.312 inches
2. Travel #1 = 1.414 x offset
   = 1.414 x 10 in.
   = 14.14 (14 1/8) inches
3. Travel #2 = shorter travel + (2 x difference)
   = 14.14 in + (2 x 3.312 in.)
   = 14.14 in. + 6.624 in.
   = 20.764 (20 5/8) inches

Metric Calculations:

1. Difference in length = spread x 0.414
   = 203.2 x 0.414
   = 84.12 mm
2. Travel #1 = 1.414 x offset
   = 1.414 x 254 mm
   = 359.16 mm
3. Travel #2 = shorter travel + (2 x difference)
   = 359.16 mm + (2 x 84.12 mm)
   = 359.16 mm + 168.24 mm
   = 527.4 mm

Rolling Offsets

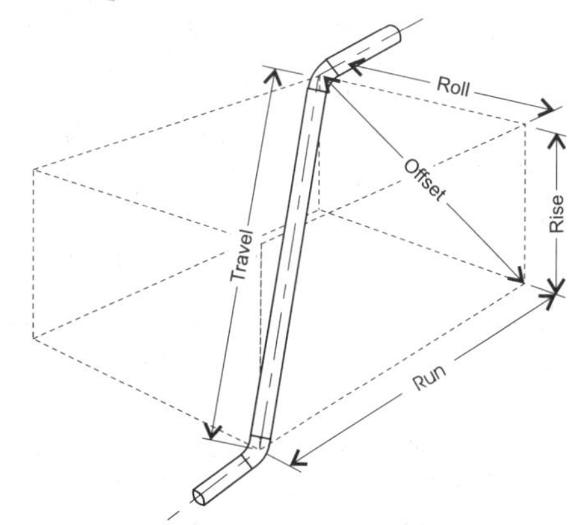

Illustration #124 - Rolling Offset Terminology

Rolling Offsets

When calculating rolling pipe offsets, it is helpful to visualize the piping offset within a three dimensional box. The pipe enters the box at a lower corner and angles diagonally across the box exiting from the opposite upper corner. Illustration #124 displays a typical rolling offset arrangement and the terminology used to describe the various parts of the rolling offset. A rolling offset can be made with any two fittings having the same fitting angle.

Rolling Offset Calculation

To calculate the travel in a rolling offset, the offset is multiplied by a constant:

Travel = Offset x Constant

The constants for typical fitting angles used in rolling offsets are given in table #68.

The offset length can be determined by using either the formula method or by the simpler steel square layout method shown in illustration #125.

$$\textbf{\textit{Offset}} = \sqrt{\textbf{\textit{Roll}}^2 + \textbf{\textit{Rise}}^2}$$

CONSTANTS AND FORMULAS FOR CALCULATING COMMON ROLLING OFFSETS

	Elbow Fitting Angle						
Formula	**90°**	**72°**	**60°**	**45°**	**30°**	**22 1/2°**	**1 1/4°**
Travel = Offset x	1.000	1.052	1.155	1.414	2.000	2.613	5.126
Run = Offset x	0	0.325	0.577	1.000	1.732	2.414	5.027

Table #68 - Rolling Offset Constants and Formulas

Rolling Offset Calculation

Example:

Find the travel and run for a 45 degree rolling offset with a roll of 11.5 inches (292.1 mm) and a rise of 15.5 inches (393.7 mm).

Solution Steps:

1. Use these formulas:
   a. Travel = Offset x Constant
   b. Run = Offset x Constant
2. Determine the offset by using the steel square or formula method.
3. The formula method for finding the offset is as follows:

Imperial Calculation:

$$Offset = \sqrt{Roll^2 + Rise^2}$$

$$Offset = \sqrt{11.5\,in.^2 + 15.5\,in.^2}$$

$$Offset = \sqrt{132.25\,in. + 240.25\,in.}$$

$$Offset = \sqrt{372.5\,in.} = 19.3\ (19\tfrac{5}{16})\ inches$$

Metric Calculation:

$$Offset = \sqrt{292.1mm^2 + 393.7\,mm^2}$$

$$Offset = \sqrt{85322.41mm + 154999.69\,mm}$$

$$Offset = \sqrt{240322.1mm} = 490.23\,mm$$

4. Select the travel formula and constant for a 45 degree elbow from table #68:

Imperial Calculation:

Travel = Offset x Constant
= 19.3 in. x 1.414 = 27.29 inches

Metric Calculation:

Travel = Offset x Constant
= 490.23 mm x 1.414 = 693.2 mm

5. Select the run formula and constant for a 45 degree elbow from table #68:

Imperial Calculation:

Run = Offset x Constant
= 19.3 in. x 1 = 19.3 inches

Metric Calculation:

Run = Offset x Constant
= 490.23 mm x 1 = 490.23 mm

Steel Square Method for Offsets

When using the steel square method, small offsets can be found by laying out the offset directly on the square.

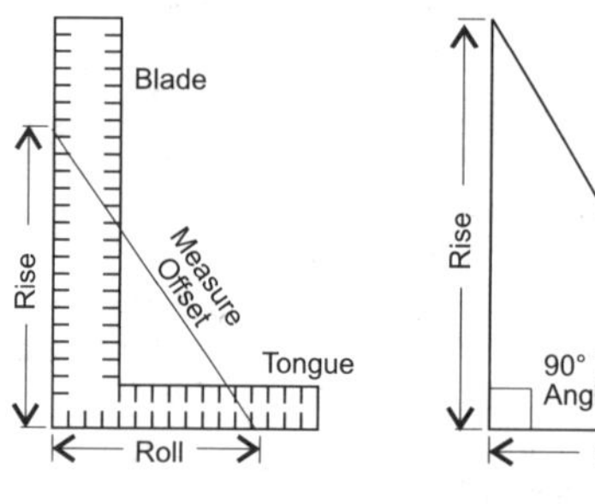

A. Direct Layout Method

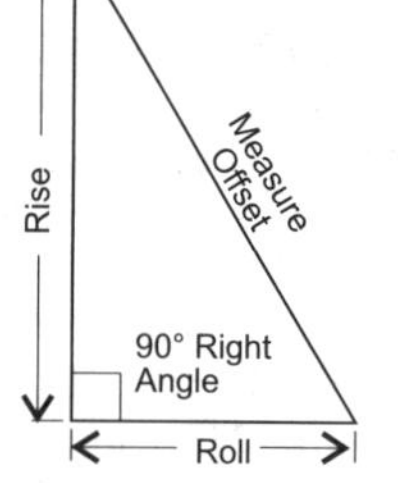

B. Triangle Layout Method

Illustration #125 - Rolling Offset Using Steel Square

In this method, the rise length of the offset is usually located on the blade side of the square and the roll on the tongue side. The distance measured diagonally across these two points establishes the offset. This method of establishing an offset is displayed in illustration #125A.

Triangle Method for Offsets

For larger offsets that do not fit on a steel square, a triangle can be laid out using the square to ensure a right angle. The rise dimension is placed on one side of the triangle and the roll on the other side. The travel is measured diagonally across these two points, see illustration #125B.

Note: The formula method gives a slightly more accurate calculation than does the use of a steel square. However, the square method is simpler and often more practical on job situations because it requires no involved math calculations.

Piping Offsets Around Square Corners

The starting point must be determined when making a 45 degree offset around a square obstruction of any type. The starting point is the distance from the obstruction to the center of fitting making the offset, see illustration #126. This point provides for an orderly pipe turn that does not run into, or over-protrude the corner, thereby becoming an obstruction in itself. The starting point of the elbow is calculated by the following formula:

Starting Point = A + (B x 1.414)

A = Distance from the center of straight run of pipe to end of wall

B = Distance perpendicular from the cen ter of the piping offset to corner

Example:

Find the starting point distance for a 45 degree offset from a square obstruction given A = 15 inches (381 mm) and B = 4 inches (101.6 mm).

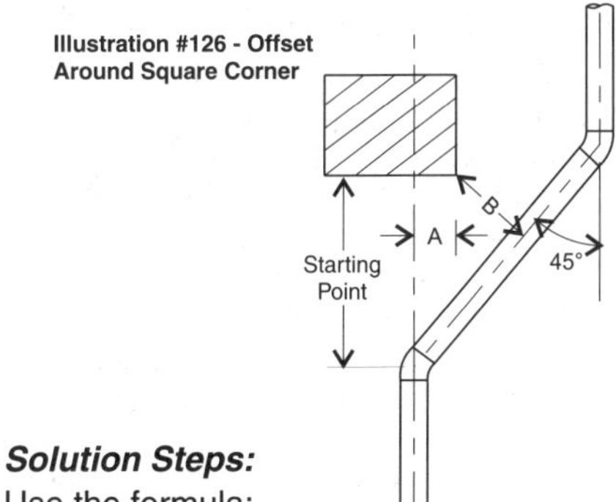

Solution Steps:

Use the formula:

Starting point = A + (B x 1.414)

Imperial Calculation:

Starting point = 15 in. + (4 in. x 1.414)

Starting point = 20.656 (20 $^{5}/_{8}$) inches

Metric Calculation:

Starting point = 381 mm + (101.6 mm x 1.414)

Starting point = 524.66 mm

Equal Spacing of Pipe Around Circular Objects

In order to provide for equal side spacing distance for a pipe offset around circular objects, the starting locations of the offset must be determined.

This method of equal side spacing of a pipe offset is shown in illustration #127.

The starting location is calculated using the following formula and constant from table #69:

Starting Point "A" = Distance from Pipe Center to Object Center x Constant

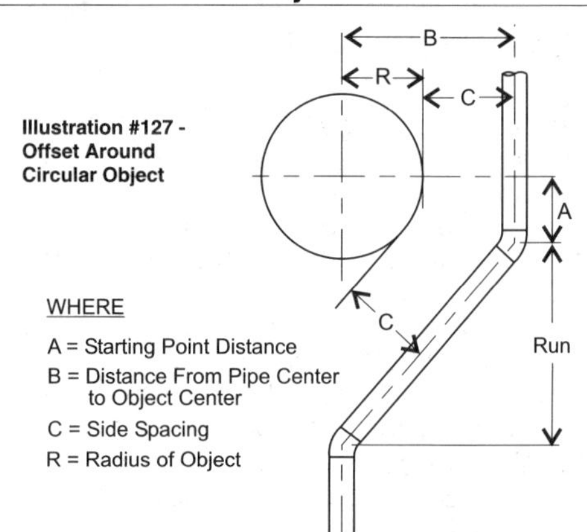

Illustration #127 - Offset Around Circular Object

CONSTANTS FOR CALCULATING STARTING POINT DISTANCE AROUND CIRCULAR OBJECTS				
Fitting or Bend Angles	**60°**	**45°**	**30°**	**22°**
Constants	0.5773	0.4142	0.2679	0.1989

Table #69

The distance from the pipe center to the circular object center can be determined by adding the radius of the object to the required side spacing distance between the pipe and the object:

Pipe Center to Object Center = Radius of Object + Required Side Spacing

Example:

Find the starting point length "A", for a 45 degree offset requiring equal side spacing around a circular object. The diameter of the circular object is 28 inches (711.2 mm) and the required equal side spacing distance between the pipe and object is 6 inches (152.4 mm).

Solution Steps:

Use the formula:

1. Starting point "A" = distance from pipe center to object center x constant
2. Calculate the distance from pipe center to object center:
3. Pipe center to object center "B" = radius of object + required side spacing

Imperial Calculation:

"B" = 14 in. + 6 in.
 = 20 inches

Metric Calculation:

"B" = 355.6 mm + 152.4 mm
 = 508 mm

Note: Radius = Diameter Divided by 2

4. Select the constant for a 45 degree elbow from table #69:
 Constant = 0.4142
5. Calculate starting point "A".

Imperial Calculation:

Starting point = 20 in. x 0.4142
 = 8.284 (8 5/16) inches

Metric Calculation:

Starting point = 508 mm x 0.4142
 = 210.4 mm

Calculating Pipe Runs Inside and Outside of Circular Tanks

Piping is often required to be installed so that it follows the contour of a circular tank using standard degree bends of fittings. Illustration #128 shows a typical example. The center to center measurement between each bend or fitting can be calculated using the following formula:

Center to Center Measurement = Piping Arrangement Radius x Constant

The piping arrangement radius is determined by:

a. Piping Runs Inside Tank = Tank Radius - Distance Between Tank and Piping

b. Piping Runs Outside Tank = Tank Radius + Distance Between Tank and Piping

Note: The radius measurement is to the center of the fitting.

Constant numbers for common fitting angles used for calculating center to center piping runs that follow the contour of circular tanks are given in table #70.

CONSTANTS FOR CALCULATING PIPE RUNS INSIDE AND OUTSIDE OF CIRCULAR TANKS							
Fitting or Bend Angles	**90°**	**60°**	**45°**	**30°**	**22½°**	**11¼°**	**5 ⅝°**
Constants	1.4142	1.0000	0.7653	0.5176	0.3902	0.1960	0.0981
# of Pipe Sections	4	6	8	12	16	32	64

Table #70 - Circular Tank Piping Constants

Piping Inside and Outside Tanks

Example:

Find the center to center measurements for piping runs to be installed following the inside and outside contour of a circular tank. The following information applies to this installation:

a. Tank diameter = 250 in. (6350 mm).

b. Distance (inside and out) between tank wall and piping = 10 in. (254 mm).

c. 45 degree elbows are to be used.

Solution Steps:

1. Change diameter of tank to radius:

Imperial Calculations:

Radius = diameter ÷ 2

= 250 in. ÷ 2 = 125 inches

Metric Calculations:

Radius = diameter ÷ 2

= 6350 mm ÷ 2 = 3175 mm

2. Calculate inside and outside piping arrangement radius.

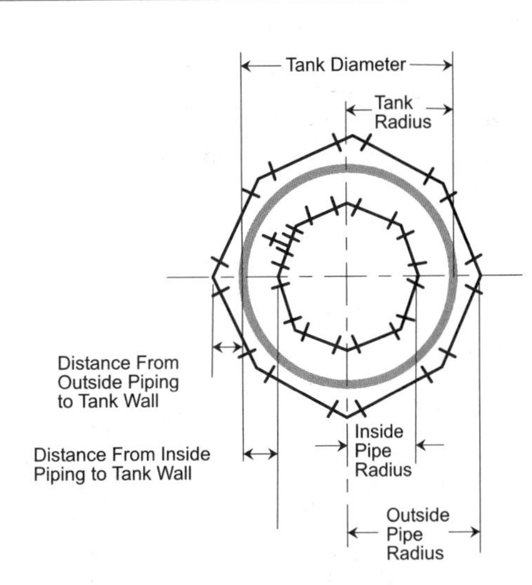

Illustration #128 - Contour Piping for Circular Tank

3. Piping inside radius = tank radius - distance between tank and piping

Imperial Calculations:

Piping inside radius = 125 in. - 10 in.
= 115 inches

Metric Calculations:

Piping inside radius = 3175 mm - 254 mm
= 2921 mm

4. Piping outside radius = tank radius + distance between tank and piping

Imperial Calculations:

Piping outside radius = 125 in. + 10 in.
= 135 inches

Metric Calculations:

Piping outside radius = 3175 mm + 254 mm
= 3429 mm

5. Select the constant for a 45 degree elbow and the number of pipes required to make the piping arrangement from table #70.

Constant for 45 degrees = 0.7653
Number of pipes required = 8

6. Determine the center to center measurements for the inside and outside piping sections using the preceding calculated information and the following formula:

Center to Center Measurement = Piping Arrangement Radius x Constant

7. Inside piping sectional center to center measurements (C to C)

Imperial Calculations:

C to C measurement = 115 in. x 0.7653
= 88 inches

Metric Calculations:

C to C measurement = 2921 mm x 0.7653
= 2235.44 mm

8. Outside piping sectional center to center measurements (C to C)

Imperial Calculations:

C to C measurement = 135 in. x 0.7653
= 103.32 (103 5/16) inches

Metric Calculations:

C to C measurement = 3429 mm x 0.7653
= 2624.21 mm

SECTION SEVEN

TRIGONOMETRY

Pythagorean Theorem

In any right-angle triangle, the square of the hypotenuse equals the sum of the square of the other two sides. This is shown in the formula in illustration #129.

It follows that if the hypotenuse and either side of a right-angle triangle is known, the other unknown side's length can be found by transposing the Pythagorean formula. This method of finding the unknown side is shown in the following:

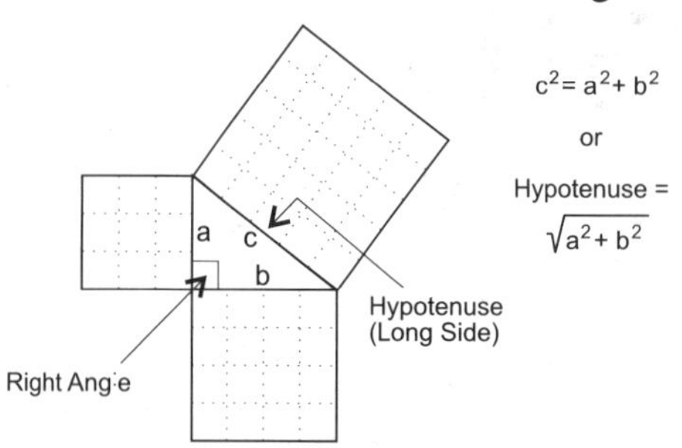

Illustration #129 - Pythagorean Theorem

$c^2 = a^2 + b^2$ $\quad$ $c = \sqrt{a^2 + b^2}$

$a^2 = c^2 - b^2$ $\quad$ OR $\quad$ $a = \sqrt{c^2 - b^2}$

$b^2 = c^2 - a^2$ $\quad$ $b = \sqrt{c^2 - a^2}$

Example #1:

Find the hypotenuse length C, for right-angle triangle, side a = 16 inches (406.4 mm) and side b = 30 inches (762 mm).

Solution:

Imperial Calculations:

Hypotenuse (c) $= \sqrt{a^2 + b^2}$

$= \sqrt{16^2 + 30^2} = \sqrt{(16 \times 16) + (30 \times 30)}$

$= \sqrt{256 + 900} = \sqrt{1156}$

= 34 in. or 2 ft. 10 in.

Metric Calculations:

$= \sqrt{406.4^2 + 762^2}$

$= \sqrt{165160.95 + 580644} = \sqrt{745804.96}$

= 863.6 mm

Trigonometry

The solution to a great many piping problems can be solved using trigonometry. Trigonometry is the study of triangle measurements and angles. Most piping offsets are actually triangle type layouts, especially right-angle triangles. Right-angle triangles consist of three sides and three angles (one angle being 90 degrees) formed by these sides.

Finding Angles for Right Angle Triangles

If any one side and any two parts (sides or angles) of the triangle are known, the other three parts of the triangle can be determined. Assuming two angles of a triangle are known, the third angle can be determined by the fact that the angles of a triangle ALWAYS equal 180 degrees.

Angles A + B + C = 180 degrees

One angle of a right angle triangle will always be 90 degrees.

Right Angle Triangle Example:

Angle C = 90 Degrees
Angle A + B = 90 degrees
Angle A = C (90 degrees) - Angle B
Angle B = C (90 degrees) - Angle A

In illustration #130, the parts of a right-angle triangle are labeled:

Letters A and B are the acute angles (angles of less than 90 degrees). Letter C indicates the right angle (90 degree angle). Small letters “a”, “b” and “c” indicate the sides OPPOSITE the angles (side “a” is opposite angle A, etc.).

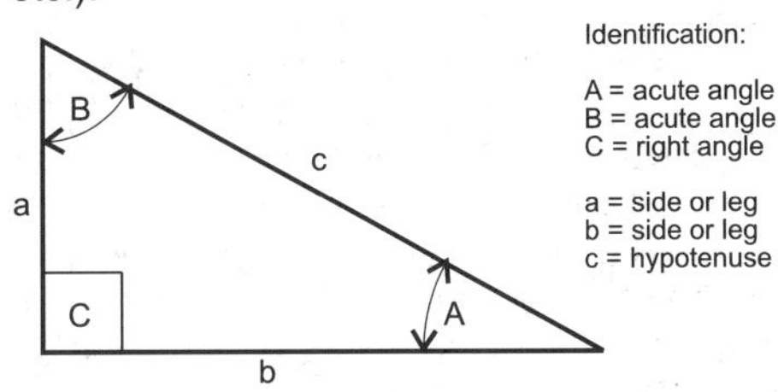

Illustration #130 - Angles and Sides of Right Angle Triangle

Right Angle Triangle Ratios or Functions

- ***Sine A*** - If the length of side "a" is divided by the length of side "c", the ratio obtained is called the sine of angle A.
- ***Cosine A*** - If the length of side "b" is divided by the length of side "c", the ratio is called the cosine of angle A.
- ***Tangent A*** - If the length of side "a" is divided by the length of side "b", the ratio is called the tangent of angle A.

These ratios between the sides are the same for all right angle triangles which have the same acute angles, regardless of the lengths of the sides. Ratios can be used to find the length of sides of any other right angle triangle with the same acute angle, as long as the length of one of its sides is known.

In defining these ratios, the legs or sides of the triangle are usually referred to as either "opposite" the acute angle, or "adjacent" to the angle.

Opposite side is located directly opposite the acute angle. Adjacent side is positioned next to or adjacent to the acute angle. The "hypotenuse" is always labeled the same (see illustration #131) and is positioned directly across from the 90 degree angle.

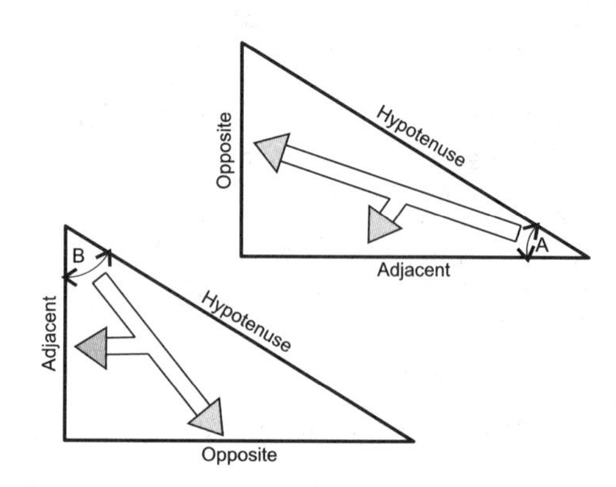

Illustration #131 - Opposite and Adjacent Sides

Right Angle Triangle Ratios or Functions

For each of the acute angles, we can compute six ratios or functions. The six ratios are given in table #71. Of the six angle functions, sine, cosine and tangent are the fundamental ones needed in right-angle triangle calculations.

Ratios of these sides make it possible to find any unknown angle or side. The information and formulas needed to find unknown angles or sides for right-angle triangles are given in table #72.

The calculated ratios or trigonometric functions are listed in table #73.

Example #2:

In illustration #132, calculate the length of the hypotenuse needed to connect two pipes that are running parallel 10 inches (254 mm) apart using 30 degree elbows.

Solution Steps:

1. Select relevant formula given information known:

side “a” is known = 10 inches (254 mm)

angle “A” is known = 30 degrees

2. From table #72, a formula is found:

side “c” = side “a” ÷ sine A

3. From table #73, the sine for a 30 degree angle = .50000

Imperial Calculations:

$$\text{“c”} = \frac{10}{.5000}$$

“c” = 20 inches = 1 ft. 8 in.

Metric Calculations:

$$\text{“c”} = \frac{254}{.5000}$$

“c” = 508 mm

Example #3:

In illustration #132, calculate the length of side "b" in the pipe offset.

Solution Steps:

1. Select relevant formula given information known:
   side "a" is known = 10 inches (254 mm)
   angle "A" is known = 30 degrees
2. From table #72, a formula is found:
   side "b" = side "a" ÷ tangent A
3. From table #73, the tangent for a 30 degree angle = .57735

Imperial Calculations:

$$\text{"b"} = \frac{10}{.57735}$$

"b" = 17.32 inches = 1 ft. 5 5/16 in.

Metric Calculations:

$$\text{"b"} = \frac{254}{.57735} = 439.9 \text{ mm}$$

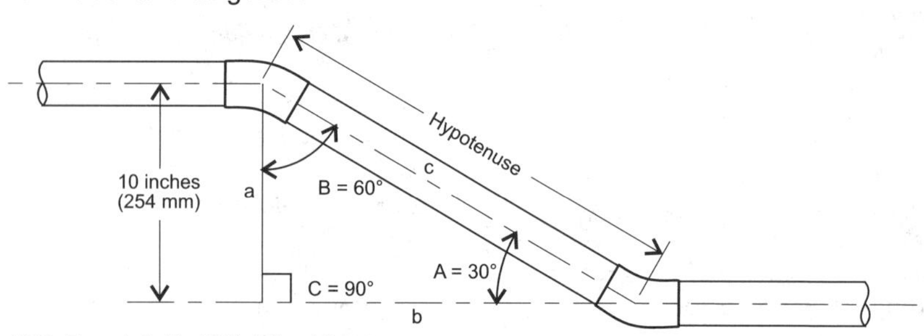

Illustration #132 - Example to Find Side "b" and Hypotenuse

TRIGONOMETRIC FUNCTIONS				
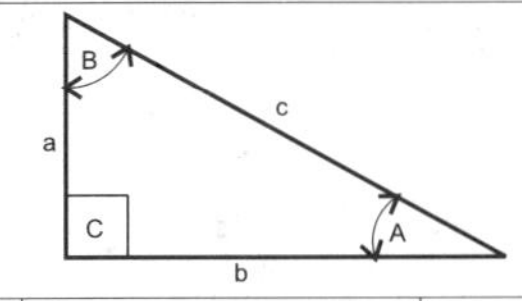				
Ratios		**Abbreviation**	**For Angle A**	**For Angle B**
Sine	$= \frac{\textit{Opposite Side}}{\textit{Hypotenuse}}$	$\sin = \frac{Opp}{Hyp}$	$\sin A = \frac{a}{c}$	$\sin B = \frac{b}{c}$
Cosine	$= \frac{\textit{Adjacent Side}}{\textit{Hypotenuse}}$	$\cos = \frac{Adj}{Hyp}$	$\cos A = \frac{b}{c}$	$\cos B = \frac{a}{c}$
Tangent	$= \frac{\textit{Opposite Side}}{\textit{Adjacent Side}}$	$\tan = \frac{Opp}{Adj}$	$\tan A = \frac{a}{b}$	$\tan B = \frac{b}{a}$
Cotangent	$= \frac{\textit{Adjacent Side}}{\textit{Opposite Side}}$	$\cot = \frac{Adj}{Opp}$	$\cot A = \frac{b}{a}$	$\cot B = \frac{a}{b}$
Secant	$= \frac{\textit{Hypotenuse}}{\textit{Adjacent Side}}$	$\sec = \frac{Hyp}{Adj}$	$\sec A = \frac{c}{b}$	$\sec B = \frac{c}{a}$
Cosecant	$= \frac{\textit{Hypotenuse}}{\textit{Opposite Side}}$	$\csc = \frac{Hyp}{Opp}$	$\csc A = \frac{c}{a}$	$\csc B = \frac{c}{b}$

Table #71 - Trigonometric Functions

TRIGONOMETRY FORMULAS FOR FINDING UNKNOWN SIDES AND ANGLES OF RIGHT-ANGLE TRIANGLES			
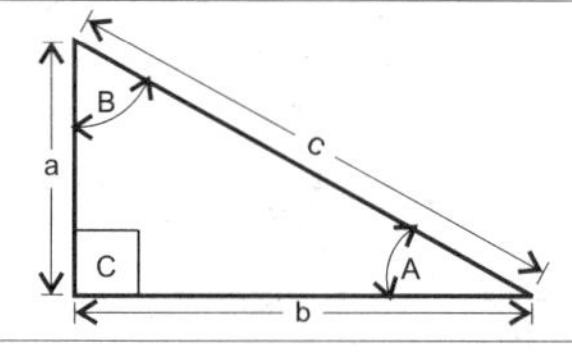			
Information Known Sides and Angles	**Formulas to Find Unknown Sides and Angles**		
	Sides		**Angles**
Side c and Angle A	a = c x sin A	b = c x cos A	B = 90 degrees - A
Side c and Angle B	a = c x cos B	b = c x sin B	A = 90 degrees - B
Side a and Angle A	$c = \dfrac{a}{\sin A}$	$b = \dfrac{a}{\tan A}$	B = 90 degrees - A
Side a and Angle B	$c = \dfrac{a}{\cos B}$	b = a x tan B	A = 90 degrees - B
Side b and Angle A	$c = \dfrac{b}{\cos A}$	a = b x tan A	B = 90 degrees - A
Side b and Angle B	$c = \dfrac{b}{\sin B}$	$a = \dfrac{b}{\tan B}$	A = 90 degrees - B

Table #72 - Trigonometric Formulas

TRIGONOMETRY TABLE							
Degree	Min.	Sine	Cosine	Tangent	Cotangent	Secant	Cosecant
0	00	.00000	1.00000	.00000	Infinite	1.0000	Infinite
1		.01745	.99985	.01745	57.290	1.0001	57.299
1	30	.02618	.99966	.02618	38.188	1.0003	38.201
2		.03490	.99939	.03492	28.636	1.0006	28.654
2	30	.04362	.99905	.04366	22.904	1.0009	22.925
3		.05234	.99863	.05241	19.081	1.0014	19.107
3	30	.06105	.99813	.06116	16.350	1.0019	16.380
4		.06976	.99756	.06993	14.301	1.0024	14.335
4	30	.07846	.99692	.07870	12.706	1.0031	12.745
5		.08715	.99619	.08749	11.430	1.0038	11.474
5	30	.09584	.99540	.09629	10.385	1.0046	10.433
6		.10453	.99452	.10510	9.5144	1.0055	9.5668
6	30	.11320	.99357	.11393	8.7769	1.0065	8.8337
7		.12187	.99255	.12278	8.1443	1.0075	8.2055
7	30	.13053	.99144	.13165	7.5957	1.0086	7.6613
8		.13917	.99027	.14054	7.1154	1.0098	7.1853
8	30	.14781	.98901	.14945	6.6911	1.0111	6.7655
9		.15643	.98679	.15838	6.3137	1.0125	6.3924
9	30	.16505	.98628	.16734	5.9758	1.0139	6.0538
10		.17365	.98481	.17633	5.6713	1.0154	5.7588
10	30	18223	.98325	.18534	5.3955	1.0170	5.4874
11		.19081	.98163	.19438	5.1445	1.0187	5.2408
11	30	.19937	.97972	.20345	4.9151	1.0205	5.0158
12		.20791	.97815	.21256	4.7046	1.0223	4.8097
12	30	.21644	.97630	.22169	4.5170	1.0243	4.6201
13		.22495	.97437	.23087	4.3315	1.0263	4.4454
13	30	.23344	.97237	.24008	4.1653	1.0284	4.2836
14		.24192	.97029	.24933	4.0108	1.0306	4.1336
14	30	.25038	.96815	.25862	3.8667	1.0329	3.9939
15		.25882	.96592	.26795	3.7320	1.0353	3.8637

Table #73A - Trigonometry Table

TRIGONOMETRY TABLE							
Degree	Min.	Sine	Cosine	Tangent	Cotangent	Secant	Cosecant
15	30	.26724	.96363	.27732	3.6059	1.0377	3.7420
16		.27564	.96126	.28674	3.4874	1.0403	3.6279
16	30	.28401	.95882	.29621	3.3759	1.0429	3.5209
17		.29237	.95630	.30573	3.2708	1.0457	3.4203
17	30	.30070	.95372	.31530	3.1716	1.0485	3.3255
18		.30902	.95106	.32492	3.0777	1.0515	3.2361
18	30	.31730	.94832	.33459	2.9887	1.0545	3.1515
19		.32557	.94552	.34433	2.9042	1.0576	3.0715
19	30	.33381	.94264	.35412	2.8239	1.0608	2.9957
20		.34202	.93969	.36397	2.7475	1.0642	2.9238
20	30	.35031	.93667	.37388	2.6746	1.0676	2.8554
21		.35837	.93358	.38386	2.6051	1.0711	2.7904
21	30	.36650	.93042	.39391	2.5386	1.0748	2.7285
22		.37461	.92718	.40403	2.4751	1.0785	2.6695
22	30	.38268	.92388	.41421	2.4142	1.0824	2.6131
23		.39073	.92050	.42447	2.3558	1.0864	2.5593
23	30	.39875	.91706	.43481	2.2998	1.0904	2.5078
24		.40674	.91354	.44523	2.2460	1.0946	2.4586
24	30	.41469	.90996	.45573	2.1943	1.0989	2.4114
25		.42262	.90631	.46631	2.1445	1.1034	2.3662
25	30	.43051	.90258	.47697	2.0965	1.1079	2.3228
26		.43837	.89879	.48773	2.0503	1.1126	2.2812
26	30	.44620	.89493	.49858	2.0057	1.1174	2.2411
27		.45399	.89101	.50952	1.9626	1.1223	2.2027
27	30	.46175	.88701	.52057	1.9210	1.1274	2.1657
28		.46947	.88295	.53171	1.8807	1.1326	2.1300
28	30	.47716	.87882	.54295	1.8418	1.1379	2.0957
29		.48481	.87462	.55431	1.8040	1.1433	2.0627
29	30	.49242	.87035	.56577	1.7675	1.1489	2.0308
30		.50000	.86603	.57735	1.7320	1.1547	2.0000

Table #73B - Trigonometry Table

TRIGONOMETRY TABLE							
Degree	Min.	Sine	Cosine	Tangent	Cotangent	Secant	Cosecant
30	30	.50754	.86163	.58904	1.6977	1.1606	1.9703
31		.51504	.85717	.60086	1.6643	1.1666	1.9416
31	30	.52250	.85264	.61280	1.6318	1.1728	1.9139
32		.52992	.84805	.62487	1.6003	1.1792	1.8871
32	30	.53730	.84339	.63707	1.5697	1.1857	1.8611
33		.54464	.83867	.64941	1.5399	1.1924	1.8361
33	30	.55191	.83388	.66188	1.5108	1.1992	1.8118
34		.55919	.82904	.67451	1.4826	1.2062	1.7883
34	30	.56641	.82413	.68728	1.4550	1.2134	1.7655
35		.57358	.81915	.70021	1.4281	1.2208	1.7434
35	30	.58070	.81411	.71329	1.4019	1.2283	1.7220
36		.58778	.80902	.72654	1.3764	1.2361	1.7013
36	30	.59482	.80386	.73996	1.3514	1.2442	1.6812
37		.60181	.79863	.75355	1.3270	1.2521	1.6616
37	30	.60876	.79335	.76733	1.3032	1.2605	1.6427
38		.61566	.78801	.78128	1.2799	1.2690	1.6243
38	30	.62251	.78261	.79543	1.2572	1.2778	1.6064
39		.62932	.77715	.80978	1.2349	1.2867	1.5890
39	30	.63608	.77162	.82434	1.2131	1.2960	1.5721
40		.64279	.76604	.83910	1.1917	1.3054	1.5557
40	30	.64945	.76041	.85408	1.1708	1.3151	1.5398
41		.65606	.75471	.86929	1.1504	1.3250	1.5242
41	30	.66262	.74895	.88472	1.1303	1.3352	1.5092
42		.66913	.74314	.90040	1.1106	1.3456	1.4945
42	30	.67559	.73728	.91633	1.0913	1.3563	1.4802
43		.68200	.73135	.93251	1.0724	1.3673	1.4663
43	30	.68835	.72357	.94896	1.0538	1.3786	1.4527
44		.69466	.71934	.96569	1.0355	1.3902	1.4395
44	30	.70091	.71325	.98270	1.0176	1.4020	1.4267
45		.70711	.70711	1.00000	1.0000	1.4142	1.4142

Table #73C - Trigonometry Table

TRIGONOMETRY TABLE							
Degree	Min.	Sine	Cosine	Tangent	Cotangent	Secant	Cosecant
45	30	.71325	.70091	1.0176	.98270	1.4267	1.4020
46		.71934	.69466	1.0355	.96569	1.4395	1.3902
46	30	.72357	.68835	1.0538	.94896	1.4527	1.3786
47		.73135	.68200	1.0724	.93251	1.4663	1.3673
47	30	.73728	.67559	1.0913	.91633	1.4802	1.3563
48		.74314	.66913	1.1106	.90040	1.4945	1.3456
48	30	.74895	.66262	1.1303	.88472	1.5092	1.3352
49		.75471	.65606	1.1504	.86929	1.5242	1.3250
49	30	.76041	.64945	1.1708	.85408	1.5398	1.3151
50		.76604	.64279	1.1917	.83910	1.5557	1.3054
50	30	.77162	.63608	1.2131	.82434	1.5721	1.2960
51		.77715	.62932	1.2349	.80978	1.5890	1.2867
51	30	.78261	.62251	1.2572	.79543	1.6064	1.2778
52		.78801	.61566	1.2799	.78128	1.6243	1.2690
52	30	.79335	.60876	1.3032	.76733	1.6427	1.2605
53		.79863	.60181	1.3270	.75355	1.6616	1.2521
53	30	.80386	.59482	1.3514	.73996	1.6812	1.2442
54		.80902	.58778	1.3764	.72654	1.7013	1.2361
54	30	.81411	.58070	1.4019	.71329	1.7220	1.2283
55		.81915	.57358	1.4281	.70021	1.7434	1.2208
55	30	.82413	.56641	1.4550	.68728	1.7655	1.2134
56		.82904	.55919	1.4826	.67451	1.7883	1.2062
56	30	.83388	.55191	1.5108	.66188	1.8118	1.1992
57		.83867	.54464	1.5399	.64941	1.8361	1.1924
57	30	.84339	.53730	1.5697	.63707	1.8611	1.1857
58		.84805	.52992	1.6003	.62487	1.8871	1.1792
58	30	.85264	.52250	1.6318	.61280	1.9139	1.1728
59		.85717	.51504	1.6643	.60086	1.9416	1.1666
59	30	.86163	.50754	1.6977	.58904	1.9703	1.1606
60		.86603	.50000	1.7320	.57735	2.0000	1.1547

Table #73D - Trigonometry Table

TRIGONOMETRY TABLE							
Degree	Min.	Sine	Cosine	Tangent	Cotangent	Secant	Cosecant
60	30	.87035	.49242	1.7675	.56577	2.0308	1.1489
61		.87462	.48481	1.8040	.55431	2.0627	1.1433
61	30	.87882	.47716	1.8418	.54295	2.0957	1.1379
62		.88295	.46947	1.8807	.53171	2.1300	1.1326
62	30	.88701	.46175	1.9210	.52057	2.1657	1.1274
63		.89101	.45399	1.9626	.50952	2.2027	1.1223
63	30	.89493	.44620	2.0057	.49858	2.2411	1.1174
64		.89879	.43837	2.0503	.48773	2.2812	1.1126
64	30	.90258	.43051	2.0965	.47697	2.3228	1.1079
65		.90631	.42262	2.1445	.46631	2.3662	1.1034
65	30	.90996	.41469	2.1943	.45573	2.4114	1.0989
66		.91354	.40674	2.2460	.44523	2.4586	1.0946
66	30	.91706	.39875	2.2998	.43481	2.5078	1.0904
67		.92050	.39073	2.3558	.42447	2.5593	1.0864
67	30	.92388	.38268	2.4142	.41421	2.6131	1.0824
68		.92718	.37461	2.4751	.40403	2.6695	1.0785
68	30	.93042	.36650	2.5386	.39391	2.7285	1.0748
69		.93358	.35837	2.6051	.38386	2.7904	1.0711
69	30	.93667	.35031	2.6746	.37388	2.8554	1.0676
70		.93969	.34202	2.7475	.36397	2.9238	1.0642
70	30	.94264	.33381	2.8239	.35412	2.9957	1.0608
71		.94552	.32557	2.9042	.34433	3.0715	1.0576
71	30	.94832	.31730	2.9887	.33459	3.1515	1.0545
72		.95106	.30902	3.0777	.32492	3.2361	1.0515
72	30	.95372	.30070	3.1716	.31530	3.3255	1.0485
73		.95630	.29237	3.2708	.30573	3.4203	1.0457
73	30	.95882	.28401	3.3759	.29621	3.5209	1.0429
74		.96126	.27564	3.4874	.28674	3.6279	1.0403
74	30	.96363	.26724	3.6059	.27732	3.7420	1.0377
75		.96592	.25882	3.7320	.26795	3.8637	1.0353

Table #73E - Trigonometry Table

TRIGONOMETRY TABLE							
Degree	Min.	Sine	Cosine	Tangent	Cotangent	Secant	Cosecant
75	30	.96815	.25038	3.8667	.25862	3.9939	1.0329
76		.97029	.24192	4.0108	.24933	4.1336	1.0306
76	30	.97237	.23344	4.1653	.24008	4.2836	1.0284
77		.97437	.22495	4.3315	.23087	4.4454	1.0263
77	30	.97630	.21644	4.5170	.22169	4.6201	1.0243
78		.97815	.20791	4.7046	.21256	4.8097	1.0223
78	30	.97972	.19937	4.9151	.20345	5.0158	1.0205
79		.98163	.19081	5.1445	.19438	5.2408	1.0187
79	30	.98325	.18223	5.3955	.18534	5.4874	1.0170
80		.98481	.17365	5.6713	.17633	5.7588	1.0154
80	30	.98628	.16505	5.9758	.16734	6.0538	1.0139
81		.98769	.15643	6.3137	.15838	6.3924	1.0125
81	30	.98901	.14781	6.6911	.14945	6.7655	1.0111
82		.99027	.13917	7.1154	.14054	7.1853	1.0098
82	30	.99144	.13053	7.5957	.13165	7.6613	1.0086
83		.99255	.12187	8.1443	.12278	8.2055	1.0075
83	30	.99357	.11320	8.7769	.11393	8.8337	1.0065
84		.99452	.10453	9.5144	.10510	9.5668	1.0055
84	30	.99540	.09584	10.385	.09629	10.433	1.0046
85		.99619	.08715	11.430	.08749	11.474	1.0038
85	30	.99692	.07846	12.706	.07870	12.745	1.0031
86		.99756	.06976	14.301	.06993	14.335	1.0024
86	30	.99813	.06105	16.350	.06116	16.380	1.0019
87		.99863	.05234	19.081	.05241	19.107	1.0014
87	30	.99905	.04362	22.904	.04366	22.925	1.0009
88		.99939	.03490	28.636	.03492	28.654	1.0006
88	30	.99966	.02618	38.188	.02618	38.201	1.0003
89		.99985	.01745	57.290	.01745	57.299	1.0001
90		1.00000	.00000	Infinite	.00000	Infinite	1.0000

Table #73F - Trigonometry Table

SECTION EIGHT

PIPE LAYOUT

Layout of Fabricated Fittings

There are two basic fitting types that are used in the installation of a butt welded piping system. These two categories of fittings can be classified as either:

- Factory Manufactured Fittings
- Job Fabricated Fittings

Job Fabricated Fittings

Fabricated fittings are classified as fittings that are made up on the job site or in fabrication shops. These fittings are used in circumstances where achieving the same results with a manufactured fitting is difficult, or impossible. The two methods of laying out pipe for making job fabricated fittings are:

- template development
- direct pipe layout

Template Development Method - A full sized drawing or template is developed for the required fabricated shape. The template is wrapped around the pipe and the pattern is traced on the pipe surface. The pipe is then cut, usually with an oxy-acetylene torch, following the scribed marks produced by tracing the template.

Direct layout Method - The pattern for the fabricated fitting is drawn directly on the pipe from which the fitting is to be made. Some of the more common methods of direct pipe layout and marking are explained in this section followed by detailed specific fitting fabrication methods.

Cutting Odd Angle Elbows from Manufactured Elbows

Factory manufactured butt welded elbows usually provide for standard turns of 45 and 90 degrees. These elbows and their dimensions are discussed in full in the Fittings Section of this book. Some job situations require welded elbow angles that are not available. The required elbow may be cut from a standard stock elbow using the procedure and layout dimensions given in illustration #133 and table #74.

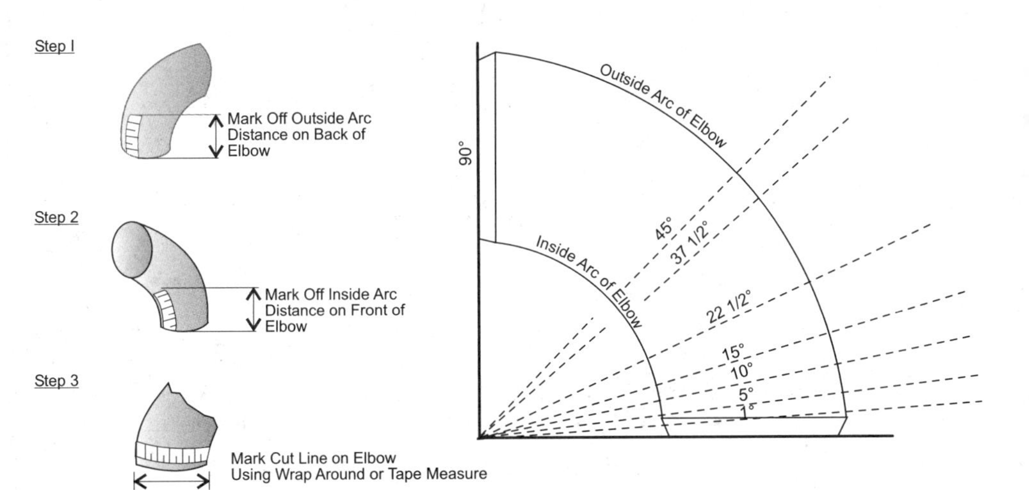

Illustration #133 - Cutting Odd Angle Elbows

ELBOWS FABRICATED FROM 90° L.R. ELBOWS (Inches)

Nom. Pipe Size	Outside Arc Distance for Required Elbow Degree							Inside Arc Distance for Required Elbow Degree						
	1°	5°	10°	15°	22½°	37½°	45°	1°	5°	10°	15°	22½°	37½°	45°
2	5/64	3/8	23/32	1 3/32	1 21/32	2 3/4	3 9/32	1/32	5/32	5/16	15/32	23/32	1 3/16	1 7/16
2 1/2	3/32	7/16	29/32	1 11/32	2 1/32	3 3/8	4 1/16	3/64	3/16	13/32	19/32	29/32	1 1/2	1 13/16
3	7/64	9/16	1 1/8	1 5/8	2 15/32	4 3/32	4 29/32	3/64	1/4	1/2	23/32	1	1 13/16	2 5/32
3 1/2	1/8	5/8	1 9/32	1 29/32	2 27/32	4 3/4	5 11/16	1/16	9/32	9/16	27/32	1 9/32	2 1/8	2 9/16
4	9/64	23/32	1 7/16	2 5/32	3 1/4	5 13/32	6 15/16	1/16	5/16	21/32	31/32	1 15/32	2 7/16	2 15/16
5	3/16	29/32	1 25/32	2 11/16	4 1/32	6 23/32	8 1/16	5/64	13/32	13/16	1 1/4	1 27/32	3 3/32	3 23/32
6	7/32	1 1/16	2 5/32	3 7/32	4 27/32	8 1/16	9 21/32	3/32	1/2	1	1 1/2	2 7/32	3 23/32	4 15/32
8	9/32	1 7/16	2 27/32	4 9/32	6 13/32	10 11/16	12 13/16	1/8	11/16	1 11/32	2	3 1/32	5 1/32	6 1/32
10	11/32	1 25/32	3 9/16	5 11/32	8	13 11/32	16	5/32	27/32	1 11/16	2 17/32	3 25/32	6 5/16	7 9/16
12	7/16	2 1/8	4 1/4	6 3/8	9 9/16	15 31/32	19 5/32	7/32	1	2 1/32	3 1/16	4 9/16	7 19/32	9 1/8
14	1/2	2 7/16	4 7/8	7 5/16	11	18 5/16	22	1/4	1 7/32	2 7/16	3 21/32	5 1/2	9 5/32	11
16	9/16	2 13/16	5 19/32	8 3/8	12 9/16	20 15/16	25 1/8	9/32	1 13/32	2 13/16	4 3/16	6 9/32	10 15/32	12 5/8
18	5/8	3 1/8	6 9/32	9 7/16	14 1/8	23 9/16	28 9/32	5/16	1 9/16	3 1/8	4 23/32	7 1/16	11 25/32	14 1/8
20	11/16	3 1/2	7	10 15/32	15 23/32	26 3/16	31 13/32	11/32	1 3/4	3 1/2	5 1/4	7 27/32	13 3/32	15 11/16
22	3/4	3 27/32	7 11/16	11 17/32	17 9/32	28 13/16	34 9/16	3/8	1 29/32	3 27/32	5 3/4	8 5/8	14 3/8	17 9/32
24	27/32	4 3/16	8 3/8	12 9/16	18 27/32	31 13/32	37 11/16	13/32	2 3/32	4 3/16	6 9/32	9 7/16	15 11/16	18 27/32
26	29/32	4 17/32	9 3/32	13 5/8	20 13/32	34 1/32	40 27/32	15/32	2 9/32	4 17/32	6 13/16	10 7/32	17 1/32	20 13/32
30	1 1/32	5 1/4	10 15/32	15 3/4	23 9/16	39 1/4	47 1/8	17/32	2 5/8	5 1/4	7 5/8	11 25/32	19 5/8	23 9/16
34	1 5/32	5 29/32	11 27/32	17 13/16	26 23/32	44 17/32	53 3/8	19/32	2 31/32	5 29/32	8 29/32	13 3/8	22 9/32	26 11/16
36	1 7/32	6 1/4	12 17/32	18 7/8	28 7/32	47	56 17/32	5/8	2 13/16	6 1/4	9 7/16	14 1/8	23 5/8	28 1/4
42	1 7/16	7 5/16	14 5/8	22	32 31/32	54 31/32	65 15/16	23/32	3 21/32	7 5/16	10 19/32	16 1/2	26 3/8	32 31/32

Table #74A - Elbows Fabricated from 90 Degree L.R. Elbows (Inches)

ELBOWS FABRICATED FROM 90° L.R. ELBOWS (Millimetres)														
Nom. Pipe Size	Outside Arc Distance for Required Elbow Degree							Inside Arc Distance for Required Elbow Degree						
	1°	5°	10°	15°	22½°	37½°	45°	1°	5°	10°	15°	22½°	37½°	45°
50	1.9	9.5	18.3	27.8	42.1	69.9	83.3	0.8	4.0	7.9	11.9	18.3	30.2	36.5
65	2.4	11.1	23.0	34.1	51.6	85.7	103.2	1.2	4.8	9.5	15.1	23.0	38.1	46.0
80	2.8	14.3	28.6	41.3	62.7	104.0	124.6	1.2	6.4	12.7	18.3	27.8	46.0	54.8
90	3.2	15.9	32.5	48.4	72.2	120.7	144.5	1.6	7.1	14.3	21.4	32.5	54.0	65.1
100	3.6	18.3	36.5	54.8	82.6	137.3	164.3	1.6	7.9	16.7	24.6	37.3	61.9	74.6
125	4.8	23.0	45.2	68.3	102.4	170.7	204.8	1.9	10.3	20.6	31.8	46.8	78.6	94.5
150	5.6	27.0	54.8	81.8	123.0	204.8	245.3	2.4	12.7	25.4	38.1	56.4	94.5	113.5
200	7.1	36.5	72.2	108.7	162.7	271.5	325.4	3.2	17.5	34.1	50.8	77.0	127.8	153.2
250	8.7	45.2	90.5	135.7	203.2	338.9	406.4	4.0	21.4	42.9	64.3	96.0	160.3	192.1
300	11.1	54.0	108.0	161.9	242.9	405.6	486.6	5.6	25.4	51.6	77.8	115.9	192.9	231.8
350	12.7	61.9	123.8	185.7	279.4	465.1	558.8	6.4	31.0	61.9	92.9	139.7	232.6	279.4
400	14.3	71.4	142.1	212.7	319.1	531.8	638.2	7.1	35.7	71.4	106.4	159.5	265.9	320.7
450	15.9	79.4	159.5	239.7	358.8	598.5	718.3	7.9	39.7	79.4	119.9	179.4	299.2	358.8
500	17.5	88.9	177.8	265.9	399.3	665.2	797.7	8.7	44.5	88.9	133.4	199.2	332.6	398.5
550	19.1	97.6	195.3	292.9	438.9	731.8	877.9	9.5	48.4	97.6	146.1	219.1	365.1	438.9
600	21.4	106.4	212.7	319.1	478.6	797.7	951.3	10.3	53.2	106.4	159.5	239.7	398.5	478.6
650	23.0	115.1	231.0	346.1	518.3	864.4	1037.4	11.9	57.9	115.1	173.0	259.6	432.6	518.3
750	26.2	133.4	265.9	400.1	598.5	997.0	1197.0	13.5	66.7	133.4	200.0	299.2	498.5	598.5
850	29.4	150.0	300.8	452.4	678.7	1131.1	1355.7	15.1	75.4	150.0	226.2	339.7	565.9	601.7
900	31.0	158.8	318.3	479.4	716.8	1193.8	1435.9	15.9	71.4	158.8	239.7	358.8	600.1	717.6
1050	36.5	185.7	371.5	558.8	837.4	1396.2	1674.8	18.3	92.9	185.7	269.1	419.1	669.9	837.4

Table #74B - Elbows Fabricated from 90 Degree L.R. Elbows (Millimetres)

Establishing Centerlines on Pipe

With Square and Level

Top, bottom and side center layout lines can be located on horizontal pipe by the use of a steel framing square. The square is leveled on the pipe with both inside edges of the square touching the pipe surface. Horizontal and vertical pipe division centerlines are located and marked off by measuring from the square's inside edges. The distance for each centerline from the square's edge equals half of the outside diameter of the pipe. This procedure is shown in illustration #134.

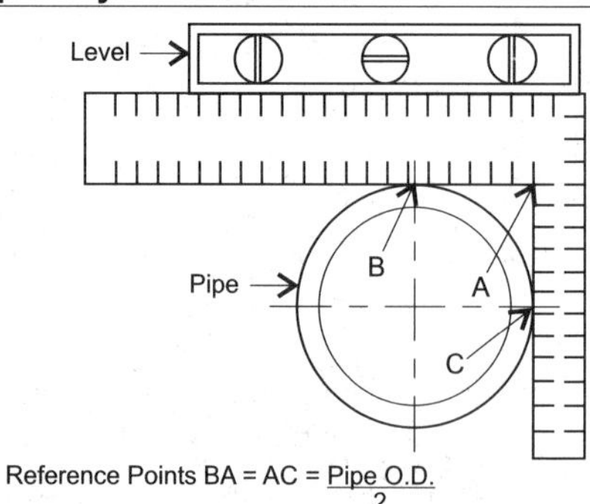

Illustration #134 - Use Square to Establish Centerlines

With Plumb Bob and Tape Measure

On larger pipe, the top and bottom center division lines can be located by the use of a plumb bob and a tape measure. A tape measure is placed across the horizontal center section of the pipe and a plumb bob is positioned over the top of the pipe.

The top of the plumb bob line is moved until the center of the line crosses the tape at a distance corresponding to half of the pipe's outside diameter. After this is completed, the pipe is marked at the top and bottom intersections of the plumb bob, see illustration #135. For side or horizontal centerlines, the pipe may be rotated 1/4 turn and marked in the same manner.

Pipe Centerlines

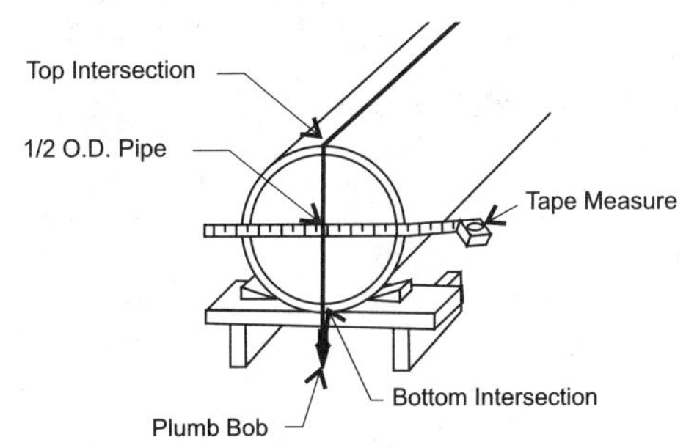

Illustration #135 - Use Plumb Bob for Centerlines

Extending Straight Lines on Pipe

Straight lines along the length of pipe can be easily extended from established center division lines by the use of angle iron. A piece of angle iron is placed on the pipe and is used as a straight edge for extending the centerlines along the outside length of the pipe.

It is important that the angle iron be held squarely and solidly in place without movement. This procedure will ensure accurate parallel lines extending along the length of the pipe form each center division line or reference point. This method of line extension is shown in illustration #136.

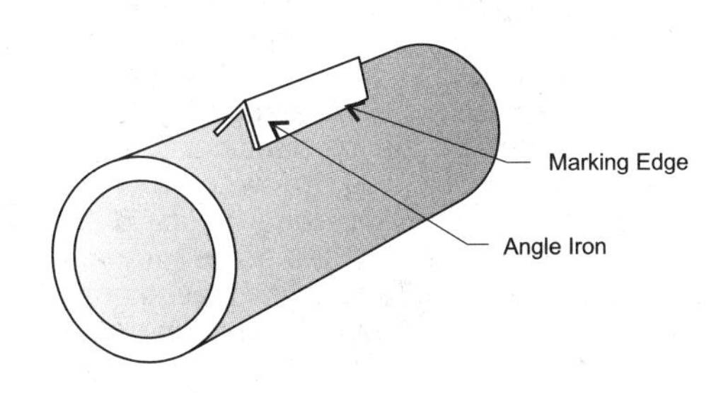

Illustration #136 - Use Angle Iron for Straight Edge

Wrap-Around for Circumference Line Layout

Circumference lines can be easily and precisely marked on pipe by the use of a "wrap-around". A wrap-around is a flexible wrap made from gasket-like material which is used as a guide for marking pipe. The wrap-around is placed around the pipe and pulled tight to ensure complete surface contact with the pipe. It is wrapped around the pipe, overlapping by at least 1/4 of the pipe circumference. This overlapping provides for a circumference line perpendicular to the pipe centerlines. When the wrap-around is positioned correctly, the square edge of the wrap-around is used as the guide for marking. Illustration #137 shows the placement of a wrap-around on a pipe for marking a circumference line.

Sectoring Pipe Into Equal Divisions

Pipe may be divided into four equal parts by the square and level method described previously, or on smaller pipe sizes by the paper folding method.

In the paper folding method, a strip of paper or similar material is securely wrapped around the piece of pipe that is to be divided. The overlapping end of the paper is cut so that both ends of the paper are just touching each other. The paper is folded in half. The two halves are then folded in half again. See illustration #138. These folded creases and paper ends, when placed on the pipe, provide the location points for marking the pipe into quarters.

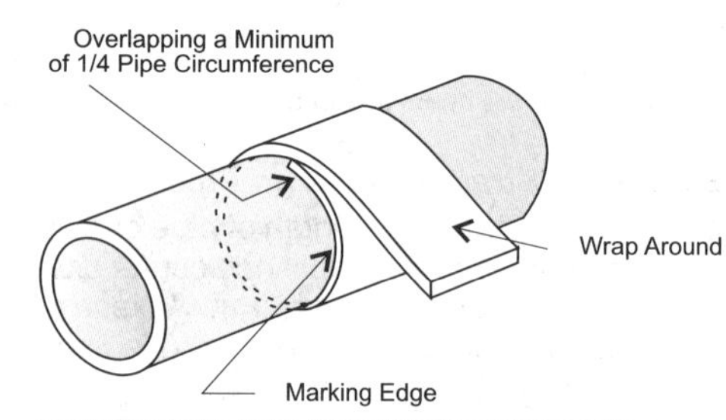

Illustration #137 - Wrap-Around for Circumference Line

Sectoring Pipe into Divisions

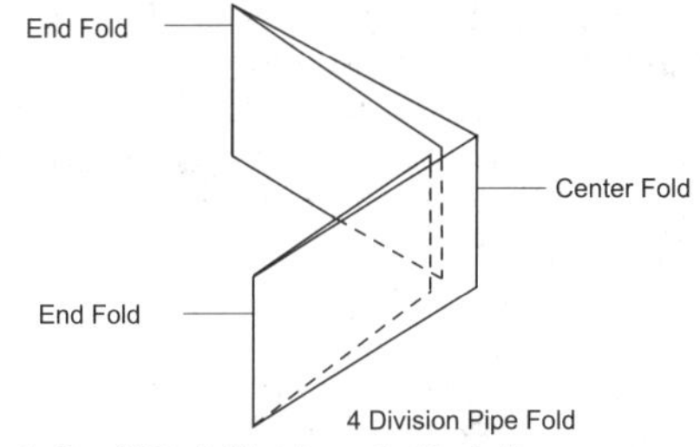

Illustration #138 - Folded Paper for Centerlines

Sectoring Pipe with a Tape Measure

A tape measure can be used to divide pipe into any number of equal sectors. The first step is to find the pipe circumference. This may be done by multiplying the pipe's outside diameter by 3.1416 (π) or by directly measuring the circumference of the pipe with a tape measure. The circumference is divided by the number of sectors needed.

The sectors or divisions are then laid out by wrapping the tape measure around the pipe and marking off the calculated division points. This procedure is shown in illustration #139.

Note: For accuracy, use a flexible tape measure.

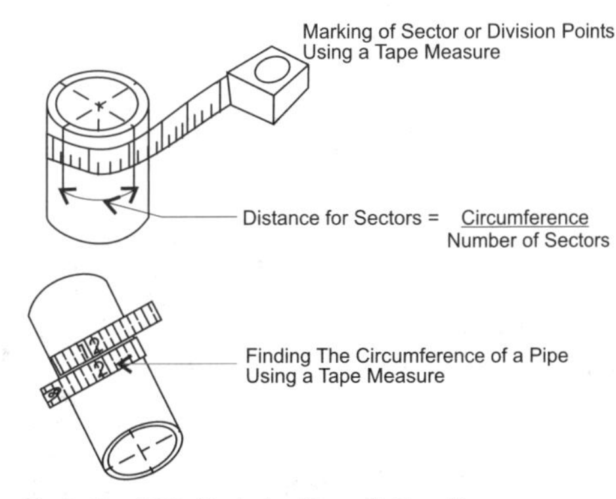

Illustration #139 - Sectoring Pipe with Tape Measure

Mitered Turns - 2 Piece

In order to fabricate a simple two piece miter turn, the following information is needed:

1. Cut Angle of Miter
2. Factor of the Cut Angle
3. Cut Back Distance

Cut Angle of Miter

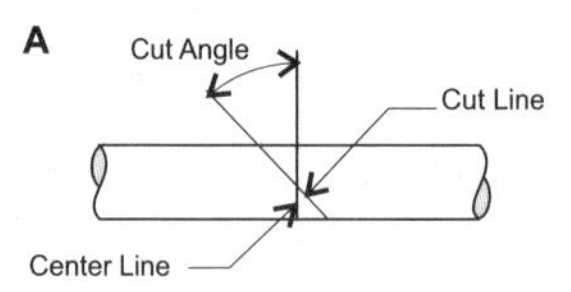

Cut Back Dimension of Miter

B

A

Cut Line

Center Line

A

A = Cut Back Dimension of Miter

Illustration #140A,B - Cut Back Distance

Cut Angle of Miter

The cut angle of a miter turn is the angle to which each piece of pipe in the turn must be cut. See illustration #140A. ***For a two piece miter turn, the cut angle can be found by dividing the angle of the turn by 2.***

Cut Angle = Angle of Turn ÷ 2

Factor of the Cut Angle

Factor numbers are used in the miter calculations for determining the cut back distance of the miter turn. Factors for the various cut angles are found in table #75.

Cut Back Distance

The cut back measurement sets the distance on each side of the centerline to start and end the miter cut. See illustration #140B. Cut back distance is determined by multiplying the pipe O.D. times the factor of the cut angle and dividing by 2:

Cut Back = Pipe O.D. x Cut Angle Factor ÷ 2

Common angles for two piece miter turns are shown in illustration #141, along with the appropriate cut angle and factor.

CUT ANGLE FACTORS							
Cut Angle	Factor	Cut Angle	Factor	Cut Angle	Factor	Cut Angle	Factor
5° 30	.09629	15° 30	.22732	25° 30	.46797	35° 30	.71329
6	.10510	16	.28674	26	.48773	36	.72654
6 30	.11393	16 30	.29621	26 30	.49858	36 30	.73996
7	.12278	17	.30573	27	.50952	37	.75355
7 30	.13165	17 30	.31530	27 30	.52057	37 30	.76733
8	.14054	18	.32492	28	.53171	38	.78128
8 30	.14945	18 30	.33459	28 30	.54295	38 30	.79543
9	.15838	19	.34433	29	.55431	39	.80978
9 30	.16734	19 30	.35412	29 30	.56577	39 30	.82424
10	.17633	20	.36397	30	.57735	40	.83910
10 30	.18534	20 30	.37388	30 30	.58904	40 30	.85408
11	.19438	21	.38386	31	.60086	41	.86929
11 30	.20345	21 30	.39391	31 30	.61280	41 30	.88472
12	.21256	22	.40403	32	.62487	42	.90040
12 30	.22169	22 30	.41421	32 30	.63707	42 30	.91633
13	.23087	23	.42447	33	.64941	43	.93251
13 30	.24008	23 30	.43481	33 30	.66188	43 30	.94896
14	.24933	24	.44523	34	.67451	44	.96589
14 30	.25862	24 30	.45573	34 30	.68728	44 30	.98285
15	.26795	25	.46631	35	.70021	45	1.0000

Table #75 - Cut Angle Factors

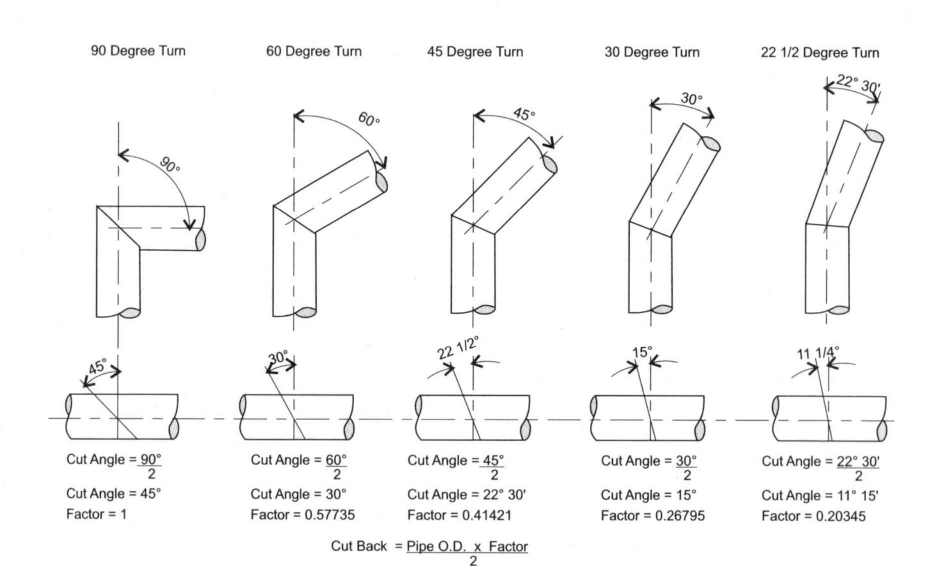

Illustration #141 - Common Two Piece Miter Turns

Two Piece Miter Layout

1. Using a wrap-around as a guide, draw a straight line around the circumference of the pipe. See illustration #142A. This line will become the center circumference line of the fitting.
2. Divide the center circumference line into 4 equal sections. Label these section lines 1, 2, 3, 4, starting at the top of the pipe and working clockwise, as shown in illustration #142A.
3. Find the cut angle of the miter and calculate the cut back distance needed.
4. Layout the calculated cut back distance on each side of the center circumference line, as shown in illustration #142B.
5. Position a wrap-around so that it forms an arc that just touches the starting cut line at the top of the pipe on division #1.

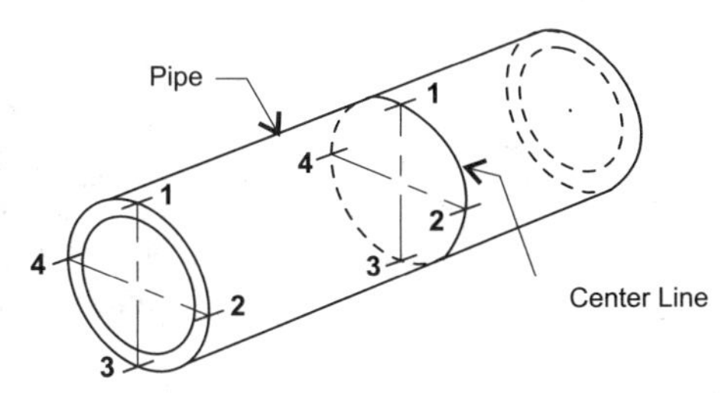

Illustration #142A - Layout Pipe Centerlines

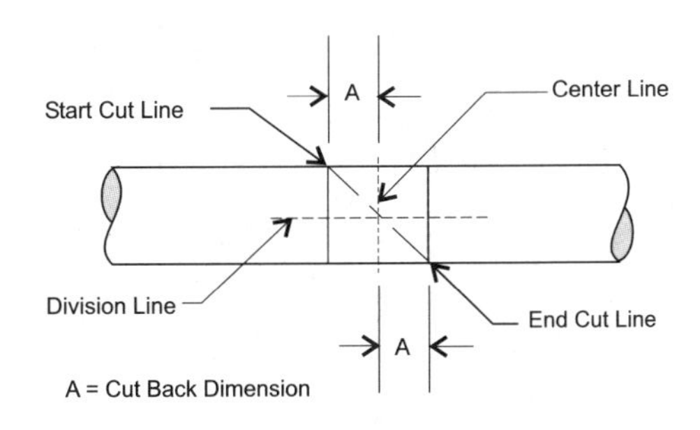

Illustration #142B - Layout Cut Back Dimension

2 Piece Miter Layout

6. The two side sleeves of the wrap-around are placed so they intersect with the side division lines #2 and #4 on the middle circumference line. Illustration #142C shows the placement of the wrap-around on these lines. Trace the curvature made by the wrap-around with chalk or a soapstone.
7. Reverse the wrap-around connecting the middle division lines #2 and #4 with the ending cut line at the bottom division line #3, once again tracing the curvature made by the warp-around.
8. If cutting the miter with an oxy-acetylene torch, the entire cut should be made in the same lane. See illustration #143. This mitering cut is then followed by the second beveling cut.

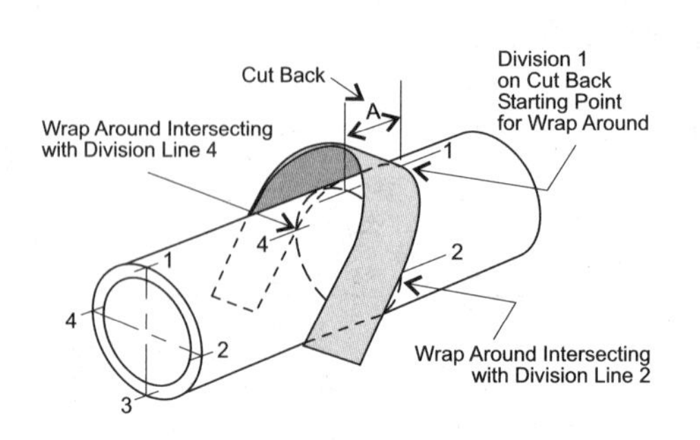

Illustration #142C - Placement of Wrap-Around

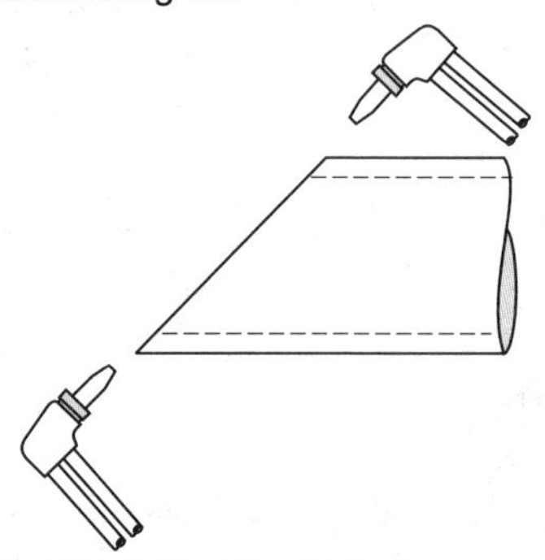

Illustration #143 - Cutting Miter with Torch

Multi-Piece 90 Degree Miter Turns (Elbows)

Most multi-piece miter turns consist of either three-piece or four-piece sectional turns, but in fact, any number of sections can be used in a miter turn. Three-piece and four-piece 90 degree miter turns are shown in illustration #144. The information needed to calculate any multi-piece turn is basically the same as that needed for simple two piece miter turns. However, the length of the pieces that make up the miter turns must also be calculated. The following is the information needed to fabricate multi-piece 90 degree turns:

a. Cut Angle of Miter
b. Factor of the Cut Angle
c. Cut Back Distance
d. Length of Sectional Pieces

Note: An ideal method of cutting miters is with a power band saw.

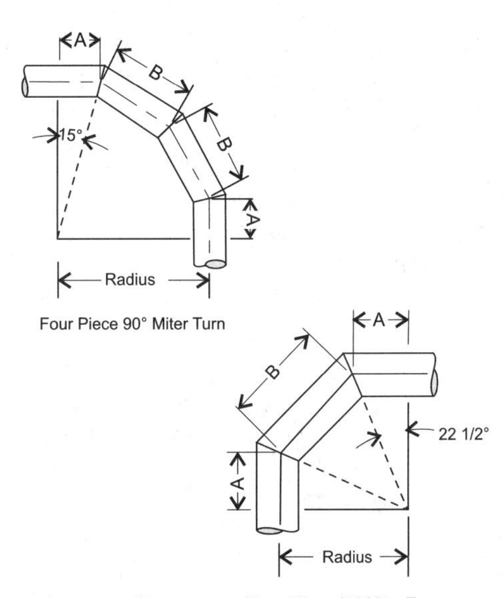

Illustration #144 - Multi-Piece Miter Turns

Cut Angle of Miter

The cut angle for a multi-piece turn can be found by dividing the angle of the turn by the number of welds needed in the fabrication times 2.

$$\text{Cut Angle} = \frac{\text{Angle of Turn}}{\text{No. of welds x 2}}$$

Note: Number of welds = number of miter pieces minus 1.

Example 1: Determine the cut angle for a 4 inch (100 mm), 4 piece 90 degree mitered elbow.

1. Number of welds = number of pieces in miter minus 1 = 4 - 1 = 3 welds.

2. $\text{Cut Angle} = \frac{\text{90 deg Elbow}}{\text{No. of welds x 2}} = \frac{\text{90 deg}}{3 \times 2}$

3. Cut Angle = 15 degrees

Factor of the Cut Angle

The factor number for each cut angle is found in table #75.

Cut Back Distance

The O.D. of the pipe x factor number divided by 2, sets the distance on each side of the centerline to start and end the miter cut.

Cut Back = Pipe O.D. x Factor of the Cut Angle $\div 2$

Example 2: Layout the cut back distance for a 4 inch (100 mm) 4 piece 90 degree miter elbow. Given a cut angle of 15 degrees.

1. In example 1, the cut angle was determined to be 15 degrees. The factor number for this cut angle is obtained from table #75. The factor number using table #75 = .26795.
2. Cut Back

$$= \frac{\text{O.D. of Pipe x Cut Angle Factor}}{2}$$

$$= \frac{\text{4.5 inches (114.3 mm) x .26795}}{2}$$

= 0.6" or 5/8 inches (15.3 mm). The calculated distance (cut back distance) is then marked-off on each side of the center line.

Length of Sectional Pieces

The lengths of the pieces that make up multi-piece miter turns are calculated by the following method:

Length A (end pieces) = radius x factor of cut angle

Length B (middle piece(s)) = Length A x 2

Example 3: Find the length needed for the end and middle pieces of a 4 inch (100 mm), 4 piece 90 degree miter elbow, given a miter radius of 24 inches (609.6 mm).

1. Find Length A for the end sections of the miter:

   Length A (end pieces) = radius x factor of cut angle = 24 inches (609.6 mm) x .26795 = 6.43 or 6 3/8 inches (163.3 mm).

Note: Cut angle and factor were used from example 1 and example 2.

2. Find Length B for the middle sections of the miter:

   Length B (middle pieces) = Length A x 2
   = 6.43 (163.3 mm) x 2
   = 12.86 or 12 7/8 inches (326.6 mm).

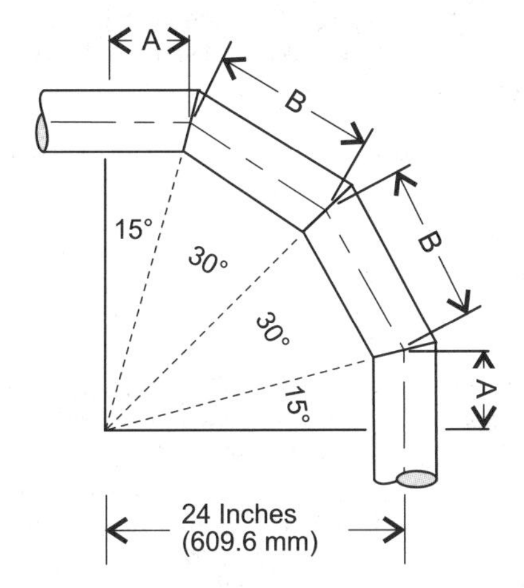

Illustration #145 - Determining Section Lengths

Large Diameter Pipe Cut Lines

Layout of the cut line on larger diameter pipe requires additional divisional and ordinate lines for accuracy. The following gives the recommended number of divisional lines for larger diameter pipe:

Max. Pipe Size	Min. # of Divisions
4" (100 mm)	4
10" (250 mm)	8
24" (600 mm)	16
42" (1050 mm)	32

Each divisional length is then calculated to provide for a guideline or ordinate mark for positioning the cut line.

The starting and ending points of the cut back line (ordinate line #1) for any number of divisions is first calculated using the following formula:

Ordinate Line #1 =

$$\frac{\textit{O.D. of Pipe x Cut Angle Factor}}{2}$$

To find the length of the remaining ordinate lines from the centerline, use one of the following formulas:

Division of 8:

Ordinate line #2 = Ordinate line #1 x .707

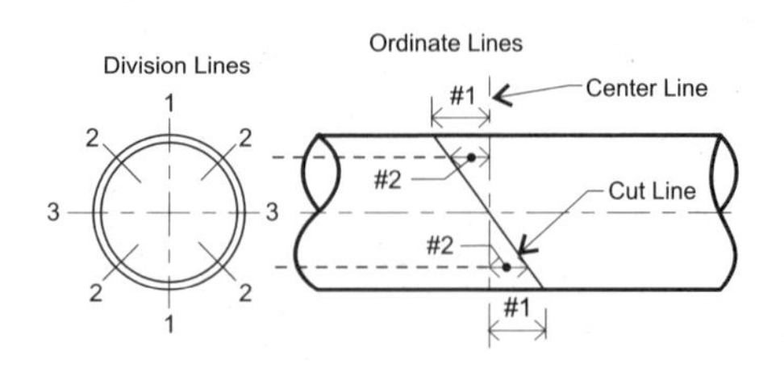

Illustration #146 - 8 Division Ordinate Lines

Multiple Ordinate Lines

Division of 16:

Ordinate line #2 = Ordinate line #1 x .924
Ordinate line #3 = Ordinate line #1 x .707
Ordinate line #4 = Ordinate line #1 x .383

Division of 32:

Ordinate line #2 = Ordinate line #1 x .978
Ordinate line #3 = Ordinate line #1 x .924
Ordinate line #4 = Ordinate line #1 x .831
Ordinate line #5 = Ordinate line #1 x .707
Ordinate line #6 = Ordinate line #1 x .558
Ordinate line #7 = Ordinate line #1 x .383
Ordinate line #8 = Ordinate line #1 x .200

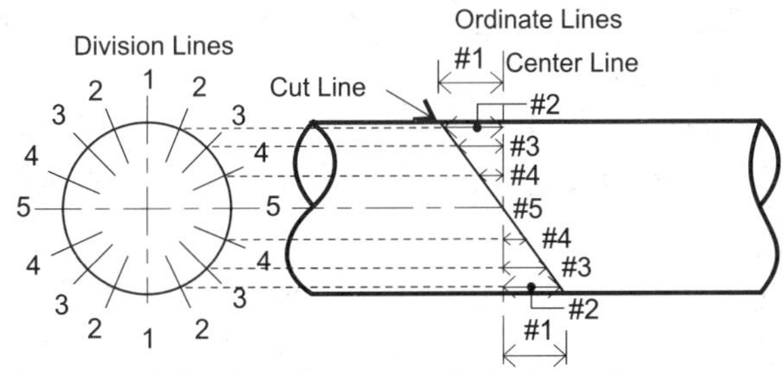

Illustration #147 - 16 Division Ordinate Lines

Multiple Ordinate Lines

Example 4: Layout the cut line for a 12 inch (300 mm) pipe which is to be mitered for the fabrication of a two piece 90 degree elbow.

1. Find the cut angle needed for the miter turn:

$$Cut\ Angle = \frac{90\ deg}{2} = 45\ degrees$$

2. Determine the factor number from table #75:

   Factor Number Using Table #75 = 1

3. Calculate the first ordinate line #1:

   Ordinate Line #1 =

$$\text{Pipe O.D. x } \frac{Cut\ Angle\ Factor}{2}$$

$$= \frac{12.75\ in.\ (324\ mm)\ x\ 1}{2}$$

   = 6.375 in. (6 $^3/_8$ in.) (162 mm)

4. Number of Divisions Needed = 16
5. Determine the length of ordinate line #2, #3, #4. The layout for these lines is shown in illustration #147.

Ordinate Lines #2 = Ordinate Line #1 x .924 = 6.375 inches (162 mm) x .924 = 5.9 in. (5 $^9/_{16}$ inches (150 mm)

Ordinate Lines #3 = Ordinate Line #1 x .707 = 6.375 inches (162 mm) x .707 = 4.5 in. (4 $^1/_2$ inches (114.5 mm)

Ordinate Lines #4 = Ordinate Line #1 x .382 = 6.375 inches (162 mm) x .382 = 2.435 in. (2 $^7/_{16}$ inches (62 mm)

Ordinate Line #5 = Center Line of Fitting

Layout for 16 Division Two Piece Miter:

1. Using a wrap-around, draw a line around the circumference of the pipe. This line will become the centerline of the fitting.
2. Divide the center circumference line into 16 equal parts using one of the methods described previously. Draw the division lines on both sides of the centerline along the length of the pipe.

3. Label these division lines 1,2,3,4,5, starting at the top of the pipe and work down in both directions to the middle of the pipe. Rotate the pipe 180 degrees and label the bottom section division lines in the same manner. See illustration #147.
4. Layout the calculated distance for the starting and ending point of the cut line (ordinate lines #1) from the centerline.
5. Layout the calculated distance for ordinate lines #2, #3, #4, from the center circumference line. Ordinate line #5 is located on the center circumference line and has no length. See illustration #147.
6. Position the wrap-around so that it forms an arc that just touches the starting cut line at the top of ordinate #1. The two sides of the wrap-around are placed intersecting along the ordinate lines #2, #3, #4, #5.
7. Trace the curvature made by the wrap-around with chalk or a soapstone onto the pipe.
8. Reverse the wrap-around or rotate the pipe, connecting the opposite ordinate lines with the ending cut line at the bottom section, ordinate line #1, once again tracing the curvature made by the wrap-around intersecting with the ordinate lines.
9. If cutting the miter with an oxy-acetylene torch, the entire cut should be made in the same plane. Refer back to illustration #143. This mitering cut is then followed by the second beveling cut.

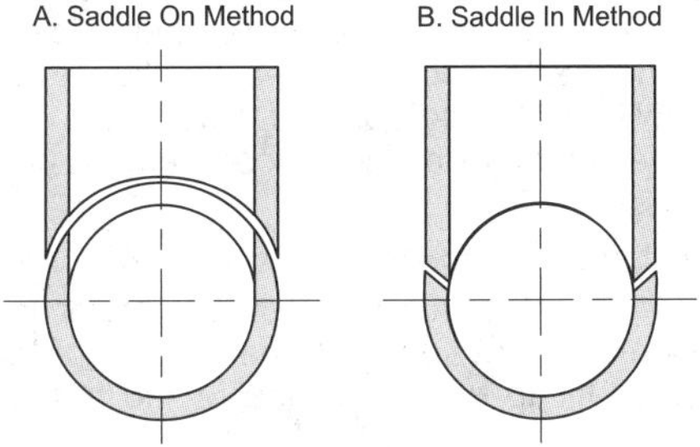

A. Saddle On Method

B. Saddle In Method

Illustration #148 - Tee Branch Connections

Tee Layout

(Equal Size Branch and Header)

The fabrication of a tee with the same branch and header size consists of two direct pipe layouts. The first layout establishes the contour line required on the branch piping in order to mate with the header surface. The second layout establishes the opening in the header necessary to receive the shaped branch piece. The tee fabrications can be laid out with either the branch sitting on the header (Saddle On Method, illustration #148A) or with the branch projecting inside of the header (Saddle In Method, illustration #148B). Both methods of layout are explained in the following procedures.

Tee Layout Procedure (Saddle In Method)

This method applies to the branch projecting inside of the header with the outside of the branch fitting to the outside of the header wall.

1. Measure and mark a distance equal to one half of the pipe's inside diameter from the end of the pipe. Using a wrap-around as a guide, draw a straight line around the circumference of the pipe at this point. This line becomes the base line for measurements.
2. Divide the center circumference line into four equal parts.
3. Number the four division lines: #1 for the top and bottom and #2 for each of the side division lines. See illustration #149A.

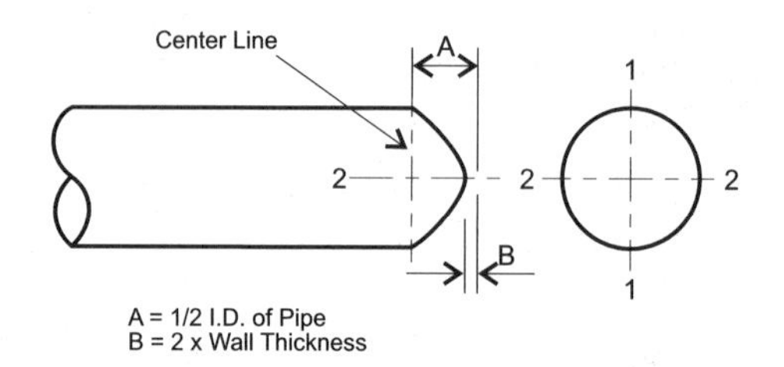

Illustration #149A - Tee Branch Layout

4. From the center circumference line around the pipe, continue both #2 division lines out to the end of the pipe.
5. Position a wrap-around so that it forms a U-shaped arc that just touches division line #1 on the center circumference line and the two sides are placed so they intersect with the side division lines #2 at the end of the pipe. Trace the curvature made by the wrap-around with chalk or a soapstone.
6. Reverse the wrap-around or rotate the pipe and connect the bottom division line #1, located on the center circumference line, with the side division lines #2, located at the end of the pipe. Once again, trace the curvature made by the wrap-around.
7. The point formed at the intersection of the wrap-around on division lines #2 can be rounded out by drawing a freehand curve.

Note: The curve should start at a location approximately two times the pipe wall thickness up from the end of the pipe. See illustration #149A.

8. When cutting the branch piece, a radial cut is made, giving a square cut edge to the pipe.

Tee Header Layout Procedure

An easy way of establishing the opening in the header for a full size tee is to use the prepared branch piece as the template for the opening. The cut branch piece is placed over the header and the header opening is traced using the branch piece as the guide, see illustration #149B.

When cutting the opening in the header, it is important to make a radial cut; that is, the cutting tip is held perpendicular to the center of the pipe. See illustration #150. The header opening is beveled after the radial cut is made.

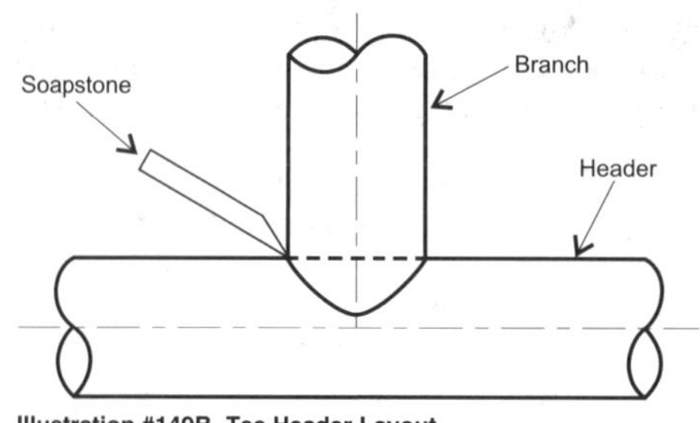

Illustration #149B -Tee Header Layout

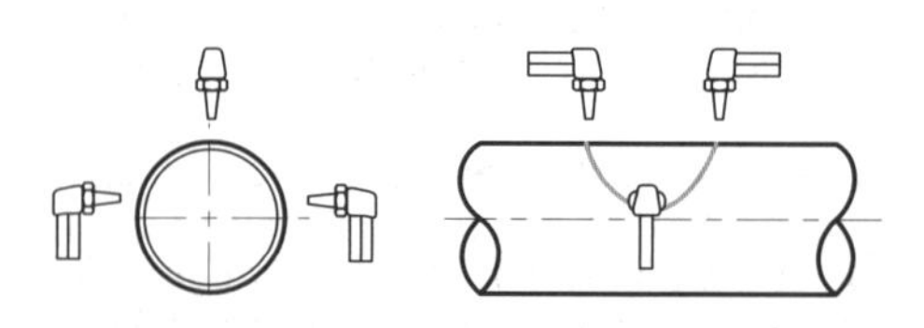

Illustration #150 - Radial Torch Cutting

Alternate Tee Header Layout

The cut line for the header may also be laid out using the following procedure.

1. Using a wrap-around as a guide, draw a straight line around the circumference of the pipe at the center of branch take off.
2. Divide the center circumference line into four equal parts and number the four division lines as #1 for the top and bottom lines and #2 for each of the side division lines. See illustration #151.
3. Measure and mark off a distance equal to one half of the branch's outside diameter to each side of the top division line #1. Use a piece of angle iron or a straight edge to extend the division line #1 to the marked off distances. Label these marked off distances as points A and B intersects.
4. At the intersection of the center circumference line and side division lines #2, label as point C.

5. Position a wrap-around so that it forms a U-shaped arc that just touches point A on division line #1. The two sides of the wrap-around are placed to intersect with points C on the side division lines #2.
6. Trace the curvature made by the wrap-around with chalk or a soapstone connecting point A with the two side points C.
7. Reposition the wrap-around connecting point B with the two side points C and trace the curvature in the same manner. See illustration #151.
8. The point formed at the intersection of the wrap-around at point C can be rounded out by drawing a freehand curve. The curve should start at a location approximately two times the pipe wall thickness up from point C. See illustration #151.
9. When cutting the opening in the header, it is important to make a radial cut; that is the cutting tip is held perpendicular to the center of the pipe. See illustration #150. The header opening is beveled after the radial cut is made.

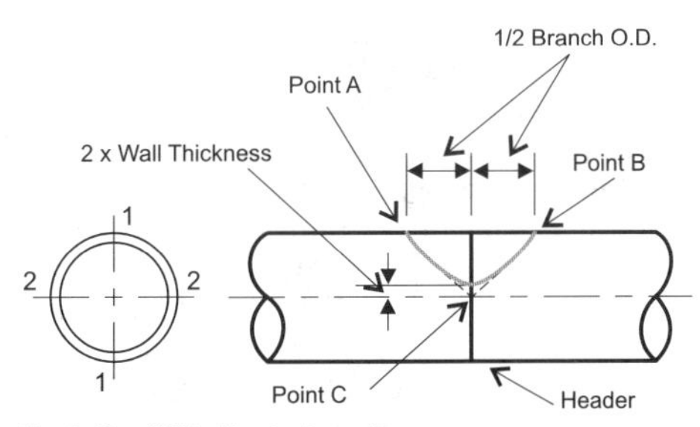

Illustration #151 - Header Layout

Reducing Tee Layout (Saddle On)

With this tee method, the inside wall of the branch is positioned on the outside wall of the header.

1. Measure and mark a distance equal to approximately one-half the branch pipe inside diameter from one end of the pipe. Using a wrap-around as a guide, draw a line around the circumference of the pipe at this point. This is the baseline.
2. Divide this line into eight equal parts for pipe sizes 4 inch to 10 inch (100 to 250 mm). Divide into 16 parts for sizes 12 inch (300 mm) and over. Only four division lines are required for sizes under 4 inch (100 mm).
3. Number the division lines as shown in illustrations #152A and #152B, depending upon the tee pipe size.
4. From table #76, select the correct branch and header sizes for the tee.

The numbers shown in the intersecting table column for lines #2, #3, #4, #5 (see illustration #152) are measured and marked from the baselines along each of the applicable division lines. This is done for both the front and back half of the header.

5. For half the pattern, position a wrap-around so that it forms a U-shaped arc that just touches division line #1 on the baseline. The wrap-around is positioned to intersect on the end points of the ordinate lines (either #2 and #3, or #2, #3, #4, and #5, depending on the pipe size), on both sides of #1.
6. Trace the wrap-around curvature on the header with a soapstone for half the pattern. Reverse the wrap-around and repeat the process for the back of the header. This becomes the cut line.
7. When cutting the branch, a radial cut is made, resulting in a square cut edge. After marking the header for cutting, the branch edge can be beveled.

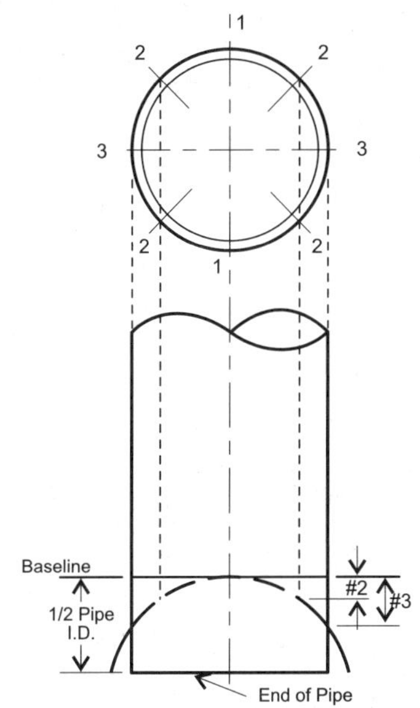

Illustration #152A - Division Lines (4 to 10 inches)

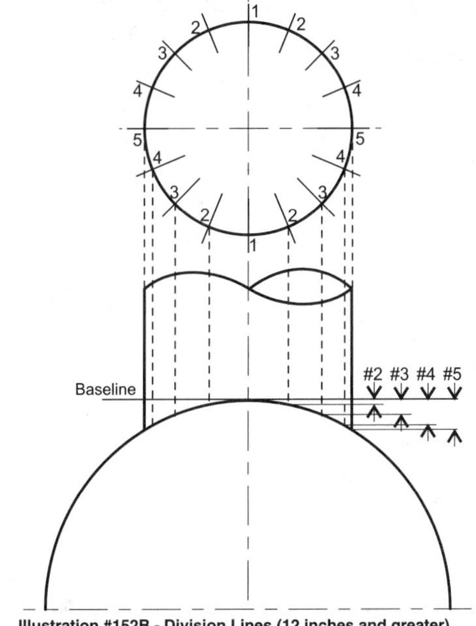

Illustration #152B - Division Lines (12 inches and greater)

ORDINATE LINE DISTANCE FOR TEE CONNECTIONS (INCHES)														
Branch Size	Line #	Header Pipe Size (Inches)												
		2	3	4	6	8	10	12	14	16	18	20	22	24
2	2	1/4	1/8	1/8	1/16	1/16	1/32	1/32	-	-	-	-	-	-
	3	9/16	5/16	1/4	3/16	1/8	1/16	1/16	-	-	-	-	-	-
3	2	-	3/8	1/4	3/16	1/8	1/8	3/32	3/32	-	-	-	-	-
	3	-	7/8	5/8	3/8	1/4	3/16	1/16	1/16	-	-	-	-	-
4	2	-	-	1/2	5/16	1/4	3/16	1/8	1/8	1/8	1/16	1/16	1/16	1/16
	3	-	-	1 1/4	11/16	1/2	3/8	5/16	1/4	1/4	3/16	3/16	3/16	1/8
6	2	-	-	-	3/4	9/16	7/16	3/8	3/8	5/16	1/4	1/4	3/16	3/16
	3	-	-	-	1 15/16	1 1/4	15/16	3/4	11/16	5/8	1/2	1/2	7/16	3/8
8	2	-	-	-	-	1 1/16	13/16	11/16	5/8	9/16	1/2	7/16	3/8	3/8
	3	-	-	-	-	2 11/16	1 3/4	1 7/16	1 1/4	1 1/16	15/16	7/8	3/4	3/4
10	2	-	-	-	-	-	1 1/4	1 1/16	1	7/8	3/4	11/16	5/8	9/16
	3	-	-	-	-	-	3 7/16	1 7/16	2 1/8	1 3/4	1 9/16	1 3/8	1 1/8	1 1/8
12	2	-	-	-	-	-	-	7/16	3/8	3/8	5/16	5/16	1/4	1/4
	3	-	-	-	-	-	-	1 5/8	1 9/16	1 3/8	1 1/16	15/16	13/16	7/8
	4	-	-	-	-	-	-	3 1/4	2 3/4	2 5/8	1 15/16	1 13/16	1 1/2	1 3/8
	5	-	-	-	-	-	-	4 1/4	3 7/16	3 1/4	2 5/16	2	1 13/16	1 5/8
14	2	-	-	-	-	-	-	-	7/16	3/8	3/8	5/16	1/4	1/4
	3	-	-	-	-	-	-	-	1 7/8	1 1/2	1 15/16	1 1/8	1 1/16	15/16
	4	-	-	-	-	-	-	-	3 9/16	2 13/16	2 3/8	2 1/16	1 7/8	1 5/8
	5	-	-	-	-	-	-	-	4 3/4	3 1/2	2 7/8	2 1/2	2 3/16	2

Table #76A - Ordinate Distances for Tee Connections (Imperial Units)

ORDINATE LINE DISTANCE FOR TEE CONNECTIONS (mm)

Branch Size	Line #	Header Pipe Size (Millimeters)												
		50	80	100	150	200	250	300	350	400	450	500	550	600
50	2	6	3	3	2	2	1	1						
	3	14	8	6	5	3	2	2						
80	2	-	10	6	5	3	3	2	2	-	-			
	3	-	22	16	10	6	5	2	2	-	-			
100	2		-	13	8	6	5	3	3	3	2	2	2	2
	3		-	32	18	13	10	8	6	6	5	5	5	3
150	2			-	19	14	11	10	10	8	6	6	5	5
	3			-	49	32	24	19	18	16	13	13	11	10
200	2				-	27	21	18	16	14	13	11	10	10
	3				-	68	45	37	32	27	24	22	19	19
250	2					-	32	27	25	22	19	18	16	14
	3					-	87	62	54	45	39	35	38	29
300	2						-	11	10	10	8	8	6	6
	3						-	41	40	35	27	24	21	22
	4						-	83	70	67	49	46	37	35
	5						-	108	87	83	68	51	46	41
350	2							-	11	10	10	8	6	6
	3							-	48	38	33	29	27	24
	4							-	90	71	60	52	48	41
	5							-	121	89	73	64	56	51

Table #76A - Ordinate Distances for Tee Connections (Metric Units)

Reducing Tee Header (Saddle On)

The method used for establishing the opening in the header is to use the prepared branch piece as the template for the opening. The cut branch piece is placed over the header and the opening for the header is traced using the branch piece as the guide. This is shown in illustration #153.

Note: When radial cutting the opening in the header, make the cut inside the cut line. This ensures that the hole in the header is not too large.

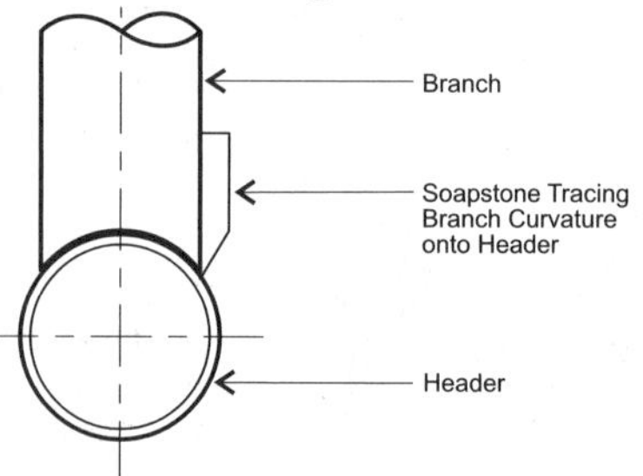

Illustration #153 - Use Branch as Template

Reducing Tee Quick Layout Method

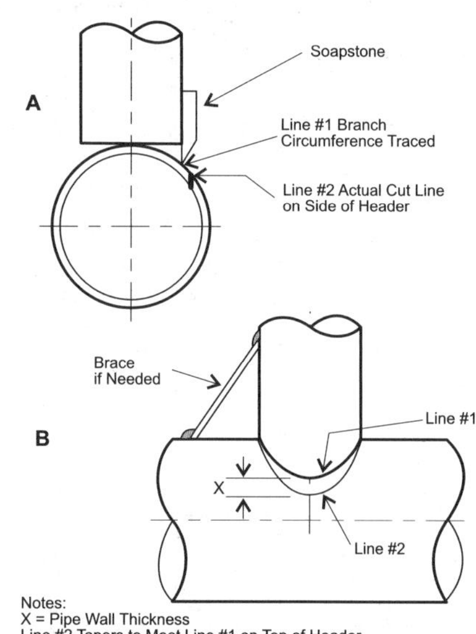

Illustration #154A,B - Header Layout

A quick method of laying out both the header and branch connection for a reducing tee is displayed in illustration #154. The cut lines for the branch and header are traced using the actual profile of the tee joint. The following are the steps needed in the layout for the header and branch.

1. Set the branch pipe squarely on the header and if necessary, hold it in place with a metal brace tacked to both branch and header.
2. Using a long sharpened soapstone held against the branch, smoothly trace the outline of the branch circumference onto the header. See illustration #154A.
3. A distance equal to the wall thickness of the pipe is measured and marked off from the traced circumference line at the sides of the header. The circumference line on the top of the pipe remains in the same position.
4. A new curved line is drawn starting at outside mark at the side and smoothly tapering to meet the top circumference line. This new line is the cut line for the header opening. See illustration #154B.

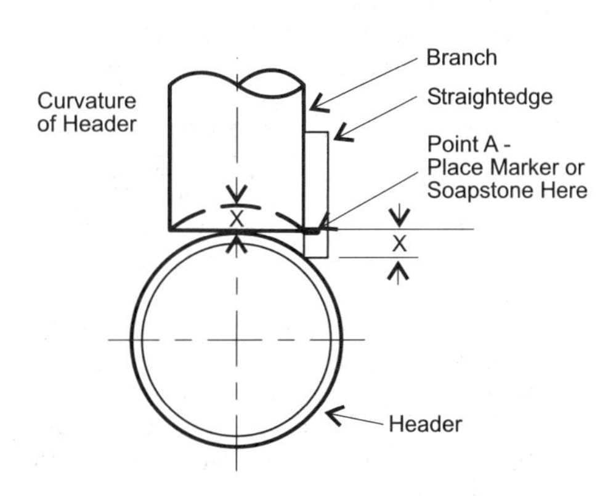

Illustration #155 - Branch Layout

Reducing Tee Branch

An easy method of marking the cut line for the branch is to slip the branch into the header opening and trace the cut line onto the branch.

Alternate Method

1. With the branch still tacked in position, establish distance X (the distance between the end of the branch pipe and the header wall). This distance is established so that when tracing the curve cut line onto the branch, it will not go off the pipe end.
2. Lay a small straight edge flat alongside the branch with the end just touching the header.

   Hold a marker or soapstone at point A on the straight edge and move the straight edge accurately around the branch. See illustration #155.The end of the straight edge must be kept against the header while tracing.
3. The established cut line on the branch is then radial cut with no bevel needed.

Lateral Layout

This method of layout establishes a full size drawing to determine the cut lines needed for lateral fabrication.

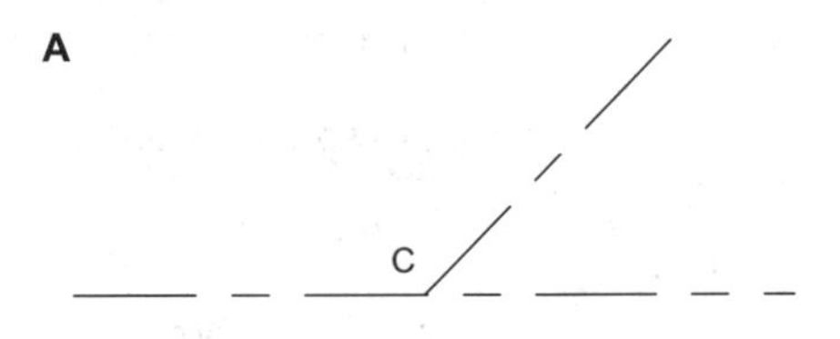

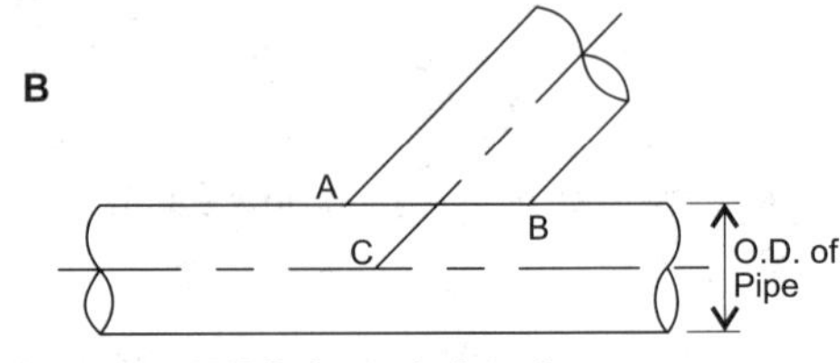

Illustration #156A,B - Layout for Lateral

1. Layout the two centerlines for the branch and header of the lateral needed on a flat sheet or surface. The branch centerline is taken off from the header centerline using the desired angle for the lateral. The intersection between these two lines is marked C. See illustration #156A.
2. Draw in the outside line of pipe around the centerlines. The distance on each side of the centerline equals 1/2 of the O.D. of the pipe. See illustration #156B.
3. Label the intersecting points of the branch with the header as point A at the back of the lateral and point B at the front of the lateral.
4. Connect point A to point C and point B to point C with straight lines on the drawing. See illustration #156C.
5. On the actual lateral header pipe, draw a center circumference line at the location of the branch take off. Divide the pipe into four equal parts and extend these division lines to both sides of the centerline on the pipe.

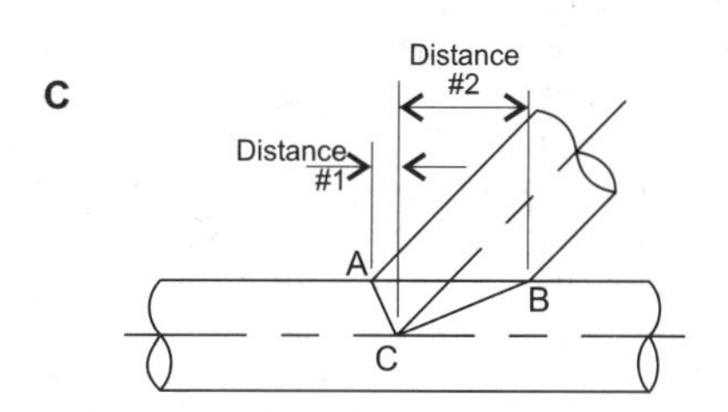

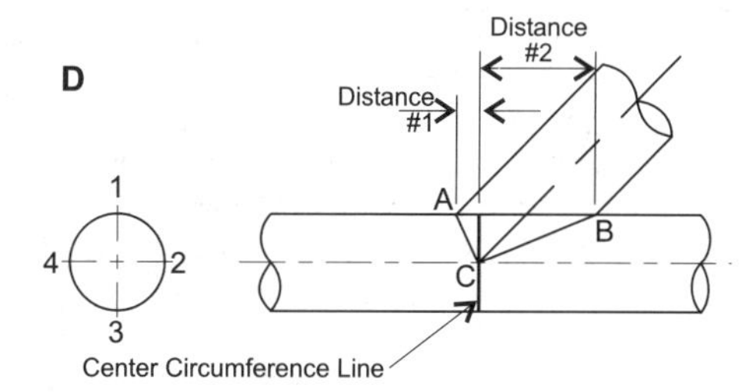

Illustration #156C,D - Layout for Lateral

Lateral Layout

6. Measure the distance horizontally from point A and from point B to the intersecting center point C of the branch on the drawing. These two distances are referred to as distance #1 and #2.
7. The two distances are marked off from the center circumference line on the actual pipe header. The distances are located along the top division line on the pipe. See illustration #156D.
8. Point C is located and marked off on the two side division lines of the actual pipe.
9. A wrap-around is used to connect point A on the top division line to the side division lines point C. The curvature made by the wrap-around is traced onto the pipe.
10. Points B and C are marked and traced in the same manner.
11. This marked curvature line becomes the cut line for the header opening.

Lateral Branch Layout

1. Divide the branch pipe into four equal sections near the pipe end.
2. Draw a right angle line at point B on the layout drawing to be used as a base line. Measure the distance between intersection point C and the base line. See illustration #157A.
3. On the actual branch pipe, mark off the distance determined in step 2 from the end of the pipe. Draw in a center circumference line at this point, which becomes the base line of the pipe.
4. Measure the distance from point A on the drawing to the base line. Mark off this distance on the top division line from the base line on the pipe. See illustration #157B.
5. A wrap-around is used to connect point A to the center division lines C at the end of the pipe. The curvature made by the wrap-around is traced onto the pipe.

6. Use a wrap-around to connect point B on the bottom division line and the center points C at the end of the pipe.
7. When cutting both the header and branch, first use a miter cut followed by a bevel cut.

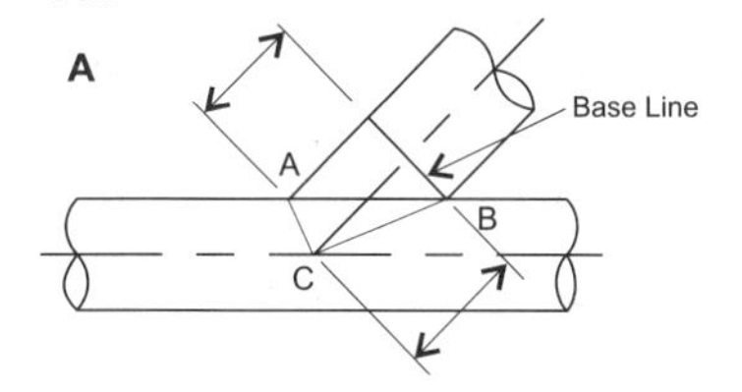

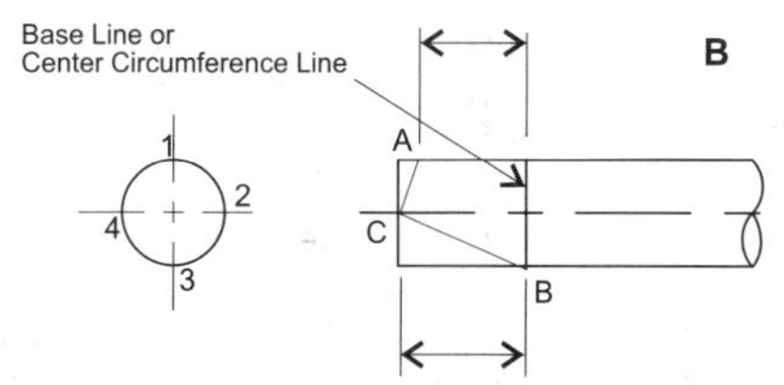

Illustration #157A,B - Branch Layout

Quick Method for Reducing Laterals

A quick method of laying out both the header and branch connection for a reducing lateral is displayed in illustration #158A & B. In this method, the cut lines for the branch and header are traced using the actual profile of the lateral joint. The following are the steps needed in the layout method for the header and branch.

1. Establish the correct angle for the branch pipe on the header and secure in place with a metal brace tacked to both the header and branch.
2. Using a long sharpened soapstone held against the branch, smoothly trace the header opening cut line using the branch pipe as the guide.
3. With the branch still tacked in position, establish distance D on both the top and bottom of the branch pipe. This will locate the starting and stopping distance so that point A will not go off the pipe end when tracing the curve cut line onto the branch.

Reducing Lateral Quick Method

4. Lay a small straight edge alongside the branch with the end just touching the header. Hold a marker or soapstone at point A on the straight edge and move the straight edge accurately around the branch, keeping the end against the header.
5. Cutting should be radial, followed by a beveling cut for the header hole.

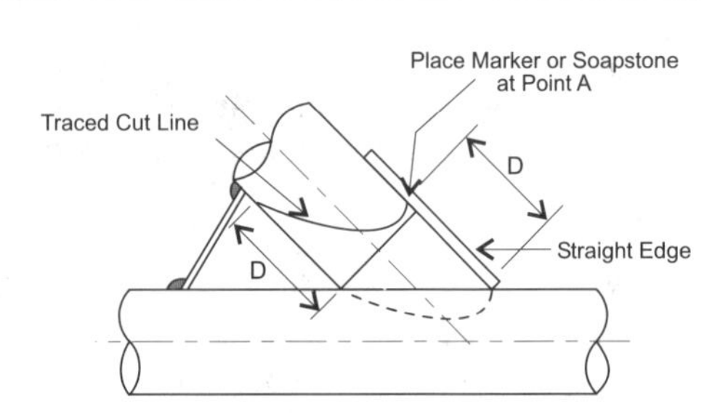

Illustration #158B - Branch Layout (Quick)

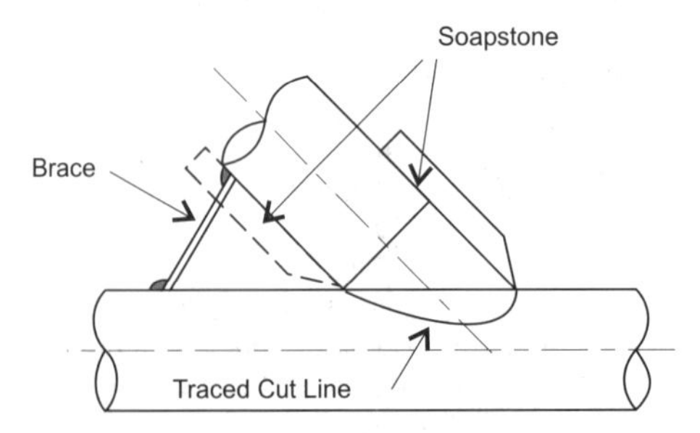

Illustration #158A - Header Layout (Quick)

Orange Peel Layout

A method of capping off a pipe end by cutting and welding the pipe end, when a manufactured cap is not available, is referred to as the orange peel method. The number of peel back sections for the layout is determined from table #77.

1. Divide the end of the pipe needing the cap into the number of sections or peels determined from table #77. Extend the division lines lengthwise along the pipe.

2. Find distance X in table #77 and mark this distance off from the end of the pipe. Place a wrap-around on the mark and draw in a center circumference line at this point.
3. A template is developed for one section of the layout using dimensions given in table #77. See illustration #159.
4. The template is aligned on each division line and traced onto the pipe surface.
5. Cut the pipe along the trace line, heat each section cherry red and bend the peels into the desired shape. Weld each section of the orange peel cap.

ORANGE PEEL LAYOUT DATA (Inches)					
Pipe Size in inches	Number of Sections	X in inches	D in inches	C in inches	B in inches
2	4	1 7/8	1 7/8	1 5/8	15/16
4	4	3 1/2	3 17/32	3 1/16	1 3/4
6	5	5 1/4	4 1/8	3 5/8	2 1/16
8	5	6 3/4	5 3/8	4 3/4	2 11/16
10	7	8 1/2	4 13/16	4 1/4	2 3/8
12	8	10	5	4 3/8	2 1/2

Table #77A - Orange Peel Data (Imperial)

ORANGE PEEL LAYOUT DATA (Millimetres)					
Pipe Size mm	Number of Sections	X in mm	D in mm	C in mm	B in mm
50	4	47.6	47.6	41.3	24.0
100	4	89.0	89.7	77.8	44.5
150	5	133.4	104.8	92.1	52.4
200	5	171.5	136.5	120.7	68.3
250	7	216.0	122.2	108.0	60.3
300	8	254.0	127.0	111.1	63.5

Table #77A - Orange Peel Data (Imperial)

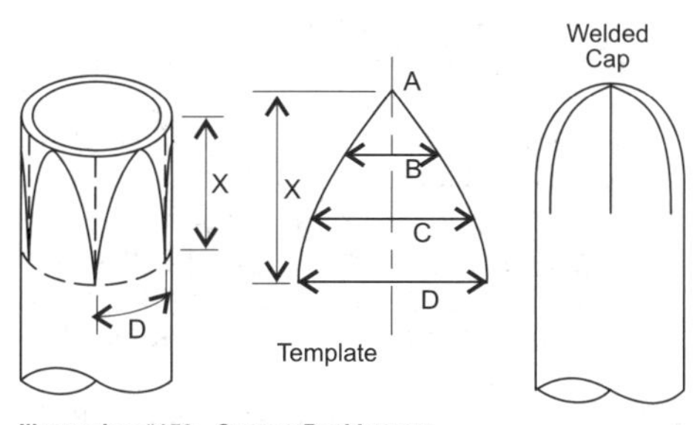

Illustration #159 - Orange Peel Layout

True Wye (Y) Equal Diameters

This method of layout establishes a full size drawing to determine the cut lines needed for the true wye fabrication.

Header Layout

1. Layout the centerlines for the wye angle needed on a flat surface. The intersection point between the header and branch lines is marked point D. See illustration #160A. The angle here is 45 degrees, but any angle can be used.
2. Draw in the outside line of pipe around the centerlines. The distance on each side of the centerline equals 1/2 of the O.D. of the pipe. See illustration #160B.
3. Label the intersecting points of the branches with the header as point A and point B. The intersecting point between the two branches is point C. See illustration #160B.
4. With straight lines, connect points A to D, B to D and C to D. See illustration #160C.
5. At point C, draw two base lines at right angles to the branch connections. Draw another base line connecting point A and point B. See illustration #160D.
6. Measure the distance E on the drawing from point D to the base line. See illustration #160D.
7. On the actual wye header pipe, draw a center circumference line at a point from the end of the pipe equal to distance E. Divide the pipe into four equal parts and extend these division lines from the end of the pipe to the center circumference line. Number these lines as shown in illustration #161.
8. Label the location where division lines #2 contact the end of the pipe as point D. Label the location where the top division line #1 contacts the center circumference line as point A and where the bottom division line #1 contacts the center circumference line as point B. See illustration #161.

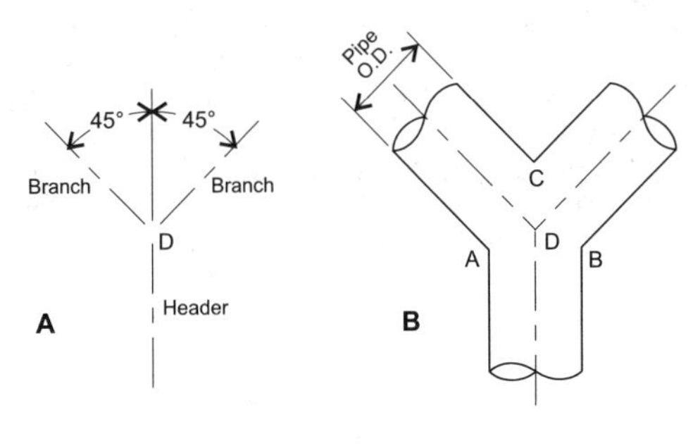

Base Lines
C
D
E
A
B

9. A wrap-around is used to connect point A on the top division line to the side division lines point D. The curvature made by the wrap-around is traced onto the pipe. See illustration #161.

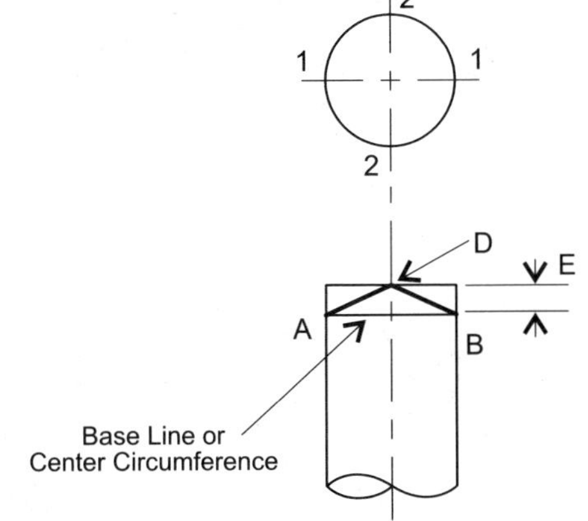

Illustration #160 - True Wye Layout

Illustration #161 - Wye Header Layout

10. The wrap-around is repositioned to connect point B on the bottom division line and the center points D at the end of the pipe. Points B and D are marked and traced in the same manner.
11. This marked curvature line becomes the cut line for the header of the wye.

True Wye Branch Layout

1. Measure the distance X on the drawing from point D to the branch base line. See illustration #162A.
2. On the actual branch pipes, draw a center circumference line at a point from the end of the pipe equal to distance X. Divide the pipe into four equal parts and extend these division lines from the end of the pipe to the center circumference line. Number these lines as shown in illustration #162B.

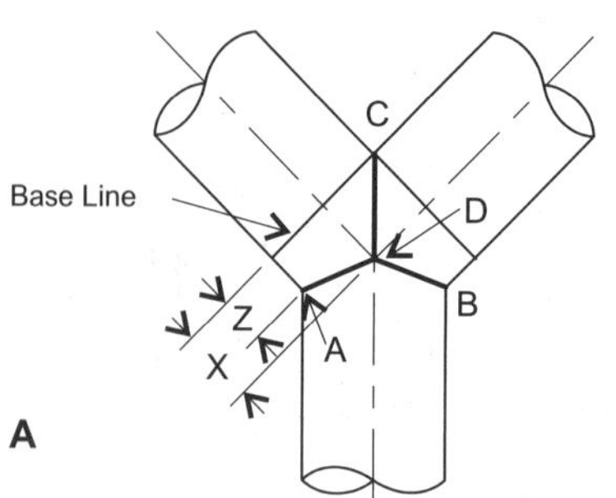

Illustration #162A - Wye Branch Layout

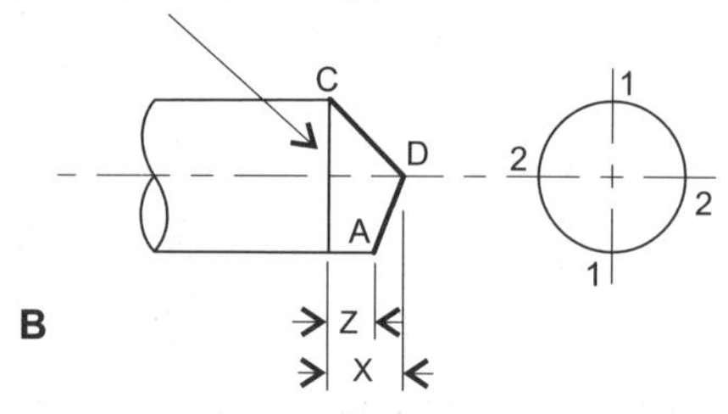

3. Label the location where division lines #2 contact the end of the pipe as point D. Label the location where the top division line #1 contacts the center circumference line as point C. See illustration #162B.
4. Measure distance Z on the drawing from point A to the branch base line. See illustration #162A.
5. On the actual branch pipes, mark off distance Z from the base line along the bottom division line #1. This mark becomes point A on the pipe.
6. A wrap-around is used to connect point C on the the top division line #1 to the side division lines at point D. The curvature made by the wrap-around is traced onto the pipe. See illustration #162B.
7. The wrap-around is repositioned to connect point A on the bottom division line and the center points D at the end of the pipe. Point B and points D are marked and traced in the same manner.
8. The branches and header pieces are miter cut and then beveled.

Template Development

If a piece of pipe was cut lengthwise and unfolded outwards until it was straight, it would take the shape of a flat piece of metal. The same piece of pipe marked for a miter cut and scribed with parallel lines evenly spaced would be projected in the flat as shown in illustration #163.

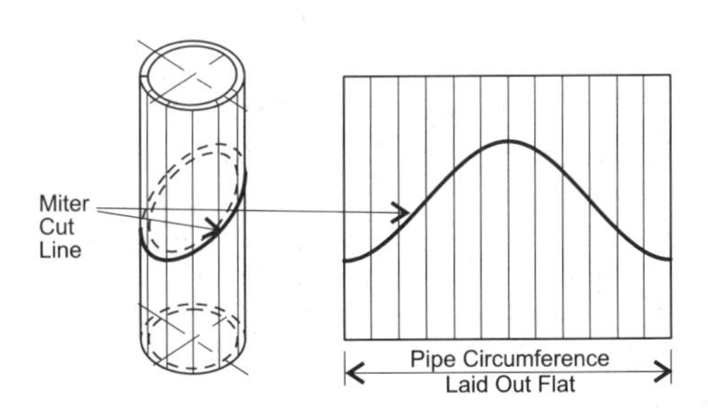

Illustration #163 - Parallel Line Development

Template Development

By using this parallel line and projection principle, flat templates can be developed for field fabrication of various fittings.

Templates are usually developed by drawing corresponding end and side views of the needed miter or fitting and projecting selected segments on to a flat surface. This flat surface then becomes the wrap around template needed to trace the fitting contour onto the pipe's surface. The template developments are generally produced on drawing paper and then transferred to thin gasket material or thin sheet metal for durability.

Development of Miter Turn Templates

Any angle of miter turn can have a template developed for tracing and marking the contour of the cut line onto the pipe. The miter angle used in this example is 45 degrees.

1. On a flat piece of paper, draw a circle with the diameter equal to the outside diameter of the pipe. See illustration #164A.
2. Directly below the circle, draw a side view of the pipe showing the miter angle needed.
3. Divide the top circle into 12 equal parts and number each dividing line as indicated.
4. Extend the division lines from the bottom half of the circle down onto the side view drawing of the pipe and miter angle, as shown in illustration #164B.
5. To one side of the drawing, establish a straight line equal to the circumference of the O.D. of the pipe. Divide this line into 12 equal spaces and number each division line, as shown in illustration #164C.

Miter Turn Template

6. From the miter angle, extend each division line horizontally until it is directly over the corresponding number on the circumference line. Extend the corresponding circumference division line up vertically until it now intersects with the horizontal division line.
7. Draw a smooth connecting curve between all of the intersection points on the flat development.

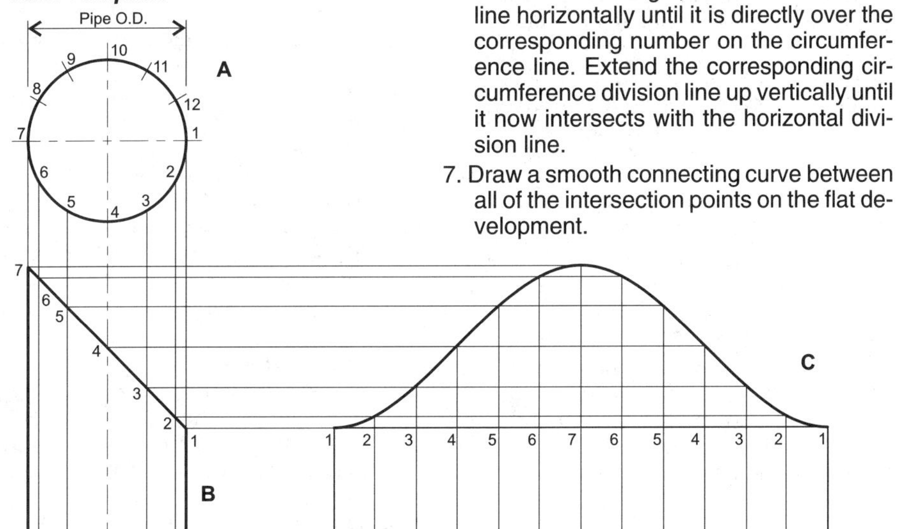

Illustration #164 - Miter Turn Layout

45 Degree Lateral Template (Equal)

1. Draw an end and side view of the lateral joint required, using the actual pipe dimensions. The two drawings should be directly across from each other on the same centerline plane. A line equal to the pipe's wall thickness is placed above the centerline, which is the lowest projection point for the development.
2. Draw semi-circles above both the side and end view drawings. Divide each semi-circle drawing into 8 equal parts and number each division line, as shown in illustration #165A and #165B.
3. Lines are extended down from the division lines on the end view drawing, stopping as they contact the round circle shaped header. These lines are then extended diagonally across onto the side drawing. See illustration #165B.
4. Lines are extended down from the semi-circle division points on the side view until they intersect with the corresponding line from the end view.
5. The intersection points are joined together with a curved line. This curved line represents the contour of the branch pipe. See illustration #165A.
6. The flat development for the branch is drawn from the side view. The length of development equals the circumference of the branch pipe.
7. Divide the flat development circumference into sixteen equal parts and number each division line.
8. Extend lines from the points located on the contour of the side view drawing until they intersect with the corresponding numbers on the flat drawing. See illustration #165D.
9. Draw a smooth, curved line on the flat development connecting the intersection points of the division lines.

45 Degree Lateral Template (Equal)

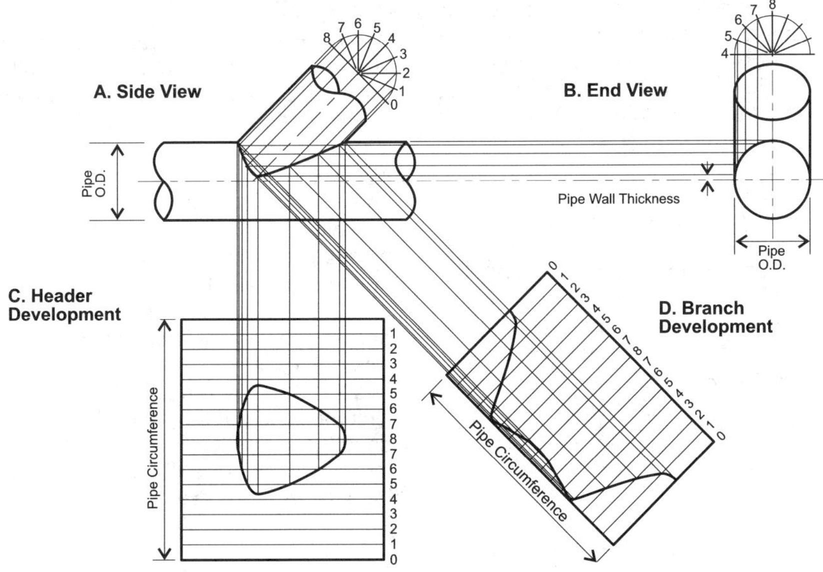

Illustration #165 - 45 Degree Lateral Layout (Equal)

45 Degree Lateral Template (Unequal)

10. The flat development for the header opening is projected directly below the side view drawing. The length of the flat header development should also equal the circumference of the header. See illustration #165C.
11. Divide the header development into sixteen equal parts and number each division as shown. Extend lines down from the side view contour points, stopping at the corresponding number on the flat drawing.
12. Connect the intersection points on the flat development to form the template opening for the header.

45 Degree Lateral Template (Unequal)

When developing lateral intersection templates with a branch connection smaller than that of the header, the same basic development procedure is used as in equal diameter laterals.

Illustration #166 shows a typical example of unequal diameter lateral development. The only difference between lateral development of equal and unequal diameters is in the header hole development.

In unequal diameter laterals, the distances between the actual flat development hole division lines are taken from measuring between the divisional lines along the circumference of the header circle. In most developments, the chord distance between the division lines is sufficient for layout, rather than the actual curvature distance.

45 Degree Lateral Template (Unequal)

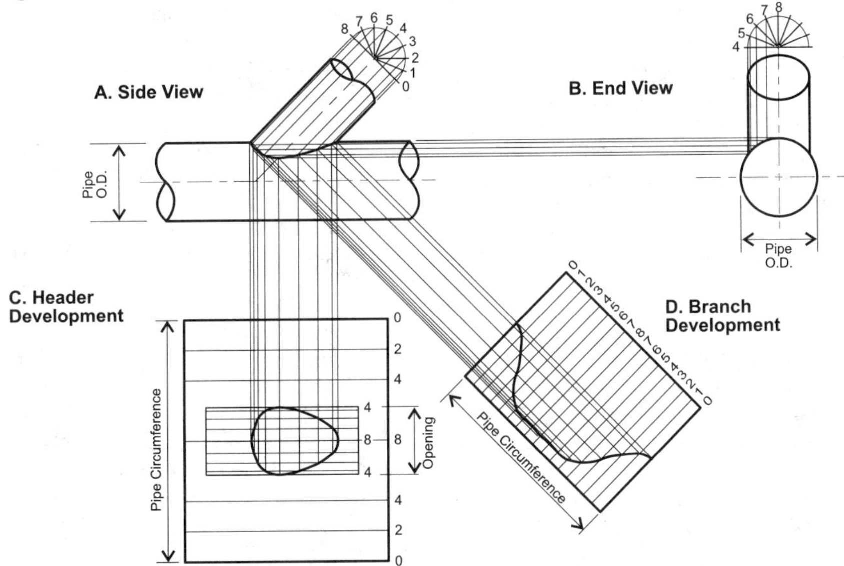

Illustration #166 - 45 Degree Lateral Layout (Unequal)

Tee Development Template (Equal)

Branch and Header

1. Draw an end and side view of the tee connection using the actual pipe dimension. The two drawings should be directly across from each other on the same centerline plane. See illustration #167.
2. Semi-circles are drawn above the branch connections on both the side and end view drawings. Divide the semi-circles into 8 equal parts and number each division line as shown in illustration #167.
3. Lines are extended down from the semi-circle division points on the end view drawing until they contact the round circle shaped header. It should be noted that the lowest projection line should remain off the centerline by a distance equal to 2 times the pipe wall thickness.
4. The division lines are then extended diagonally across onto the side view drawing.
5. Lines are extended down from the semi-circle on the side view, until they intersect with the corresponding lines brought across from the end view drawing. See illustration #167A.
6. The intersecting points of the division lines are joined together with a solid line that represents the contour line of the branch pipe.
7. The flat development pattern for the branch is drawn from the end view. The length of the flat development equals the circumference of the branch pipe. See illustration #167D.
8. Divide the flat development into sixteen equal parts and number each division line as shown.
9. Extend lines from the intersecting points located on the contour of the side view, until they intersect with the corresponding numbers on the flat development drawing. See illustration #167D.

Tee Development Template (Equal)

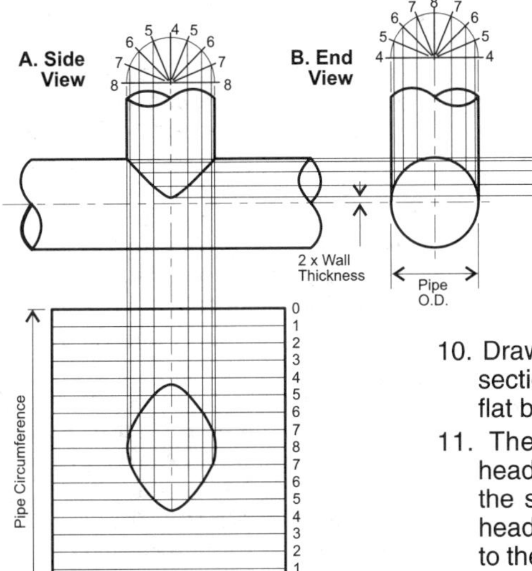

Illustration #167 - Tee Layout (Equal)

10. Draw a curved line connecting the intersection points of the divisional lines on the flat branch development.
11. The flat development pattern for the header opening is projected directly below the side view drawing. The length of the header development should also be equal to the circumference of the header. See illustration #167C.

Tee Development Template (Equal)

12. Divide the header development into sixteen equal parts and number each division as shown. Extend lines down from the side view intersection points stopping at the corresponding last number on the flat drawing.
13. Connect the intersection points on the flat development to form the template opening for the header.

Tee Development Template (Unequal)

When developing templates for unequal diameter tees, the same basic development procedure is followed as that used for equal diameter tees.

Illustration #168 shows a typical example of unequal tee development where the branch is a smaller diameter than that of the header.

The only difference between the two procedures is in the development of the header opening.

In the development of unequal diameter tees, the distance between division lines on the flat header opening is established by measuring the arc distance between the divisional lines on the circumference of the header circle. In most developments, the chord distance between the division lines is sufficient for layout, rather than the actual curvature distance.

Note: Either the inside or outside of the branch may be used in the development of the header hole for the tee. The template may be designed to let the branch sit on the header (saddle on) or it may be designed to let the branch project into the header (saddle in). These two methods are shown in illustration #169.

Tee Development Template (Unequal)

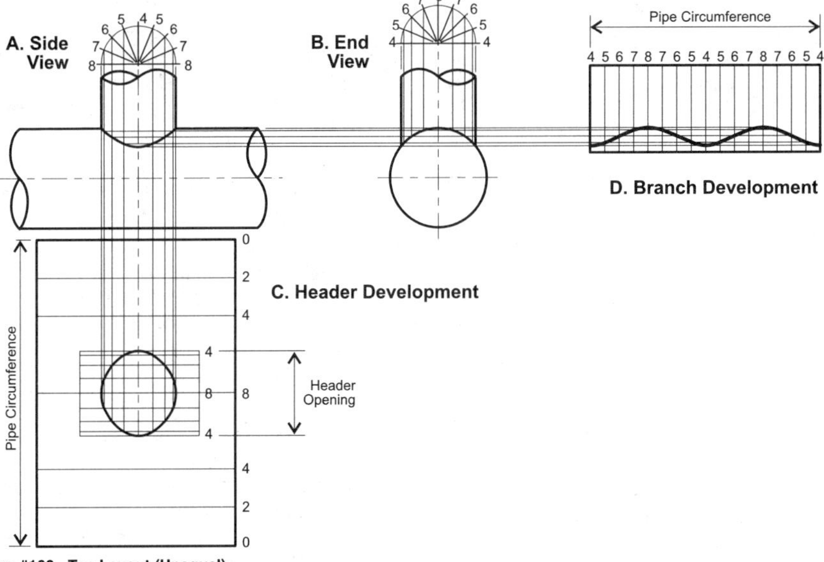

Illustration #168 - Tee Layout (Unequal)

Saddle In or On Development

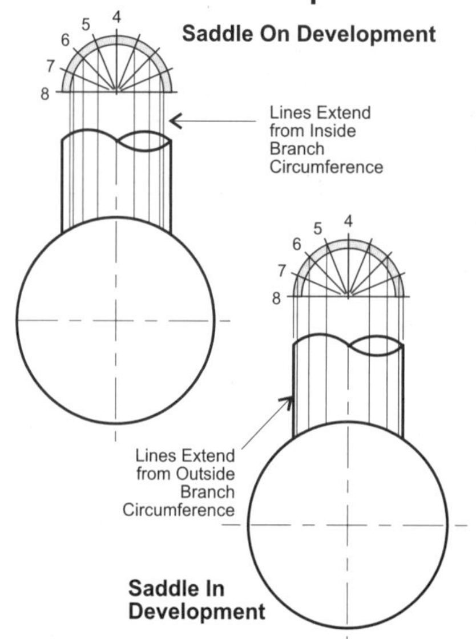

Illustration #169 - Saddle On/Saddle In

Wye Template Development

1. Draw a side elevation view of the wye needed, using the actual pipe dimensions. The centerlines of the wye should be drawn first, and then the outside lines of the fitting should be added. See illustration #170A.
2. Sectional lines are added to the drawing to divide the header from the branch pieces. See illustration #170A.
3. Semi-circles are drawn over one of the branches and below the header piece. Each semi-circle is divided into 6 equal parts and the division lines for the semi-circles are numbered as shown in illustration #170B.
4. Extend the division lines from the semi-circles until they meet the sectional lines of the wye.
5. The flat development for the branch is projected in line with the branch angle. The length of the branch development should equal the circumference of the branch.

6. Divide the branch development into twelve equal parts and number each division as shown. Extend the division lines diagonally across from the wye drawing, stopping in line with the corresponding numbers on the flat development drawing.
7. Connect the intersection points of the flat development to form the template of the branch pieces.
8. The flat development for the header is projected directly to the side of the header drawing, following the same procedures as that of the branch. See illustration #170B.

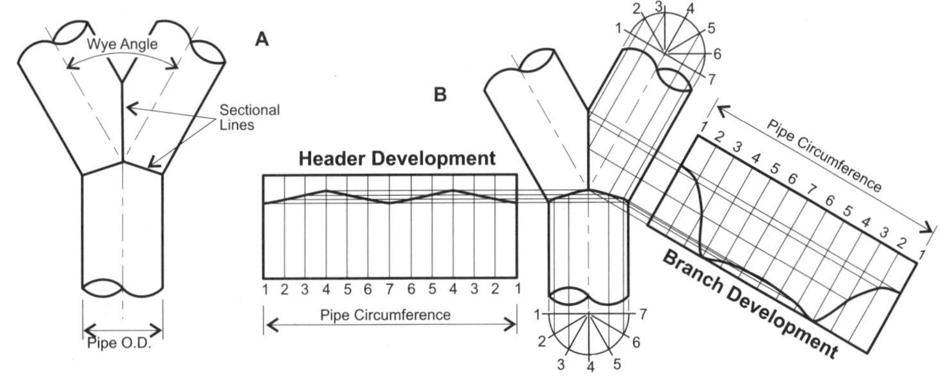

Illustration #170 - True Wye Layout

SECTION NINE

PIPING PRINTS

Introduction

The purpose of this section is to assist in the interpretation of piping drawings and blueprints.

Within this section, and as a rule throughout the industry, drawings or blueprints have an interchangeable meaning and refer to copies of the engineering drawings.

Drawings and their blueprint copies are the principle language used to convey piping information from engineers and designers to the fabrication and construction personnel. This transfer of information using drawings enables the fabrication, building and maintenance of piping systems in a wide and ever growing field of applications. Piping drawings primarily represent piping layouts by the use of both single and double line drawings, symbols, and special notations.

Lines and Their Functions

Lines play an important role in communicating necessary information on blueprints. Every line on a drawing has a special purpose and meaning that must be understood in order to correctly interpret and use blueprints.

The various line types (often referred to as the alphabet of lines) are shown in illustration #171, preceded by a brief description and specific use for each.

Piping in blueprints is generally represented by a solid line, but when it is necessary to differentiate between dissimilar lines of pipe, lettering and symbols are often used. Illustration #172A and #172B display common graphic symbols used in air conditioning, heating, instrumentation, plumbing, and sprinkler system prints to distinguish between the dissimilar lines of pipe.

Line Description and Function

1. ***Break Lines*** - Three types of break lines are frequently used on drawings to shorten continual long objects or sections in drawings. The short break line is used when a break is required across a small space. It is represented by a thick wavy line. The long break line is drawn thin with a horizontal Z placed between the line showing the break. For round objects or pipe, a thick S break is usually used.

2. ***Centerline*** - A thin line composed of long dashes alternating with short dashes used to represent the center of fittings, pipe and equipment.

3. ***Cutting Plane Line*** - A thick line used to indicate where a difficult to show object or section of the drawing is cut through for viewing. A drawing of the view is then shown through the cut plane, typically distinguished by sectional letters.

4. ***Dimension and Extension Lines*** - Thin lines that are typically used together to represent the extended features of an object and the dimension of those features. Dimension lines typically use arrow heads placed against the extension lines to limit or show the dimension features.

5. ***Flow Lines*** - Primary and secondary flow lines are used on flow diagrams and P&ID (Piping and Instrumentation Diagram) drawings to show flow paths. Primary lines are shown as thicker lines and secondary lines as medium to thick lines.

6. ***Hidden Line*** - Medium weight dash lines used to show features hidden from the particular view on a drawing.

7. ***Match and Boundary Line*** - A thick line used to show where two drawings align or project boundaries begin.

8. ***Object or Visible Line*** - Continuous line used to represent all visible object surfaces that are in view.

9. ***Section Line*** - Medium or thin weight lines usually drawn at 45 degrees indicate the object or surface has been cut through along this point or plane. The nature of material used in the object can be identified by the type of cross hatching line used in showing the cut surface. The cross hatching symbols for various types of material are shown in illustration #173.

10. ***Phantom Line*** - A thin line consisting of long dashes alternating with a pair of short dashes. The phantom line indicates: alternating positions or movable parts, future location of equipment or piping, and/or repeated detail.

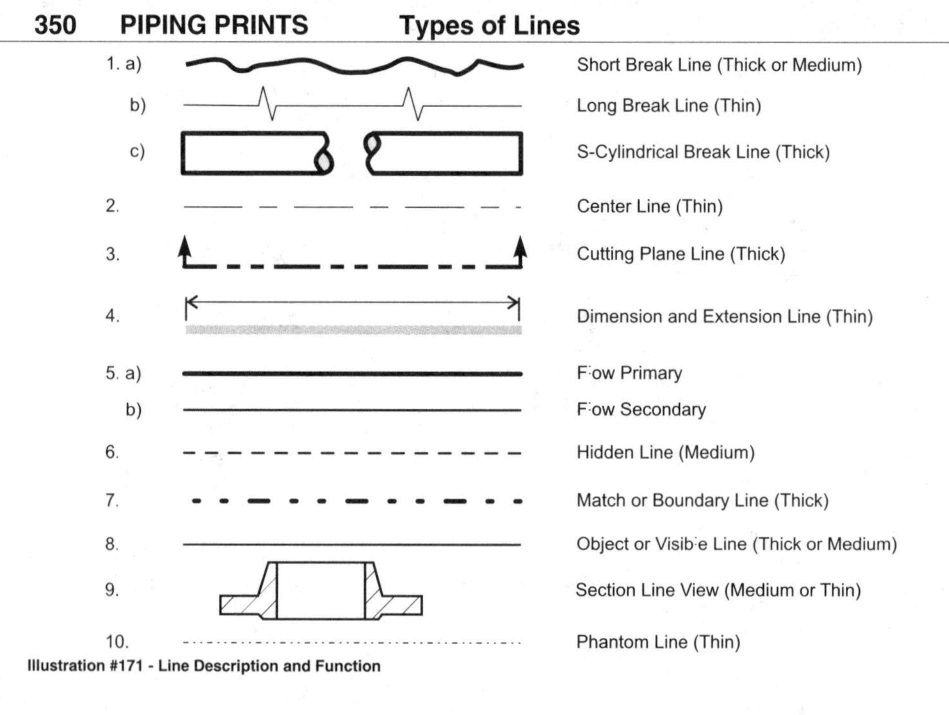

Illustration #171 - Line Description and Function

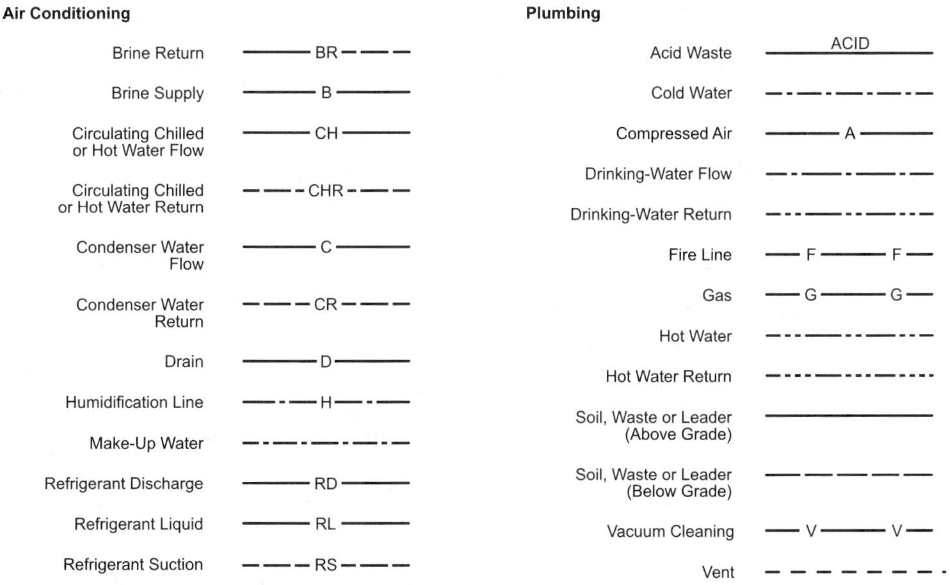

Air Conditioning

Brine Return	BR
Brine Supply	B
Circulating Chilled or Hot Water Flow	CH
Circulating Chilled or Hot Water Return	CHR
Condenser Water Flow	C
Condenser Water Return	CR
Drain	D
Humidification Line	H
Make-Up Water	
Refrigerant Discharge	RD
Refrigerant Liquid	RL
Refrigerant Suction	RS

Plumbing

Acid Waste	ACID
Cold Water	
Compressed Air	A
Drinking-Water Flow	
Drinking-Water Return	
Fire Line	F F
Gas	G G
Hot Water	
Hot Water Return	
Soil, Waste or Leader (Above Grade)	
Soil, Waste or Leader (Below Grade)	
Vacuum Cleaning	V V
Vent	

Illustration #172A - Piping Line Symbols

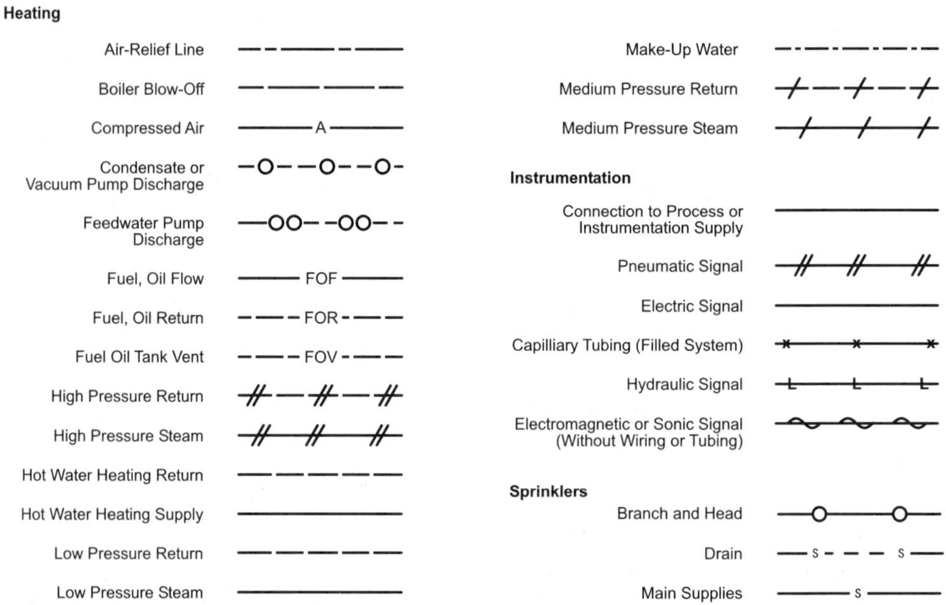

Heading

Air-Relief Line

Boiler Blow-Off

Compressed Air — A —

Condensate or Vacuum Pump Discharge

Feedwater Pump Discharge

Fuel, Oil Flow — FOF —

Fuel, Oil Return — FOR —

Fuel Oil Tank Vent — FOV —

High Pressure Return

High Pressure Steam

Hot Water Heating Return

Hot Water Heating Supply

Low Pressure Return

Low Pressure Steam

Make-Up Water

Medium Pressure Return

Medium Pressure Steam

Instrumentation

Connection to Process or Instrumentation Supply

Pneumatic Signal

Electric Signal

Capilliary Tubing (Filled System)

Hydraulic Signal

Electromagnetic or Sonic Signal (Without Wiring or Tubing)

Sprinklers

Branch and Head

Drain — S — — — S —

Main Supplies — S —

Illustration #172B - Piping Line Symbols

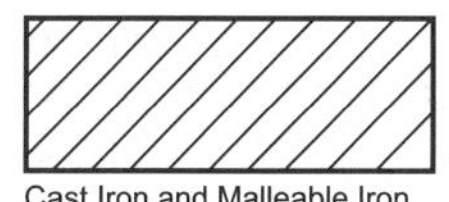
Cast Iron and Malleable Iron (Also Used for all Materials)

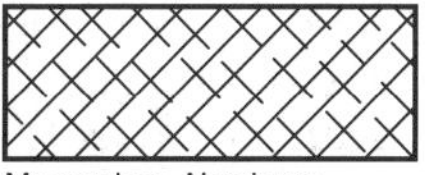
Magnesium, Aluminum, and Aluminum Alloys

Electric Windings, Electromagnets, Resistors, etc.

Rock

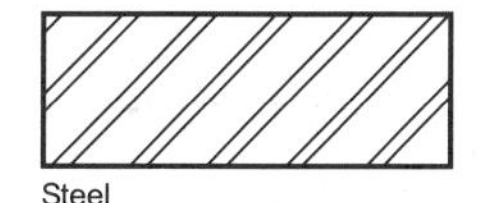
Steel

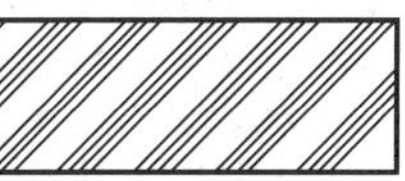
Rubber, Plastic, Electrical Insulation

Marble, Slate, Glass, Porcelain, etc.

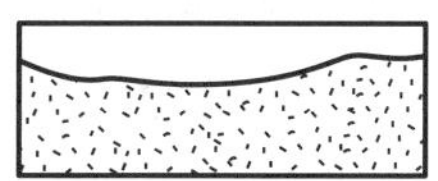
Sand

Brass, Bronze and Compositions

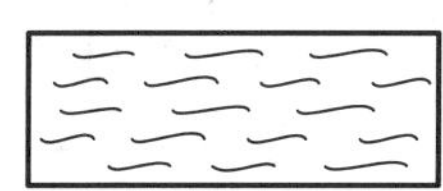
Cork, Felt, Fabric, Leather, Fibre

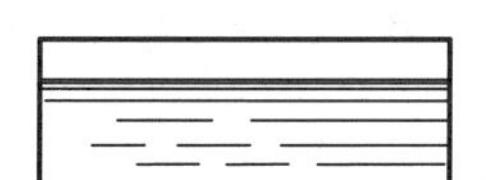
Water and Other Liquids

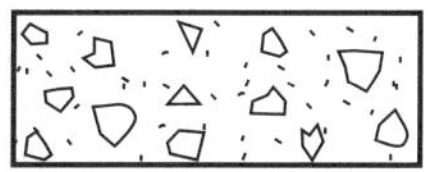
Concrete

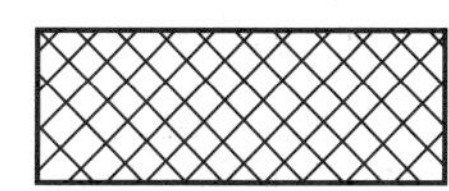
White Metal, Zinc, Lead Babbitt and Alloys

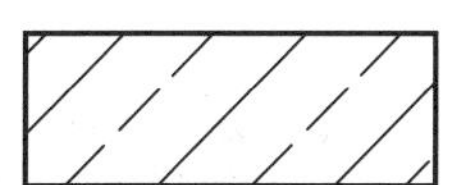
Firebrick and Refactory Material

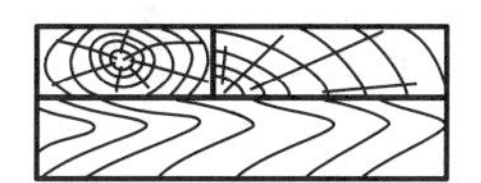
Wood, (Across Grain /With Grain)

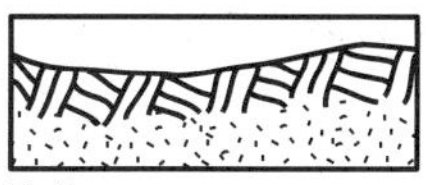
Earth

Illustration #173 - Section Line Symbols

Pipe and Fitting Representation

Pipe and fittings are shown on drawings and blueprints as either single or double lines, see illustration #174. Because single line pipe drawings are faster to draw, most prints use this method of showing pipe and fittings.

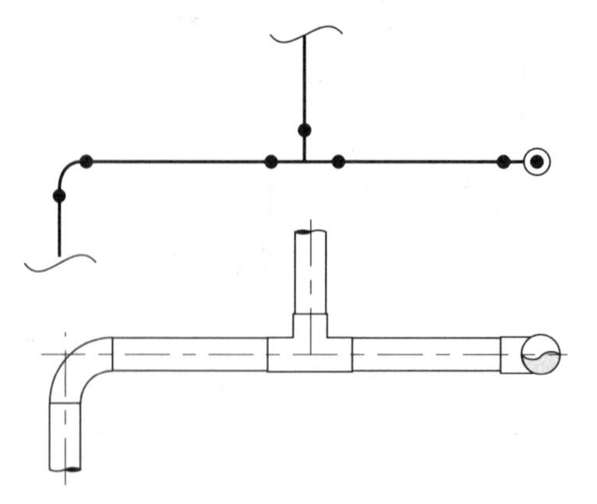

Illustration #174 - Single and Double Line Drawings

Double line pipe drawings are usually only used on pipe sizes over 12 inches (30 mm), and where it is important to show the clearance or relationship between lines and/or equipment.

Line Identification Numbers

Line identification numbers or codes are used in industry to differentiate between the various pipe runs throughout a system. The line number is typically placed beside the line or located directly in the pipe line drawing as shown in illustration #175.

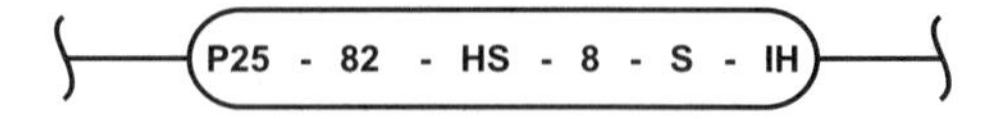

Illustration #175 - Pipe Identification Number Symbol

Even though the line identification codes are not standardized, most companies provide the same fundamental information within each line number.

Line Identification Numbers

The following gives an example of the information that may be found in a typical line number:

P25-82-HS-8-S-IH

P25 - sequential number of the line (25th line in process)

82 - denotes the area number of the line (i.e. Area 82 in plant)

HS - represents the medium of conveyed fluid (High Pressure Steam)

8 - line pipe size (8 inch/400 mm)

S - pipe material specification (Standard / Carbon Steel Pipe to API 5L Grade B or A, Flanges to ASTM A181 Grade B)

IH - insulation type (Hot Insulation)

Note: It should be noted that the line number remains with that line, except where the size changes or the line runs through a major piece of equipment. Branch lines taken off the continuous line will have different line numbers regardless of branch size. The sequence of number or letter identification for line specification are determined by the company and vary from one company to another.

Standard Piping Symbols

Standard piping symbols for fittings, valves, and line designations on drawings as a rule correspond to ANSI Z32.23 standard. Commonly used single and double line fittings and valve symbols are shown in illustrations #176A and #176B. Companies often modify these symbols, but typically the variations are slight and easily recognized.

	Fitting or Valve Types	Flanged	Screwed	Bell and Spigot	Welded X or ●	Soldered	Double Line
VALVES	Gate Valve						
	Globe Valve						
	Motor Op. Gate Valve						
	Motor Op. Globe Valve						
	Gate Hose Valve						
	Globe Hose Valve						
	Diaphragm Valve						
	Lockshield Valve						
	Motor Control Valve						
	Quick Opening Plug						
	Solenoid Valve						
	Safety Relief Valve						

Illustration #176A - Commonly Used Fitting and Valve Symbols

	Fitting or Valve Types	Flanged	Screwed	Bell and Spigot	Welded X or ●	Soldered	Double Line
VALVES	Angle Check Valve						
	Angle Gate (Elevation)						
	Angle Gate (Plan)						
	Angle Globe (Elevation)						
	Angle Globe (Plan)						
	Angle Hose Valve						
	Ball Valve						
	Butterfly Valve						
	Straightway Check Valve						
	Cock or Plug Valve						
	Y-Valve						
	Strainer						

Illustration #176B - Commonly Used Fitting and Valve Symbols

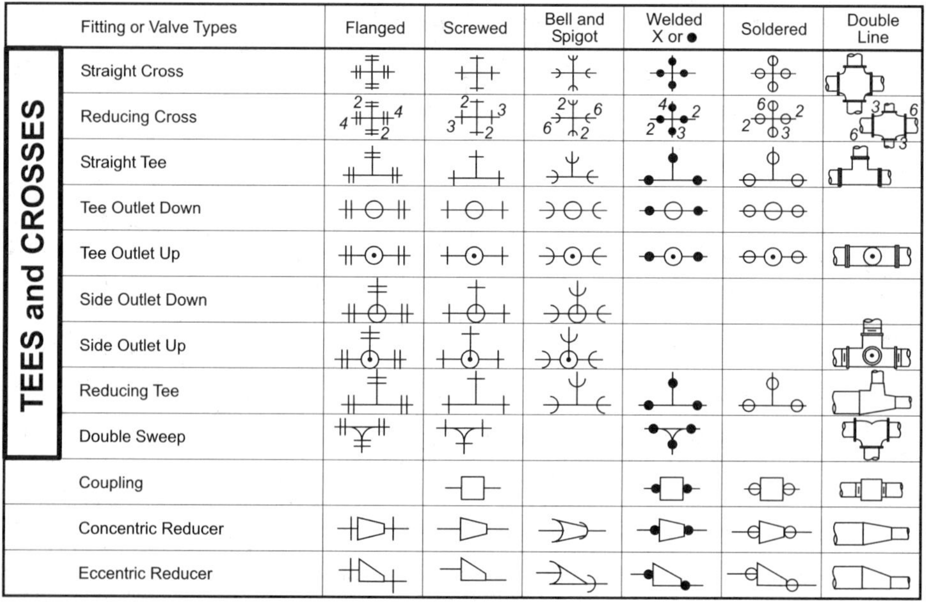

	Fitting or Valve Types	Flanged	Screwed	Bell and Spigot	Welded X or ●	Soldered	Double Line
TEES and CROSSES	Straight Cross						
	Reducing Cross	2 4 4 2	2 3 3 2	2 6 6 2	4 2 2 3	6 2 2 3	3 6 6 3
	Straight Tee						
	Tee Outlet Down						
	Tee Outlet Up						
	Side Outlet Down						
	Side Outlet Up						
	Reducing Tee						
	Double Sweep						
	Coupling						
	Concentric Reducer						
	Eccentric Reducer						

Illustration #176C - Commonly Used Fitting and Valve Symbols

	Fitting or Valve Types	Flanged	Screwed	Bell and Spigot	Welded X or ●	Soldered	Double Line
ELBOWS	45° Elbow						
	90° Elbow						
	Elbow Turned Down						
	Elbow Turned Up						
	Base Elbow						
	Long Radius Elbow						
	Reducing Elbow	4 2	4 2	4 2	4 2	4 2	4 2
	Side Outlet (Turned Down)						
	Side Outlet (Turned Up)						
	Elbowlet						
	Bushing						
	Union						

Illustration #176D - Commonly Used Fitting and Valve Symbols

Fitting or Valve Types	Flanged	Screwed	Bell and Spigot	Welded X or ●	Soldered	Double Line
Connecting Pipe Joint						
Expansion Joint						
Lateral Joint						
Sleeve						
Orifice Flange						
Reducing Flange						
Socket Weld Flange						
Weld Neck Flange						
Blind Flange						
Bull Plug						
Pipe Plug						
Cap						

Illustration #176E - Commonly Used fitting and Valve Symbols

Notes and Specifications

Any information that is not easily interpreted from a drawing or blueprint is usually noted for recognition. Piping notes can be classified as either general or local.

General notes, when used, apply to the entire drawing and are normally placed close to the edge of the drawing or near the title block. Local or specific notes take priority over general notes and convey precise information about one part or area of the drawing. Items such as reducer sizes and valve types are locally noted or called out by size and, in the case of valves, given a number to represent a specific type. Examples of a common local note and callout are presented in illustration #177A and #177B.

Specification or Specs, as they are often referred to, are the written guide lines or project standard to which the job is to be constructed.

Pipe material specifications deal with the requirements for pipe, valves, fittings, flanges, and gaskets which are not normally found on the piping drawing. Specifications for a particular project are found in the project standards or project specification book.

Most notes use abbreviations extensively to save time and space in preparing the drawing. Common abbreviations used on pipe drawings are listed in table #78.

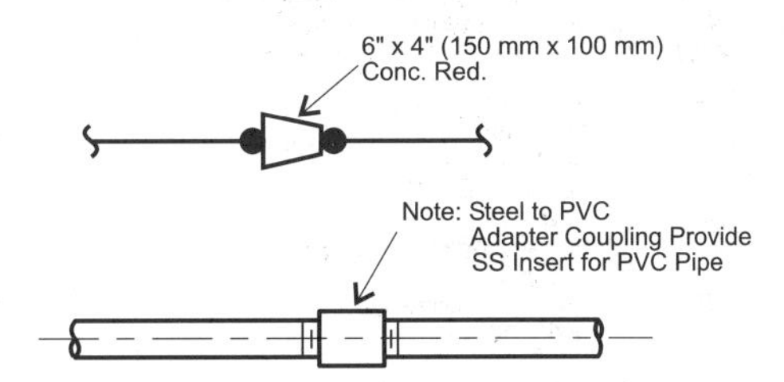

Illustration #177 - Typical Local Note and Callout Information Found on Blueprints

PIPING DRAWING ABBREVIATIONS			
A	Air	CS	Carbon Steel
	Absolute		Cold Spring
AC	Air to Close	CSO	Car-Seal Open
	Combustion Air	CTR	Center
AGA	American Gas Association	CV	Control Valve
AI	Instrument Air		
AISC	American Institute of Steel Construction	DIM	Diameter
ANSI	American National Standard Institute	DS	Dummy Support
AO	Air to Open	DWG	Drawing
AP	Plant Air		
API	American Petroleum Institute	E	East
ASA	American Standard Association	E-E	End to End
ASME	American Society of Mechanical Engineers	EL	Elevation
ASTM	American Society of Testing Materials	ELB	Elbowlet
AWS	American Welding Society	ELL	Elbow
AWWA	American Waterworks Association	ERW	Electric Resistance Welding
BBE	Beveled Both Ends	F	Fahrenheit
BC	Bolt Circle	FAB	Fabrication
BF	Blind Flange	F&D	Face and Drilled
BLE	Beveled Large End	FF	Face to Face
BLVD	Beveled		Flanged Face
BOF	Bottom of Face of Flange		Flat Face
BOP	Bottom of Pipe		Full Face (of gasket)
BSE	Beveled Small End	FLG	Flange
BW	Butt Weld	FOB	Flat on Bottom
		FOT	Flat on Top
C-C	Center to Center	FS	Far Side
CFM	Cubic Feet Per Minute		Field Support
CI	Cast Iron	FTG	Fitting
Cr	Chromium	FW	Field Weld

Table #78A - Piping Drawing Abbreviations

PIPING DRAWING ABBREVIATIONS			
G	Gas	LT	Level Transmitter
GALV	Galvanized		Low Point
GPM	Gallon Per Minute		Light-wall (of pipe)
GR	Grade	M	Meter
		Mo	Molybdenum
HDR	Header	MS	Mild Steel
HEX	Hexagon		
Hg	Mercury	N	North
HPT	Hose Pipe Thread	NC	Normally Closed
HP	High Point	NO	Normally Open
	High Pressure	NPS	National Pipe Size
HTR	Heater		Nominal Pipe Size
HVAC	Heating, Venting & Air Conditioning	NPT	National Pipe Thread
		NRS	Non Rising Stem
ID	Inside Diameter	NS	Near Side
IE	Invert Elevation	NTS	Not to Scale
IMP	Imperial		
INS	Insulation	OD	Outside Diameter
INST	Instrument	OS&Y	Outside Screw and Yoke
IPS	Iron Pipe Size		
IS&Y	Inside Screw and Yoke	P&ID	Piping & Instrumentation Diagram
ISO	Isometric Drawing	PBE	Plain on Both Ends
		PC	Pressure Controller
LA	Level Alarm	PCV	Pressure Control Valve
LC	Level Controller	PE	Plain End
LG	Level Gage	PI	Pressure Indicator
LI	Level Indicator	POE	Plain on One End
LP	Low Pressure	PR	Pressure Regulator
	Low Point	PRV	Pressure Reducing Valve
LR	Long Radius	PS	Pipe Support
LS	Level Switch	PSI	Pounds Per Square Inch
		PSV	Pressure Safety Valve

Table #78B - Piping Drawing Abbreviations

PIPING DRAWING ABBREVIATIONS			
R	Radius	TBE	Threaded on Both Ends
RED	Reducer or Reducing	TC	Temperature Controller / Test Connection
REF	Reference	TE	Threaded End
REQ	Required	TEF	Teflon
RF	Raised Face	TOC	Top of Concrete
RJ	Ring Joint	TOL	Threadolet
R/L	Random Length	TOP	Top of Pipe
RS	Rising Stem	TOS	Top of Support / Top of Steel
		TPI	Threads Per Inch
S	South / Steam Pressure	TYP	Typical
SC	Sample Connection / Steam Condensate		
SCH	Schedule	UNC	Unified Course Thread
SCFH	Standard Cubic Feet Per Hour	UNF	Unified Fine Thread
SCFM	Standard Cubic Feet Per Minute		
SECT	Section	V	Valve
SMLS	Seamless	VC	Vitrified Clay
SO	Slip-On	VERT	Vertical
SOL	Sockolet		
SP	Sample Point / Steam Pressure	W	West / Water
SPEC	Specification	W/	With
SQ	Square	WC	Water Column
SR	Short Radius	WE	Welded End
SS	Stainless Steel	WN	Welded Neck
ST	Steam Tracing	WOG	Water, Oil, Gas
STD	Standard	WOL	Weldolet
SW	Socket Weld	WP	Working Pressure / Working Point
SWG	Swage	WT	Weight
SWP	Standard Working Pressure		
		XH	Extra Heavy
T	Threaded / Steam Trap / Temperature	XS	Extra Strong
		XXH	Double Extra Heavy
		XXS	Double Extra Strong

Table #78C - Piping Drawing Abbreviations

Reading Scale Drawings

Whenever possible, needed measurements from blueprints should be taken from the dimensions given on the blueprint. Because blueprints can shrink or expand for various reasons (reproduction or copying for example), scaling from a blueprint should only be done when there is no other way of determining the dimensions needed.

When it is necessary to use a scale rule to determine a dimension from a blueprint, the proper scale or ratio of the rule must match the blueprint scale used.

The scale used on a drawing is always given in the title block of the print.

The principle of scale measurement used on most drawings is that a smaller unit dimension represents a larger unit dimension. Common scales found in the piping field for imperial measurements are 1/4 in. = 1 ft. and 3/8 in. = 1 ft.. In the metric system 1 mm = 50 mm and 1 mm = 20 mm are the most commonly used. Table #79 gives the standard scales and ratios found on an architect's scale and which may be used on blueprints. Examples of both imperial and metric scale usage are shown in illustration #178.

SCALE AND RATIO			
Imperial		**Metric**	
3/32 inch = 1 ft.	1:128 ratio	1 mm = 2 mm	1:2 ratio
1/8 inch = 1 ft.	1:96 ratio	1 mm = 5 mm	1:5 ratio
3/16 = 1 ft.	1:64 ratio	1 mm = 10 mm	1:10 ratio
1/4 inch = 1 ft.	1:48 ratio	1 mm = 20 mm	1:20 ratio
3/8 inch = 1 ft.	1:32 ratio	1 mm = 30 mm	1:30 ratio
1/2 inch = 1 ft.	1:24 ratio	1 mm = 50 mm	1:50 ratio
3/4 inch = 1 ft.	1:16 ratio	1 mm = 100 mm	1:100 ratio
1 inch = 1 ft.	1:12 ratio	1 mm = 200 mm	1:200 ratio
		1 mm = 500 mm	1:500 ratio
		1 mm = 1000 mm	1:1000 ratio

Table #79 - Scale and Ratio

Reading Scale

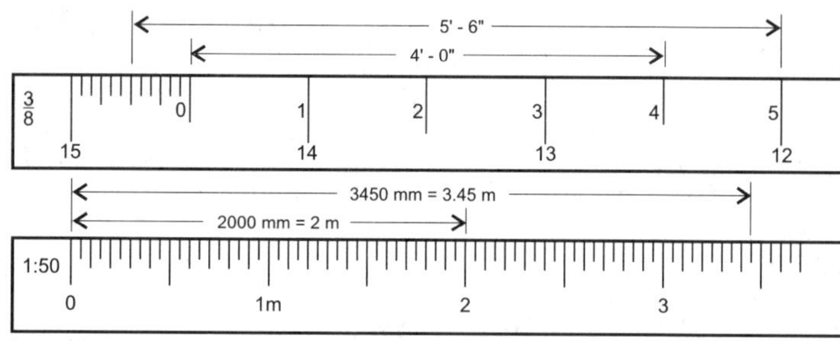

Illustration #178 - Reading Scales

Development of Pipe Drawings

Pipe drawings and blueprints range from general information schematic and flow diagrams to the detailed piping isometric and spool sheets needed for fabrication of pipe sections. The development of piping drawings usually progress in the following sequence:

1. Schematic Diagram
2. Flow Diagram
3. Piping and Instrumentation Diagram (P&ID)
4. Piping Drawing
5. Isometric Drawings
6. Spool Drawings

Note: Some of the drawings may be grouped together or left out depending on the project size and type of piping system.

Schematic Diagram

The schematic diagram or drawing is a theoretical layout of the system and its operation. It is only used in the initial planning stage as a basic guide for the development of the flow diagram. This diagram uses non scale single line flow paths with rectangles or circles representing general system operations and process equipment.

Flow Diagram

The flow diagram is a more sophisticated schematic drawing showing typical equipment layout and the flow of fluids through the system. An example of a typical process flow diagram is represented in illustration #179.

The diagram provides an overall perspective of the entire system operation or a specified plant process using basic symbols, flow arrows and single line process runs.

Flow diagrams are not drawn to scale and may be presented as either an elevation or plan view drawing. Installations covering large areas are usually shown as plan view flow diagrams, where less complicated systems are usually shown from an elevation view.

Flow Diagram

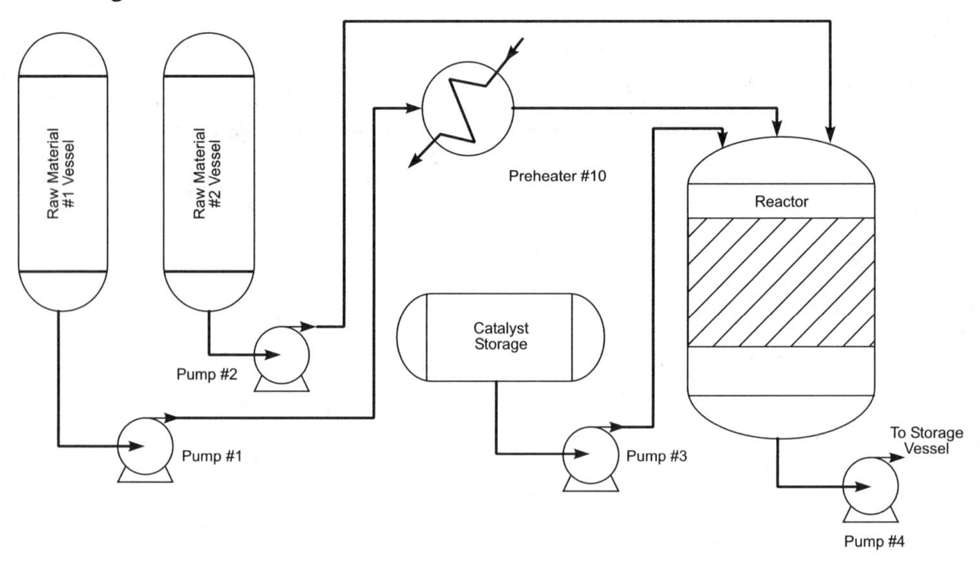

Illustration #179 - Typical Process Flow Diagram

Piping and Instrumentation Diagrams

P&ID drawing is the term frequently given to the piping and instrumentation diagram or occasionally referred to as the process and instrumentation diagram. The drawing is essentially a detailed mechanical or process flow diagram in schematic form.

The P&ID drawing is not drawn to scale and equipment on the drawing is located as to allow major process flow runs form left to right on the drawing. The content of a typical P&ID generally includes:

- Major equipment and needed valves
- All process lines and pipe sizes for each pipe line.
- Line numbers of codes along with designated flow direction.
- Instrumentation and control devices.

Illustration #180 shows one section of the previous flow diagram (raw material vessel) as it would be developed in a P&ID format.

Instrumentation on P&ID

Most instrumentation on piping and instrumentation diagrams use standardized ISA (Instrumentation, Systems and Automation Society) symbols and identification methods. Balloons or bubbles (circular shaped symbols) are generally used in identifying instruments (see illustration #181) with the instrument functions specified by the use of abbreviations within the balloons.Generally, each instrument is first given a functional letter identification followed by a loop identification number. The first letter in the abbreviation typically identifies the measured or variable controlled by the instrument, see table #80.

Piping and Instrumentation Diagram

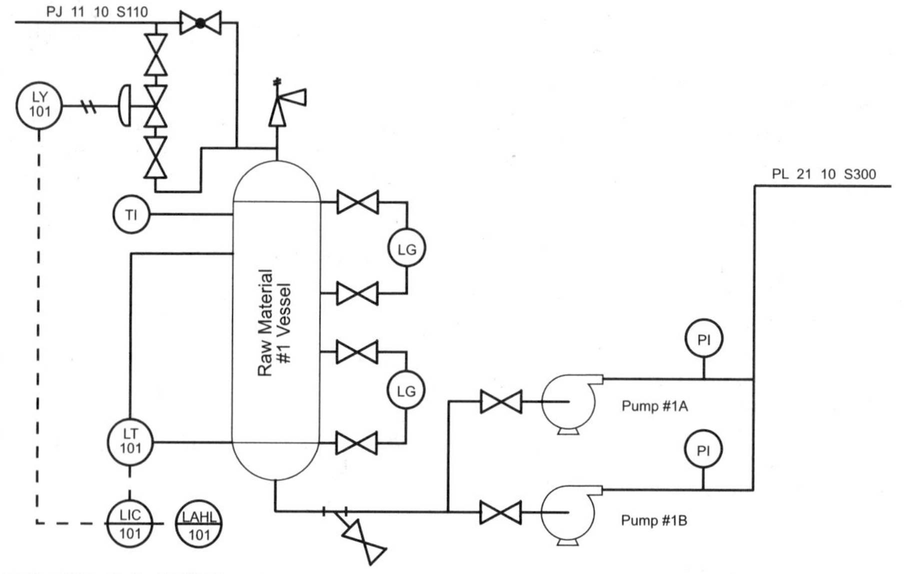

Illustration #180 - Typical P&ID Diagram

Instrumentation on P&ID

Locally or Field Mounted Instrument

Board Mounted (Control Room) Instrument

Mounted Behind Board

Instruments Sharing a Common Housing

Illustration #181 - Instrumentation Balloons

In instrumentation function letter identification, most often the first letter will represent one of the following:

F - Flow, T - Temperature, L - Level, or P - Pressure. Table #80 shows succeeding letters used in the abbreviation identify the actual function performed by the instrument.

INSTRUMENT FUNCTIONS	
Flow	
FI	= Flow Indicator
FR	= Flow Recorder
FC	= Flow Controller
FT	= Flow Transmitter
Level	
LI	= Level Indicator
LR	= Level Recorder
LC	= Level Controller
LT	= Level Transmitter
Temperature	
TI	= Temperature Indicator
TR	= Temperature Recorder
TC	= Temperature Controller
TT	= Temperature Transmitter
Pressure	
PI	= Pressure Indicator
PR	= Pressure Recorder
PC	= Pressure Controller
PT	= Pressure Transmitter

Table #80 - Instrument Functions

Piping Drawing (Orthographic Projection)

The piping drawing is the detailed outline to which the piping system is to be fabricated. The true shape and dimensions of equipment are represented on the piping drawings. Most piping drawings are developed using the orthographic projection method.

True orthographic projection drawings consist of six distinct views: top, front, right side, left side, bottom, and back. However, most drawings find it necessary only to show the top, side and front views. This three view method of orthographic projection is shown in illustration #182B.

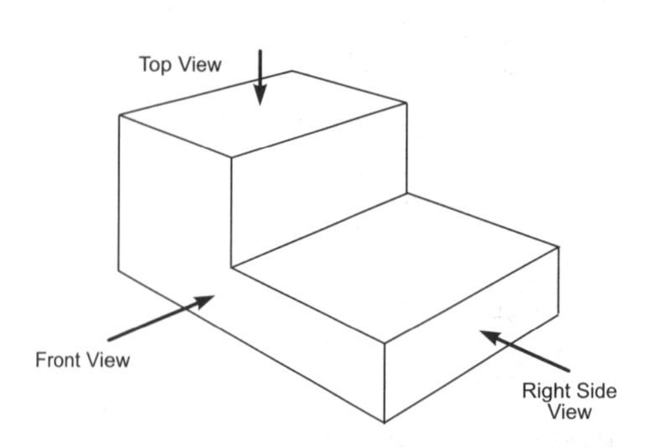

Illustration #182A - Actual View of Object

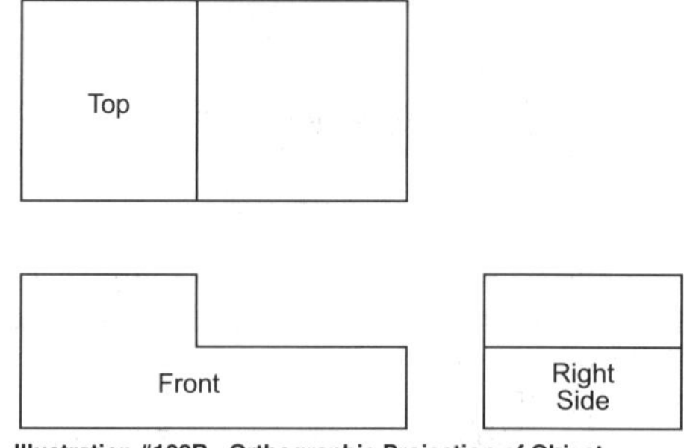

Illustration #182B - Orthographic Projection of Object

Piping Drawing (Orthographic Projection)

In pipe drawing, orthographic projections are referred to as plan, elevation, and sectional views. This type of projection drawing may be used for large and complicated piping systems, piping in buildings, and for some small spool piece drawings. A simple double line orthographic spool sheet drawing showing a plan and sectional view is shown in illustration #183.

Dimensions, Elevation and Coordinates

Most pipe orthographic drawings are drawn to some scale; with the exception being the spool sheet drawing, which as a rule, relies on written dimensions. Vertical dimensions on large scale plant drawings are specified in elevation designations rather than using dimension lines. The elevation at grade is the normal starting point and typically given an arbitrary elevation of 100 ft. 00 in. (100 m 000 mm).

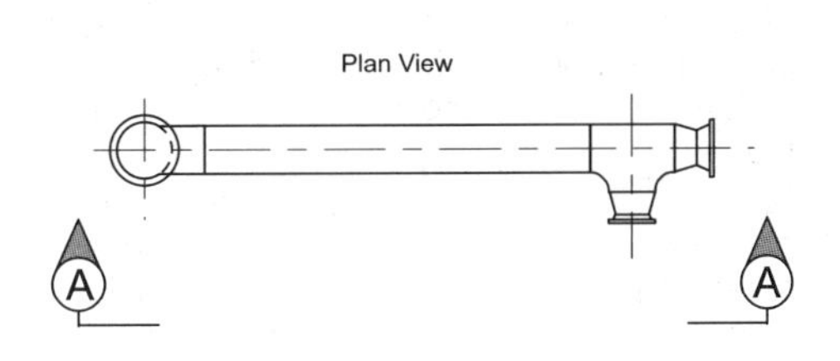

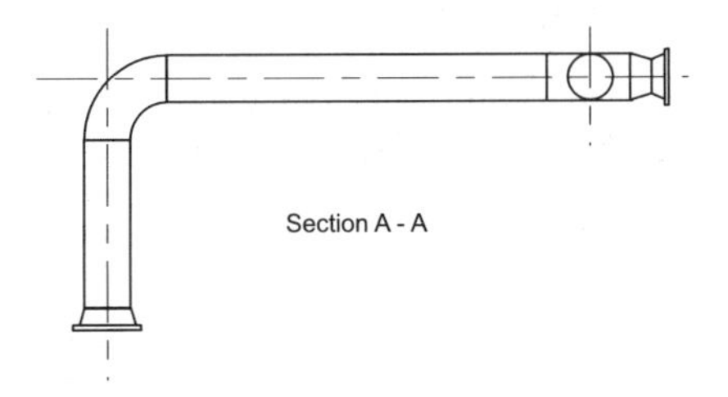

Illustration #183 - Double Line Orthographic Spool Drawing

Dimensions, Elevation and Coordinates

This dimensioning method provides for a positive number for underground service elevation dimensions and an even number starting point.

In plan view orthographic drawings, coordinates are often used to locate structural steel, vessels, tanks and major equipment. Coordinates, as a rule, start from an established reference point at the South-West corner of the project. As the distance increases (going north and east from the reference point), the coordinates get larger. In buildings and outlined structures, dimensions are usually given from the structure steel of columns.

Pictorial Piping Drawing

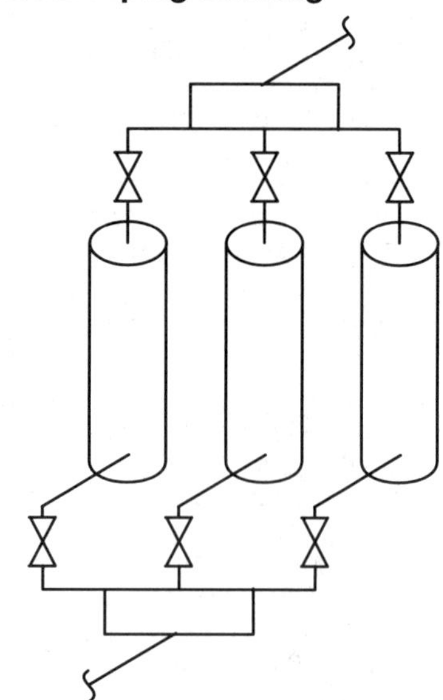

Illustration #184 - Oblique Piping Drawing

Pictorial Piping Drawing

Piping drawings are often shown in a pictorial type of representation for clarity and ease of interpretation, particularly on intricate drawings. The two most common types of piping pictorial drawings are the isometric and the oblique projections.

The oblique drawing is the less common of the two drawings but may be used when an emphasized front view is desirable for clarity. Oblique drawings show the plan or front face of the piping drawing in its true form or plane.

All other parts and piping from this plane are then projected back from this view on an angle of 60 degrees or at 45 degrees on some drawings. Illustration #184 shows a parallel tank and piping layout using an oblique projection.

Isometric or Iso Drawings

The most used pictorial drawing in the piping industry is the isometric projection. The major advantage of an isometric drawing is that three sides of the object are displayed in one practical easy to read view. This ease of interpreting an isometric drawing as compared to an orthographic drawing is shown in illustration #185A and #185B. Some companies use isometric projection as their major piping working print, but because of the complexity involved in drawing overlapping multiple pipe runs, it is often only used for fabrications and detail piping work.

Isometric drawing construction uses three axis that are equally spaced at 120 degrees form each other. All horizontal lines are drawn at angles of 30 degrees, while all vertical lines remain vertical, see illustration #186.

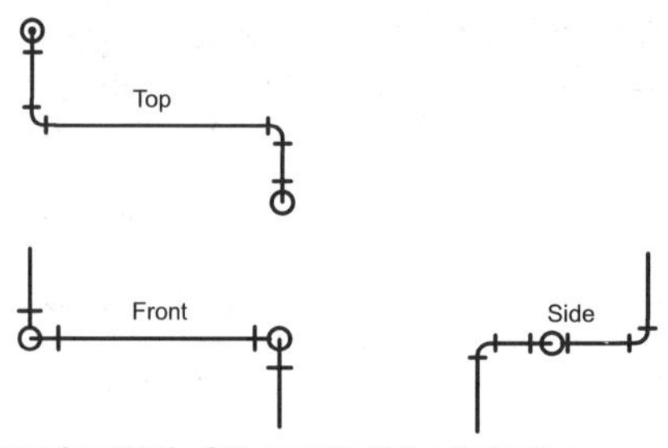

Illustration #185A - Orthographic Piping Projection

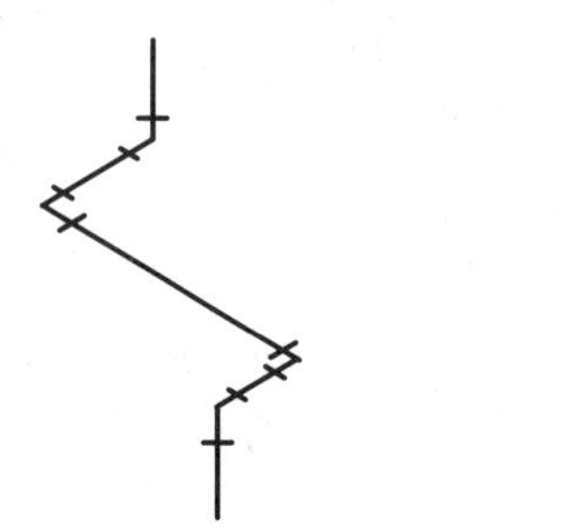

Illustration #185B - Isometric Drawing of the Same System

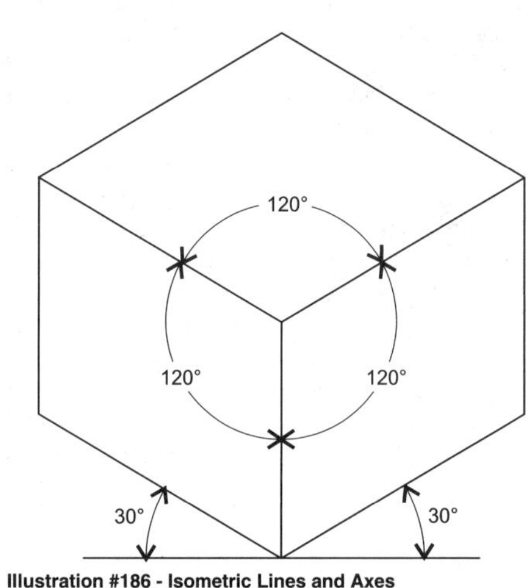

Illustration #186 - Isometric Lines and Axes

Isometric or Iso Drawings

In piping systems, not all lines run at right angles to each other and when diagonal lines are needed on isometric drawings, they are shown by framing the diagonal line. The frame is represented by an isometric square or rectangle in the same plane as the offset, shown in illustration #187.

Isometric Dimensions

Dimensions on isometric drawings are normally indicated and should not be scaled from the drawing. When drawing isometric piping, the tendency is to give priority to indicating and positioning fittings and valves for clarity rather than scale. The dimensions given are center-to-center for most fittings and face-to-face for flanges and valves.

Hash marks or parallel extension lines used inside dimension lines on flanges and valves indicate that the face-to-face dimension includes the gasket dimension. Illustration #188 shows a typical isometric drawing and the basic information that may be found on it.

Isometric or Iso Drawings

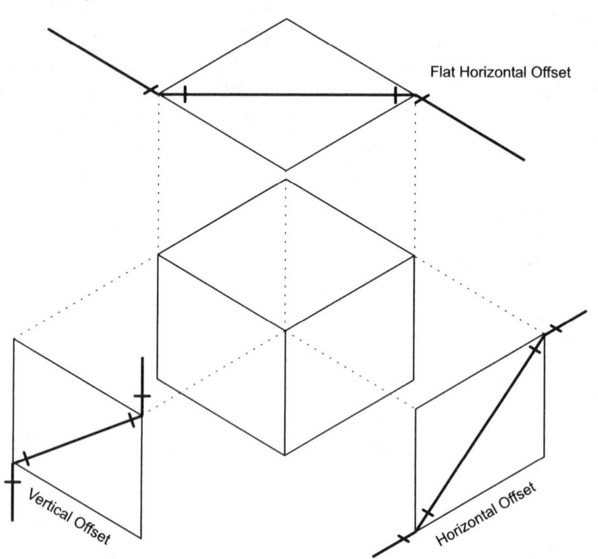

Illustration #187 - Isometric Rectangle Example

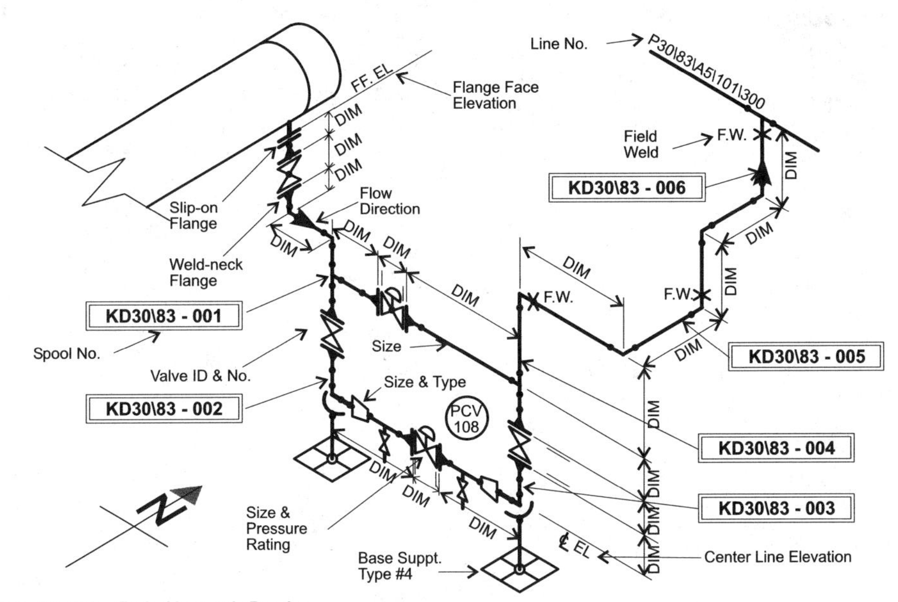

Illustration #188 - Typical Isometric Drawing

Pipe Spools

The prefabrication of a particular piping section including fittings and flanges is referred to as a spool. The physical size is usually limited to the type of transportation used to deliver the spool from the fabrication location to the actual job site.

Spool drawings or shop fabrication drawings, as they are sometimes referred to, are made up from piping drawings or from detailed isometric drawings. They are separate drawings incorporating all the dimensions, material specifications and information needed for the complete fabrication of the spool piping.

The spool drawings can be done in orthographic or isometric projection depending on company preference. Isometric spool drawings are generally done in single line and orthographic drawings in either single or double line.

Double line orthographic drawings are usually preferred to the single line projection. Illustration #189A, #189B and #189C show a comparison between various spool drawing methods.

Each pipe spool drawing includes a spool number or mark number which is transferred to the finished fabricated pipe spool for identification. The number usually matches the line number on the piping drawing or isometric drawing to which it belongs. A letter or number is usually added to the end of line number to indicate the position of the spool in the field piping fabrication.

Double Line Orthographic Spool Drawing

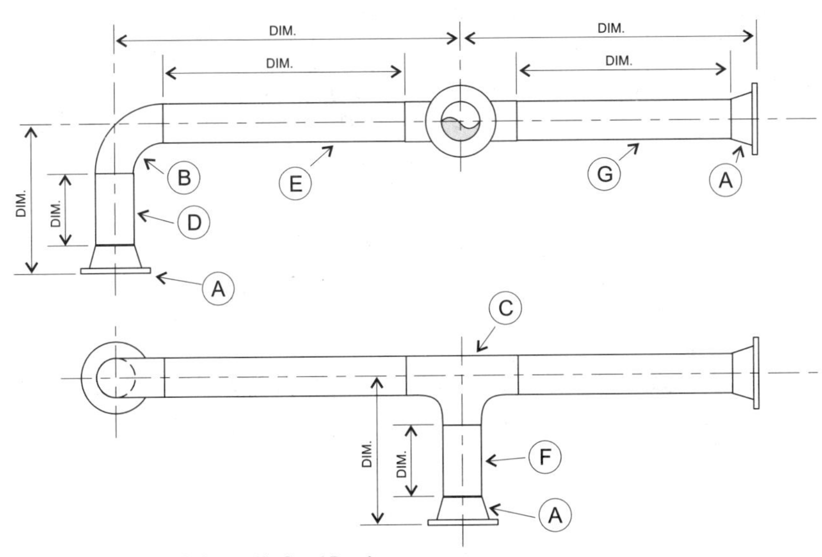

Illustration #189A - Double Line Orthographic Spool Drawing

Single Line Orthographic Spool Drawing

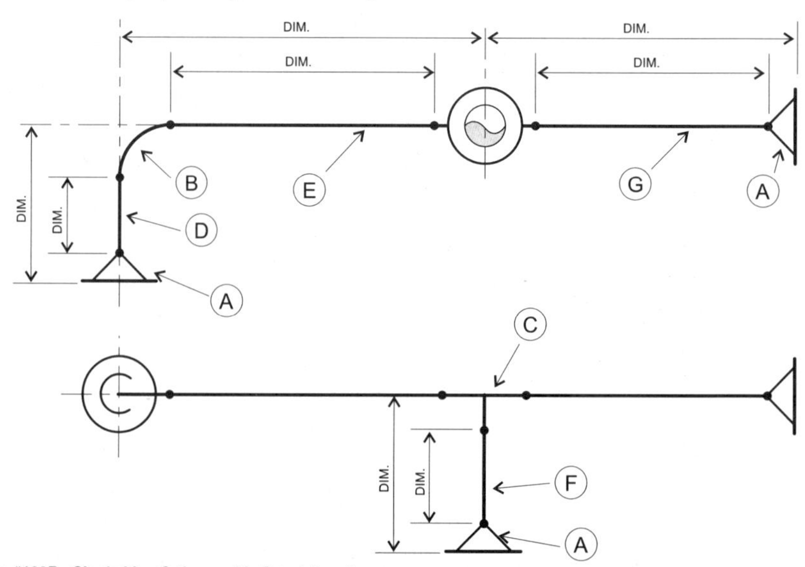

Illustration #189B - Single Line Orthographic Spool Drawing

Single Line Isometric Spool Drawing

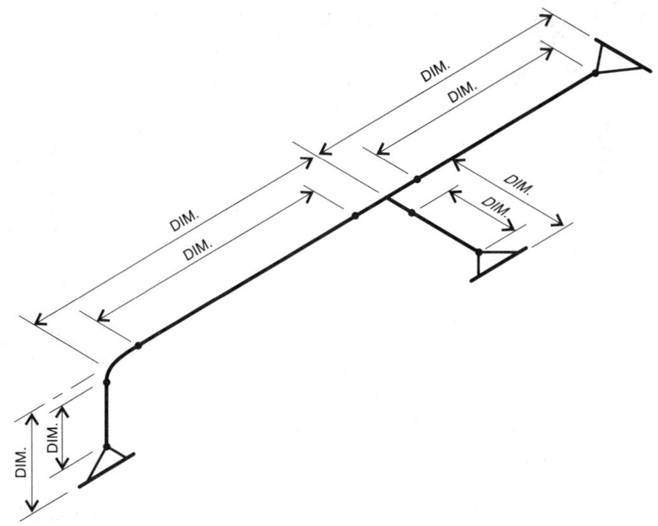

Illustration #189C - Single Line Isometric Spool Drawing

Bill of Material

Each spool drawing includes a bill or list of material usually specifying the quantity and type of fittings, flanges, bolts, gaskets and pipe to use in the spool fabrication. An example of the bill of material for the spool drawing in illustration #189 is shown in illustration #190.

Bill of Materials for Spool Drawing No. KD30\83 - 004		
Item No.	**Quantity**	**Description**
		Flanges
A	3	6" (150 mm) 150# R.F. W.N. (STD BORE)
		Fittings
B	1	6" (150 mm) STD. L.R. 90 DEG. ELL
C	1	6" (150 mm) STD. TEE
		Pipe
D	1	6" X 1' - 4" (150 mm X 406.5 mm) SCH 40
E	1	6" X 6' - 10" (150 mm X 2083 mm) SCH 40
F	1	6" X 2' - 0" (150 mm X 609.6 mm) SCH 40
G	1	6" X 5' - 9" (150 mm X 1752.6 mm) SCH 40
		Other

Illustration #190 - Bill of Material

SECTION TEN

PIPE WELDING

Hazardous Environment Tips

1. Do not cut or weld tanks, vessels, containers, or piping which has previously been in contact with hazardous substances unless they have been thoroughly cleaned.
2. Use extreme caution when cutting or welding components which have been painted, plated, or coated. Highly toxic fumes or gases can be released. Use proper ventilation and safe breathing apparatus.
3. Working in a confined space will require a special entry procedure and proper ventilation.
4. Be careful of electrical shock from welding machines.
   a. Properly ground the machine.
   b. Use dry gloves with no holes. Do not weld if wet.
   c. Use proper size cables in good condition.
   d. Do not wrap cables around the body.

Clothing

The best possible material for welders clothing is tanned leather. However leather protective clothing is expensive, hot and heavy.

Cotton denims are possibly the most popular dress. Denim is inexpensive and sheds sparks fairly well.

The operators' clothing should conform to the following:

- Pants should have no cuffs.
- Shirt pockets should have flaps.
- Steel toed boots should be worn.
- A cap or beanie should be worn to protect the head and hair.
- Gauntlet type gloves should be worn.

Note: Do not wear light, shoddy, frayed, cottons or flannelette as these materials tend to flash easily and quickly. Do not wear oily clothing while welding!

Welding Lenses

The welder needs eye protection against visible light rays, ultra-violet light rays, infra red rays, heat rays, flying metal particles, sparks and slag.

Lenses are available in three basic colors: amber, green and cobalt blue. The shade of the lens is indicated by numbers. The greater the number the darker the shade.

The lens must eliminate glare but still allow the work to be seen distinctly.

Selecting Correct Lenses

After welding for a few minutes, lift the goggles or helmet. If light spots are seen, a darker lens is needed. If dark spots are seen, a lighter shade is needed.

WELDING FILTER LENSES	
Lens #	**Uses**
Shade 4	Light oxy-acetylene welding and cutting
Shade 5	Optional to above
Shade 6	General oxy-acetylene welding
Shade 8	Sheet metal arc welding
Shade 9	Light arc welding
Shade 10	Medium arc welding
Shade 11	Heavy arc welding
Shade 12	Used with heavy electrodes

Table #81 - Welding Filter Lenses

Note: For additional welding tips and information, see IPT's Metal Trades Handbook.

Welding Lenses

To protect the welding filter lens from sparks, the filter lens should have two clear glass cover lenses, one positioned in front and one behind, see illustration #191. These two glass cover lenses are in addition to the one plastic lens in the fixed position in the shield.

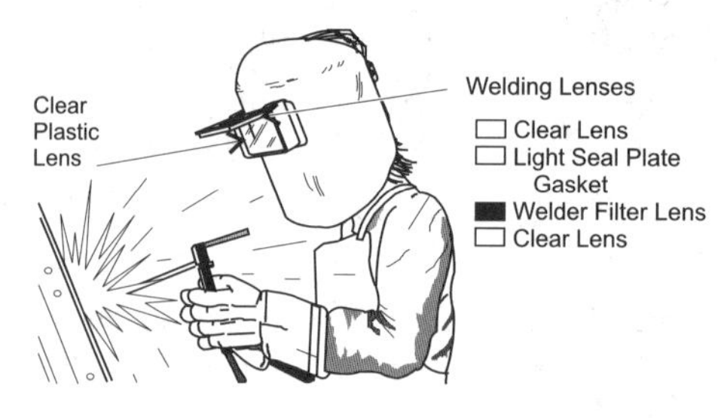

Illustration #191 - Welding Lens Positions

Arc Flash

Arc flash is the term used to describe eyes damaged by ultra-violet rays from welding. The arc has burned the eyeballs and they are covered with small water blisters. This condition is described as having sand in the eyes. The result is that the eyelids flutter and tears are profuse.

The following medical aids will help the person with an arc flash. ***A Doctor's prescription is strongly recommended.***

- 1% Pontocain drop (local anaesthetic)
- 2% Butyn drops (local anaesthetic)
- 1% Holocaine (local anaesthetic)

Do not use these medical aids for a prolonged length of time (12 hours, maximum).

Note: Do not work with eyes frozen, and if the eyes are not noticeably better after 12 hours see a doctor.

Ventilation

Welders must be provided with good ventilation. Welding shops should have a minimum of four air changes per hour.

Screens should be arranged so that they do not restrict ventilation.

When welding on non-ferrous or galvanized metals, extra ventilation will be required. Under some circumstances, it may be necessary for the operator to use a respirator or mask.

Exposure to zinc fumes from galvanized steel may result in metal fume fever, commonly called "zinc chills".

"Zinc chills" are self eliminating, and at the time of writing are without any known complications or after effects.

Cadmium and lead fumes are definitely dangerous. The welder must be provided with an air line respirator approved for use with these fumes.

Special precautions must be taken when welding with either tungsten or consumable electrodes using an argon shield. Argon is heavier than air and will sink to the bottom of confined areas and has the net result of actually drowning the welder in the argon atmosphere.

Welding Fire Prevention

Due to the high temperature involved in welding, and inevitable production of hot metal accompanied by sparks, it is most essential that fire prevention precautions be rigorously observed wherever these operations are performed. Obviously, such precautions are particularly necessary when the work is carried out in combustible surroundings.

Keep fire extinguishers at hand and ready for immediate use. When work has to be done in the vicinity of combustible materials, someone should stay in attendance for at least a half hour after the work is finished to insure that no fires break out.

Oxygen

Oxygen is a colorless, odorless, tasteless gas at ordinary temperatures; it is slightly heavier than air, and while not flammable, it must be considered a potential danger since it combines readily, and in some cases very violently, with many other materials.

Oxygen (O_2) is the element in air that supports life and combustion. In general, materials that burn in air, burn much more violently in O_2, and some materials, not considered combustible in air, will burn readily in O_2.

At normal temperatures, oil and grease is not considered highly flammable, but if either one is brought in contact with pure O_2 under high pressure or friction, an explosion may occur.

Note: Keep oxy-acetylene equipment away from oil or grease. Never oil regulators or torch parts.

Never Use Oxygen:

- To operate pneumatic tools.
- In the presence of oil or grease.
- As a compressed air substitute.
- In oil pre-heating burners.
- To start or run internal combustion engines.
- To blow out pipe or tubing lines.
- To create pressure.
- For ventilation purposes.
- To dust off clothing or work areas.

Oxygen Cylinders

An oxygen cylinder is a pressure vessel and must be treated as such. The top is hemispherical and the valve hole is reinforced.

Do not attempt any repairs to an oxygen cylinder and if a cylinder leaks, remove it from any building and place it in the open. Tag it indicating the type of fault and notify the supplier.

Oxygen Cylinders

The oxygen cylinder is protected from extreme pressure caused by heat with the use of a fusible metal rupture disc. This allows the slow controlled escape of gas.

- All oxygen connections have right hand threads.
- Oxygen cylinder valves should be opened fully while in use.
- Oxygen cylinders have no fixed draw off limit.

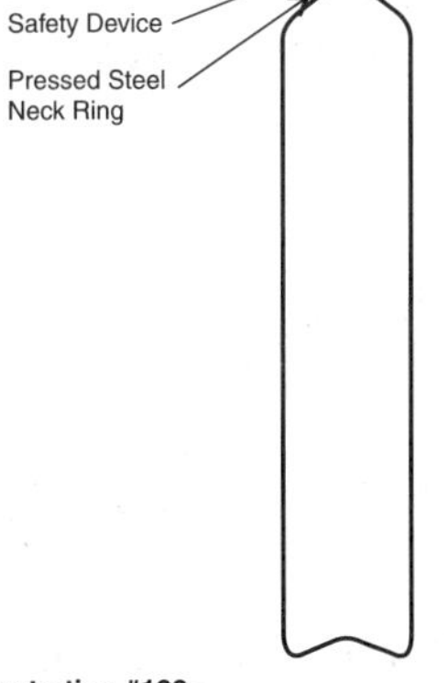

Illustration #192 - Oxygen Cylinder

Acetylene

Acetylene gas is colorless, but fortunately has a strong pungent odor. As little as 1 % in air is quite noticeable to the average person's sense of smell.

It is advisable to treat mixtures of air or oxygen and acetylene as explosive. If you can smell acetylene take no chances. Extinguish all open flames and ventilate the room.

Acetylene is an Unstable Compound.

Acetylene is not safely compressed beyond 15 psi (103.4 kPa) pressure, especially in large volumes, such as cylinders, pipelines and generators.

Acetylene Cylinders

Codes call for a minimum of one fusible plug on each end of the cylinder. The purpose of the plug, which has a melting point of about 212°F (100°C), is to melt out in case of fire, thus allowing a slow controlled escape of gas, rather than a violent explosion.

- *Acetylene cylinders must be used in the vertical position to prevent acetone from being drawn off.*
- *Store cylinders in a cool place.*
- *Never attempt to transfer acetylene from one cylinder to another.*
- *Never attempt to interchange equipment from one gas type to another.*
- *Key type acetylene valves should be opened only one and one half turns. Hand wheel type should be opened fully.*
- *All acetylene connections have left hand threads and the fittings have a groove cut around them.*
- *Use soapy water to test for leaks.*

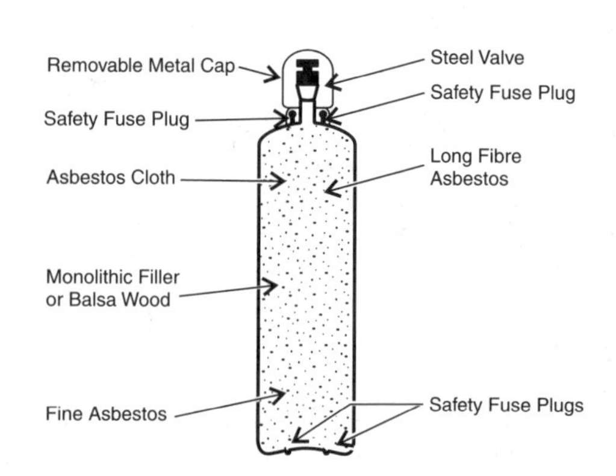

Illustration #192B - Acetylene Cylinder

Setting Up An Oxy-Acetylene Outfit

In this set-up, the equipment is being treated as though it were completely dismantled. All these steps are not necessary if only one part is being changed.

1. Stand the cylinders vertically and securely so that they will not be knocked over.
2. Uncap the cylinders.
3. Crack the cylinder valves to remove dust or dirt from the valves.
4. Attach the pressure regulators to the cylinder valves. Flat faced fittings of the type found on some acetylene regulator stems must have a fibre washer between the stem and the cylinder valve.
5. Attach the hoses to the regulators and tighten the connections.
6. Release the regulator adjusting screws by turning anti-clockwise until they turn freely.
7. Open cylinder valves slowly, allowing a slow build up of pressure within the gauges.
8. Open the regulators one at a time to purge the hoses and remove dust, dirt, and talcum powder.
9. Attach torch, mixer, tip, and tighten firmly.
10. Adjust the working pressure for both oxygen and acetylene.
11. Check connection for leaks, use soapy water around the connections and look for bubbles.

Lighting a Cutting Torch

1. If the torch is of the combination type, open the torch handle oxygen needle valve fully. Delete this step if it is a heavy duty cutting torch.
2. Open the acetylene needle valve about one half turn and light. Increase acetylene flow to desired proportions but do not attempt to take away all the smoke, or preheat will be excessive.
3. Add oxygen to preheat flames by slowly opening preheat oxygen needle valve and adjust preheat flames to neutral.
4. Depress the cutting oxygen lever or trigger, and if the preheat flames feather, readjust to neutral with the trigger still depressed.
5. Before beginning the cut, ensure that the cutting jet is perfectly straight. Clean the tip with the correct sized tip cleaner.

Shutting Down Oxy-Acetylene Equipment

Close the oxygen valve, then the acetylene torch valve when cutting or welding is stopped.

When welding or cutting is to be stopped for a considerable length of time, the following steps must be completed:

1. Close the oxygen and acetylene cylinder valves.
2. Open the torch oxygen valve to release all pressure from the hose and regulator.
3. Turn out the pressure adjusting screws of the oxygen regulator.
4. Close the torch oxygen valve.
5. Open the torch acetylene valve to release all pressure from the hose and regulator.
6. Turn out the pressure adjusting screw of the acetylene regulator.
7. Close the torch acetylene valve.

Balancing a Torch (Welding or Brazing)

The purpose of balancing a welding torch is to establish the maximum gas flow settings used for a particular welding tip. This compensates for normal regulator inaccuracies by setting the regulators under actual working pressures. A welding torch should be balanced every time a tip is selected and the regulator pressure must be adjusted accordingly. This balancing process is used for both oxy-acetylene welding and brazing.

Balancing Steps

1. Slowly open the oxygen cylinder valve until fully open.
2. Slowly open the acetylene cylinder valve until fully open (key type valves are only opened one and one half turns).
3. Turn the acetylene regulator adjusting screw clockwise until fuel starts to flow.
4. Open the torch acetylene valve one half turn. Ignite the gas, then fully open the valve.
5. Open the acetylene regulator valve until the flame just leaves the tip end.
6. Adjust the gas flow with the acetylene torch valve until the flame is back to the tip and it does not smoke.
7. Fully open the torch oxygen valve.
8. Slowly open the oxygen regulator valve until a neutral flame is obtained. See illustration #194.
9. Open the acetylene torch valve slightly and open the oxygen regulator valve to again obtain a neutral flame.
10. Repeat step #9 until the torch acetylene valve is fully open and there is a bright neutral flame.

Note: When the oxygen and acetylene torch valves are fully open and a neutral flame is obtained, the torch is balanced for the selected tip.

11. Slowly turn down both valves until the flame suitable for the job is obtained. The torch can be turned off by shutting off the torch acetylene first, then the oxygen.

Note: The ideal tip flow is a medium rate.

Types of Oxy-Acetylene Flames

- ***Carbonizing Flames*** - A flame that is rich in acetylene, see illustration #193.
- ***Neutral Flame*** - A flame that has no excess of acetylene or oxygen, see illustration #194.
- ***Oxidizing Flame*** - A flame that is rich in oxygen, see illustration #195.

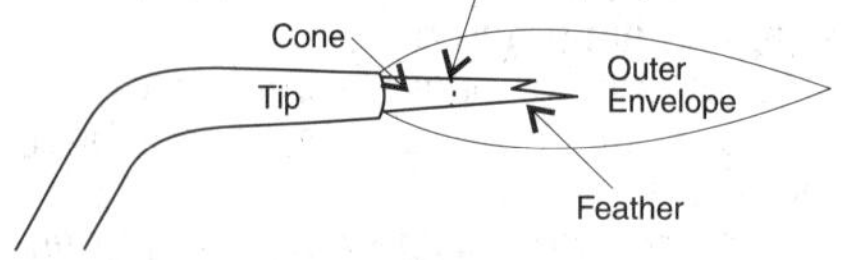

Illustration #193 - Carbonizing Flame

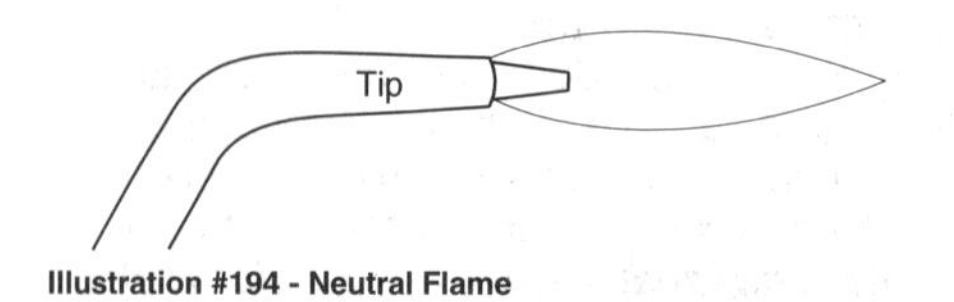

Illustration #194 - Neutral Flame

Illustration #195 - Oxidizing Flame

Cutting Torch Tip Hole Alignment

The correct preheat hole alignment for cutting is indicated in illustration #196.

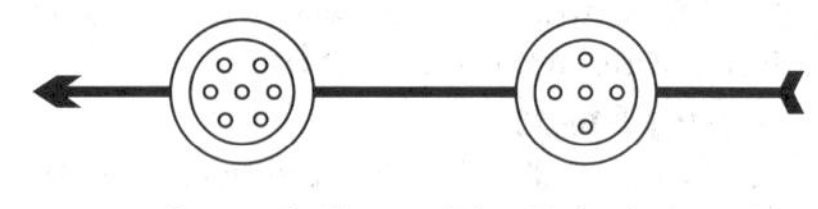

Correct For Square Edge 90 Cutting

Correct For Bevel Cutting

Illustration #196 - Torch Preheat Hole Alignment

Welding Symbols

Welding symbols provide a method of conveying complete welding information from the designer to the welder.

Fillet	Plug or Slot	Spot or Projection	Seam

Back or Backing	Surfacing	Flange Edge	Flange Corner

Groove Welds			
Square	V	Bevel	U
J	Flare-V	Flare-Bevel	

Illustration #197A - Basic Weld Symbols

Basic Symbols

Arc and gas welding symbols are indicated in illustration #197A.

Supplementary Symbols

Supplementary symbols used in connection with basic weld symbols are indicated in illustration #197B.

Weld All Around	Field Weld	Back or Backing	Melt Through

Contour		
Flush	Convex	Concave

Illustration #197B - Supplementary Welding Symbols

Welding Symbol Construction

Basic Rules

- ***When drawing the fillet, J, and bevel groove weld symbols, the perpendicular leg must always be on the left side, as in illustration #198A.***

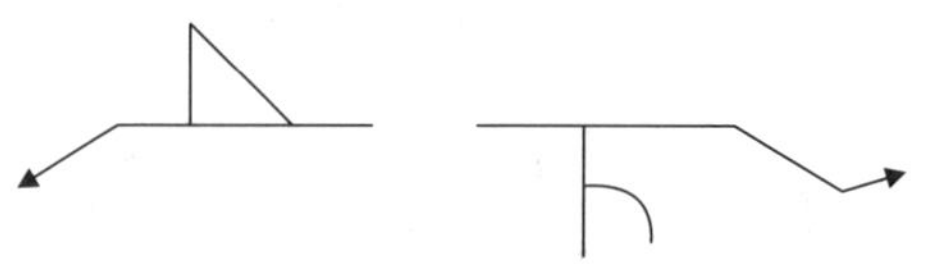

Illustration #198A - Perpendicular Weld Symbol

- ***Arrow Side - When the weld is to be made on the arrow side of the joint, the symbol is located on the bottom of the reference line, regardless of whether the arrow points up or down, as in illustration #198B.***

Illustration #198B - Arrow Side

- ***Other Side - When reference is made to the other side of the joint, the weld symbol is located on the top of the reference line, as in illustration #198C.***

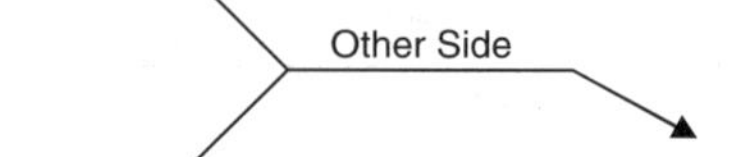

Illustration #198C - Other Side

Note: If the reference line is perpendicular, turn the welding symbol 1/4 turn clockwise to determine arrow side and other side.

Welding Symbol Construction

- ***Both Sides - When reference is made to both sides of the joint, the weld symbol is placed on the top and bottom of the reference as in illustration #198D.***

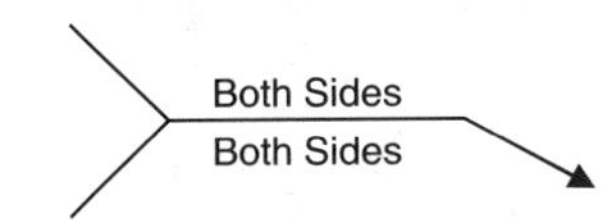

Illustration #198D - Both Sides

- ***Broken Arrow - When using a bevel or J-groove weld symbol, the arrow line must have a definite break toward the member of the joint which is to be prepared, as in illustration #198E.***

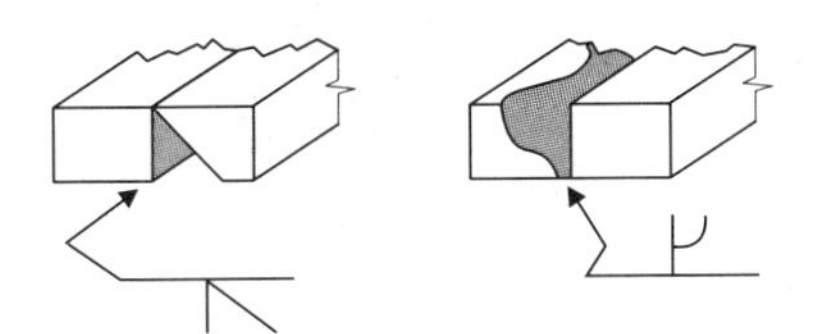

Illustration #198E - Broken Arrow

WELDING PROCESS DESIGNATION LETTERS		
	Welding Process	**Letter Designation**
Brazing	Torch Brazing	TB
Gas Welding	Oxy-acetylene Welding	OAW
Resistance Welding	Resistance Spot Welding	RSW
Arc Welding	Stud Welding	SW
	Plasma-Arc Welding	PAW
	Submerged Arc Welding	SAW
	Gas Tungsten-Arc Welding (T.I.G.)	GTAW
	Gas Metal-Arc Welding (M.I.G.)	GMAW
	Flux Cored Arc Welding	FCAW
	Shielded Metal-Arc Welding (Stick Electrode)	SMAW
	Cutting Process	
Arc Cutting		AC
	Air Carbon-Arc Cutting	AAC
	Plasma-Arc Cutting	PAC
Oxygen Cutting		OC
	Metal Powder Cutting	POC

Table #82 - Welding Process Designation Letters

Elements of a Welding Symbol

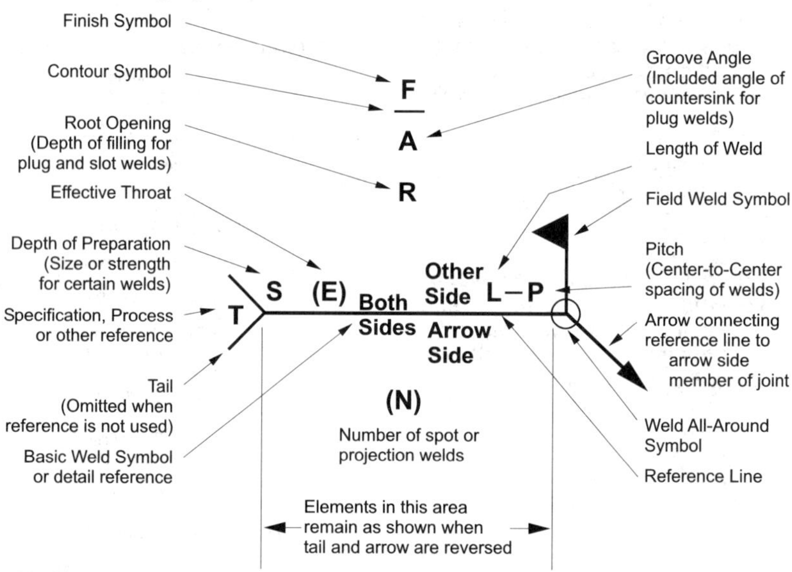

Illustration #199 - Elements of a Weld Symbol

Position of Groove Welds

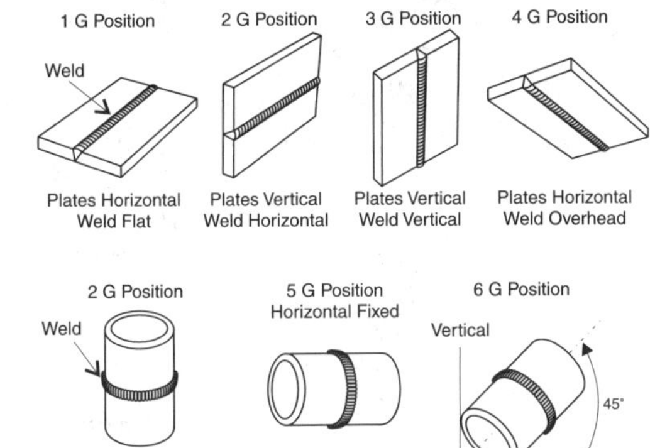

Illustration #200 - Groove Weld Positions

Position of Fillet Welds

The "F" in the numbering system stands for fillet weld.

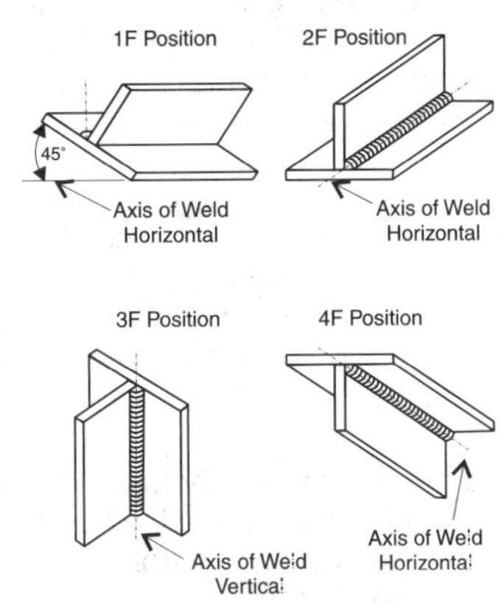

Illustration #201 - Fillet Weld Positions

Preparation of Pipe Test Coupons

Most pipe test coupons should be tacked in the horizontal position. Bend a piece of wire into a V-shape having a diameter equal to the width of the required root opening.

Place this gap rod between the two coupons to act as a spacer. Align the coupons on the inside.

With the pipe coupons properly aligned, one tack is made in the root of the joint, as in illustration #202A. The gap rod is then moved so that only the end is between the coupons. When the gap is correct, make the second tack 180° from the first tack, see illustration #202B.

Remove the gap rod and adjust the gap until the openings are equalized. The third and fourth tacks are made 90° from the first two tacks, see illustration #202C.

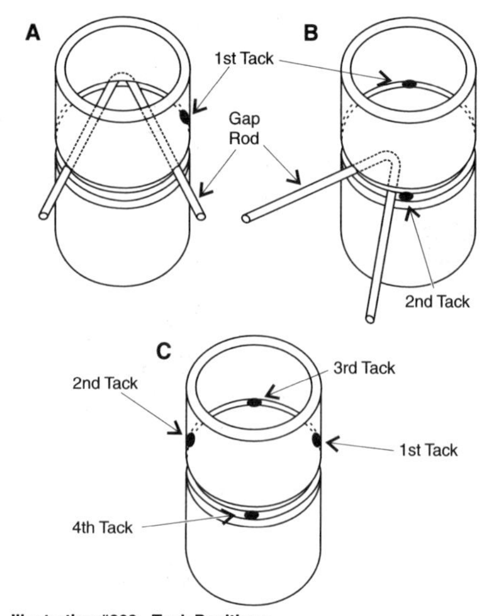

Illustration #202 - Tack Positions

RECOMMENDED ROOT PASS VARIABLES								
Weld Type	Pipe & Tube Diameter	Wall Thickness	Root Face Land	Gap Space	Polarity	Length of Tack	Number of Tacks	Shielding Gas Pressure
SMAW 1/8" (3.2 mm) E-6010, E-7010	all diameters	ALL	1/8" (3.2 mm)	3/32" - 1/8" (2.5 - 3.2 mm)	reverse	1/2" - 1" (13-25.4 mm)	4	
SMAW 3/32" (2.5 mm) E-6010, E-7010	all diameters	ALL	3/32" (2.5 mm)	3/32" (2.5 mm)	reverse	1/2" - 1" (13-25.4 mm)	4	
SMAW 3/32" (2.5 mm) E-7018, E-8018	all diameters	ALL	3/32" (2.5 mm)	3/32" (2.5 mm)	reverse	1/2" - 1" (13-25.4 mm)	4	
SMAW 3/32" (2.5 mm) stainless steel	all diameters	ALL	3/32" (2.5 mm)	3/32" (2.5 mm)	reverse	1/2" - 1" (13-25.4 mm)	4	
GTAW finger rest process, 2% thoriated tungsten 3/32" (2.5 mm) filler wire	all diameters	ALL	no land	5/32" (4.0 mm)	straight	1/2" - 1" (13-25.4 mm)	4	15 cu. ft.
GTAW finger rest process, 2% thoriated tungsten 3/32" (2.5 mm) filler wire	all diameters	ALL	no land	3/32" (2.5 mm)	straight	1/2" - 1" (13-25.4 mm)	4	15 cu. ft.

Table #83 - Recommended Root Pass Variables

Polarity Check

Use an E-6010 electrode in the normal heat range to determine the polarity of a welding machine if the terminals are not marked.

If the arc blows wildly, the electrode is fastened to the negative pole and the polarity would be straight.

If the arc reacts normally, the electrode is fastened to the positive pole and the polarity would be reverse.

Note: Use only E-6010 or E-7010 for this test.

Welding Electrodes

A consumable electrode is a filler metal for welding. The American Welding Society (A.W.S.) has developed specifications for these filler metals.

For the common E-60 and E-70 series of electrodes, the core wire is generally from the same wire stock.

It is S.A.E. 1010 carbon steel, having a carbon range of 0. 05% to 0. 15%.

Identification of Electrodes

For identification, each class of electrode is designated by the letter E, followed by a four or five (five digit metric) digit number.

The first two or three digits represent the first factor (minimum tensile strength of the deposited weld metal) expressed in thousands of pounds per square inch. For example E60xx indicates 60,000 psi minimum tensile strength (metric - three digits representing megapascals).

The second last digit refers to the welding position in which the electrode can be used:

E-xx1x - All positions
E-xx2x - Flat and horizontal positions
E-xx3x - Flat position only
E-xx4x - Vertical down only

Identification of Electrodes

The final digit indicates the type of electrode coating, but may also be used to identify the type of current used, as well as the type of electrode coating. For full identification, however, the last two digits (sometimes called the usability identification) must be read together.

E-xx10 - DC reverse polarity only (cellulose sodium)

E-xx11 - AC or DC reverse polarity (cellulose potassium)

E-xx12 - DC straight polarity or AC (rutile sodium)

E-xx13 - AC or DC either polarity (rutile potassium)

E-xx14 - DC either polarity or AC (rutile iron powder)

E-xx15 - DC reverse polarity only (lime sodium)

E-xx16 - DC reverse polarity or AC (lime potassium)

E-xx18 - DC reverse polarity or AC (lime iron powder, low hydrogen)

E-xx20 - DC either polarity or AC (mineral)

E-xx24 - DC either polarity or AC (rutile iron powder)

E-xx27 - DC either polarity or AC (mineral iron powder)

E-xx28 - DC reverse polarity or AC (lime iron powder, low hydrogen)

E-xx30 - DC either polarity or AC (mineral)

E-xx28 - AC or DC reverse polarity (iron powder, low hydrogen)

WELDING ELECTRODE SELECTION (MILD STEEL)								
	Electrode Classification							
	E-6010	E-6011	E-6012	E-6013	E-7014	E-7018	E-7024	E-6027
Groove butt welds, flat (> 1/4")	4	5	3	8	9	9	9	10
Groove butt welds, all positions (> 1/4")	10	9	5	8	6	6	(b)	(b)
Fillet welds, flat or horizontal	2	3	8	7	9	9	10	9
Fillet welds, all positions	10	9	6	7	7	10	(b)	(b)
Current (c)	DCR	AC	DCS	AC	DC	DCR	DC	DC
		DCR	AC	DC	AC	AC	AC	AC
Thin material (> 1/4 in.)	5	7	8	9	8	2	7	(b)
Heavy plate or highly restrained joint	8	8	6	8	8	10	7	8
High-sulphur or off-analysis steel	(b)	(b)	5	3	3	10	5	(b)
Deposition rate	4	4	5	5	6	6	10	10
Depth of penetration	10	9	6	5	6	7	4	8
Appearance, undercutting	6	6	8	9	9	10	10	10
Soundness	6	6	3	5	7	10	8	9
Ductility	6	7	4	5	6	10	5	10
Low spatter loss	1	2	6	7	9	8	10	10
Poor fit-up	6	7	10	8	9	4	8	(b)
Welder appeal	7	6	8	9	10	10	10	10
Stag removal	9	8	6	8	8	7	9	9

(a) Rating is on a comparative basis of same size electrodes with 10 as the highest value. Ratings may change with size.
(b) Not recommended.
(c) DCR - direct current reverse, electrode positive DCS - direct current straight, electrode negative; AC - Alternating current; DC - direct current, either polarity

Table #84 - Mild Steel Welding Electrode Selection

Stainless Steel Electrodes

Metallurgists have classified the more common type of stainless steel in use today as martensitic, ferritic or austenitic. The martensitic type is hard and brittle, the ferritic type by contrast is soft and ductile, while austenitic is strong, resistant to impact and ductile. It is evident that a different welding method is required for each.

Chromium is the major alloy in stainless steel, although austenitic steel has both chromium and nickel. However, carbon is found in varying degrees, ranging from less than 0.35% in the E.L.C. (extra low carbon) grade to approximately 1.0% in some of the martensitic-type stainless steels.

Enhanced mechanical and corrosion-resistant properties are obtained in some types of stainless steels by the addition of columbium, molybdenum, titanium, etc.

The martensitic hardenable stainless steels generally have chromium contents between 11% and 13%. Some typical examples of these kinds are the designation of E-403, E-405, E-410, E-420 and E-440.

When using martensitic stainless steel electrodes, preheating and postheating are required so that the weld metal will not be weaker than the base metal.

The ferritic nonhardenable stainless steels have chromium contents greater than 13%. The higher chromium content makes those steels nonhardenable. Some typical examples of these stainless electrodes are E-430, E-436, E-442 and E-446. These stainless steels require preheating and postheating.

The austenitic nonhardenable stainless steels have at least 11% or 12% chromium and range up to about 26% chromium with additions of nickel from 3.5% to 22%.

Stainless Steel Electrodes

The austenitic stainless steels are designated as the 200 and 300 series. In the 200 series of steels, manganese is used to replace some of the nickel. Some common examples of these stainless electrodes are: E-302, E-304, E-308, E-310, E-316, E-321, E-347, E-201 and E-202. Preheating and postheating are usually not required, but preheating may be used to remove the chill.

All stainless steel electrodes are furnished with an extruded type coating. These electrodes are available with a lime type coating for use with DC reverse polarity or with a titanium type coating for use with either AC or DC reverse polarity. The straight chromium electrodes are obtainable only with a lime type coating for DC reverse polarity. The lime coating is referred to as Type 15, as in 308-15.

Gas Tungsten Arc Welding (GTAW)

Gas tungsten arc welding is often referred to as "Tig" or "Heliarc" welding. It is a process where a joining of metals is produced by heating with an arc between a non-consumable tungsten electrode and the work. The electrode, arc, weld puddle, and adjacent heated area of the work piece is protected from the atmosphere by a gaseous shield, as indicated in illustration #203.

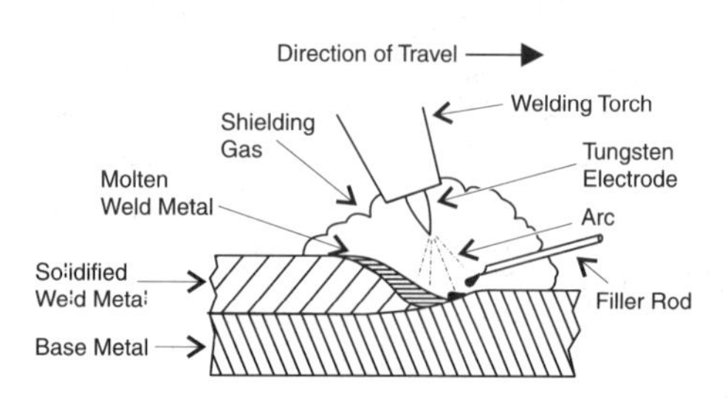

Illustration #203 - Gas Tungsten Arc Welding

Gas Tungsten Arc Welding

There are specific advantages compared to other welding processes, such as: the elimination of flux, less distortion, welds thin material in all positions, and creates very little smoke and sparks.

"Tig" welding equipment can be connected to almost any type of welding machine. The selection of power supply and equipment depends upon the material to be welded. The typical components required for gas tungsten arc welding are shown in illustration #204.

Welding Machines for "Tig" Welding

A specially designed welding machine can be used for "Tig" welding. These types of machines usually contain a high-frequency generator which is used to aid arc starting when welding with alternating current. Alternating current is used for welding aluminum and magnesium.

Other machines used for "Tig" welding are the AC, DC or rectified constant current welding machines, as used for shielded metal arc welding.

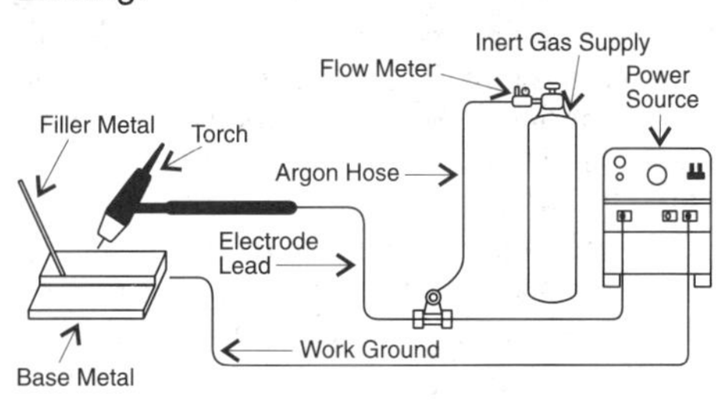

Illustration #204 - Equipment for "Tig" Welding

"Tig" Torches

"Tig" torches direct the shielding gas and hold the tungsten electrode. They are sized by current capacities. Some torches are water cooled.

Gas Tungsten Arc Welding

Tungsten Electrodes

Tungsten electrodes come in a variety of sizes and lengths. They may be either pure tungsten or tungsten alloys, either thorium or zirconium oxide. They are manufactured to AWS standards. Tungsten electrodes have a color identification band on the end, as indicated in Table #85.

When preparing an electrode for use with direct current, the end of the tungsten is ground to a sharp point, as indicated in illustration #205A.

TUNGSTEN IDENTIFICATION	
TYPE	**COLOR**
Pure Tungsten	green band
Thoriated 1%	yellow band
Thoriated 2%	red band
Thoriated striped	blue band
Zirconium	brown band

Table #85 - Tungsten Color Identification

When preparing an electrode for use with alternating current, use a hemispherical balled end. As a result of the heat and the AC current, the ball will remain at 1½ times the electrode diameter.

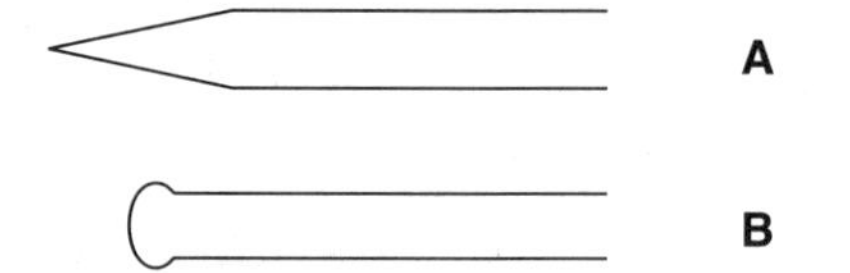

Illustration #205 - Tungsten Preparation

GTAW Shielding Gas

Shielding gas excludes the atmosphere from the welding area. The most commonly used shielding gases are argon and helium.

GTAW Filler Metals

The selection of a filler metal primarily depends on the type of base metal being joined. The filler wire is multi-dioxidized and is not subject to corrosion resulting from carbide precipitation.

GAS TUNGSTEN ARC WELDING PROCEDURES				
Metal	**Special Procedures**	**Type of Current**	**Shielding Gas**	**Type of Tungsten**
Carbon steel plate & pipe	includes stainless	DCSP	Argon or argon & helium	Thoriated pointed end
Nickel, monel, inconel		DCSP	Argon or argon & helium	Thoriated pointed end
Titanium		DCSP	Argon	Thoriated pointed end
Aluminum	sheets, plates, castings	ACHF	Helium or argon & helium	Zirconium balled end
Copper & copper alloys	sheets & pipe	DCSP	Argon or argon & helium	Thoriated pointed end
	very thin material	ACHF	Argon	Zirconium balled end
DCSP ACHF	direct current straight polarity alternating current with a high frequency			

Table #86 - Gas Tungsten Arc Welding Procedures

Pipe or Tube Purging

Mild steel pipe can be welded by the GTAW process without the use of purging, as a sufficient quantity of inert gas reaches the bottom of the joint to provide protection from the oxygen.

Steel alloys containing chrome in percentages higher than 2½% or nickel and nickel alloys must have extra protection at the bottom of the joint. This is done by filling the inside of the pipes in the region of the joint with an inert gas.

Several methods can be used to contain the inert gas in the pipe: rubber pistons, plastic balloons, water soluble paper or bread.

The purge gas can be supplied either through the purge dams or through the joint, see illustration #206.

The purge tube can be a copper tube or a hose with a football needle fitted to the end. As argon is heavier than air, a vent must be provided to allow the air to escape. The pipe joint is taped shut and this tape is removed in sections just ahead of the weld.

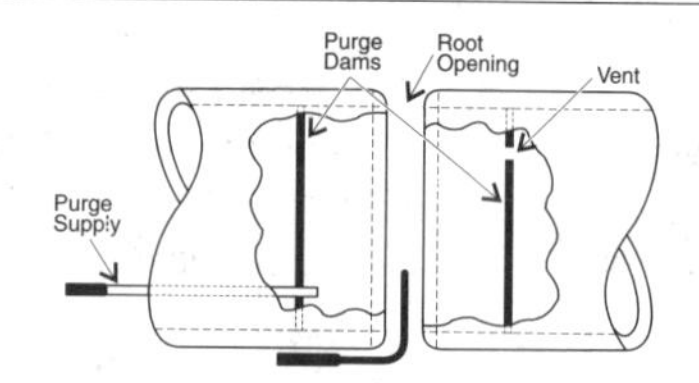

Illustration #206 - Pipe or Tube Purging

Preheating for Welding

Preheating before welding is a well-recognized preventive measure against cracking. It was used for many years before scientific reasons were advanced to explain its function in preventing cracks. The importance of preheating is shown by the fact that recommendations for preheating are written into modern welding specifications whenever there is the slightest risk of weld defect being present.

Note: Use caution, as preheat cannot be used with some metals due to the intended use.

PREHEAT FOR WELDING		
Metal Group	**Metal Designation**	**Recommended Preheat**
Plain Carbon Steels	Plain Carbon Steel	Up to 200°F
Carbon Moly Steels	Carbon Moly Steel	400°F - 600°F
Manganese Steels	Silicon Structural Steel	300°F - 500°F
	Medium Manganese Steel	300°F - 500°F
	SAE T 1330 Steel	400°F - 600°F
	SAE T 1340 Steel	500°F - 800°F
	SAE T 1350 Steel	600°F - 900°F
	12% Manganese Steel	Usually Not Required
High Tensile Steels	Manganese Moly Steel	300°F - 500°F
	Jalten Steel	400°F - 600°F
	Manten Steel	400°F - 600°F
	Armco High Tensile Steel	Up to 200°F
	Double Strength #1 Steel	300°F - 600°F
	Double Strength #1A Steel	400°F - 700°F
	Otiscoloy Steel	200°F - 400°F
	Nax High Tensile Steel	Up to 300°F
	Cromansil Steel	300°F - 400°F
	A.W. DYN-EL	Up to 300°F
	Corten Steel	200°F - 400°F
	Chrome Copper Nickel Steel	200°F - 400°F
	Chrome Manganese Steel	400°F - 600°F
	Yoloy Steel	200°F - 600°F
	Hi-Steel	200°F - 500°F

Table #87A - Preheat for Welding

PREHEAT FOR WELDING		
Metal Group	**Metal Designation**	**Recommended Preheat**
Nickel Steels	SAE 2015 Steel	Up to 300°F
	SAE 2115 Steel	200°F - 300°F
	2 1/2% Nickel Steel	200°F - 400°F
	SAE 2315 Steel	200°F - 500°F
	SAE 2320 Steel	200°F - 500°F
	SAE 2330 Steel	300°F - 600°F
	SAE 2340 Steel	400°F - 700°F
Medium Chrome Moly Steels	5% Cr. - 1/2% Mo. Steel	600°F - 900°F
	8% Cr. - 1% Mo. Steel	600°F - 900°F
Plain High Chromium Steels	12 - 14% Cr. Type 410	300°F - 500°F
	16 - 18% Cr. Type 430	300°F - 500°F
	23 - 30% Cr. Type 446	300°F - 500°F
High Chrome Nickel Stainless Steels	18% Cr. 8% Ni. Type 304	Usually do not require preheat but it may be desirable to remove chill
	25 - 12 Type 309	
	25 - 20 Type 310	
	18 - 8 Cb. Type 347	
	18 - 8 Mo. Type 316	
	18 - 8 Mo. Type 317	
NOTE: See Appendix D to convert degrees F to degrees C preheat temperatures.		

cont...

Table #87B - Preheat for Welding

PREHEAT FOR WELDING		
Metal Group	**Metal Designation**	**Recommended Preheat**
Medium Nickel Chromium Steels	SAE 3115 Steel	200°F - 400°F
	SAE 3125 Steel	300°F - 500°F
	SAE 3130 Steel	400°F - 700°F
	SAE 3140 Steel	500°F - 800°F
	SAE 3150 Steel	600°F - 900°F
	SAE 3215 Steel	300°F - 500°F
	SAE 3230 Steel	500°F - 700°F
	SAE 3240 Steel	700°F - 1000°F
	SAE 3250 Steel	900°F - 1100°F
	SAE 3315 Steel	500°F - 700°F
	SAE 3325 Steel	900°F - 1100°F
	SAE 3435 Steel	900°F - 1100°F
	SAE 3450 Steel	900°F - 1100°F
Moly Bearing Chromium and Chromium Nickel Steels	SAE 4140 Steel	600°F - 800°F
	SAE 4340 Steel	700°F - 900°F
	SAE 4615 Steel	400°F - 600°F
	SAE 4630 Steel	500°F - 700°F
	SAE 4640 Steel	600°F - 800°F
	SAE 4820 Steel	600°F - 800°F
Low Chrome Moly Steels	2% Cr. - 1/2% Mo. Steel	400°F - 600°F
	2% Cr. - 1/2% Mo. Steel	500°F - 800°F
	2% Cr. - 1% Mo. Steel	500°F - 700°F
	2% Cr. - 1% Mo. Steel	600°F - 800°F
NOTE: See Appendix D to convert degrees F to degrees C pre-heat temperatures.		

Table #87C - Preheat for Welding

Pipe and Fitting Alignment

Before butt welded fittings, valves, or pipe can be tacked and welded together, proper fit-up or alignment must be achieved. If alignment is done incorrectly, welding is more difficult with a greater possibility of weld failure and system malfunction. Proper pipe and fitting alignment should include:

- Pipe end preparation.
- Joint assembling and gapping.
- Tacking the joint.
- Alignment of pipe and fittings to other parts of the piping spool.

Preparation for butt welding pressure pipe or matching fitting ends consists of V-beveling the ends to an angle of $37^{1}/_{2}$ degrees. For wall thickness greater than $^{7}/_{8}$ in. (22.2 mm), a compound or double angle bevel is used. See illustration #207A and #207B. The land or root face is approximately $^{1}/_{16}$ to $^{1}/_{8}$ of an inch (1.6 to 3.2 mm). This land helps to prevent the sharp edges of the bevel from burning off in the welding process.

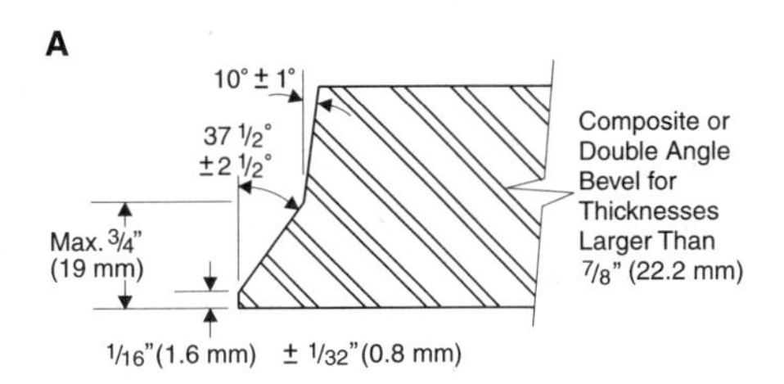

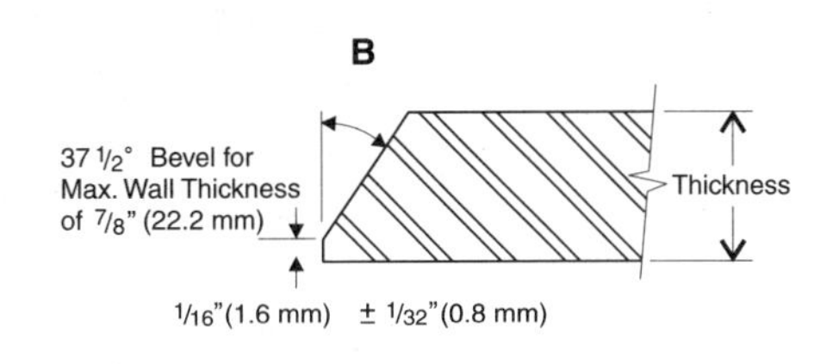

Illustration #207A,B - Bevel Preparation

Welded pipe fittings and pipe lengths are supplied with standard beveled ends and only require cleaning. If the pipe has been cut or is not beveled, the bevel is usually placed on the pipe by the use of a pipe cutting and beveling machine or hand cut with an oxy-acetylene torch. Hand cut pipe bevels usually require smoothing with a grinder or file. The land or root face of the joint is also made with a grinder or file.

When it is necessary to butt weld pipe of unequal wall thicknesses, the pipe ends should be evenly tapered. The length of taper between the ends should be four times the offset thickness. The two recommended methods of joint preparation for unequal wall thicknesses are displayed in illustration #208. For valves and fittings with larger wall thickness than the pipe, the joint should be prepared as shown in illustration #209.

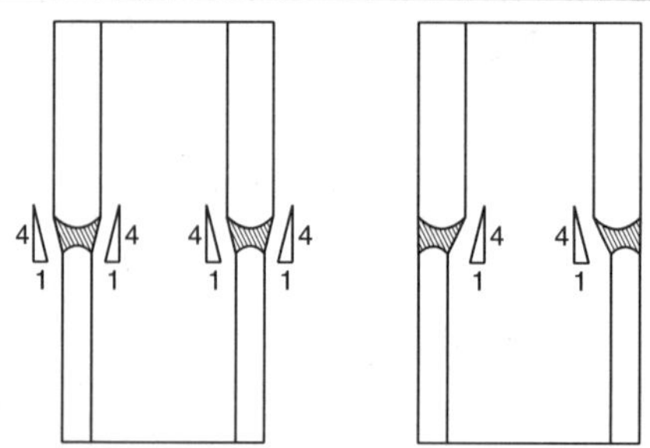

Illustration #208A,B - Taper of Unequal Thickness Pipe

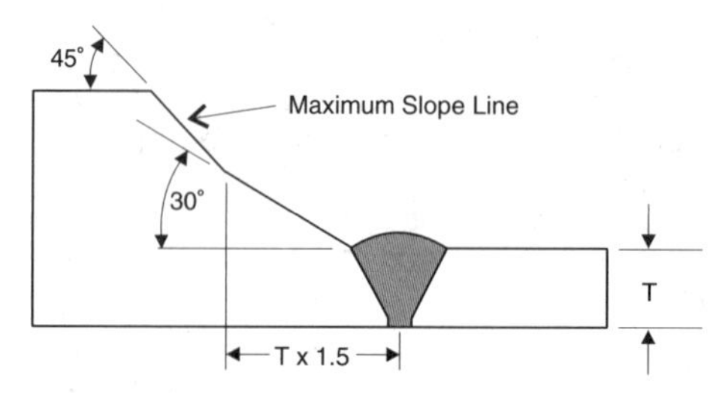

Illustration #209 - Preparation for Valves & Fittings

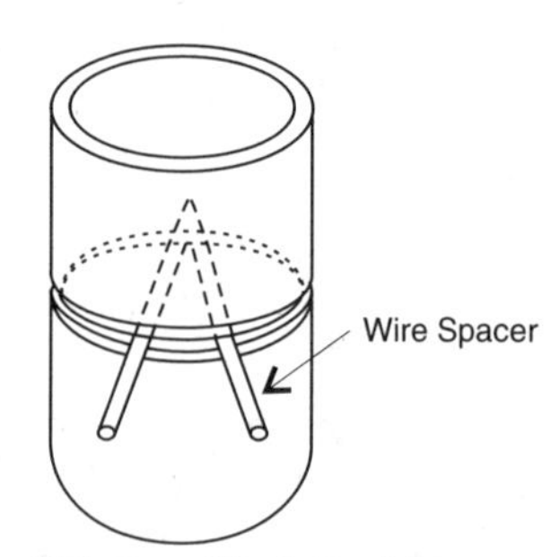

Illustration #210 - Use of Wire to Gap Joint

Joint Assembly Methods

The pipe and/or fitting(s) to be welded together must be accurately assembled and gapped before tacking can take place. Gap or the root opening between the two ends must be evenly spaced with a distance of approximately 1/16 to 1/8 of an inch (1.6 to 3.2 mm). This space can be fixed by placing a piece of bent wire (V or U shaped) the same diameter as the gap needed between the ends to be joined. See illustration #210.

Note: Joint preparation should include cleaning joint ends of any substance such as oil, grease, rust, paint, or scale that may interfere with the welding process.

Not only must the gap between the beveled ends be maintained, but both the inside and outside surfaces of the joint must match evenly without high or low spots (referred to as hi-low). To maintain proper joint gap and hi-low alignment, clamps are often used. Two types of pipe and fitting clamps are shown in illustration #211A and #211B.

Proper hi-low and alignment on smaller pipe and fitting sizes can be maintained simply by the use of a piece of angle iron, see illustration #212A.

On larger pipe and fittings, pipe with thin wall or pipe slightly out of round, dogs can be made up to help in the alignment of hi-low. Using dogs to align pipe ends is displayed in illustration #212B.

Backing Rings

Backing rings are sometimes used to help maintain gap and alignment. These rings are placed inside the pipe and have nodules in the joint to maintain the gap, see illustration #213. The backing rings help prevent root pass burn through in light wall pipe, and prevent slag or spatter from entering the pipe. This ring becomes part of the piping system and may cause flow restriction. Some backing rings are described as consumable and are designed to melt under welding heat. This type of ring is typically used with the gas tungsten arc welding (GTAW) process.

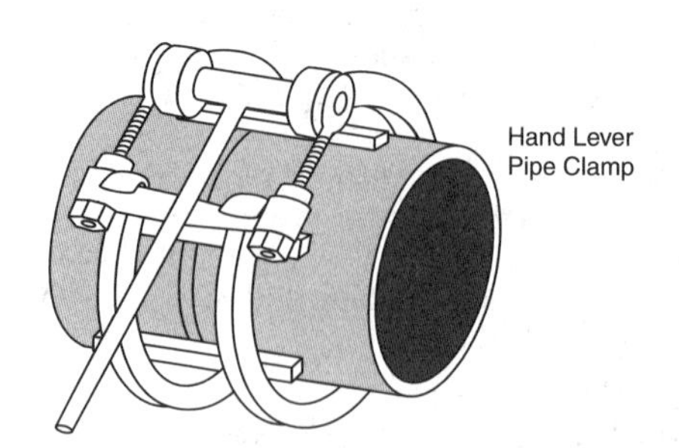

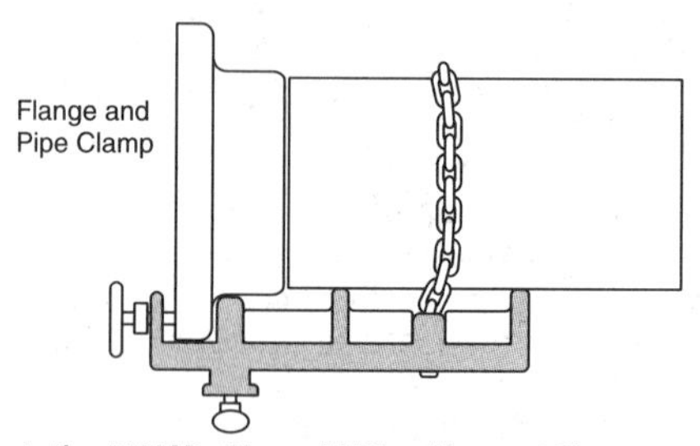

Illustration #211AB - Pipe and Fitting Alignment Clamps

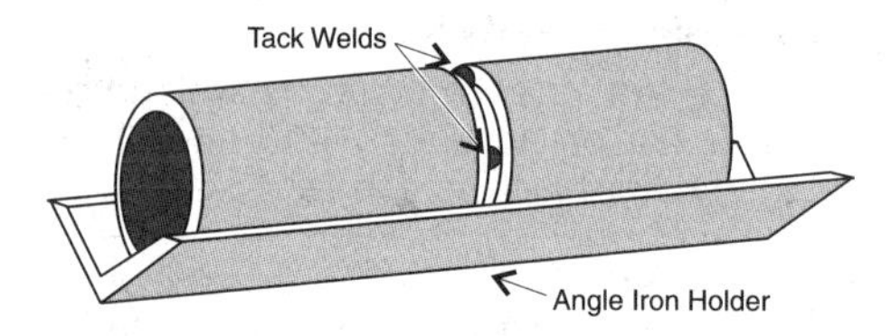

Illustration #212A - Pipe Alignment with Angle Iron

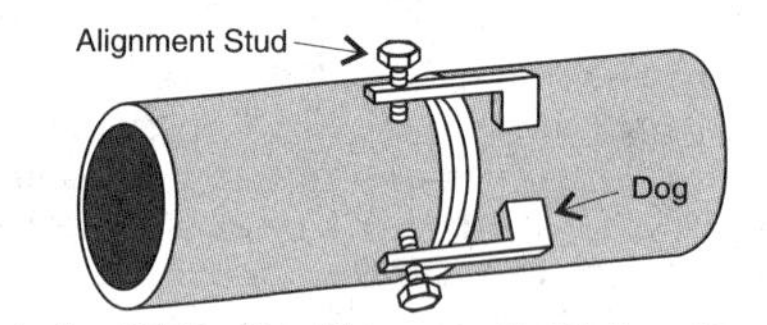

Illustration #212B - Pipe Alignment with Fabricated Dogs

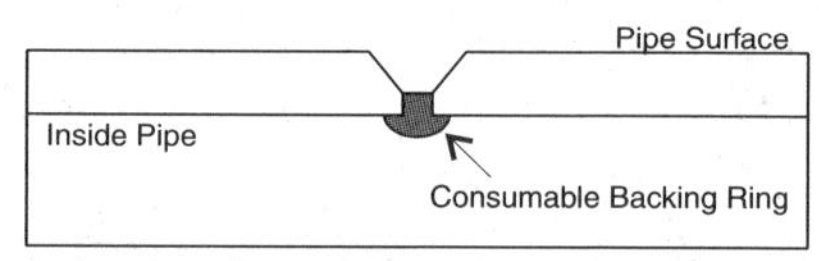

Illustration #213 - Consumable Backing Rings

Tacking the Joint

After the pipe and/or fitting(s) are aligned and properly gapped, four tack welds should be evenly spaced around the joint. The tacks maintain the spacing and alignment during the welding of the joint. The length of the tack weld should be approximately three times the pipe wall thickness.

When using a spacer wire, the wire should be removed after the first tack is welded. The second tack weld is made on the opposite side from the first, 180 degrees opposite.

The third tack is made 90 degrees from either the first or second tack. Adjust the gap until the openings are equalized, or if one side of the root opening is slightly wider, then the third tack weld should be placed there. Any shrinkage in the third tack weld will even out the root opening space. The fourth tack weld is placed 180 degrees from the third. Common tack welding locations for pipe and fittings are outlined in illustration #214A, B, C.

Note: After tacking, the alignment and gap spacing should be re-checked for accuracy. Allowances for shrinkage must be considered before making the 2nd and 3rd tack.

There are many methods of aligning pipe and fittings to other spool parts or to blueprint specifications. The following information and illustrations #215 - #221 are not intended to be taken as the only methods of alignment, but are given as alternates to the many methods that may be used.

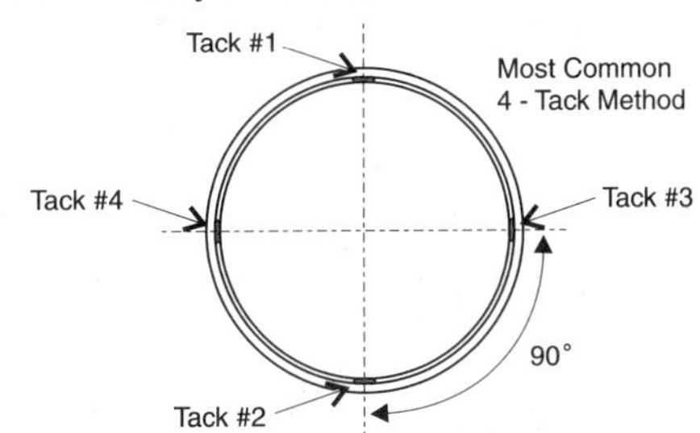

Illustration #214A - Four Weld Tacks

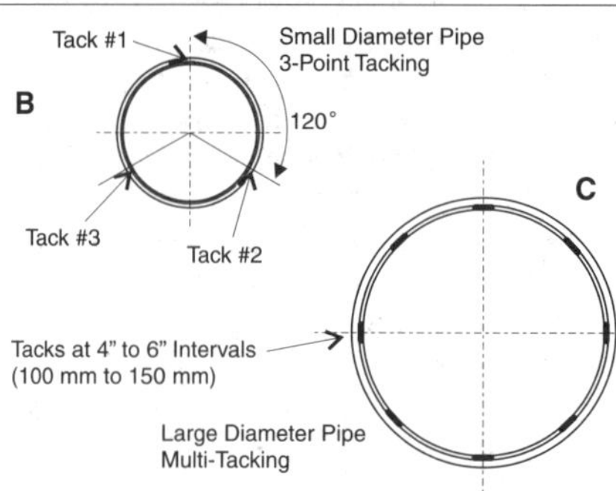

Illustration #214B,C - Three & Multiple Tack Welds

Tee To Pipe Fit-Up

1. Align pipe and tee for gap and hi-low.
2. Tack on top.
3. Again align, then open up bottom slightly to allow for shrinkage.
4. Tack on bottom.
5. Check gap on each side, tack widest side first, then opposite side.

Two methods of checking the alignment are shown in illustration #215A and #215B, one with two squares and the other by a square and tape measure.

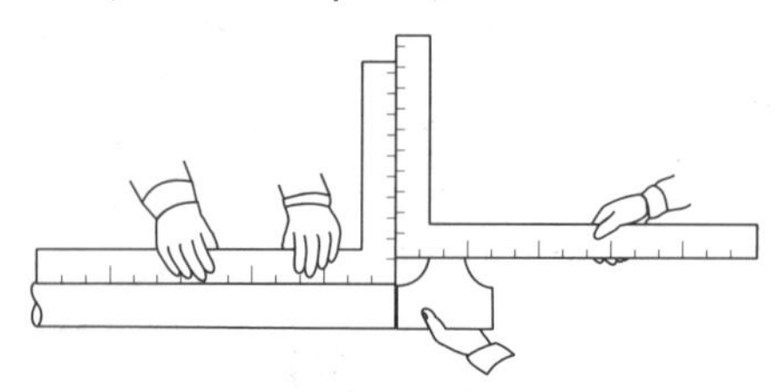

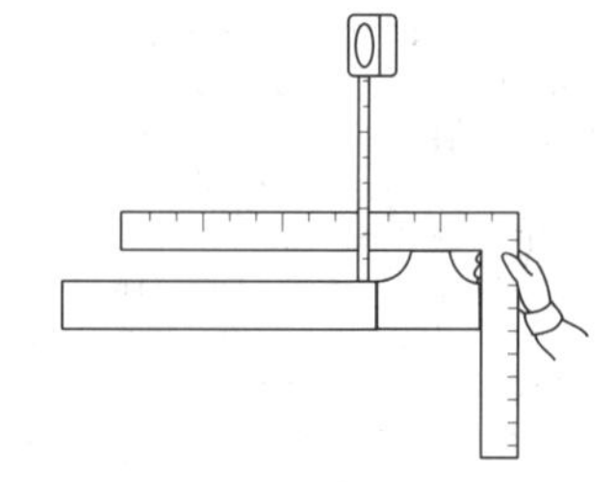

Illustration #215A,B - Fitting Pipe to Tees

Two Square Method of Pipe Alignment

1. Butt joint together with proper gap.
2. Align for straightness and hi-low at the joint.
3. Tack on one side.
4. On the side opposite the tack, again align pipe for straightness, then open the joint slightly to allow for shrinkage, then tack.
5. Roll pipe one quarter turn, check for straightness, open one side slightly and tack.
6. Tack opposite side.
7. Tacks should not be over 3/4 (19 mm) of an inch long.

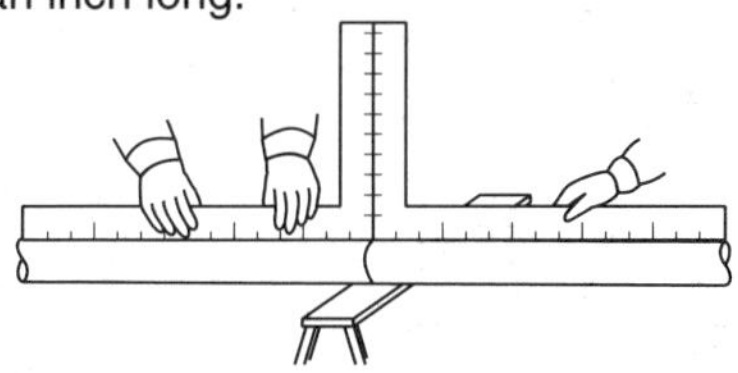

Illustration #216 - Fitting Pipe Sections

Note: Two squares can be used to check pipe for alignment, see illustration 216.

Pipe to 90° Elbow

The proper fit up of a small diameter elbow is not as critical as two pipe sections which must be straight, due to the small size and ease of allowance for the pipe which must butt onto the other end of the elbow.

The problem arises in the larger sizes. The difficulty in fitting up the pipe on the other end of the elbow seems to increase proportionately with the increase in diameter. If possible, check each pipe for a square cut end and also each fitting for squareness.

For example, a 24 inch (600 mm) pipe cut 3/16 inches (5 mm) off square is indicated in illustration #217.

Pipe to 90° Elbow (cont'd)

During fit up, the long or high side was tacked with a gap a little too wide, and the short side was tacked a little too tight. When the fitter starts to fit pipe #2, he finds that a square joint will mean one spot butted tight together and opposite it a gap at least 1/4 inch (6.5 mm) wide. This will mean time consuming work to grind down either the pipe or the elbow.

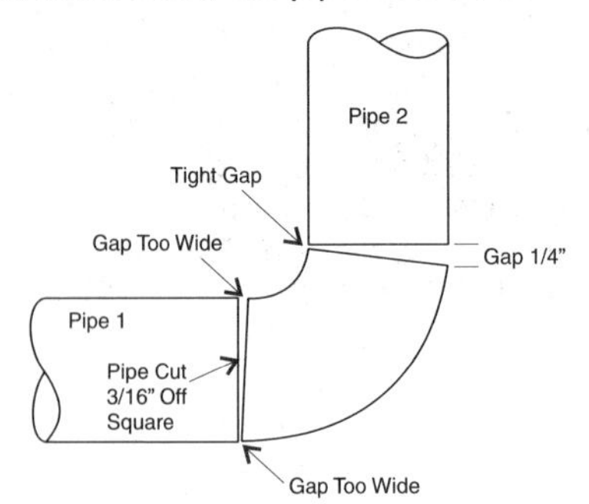

Illustration #217 - Misaligned Elbow

90° Elbow Fit-Up

1. Align pipe and elbow for gap and hi-low.
2. Tack on top.
3. Again align, then open up bottom slightly to allow for shrinkage.
4. Tack bottom.
5. Check gap on each side, tack widest side first, then opposite side.

Note: Two squares or a level can be used to check squareness, see illustration #218.

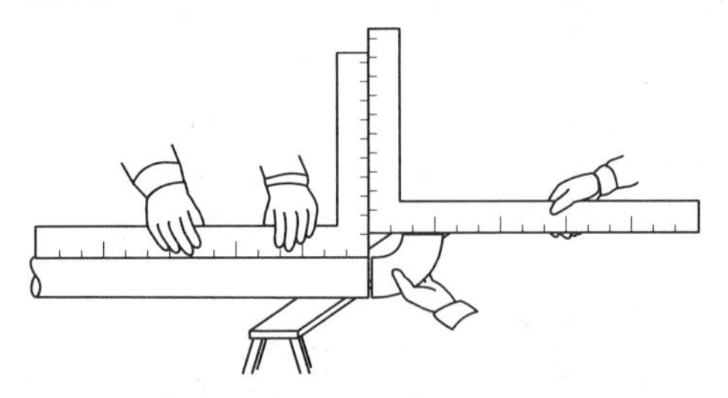

Illustration #218 - Fitting Pipe to Elbow

45° Elbow Fit-Up

The fit-up procedure for a 45° elbow is similar to that of a 90° elbow. Shown in illustration #219A and #219B are two methods of checking the alignment, one by squares and the other by spirit level.

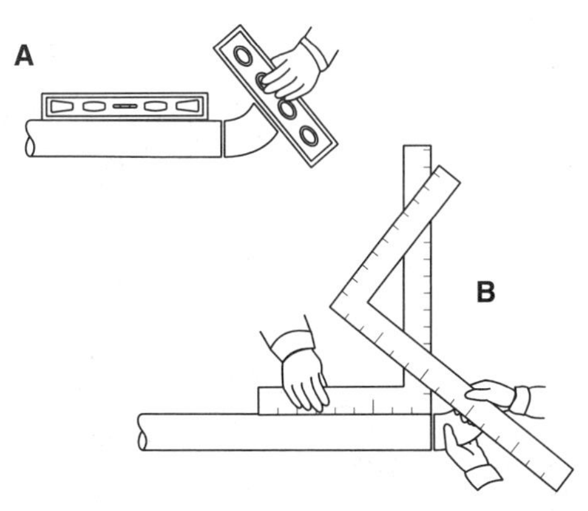

Illustration #219A,B - Fitting Pipe to 45 Degree Elbow

Pipe To Flange Fit-Up

1. Butt up flange to pipe. Three things have to be watched at the same time, that is gap, hi-low and level of holes (two-holing the flange), see illustration #220.
2. Tack on top.
3. Align bottom, gap, then open slightly to allow for shrinkage. A square and tape measure or a vertical spirit level can be used to check squareness, see illustration #221.
4. Tack bottom.
5. Using a square on the sides, check gap and squareness, then tack both sides.

Pipe To Flange Fit-Up (cont'd)

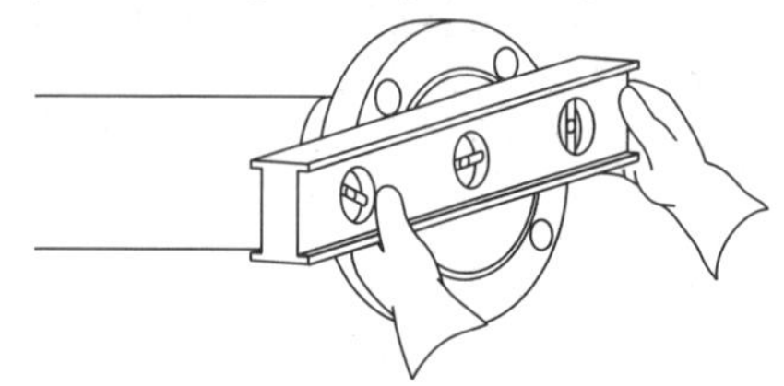

Illustration #220 - Two Holing Flange

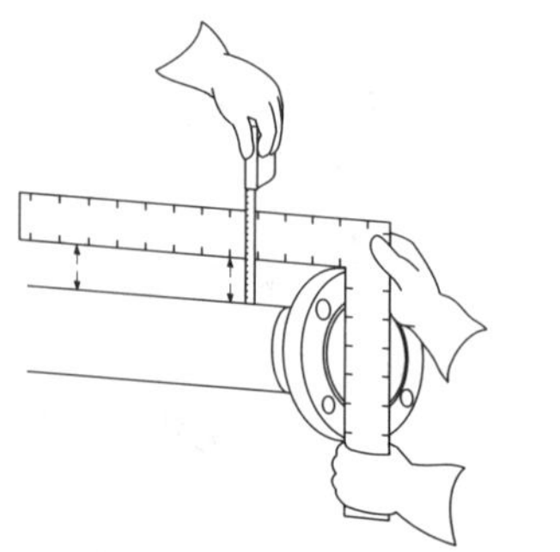

Illustration #221 - Check Alignment of Pipe to Flange

Small diameter Pipe Jig

A jig made of channel is helpful in aligning small diameter pipe and elbows. Layout a 90" notch on both sides of a channel and cut out to form a "V". Heat, bend to a 90° angle, and weld. See illustration #222.

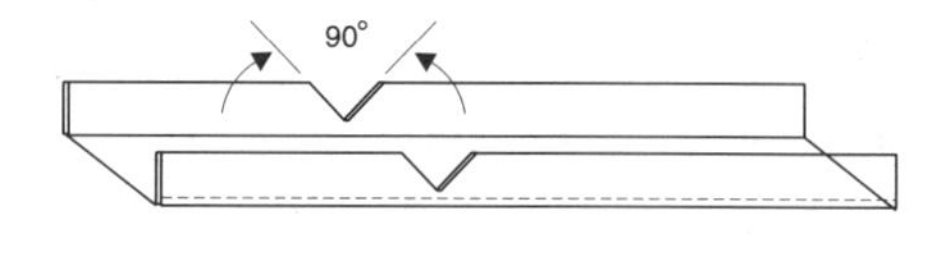

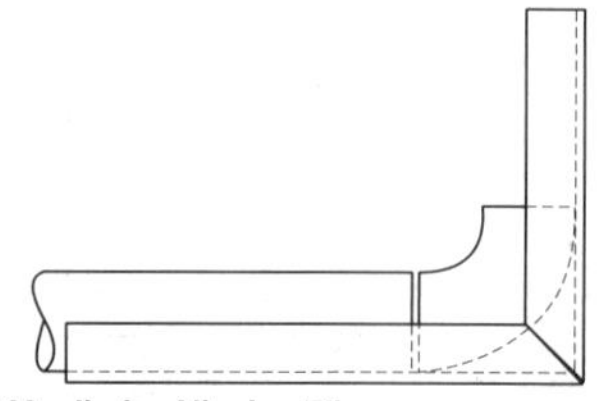

Illustration #222 - Jig for Aligning Elbows

SECTION ELEVEN

RIGGING

Fiber Rope

The most common type of natural fiber rope used in the construction industry is Number One Grade Manila. The other types of natural fiber rope are not strong enough or deteriorate too quickly.

Manila Rope

Number One Natural Manila is strong and durable. It is recognized by its light yellow color. As the grade and strength decreases, the color darkens. A low grade Manila is dark brown. The minimum breaking strength of a one inch diameter Manila rope is 9,000 pounds (4 082 kg).

Nylon Rope

Nylon is the strongest rope available. It will absorb greater shock load than any other and outlast all natural fiber ropes by a wide margin. Nylon is flexible, has high abrasion resistance, can be stored wet, resists most alkalis and organic solvents. It will not rot.

Rope Design Factors

For most uses, the design factor of all rigging equipment, including fiber rope, is 5:1.

When rigging equipment (including fiber rope) is used to hoist personnel or hazardous loads, the design factor must be increased to 10:1.

Safety factors are applied to compensate for the reduced capacity of rope for such reasons as: normal wear, exposure to sun and moisture; extra load imposed by jerky lifting and stopping; excessive sling angle; unknown load weight.

Manila rope will fail with a static load as the rope fibers do not run the full length of the rope. They are intertwined and will pull apart under the load. If loaded within 50% of its ultimate load (breaking strength), the rope will fail in several hours, but if loaded within 75% of ultimate load, it will fail within minutes. This is due to creepage of the fibers.

Knots and Hitches

Most rigging applications involve the tying of knots and hitches in a method to ensure:

- The knot or hitch can be tied and untied quickly and efficiently.
- The knot or hitch when tied holds without slippage.
- The knot or hitch is suitable and retains acceptable degree of rope strength.
- The knot or hitch ensures the safety of people and equipment.

Approximate WLL of New Fiber Rope (3-Strand Ropes) Design (Safety) Factor of 5												
Rope Diameter		Manila		Nylon		Polypropylene		Polyester		Polyethylene	
Inch	(mm)	lbs	kg	lbs	kg	lbs	kg	lbs	kg	lbs	kg
3/16	(4.82)	100	45.3	200	90.7	150	68.0	200	90.7	150	68.0
1/4	(6.35)	120	54.4	300	136.1	250	113.4	300	136.0	250	113.4
5/16	(7.87)	200	90.7	500	226.8	400	181.4	500	226.8	350	158.8
3/8	(9.53)	270	122.5	700	317.5	500	226.8	700	317.5	500	226.8
1/2	(12.7)	530	240.4	1250	566.9	830	376.5	1200	544.3	800	362.9
5/8	(15.88)	880	399.2	2000	907.2	1300	589.7	1900	861.8	1050	476.3
3/4	(19.05)	1080	489.9	2800	1270.0	1700	771.1	2400	1088.6	1500	680.4
7/8	(22.23)	1540	698.5	3800	1723.6	2200	998.0	3400	1542.2	2100	952.5
1	(25.4)	1800	816.5	4800	2177.2	2900	1315.4	4200	1905.1	2500	1133.9
1 1/8	(28.7)	2400	1088.6	6300	2857.6	3750	1701.0	5600	2540.1	3300	1496.9
1 1/4	(31.75)	2700	1224.7	7200	3265.9	4.200	1905.1	6300	2857.6	3700	1678.3

Table #88 - Working Load Limit of Fiber Rope (5:1 WLL)

Bowline

The bowline is one of the most popular knots. It never jams or slips under load and is easily untied. It has a 50% efficiency, see illustration #223.

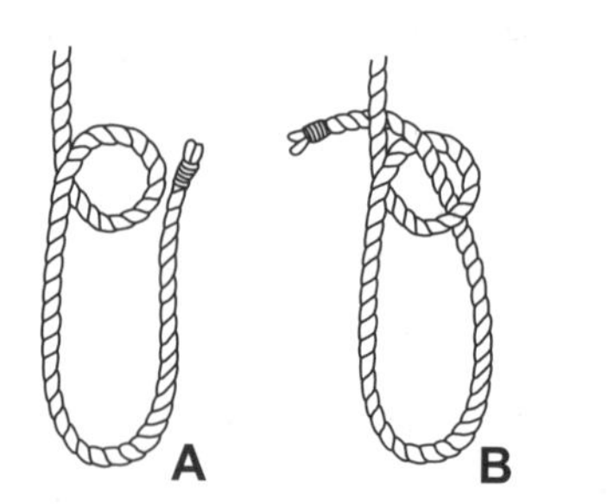

Illustration #223 - Bowline

Bowline on the Bight

The bowline on the bight is used to form a non slipping eye in the middle of a rope, see illustration #224.

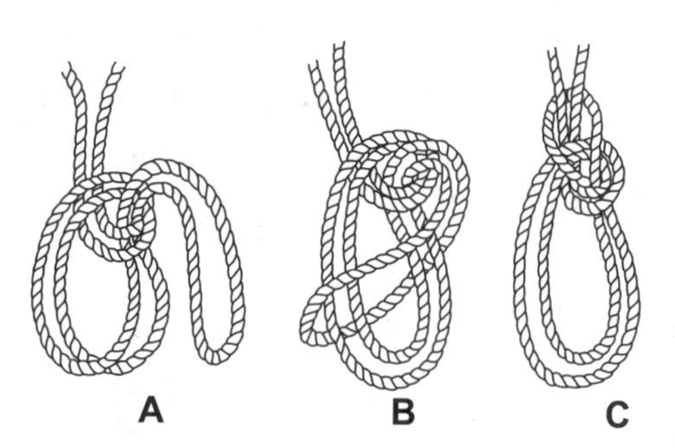

Illustration #224 - Bowline on the Bight

Self Centering Bowline

The self centering bowline is used when a knot must be tied in the center of a load and an equal leg stress is required. It complements a barrel or scaffold hitch, as in illustration #225.

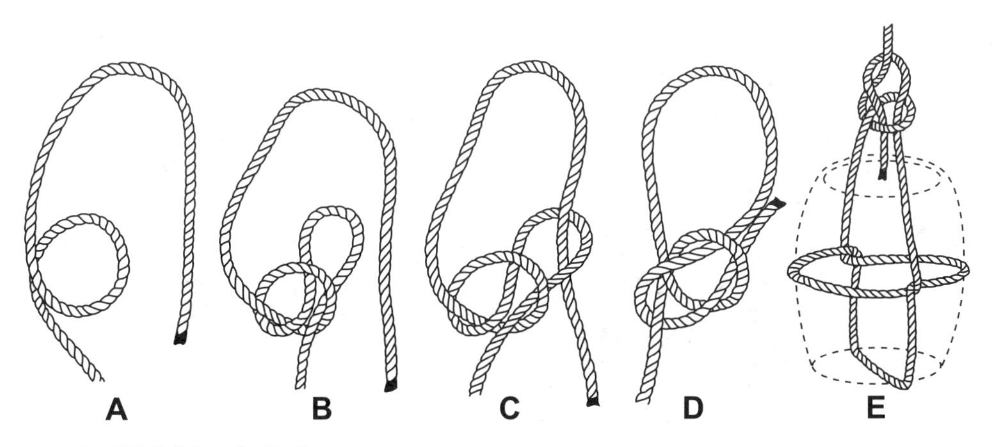

Illustration #225 - Self Centering Bowline

Barrel Hitch

A barrel hitch is used to support a barrel vertically. A self centering bowline is used to complete the knot, as in illustration #226.

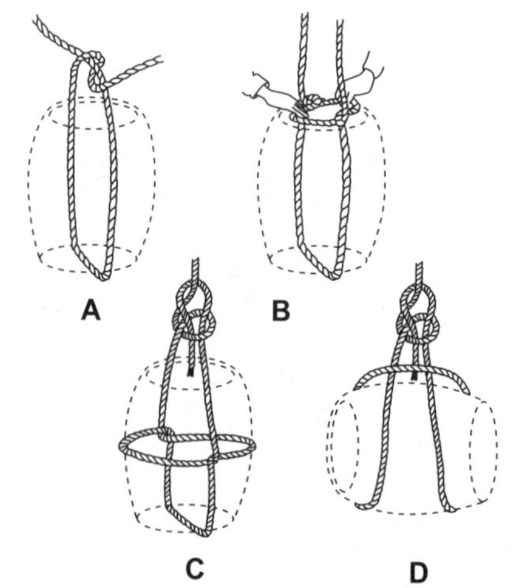

Illustration #226 - Barrel Hitch

Reef Knot

The reef knot has a reputation as a "killer knot", as it can slip when not tied properly. Use only to secure the two ends of a rope. See illustration #227.

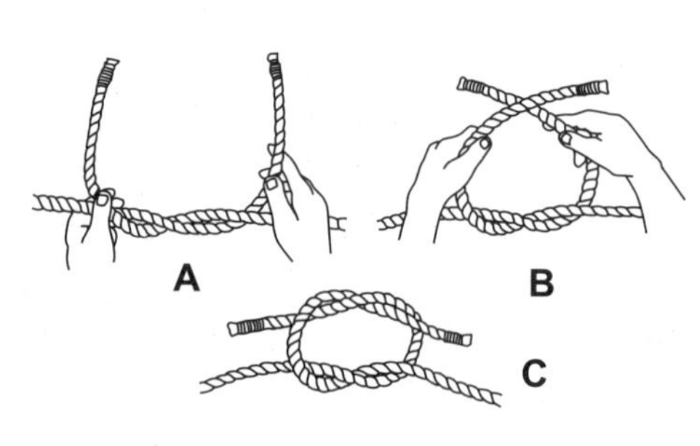

Illustration #227 - Reef Knot

Catspaw

A catspaw is used to attach a rope to a hook. Especially useful if the center of the rope is used. See illustration #228.

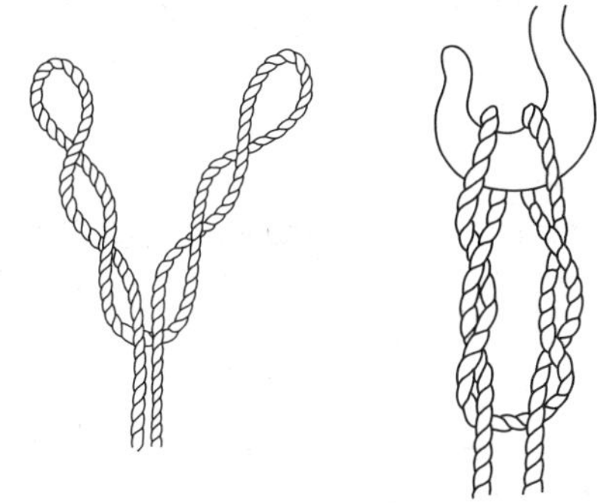

Illustration #228 - Catspaw

Clove Hitch

The clove hitch is used to tie a rope to a pipe or post. It can be tied in position or slipped over the end, as in illustration #229.

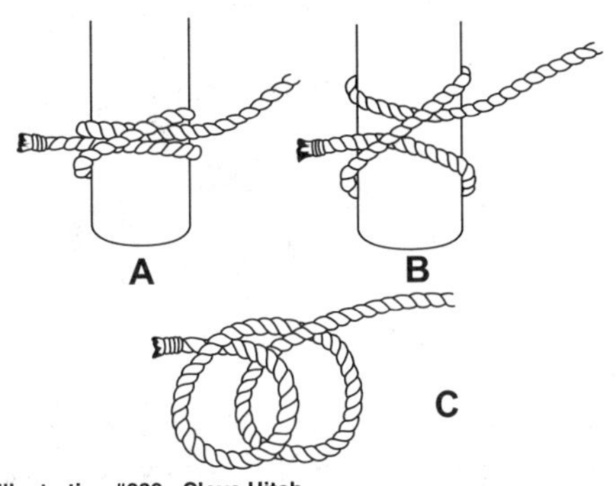

Illustration #229 - Clove Hitch

Becket Hitch

A becket hitch is used to secure the end of a line or rope to the becket on a set of rope falls, as in illustration #230.

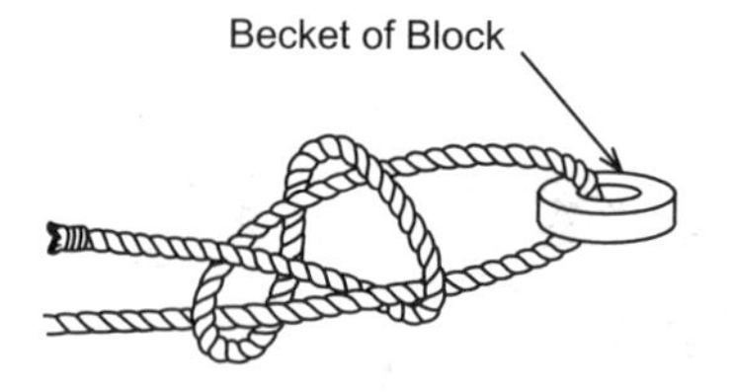

Illustration #230 - Becket Hitch

Stop Hitch

A stop hitch is used to maintain the pull on a line (wire or fiber), when the end of the line must be loosened and repositioned, as in illustration #231.

Illustration #231 - Stop Hitch

Double Half Hitch

This hitch is used to secure or tie off the end of a rope to such objects as a pipe, post or ring. See illustration #232.

Illustration #232 - Double Half Hitch

Pipe Hitch

This is used for lifting a pipe vertically or pulling horizontally. The rope is wrapped tightly around the pipe 4 - 5 times and finished off with a double half hitch. See illustration #233.

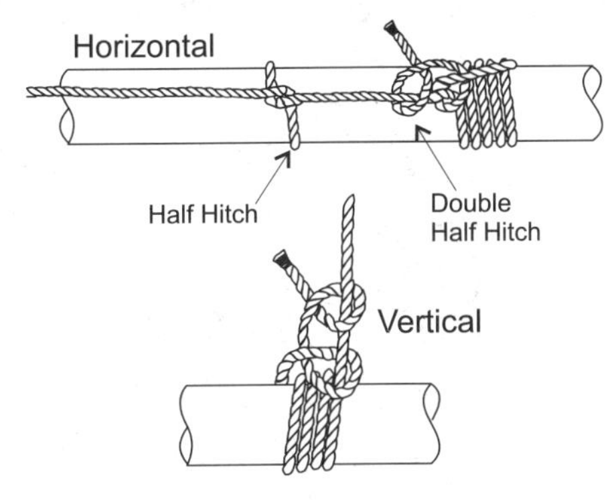

Illustration #233 - Pipe Hitch

Rope Strength With Knots, Bends and Hitches

Straight lengths of rope without knots or splices represent 100% of its strength.

When a knot is tied in a rope it loses approximately 50% of its original strength.

Illustration #234 - Rope Strength Reduction with a Knot

When a bend is tied in a rope it loses 50% of its original strength.

Illustration #235 - Rope Strength Reduction with a Bend

When a hitch is tied in a rope it loses 25% of its original strength.

Illustration #236 - Rope Strength Reduction with a Hitch

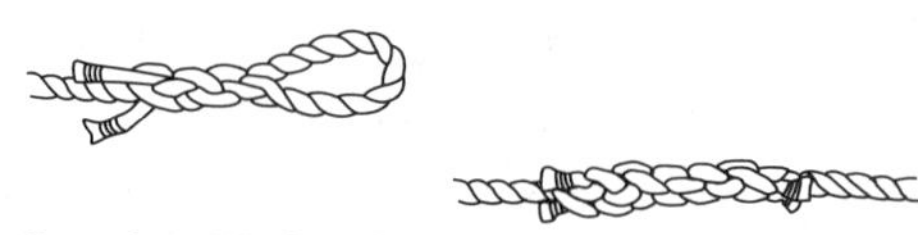

Illustration #237 - Rope Strength Reduction with a Splice

When an eye splice or a short splice is tied in a rope it loses 20% of its original strength.

When two ropes are bent in a U around each other, the strength is reduced by 50%.

Wire Rope Parts

Wire rope is made of wires laid together to form a strand (illustration #238A). These strands are formed together into a rope, usually around a central core made of either fiber or wire, as indicated in illustration #238.

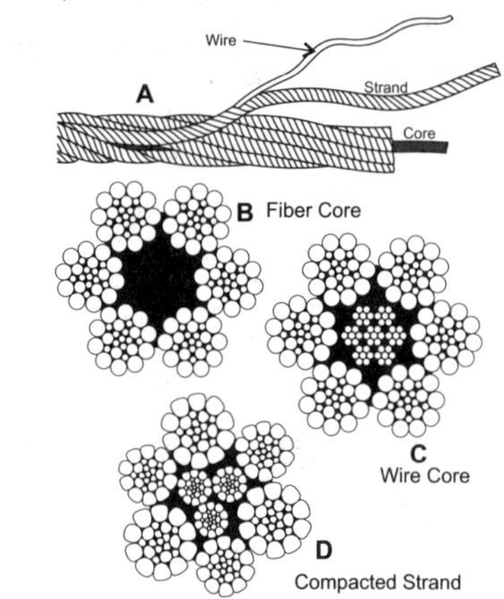

Illustration #238 - Wire Rope Components

Wire Rope Lays

The lay refers to the direction of winding of the wires in the strands, and to the type of strands in the rope. The term is usually applied to the two basic lays, which are Regular Lay and Lang Lay.

Regular Lay

The wires in the strands are laid in one direction, while the strands in the rope are laid in the opposite direction. The strand wires can withstand considerable crushing and distortion due to their short exposed length, see illustration #239A, B.

Lang Lay

The wire rope strands and the wires in the strands are laid in the same direction. Lang Lay rope is not recommended for use on single part hoisting due to its tendency to untwist. Its biggest advantage is its resistance to abrasion, see illustration #239C,D.

Note: The term lay length means the lateral distance of one strand as it makes a complete revolution.

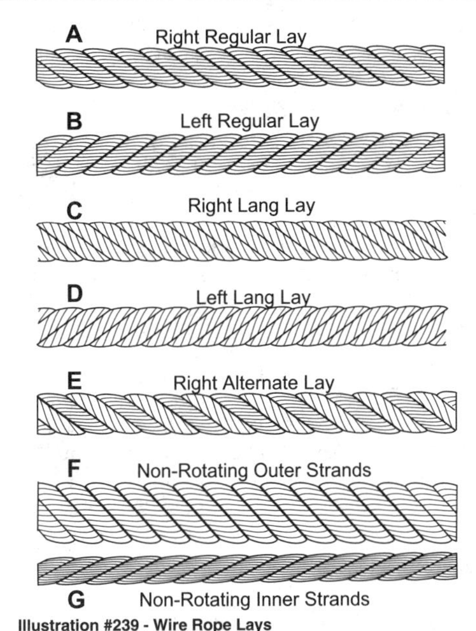

Illustration #239 - Wire Rope Lays

Illustration #239E shows a right lay rope made of alternating strands of right regular lay and right lang lay.

Rotation-Resistant

There are several categories of this type of rope, but basically the outer layer and the inner layer of strands are laid in the opposite direction. The opposing torque from the right hand lay and the left hand lay helps to prevent rope rotation under load. See illustration #239F,G.

Strand Classification

Strands are grouped according to the number of wires per strand. The number of wires and the pattern defines the rope's characteristics. The wires in the strands can all be the same size or a mixture of sizes. The four common arrangements are indicated in table #89.

Classification	No. of Strands	Wires Per Strand
6 x 7	6	3 to 14
6 x 19	6	16 to 26
6 x 37	6	27 to 49
8 x 19	8	16 to 26

Table #89 -Wire Rope Strand Classification

Note: The more wires in a strand, the more flexible the wire rope. The most common classification in the construction industry is 6 x 19.

Load Safety Terminology

"Rated Load" is the maximum working load established by a sling manufacturer. The terms "Rated Capacity", "Working Load Limit (WLL)" are used to describe "Rated Load". These terms replace "Safe Working Load". Also "Ultimate Load" replaces "Breaking Strength", and "Design Factor" replaces "Safety Factor".

Working Load Limit WLL (SWL)

The WLL of most rigging jobs is based on a 5: 1 design (safety) factor. For critical lifts or hazard to personnel the factor should be increased to 10:1. If using a wire rope chart, calculate the WLL as follows:

$$WLL = \frac{\text{Ultimate (breaking) Strength}}{\text{Design (safety) Factor}}$$

WLL Example #1:

1 inch (25.4 mm) I.W R.C.= ultimate strength of 44.9 tons (40.7 tonnes).

44.9 tons (40.7 tonnes) ÷ 5 (SF)
= 8.9 tons (8.1 tonnes) WLL.

Rule of Thumb WLL

As it is difficult to remember the ultimate strength of various wire rope sizes, a rule of thumb formula will give an approximate WLL, based on a design factor of 5.

Diameter squared x 8 = WLL

WLL Example #2:

WLL of 3/4 inch wire rope

3/4 x 3/4 x 8 = 9/16 x 8 = 72 ÷ 16
= 4.5 tons WLL

Ultimate Strength Rule of Thumb

The approximate ultimate strength of any size wire rope can be found with the formula:

Diameter squared x 45 tons (based on Improved Plow, wire core rope)

Using 3/4 inch as an example:

3/4 x 3/4 x 45 = 9/16 x 45 = 25.3 tons

Note: Rule of thumb formulas are based on an "inch-ton" ratio. They cannot be applied to the Metric system.

Rope Diameter (inches)	Maximum WLL (SWL) in Tons Design (Safety) Factor = 5			
	Grade 100/110 Plow		Grade 110/120 Plow Improved	
	Fibre Core	Steel Core	Fibre Core	Steel Core
3/16	0.26	0.28	0.30	0.32
1/4	0.48	0.52	0.54	0.58
5/16	0.76	0.82	0.82	0.88
3/8	1.08	1.16	1.20	1.28
7/16	1.40	1.50	1.60	1.72
1/2	2.00	2.14	2.20	2.36
9/16	2.34	2.50	2.66	2.86
5/8	3.00	3.22	3.30	3.54
3/4	4.30	4.62	4.76	5.10
7/8	5.66	6.08	6.40	6.89
1	7.60	8.17	8.34	8.96
1 1/8	9.70	10.42	10.60	11.38
1 1/4	12.00	12.90	13.10	14.10
1 3/8	14.70	15.80	16.20	17.40
1 1/2	17.70	19.00	19.20	20.64
1 5/8	20.60	22.14	22.60	24.28
1 3/4	23.80	25.58	26.00	27.94
1 7/8	27.60	29.66	30.40	32.68
2	30.80	33.10	33.80	36.32
2 1/4	38.60	41.48	42.00	45.14
2 1/2	47.00	50.42	52.00	55.90

Table #90 - Wire Rope Working Load Limit

Nominal Strengths Of Wire Rope 6 x 19 Classification, IWRC							
Nominal Diameter		Approximate Mass		Nominal Strength			
				Improved Plow		Extra Imp. Plow	
inches	mm	lb/ft	kg/m	tons	Metric tonnes	tons	Metric tonnes
1/4	6.4	0.12	0.17	2.94	2.67	3.40	3.08
5/16	8.0	0.18	0.27	4.58	4.16	5.27	4.78
3/8	9.5	0.26	0.39	6.56	5.95	7.55	6.85
7/16	11.5	0.35	0.52	8.89	8.07	10.2	9.25
1/2	13.0	0.46	0.68	11.5	10.4	13.3	12.1
9/16	14.5	0.59	0.88	14.5	13.2	16.8	15.2
5/8	16	0.72	1.07	17.7	16.2	20.6	18.7
3/4	19	1.04	1.55	25.6	23.2	29.4	26.7
7/8	22	1.42	2.11	34.6	31.4	39.8	36.1
1	26	1.85	2.75	44.9	40.7	51.7	46.9
1 1/8	29	2.34	3.48	56.5	51.3	65.0	59.0
1 1/4	32	2.89	4.30	69.4	63.0	79.9	72.5
1 3/8	35	3.50	5.21	83.5	75.7	96.0	871
1 1/2	38	4.16	6.19	98.9	89.7	114	103
1 5/8	42	4.88	7.26	115	104	132	120
1 3/4	45	5.67	8.44	133	121	153	139
1 7/8	48	6.50	9.67	152	138	174	158
2	52	7.39	11.0	172	156	198	180
2 1/8	54	8.35	12.4	192	174	221	200
2 1/4	57	9.36	13.9	215	195	247	224
2 1/2	64	11.6	17.3	262	238	302	274

Table #91 - Wire Rope Ultimate (Breaking) Strength

Winding Wire Rope on a Drum

To properly install a rope on a drum or winch, stand behind the drum and face it. The right hand represents Right Lay Rope and the left hand represents Left Lay Rope. Make a fist and extend the index finger.

If the rope is Right Lay, imagine the right fist as the drum, and the index finger as the rope. The wire rope will attach to the drum on the thumb side of the fist. This method is indicated in illustration #240.

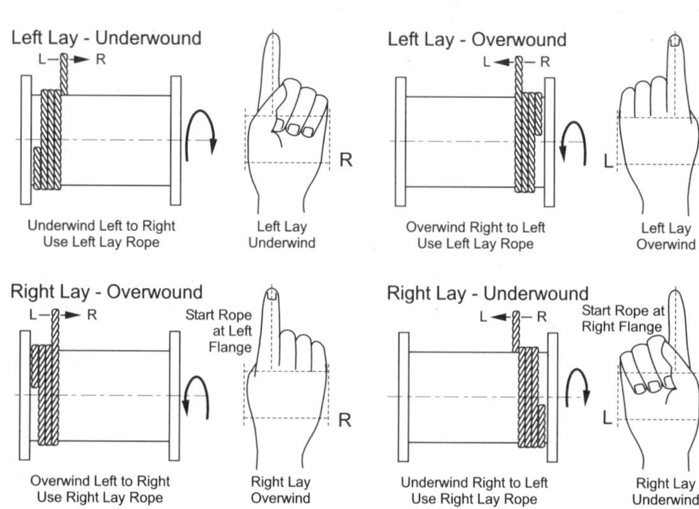

Illustration #240 - Using Hand to Indicate Proper Rope Position on Drum

Winding Wire Rope Reel to Drum

When unwinding wire rope from its storage reel to another reel or drum, the wire rope must be reeled from the top of one reel to the top of the other, or from the bottom of one to the bottom of the other, see illustration #241 . ***Do not cross wind from top to bottom.***

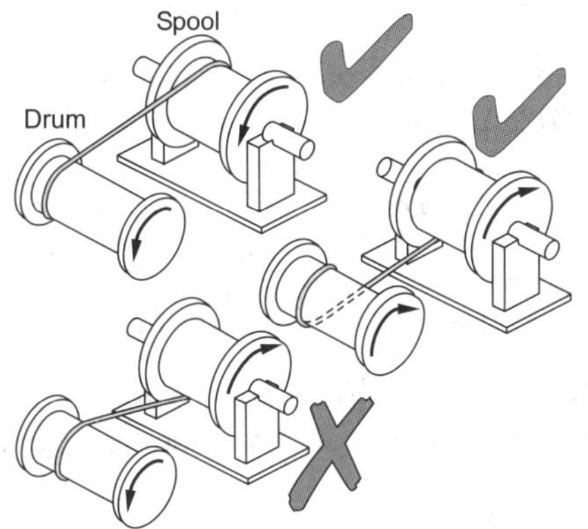

Illustration #241 - Proper Winding From Top to Top

Drum Winding Tips

- A minimum of 3 wraps should be,on the drum at all times (check with OH&S/OSHA as the requirement may be up to 5 wraps in some areas).
- The drum end of the rope should be anchored by a clamp supplied by the manufacturer.
- The flange on a drum should project a minimum of 2 diameters beyond the last layer of rope.
- Whenever possible, not more than three layers of rope should be on the drum at one time. This helps to avoid crushing of the wire rope.

Wire Rope Clips

U Bolt Type **-** When using U Bolt clips, the U bolt section must be on the dead or short end, as indicated in illustration #242. Always use a thimble to prevent the wire rope chokers from wearing the eye. Tighten the clips before tension is placed on the rope. Then tighten again after a load has been applied.

It is important to match the clip lay to the wire rope lay, otherwise the sharp ridges between the corrugations in the forging will run across and cut the strand.

Note: Remember, never saddle a dead horse. This means the U Bolt part must always be on the dead end of the wire rope.

Fist Grip Type (J Clip) **-** The fist grip clip offers a wide bearing surface for maximum strength and greater holding power. A wrench can be swung in a full arc for fast installation.

Note: Clips property attached are efficient to approximately 80% of the rope's strength.

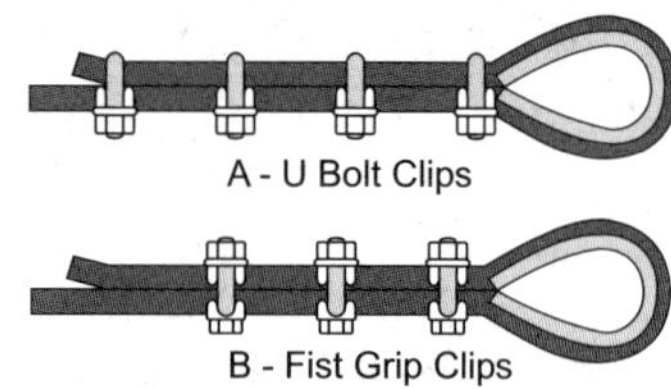

Illustration #242 - Eye Using U-Bolt or Fist Grip Clips

Number and Spacing Formulas

NOTE: Use for 7/8" and under

Number of Clips - Rule of Thumb

Number of wire rope clips = the diameter of the rope x 3 + 1 to the next whole number.

Example: 1 1/2 inch wire rope:
1 1/2 x 3 + 1 to the next whole number
= 6 clips

Spacing of Clips - Rule of Thumb

Spacing of wire rope clips = the diameter of the rope x 6

Example: 1 1/2 inch wire rope:
1 1/2 x 6 = 9 inch spacing

Rope Diam. Inches	U-Bolt and Saddle Type			Integral Saddle and Bolt Type		
	Minimum No. of Clips	Amount of Rope to turn back in inches from Thimble	Torque in lbs Foot	Minimum No. of Clips	Amount of Rope to turn back in inches from Thimble	Torque in lbs Foot
1/8	2	3 1/4	—	—	—	—
3/16	2	3 3/4	7.5	2	3 1/4	30
1/4	2	4 3/4	15	2	3 1/4	30
5/16	2	5 1/2	30	2	4	30
3/8	2	6 1/4	45	2	5	45
7/16	2	6 3/4	65	2	5 3/4	65
1/2	3	11	65	3	6 1/2	65
9/16	3	11 1/4	95	3	7 1/4	130
5/8	3	12	95	3	8	130
3/4	4	18	130	3	14	225
7/8	4	21 1/2	225	4	23	225
1	5	24	225	5	26	225
1 1/8	6	28	225	5	29	225
1 1/4	7	30	360	6	40	360
1 3/8	7	37 1/2	360	6	45	500
1 1/2	8	40 1/2	360	7	49	500
1 5/8	8	43 1/2	430			
1 3/4	8	46	590			
2	8	62	750			

Table #92 - Wire Rope Clip Application

Shackles

The screw pin anchor type is the most commonly used shackle. The safe working load of each shackle should be stamped on the bow section and is rated in tons. The shackle is sized by the diameter of the bow section, not the pin diameter.

Note: Never weld any type of rigging hardware.

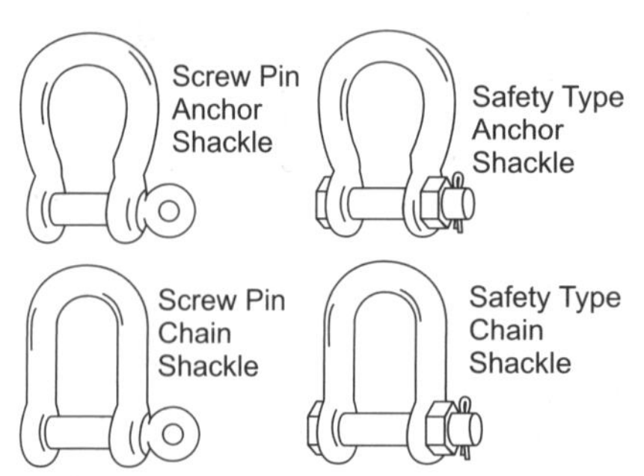

Illustration #243 - Types of Shackles

General Shackle Rules

1. Use only a shackle with an embossed rating on the bow.
2. Use only the proper manufacturer's pin, never replace it with a bolt.
3. Never use a screw pin shackle if the pin can roll under load (illustration #244A).

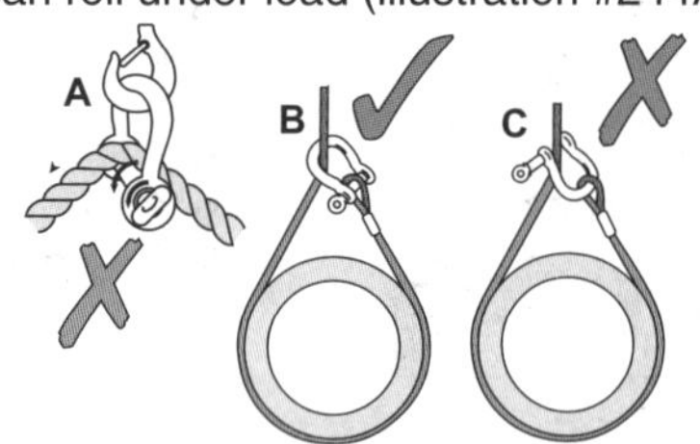

Illustration #244 - Do NOT let Rope Run Over Pin

4. Shackle pins are: $^{1}/_{16}$ inch larger in diameter than the bow on sizes up to $^{7}/_{16}$ inch; $^{1}/_{8}$ inch larger than the bow on sizes $^{1}/_{2}$ inch to 1 $^{5}/_{8}$ inch; $^{1}/_{4}$ inch larger than the bow on sizes 1 $^{3}/_{4}$ and over.

5. Shackles are designed with maximum capacity on a straight pull. See illustration #245 for capacity reduction on angled loads using screw pin and bolt type shackles. Do not side load a round pin shackle. Use a larger shackle for two slings spreads at a wide included angle.

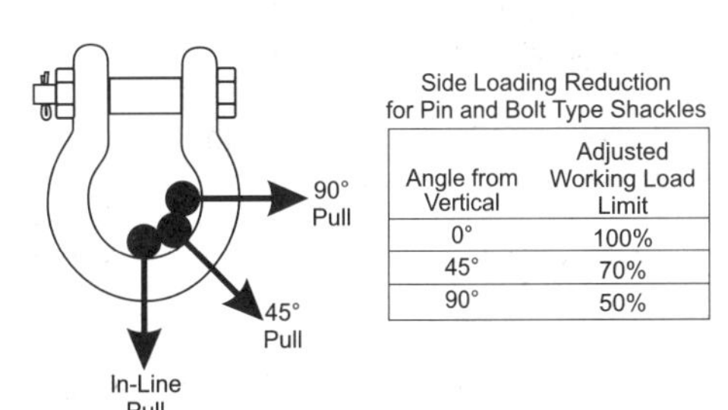

Side Loading Reduction for Pin and Bolt Type Shackles

Angle from Vertical	Adjusted Working Load Limit
0°	100%
45°	70%
90°	50%

Illustration #245 - Shackle Capacity Reduction

6. The pin of a shackle is usually hung on a hook and the load slings are placed in the body or anchor part. Washers, spacers or a spool provided by the manufacturer can be used on the pin to keep the shackle hanging evenly on a hook (see illustration #246).

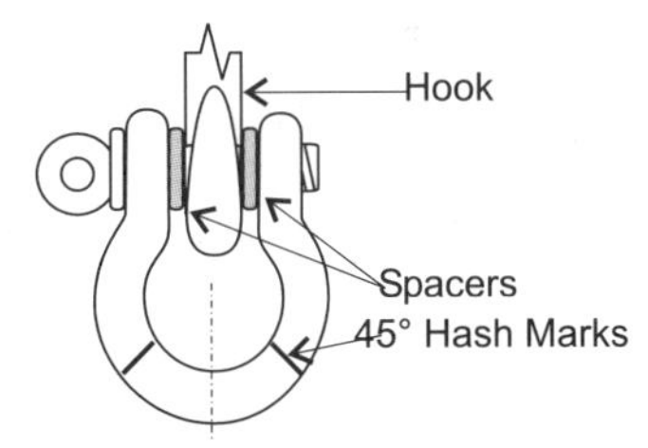

Illustration #246 - Shackle on a Hook

Note: Shackles with 45 degree hash marks on the bow are recommended. These marks indicate that slings hooked up with an included angle greater than 90 degrees have a much reduced capacity.

Shackle Safe Working Loads

SHACKLES - Weldless Construction Forged Steel		
Stock Diameter (Inches)	Inside Width At Pin (Inches)	Max. Safe Working Load Single Vertical Pull (Pounds)
3/16	3/8	665
1/4	15/32	1,000
5/16	17/32	1,500
3/8	21/32	2,000
7/16	23/32	3,000
1/2	13/16	4,000
5/8	1 1/16	6,500
3/4	1 1/4	9,500
7/8	1 7/16	13,000
1	1 11/16	17,000
1 1/8	1 13/16	19,000
1 1/4	2 1/32	24,000
1 3/8	2 1/4	27,000
1 1/2	2 3/8	34,000
1 3/4	2 7/8	50,000
2	3 1/4	70,000
2 1/2	4 1/8	100,000
3	5	150,000

Table #93 - Shackle Data

Hooks

- All hooks should be made from forged steel, except grab and sorting types.
- The safe working load should be stamped on the hooks.
- The safe working load is reduced if the load is applied anywhere between the saddle and tip of the hook. See illustration #247.
- Inspect hooks regularly and look for wear in the saddle, plus cracks and corrosion. The throat will open if the hook has been over loaded. Destroy the hook if there is any distortion.
- OH&S/OSHA regulations specify the hook must be destroyed if the throat has opened 15%, or the body has twisted 10 degrees.
- ***All hooks should be equipped with safety catches.***

Hook Tip Loading

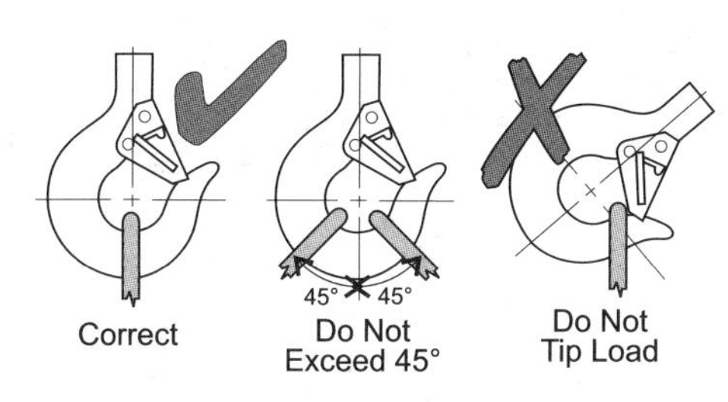

Illustration #247 - Examples of Hook Rip Loading

Turnbuckles

The working load limit is based on the outside diameter of the threaded portion. The safe working loads are indicated in table #94.

Turnbuckle Working Load Limits		
Stock Diameter (Inches)	Jaw, Eye, Stub, End Fittings (Pounds)	Hook End Fitting (Pounds)
1/4	500	400
5/16	800	700
3/8	1,200	1,000
1/2	2,200	1,500
5/8	3,500	2,250
3/4	5,200	3,000
7/8	7,200	4,000
1	10,000	5,000
1 1/4	15,200	5,000
1 1/2	21,400	7,500
1 3/4	28,000	

Table #94 - Turnbuckle Safe Working Loads

If vibration is present, a well lubricated turnbuckle can loosen; therefore, it is important to lock the frame of the end fitting (a piece of wire can be used). Do not use jam nuts if the turnbuckle didn't originally have them, as they add to the load on the thread.

Eye Bolts

The working load limit for shoulderless and shouldered eye bolts are the same for vertical loads. Angular loading is not recommended for shoulderless eye bolts.

All eye bolts used in the construction industry should be equipped with shoulders and be made of forged alloy steel.

The working load limit is reduced with angular loading. See table #95.

Note: New designs of eye bolts are available. A Swivel Hoist Ring will rotate a full 360 degrees and the bail will pivot 180 degrees. Normal eye bolt rules apply to these rings.

A Side Pull Ring is designed for a 90 degree side pull at 100% capacity.

Eye Bolt Working Load Limits - Shoulder Type Only - Forged Carbon Steel (Pounds)			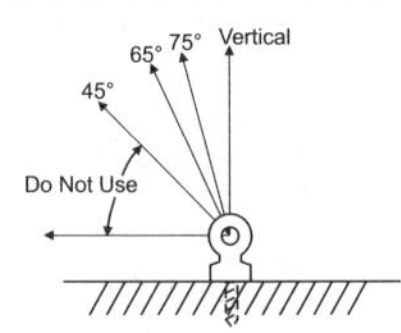	
Shank Diameter	Vertical Pull	60° Pull	45° Pull	Under 45°
1/4	650	420	195	
5/16	1200	780	360	
3/8	1550	1000	465	
1/2	2600	1690	780	
5/8	5200	3380	1560	**NOT**
3/4	7200	1680	2160	**RECOMMENDED**
7/8	10,600	6890	3180	
1	13,300	8645	3990	
1 1/4	21,000	13,600	6300	
1 1/2	24,000	15,600	7200	

Table #95 - Eye Bolt Working Load Limits

Eye Bolts

For angular lifting, the eye must be aligned as indicated. A shim may be used for correct alignment. See illustration #248.

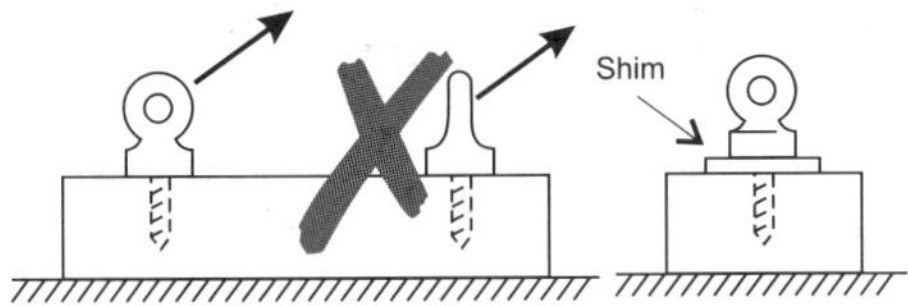

Illustration #248 - Align Eye Bolt for Angular Lifting

Do not insert the point of a hook in an eye. Always use shackles for connecting, as in illustration #249.

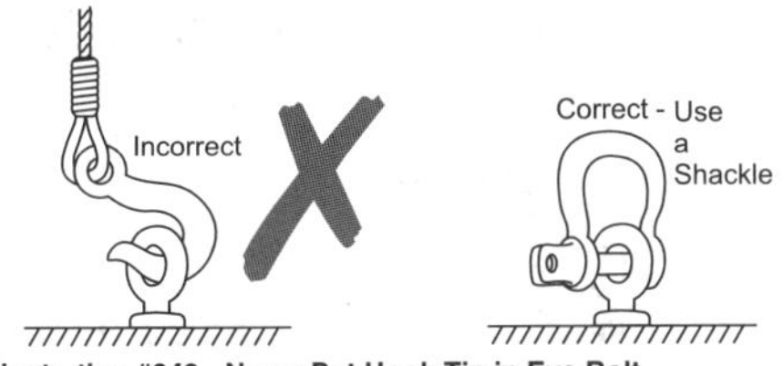

Illustration #249 - Never Put Hook Tip in Eye Bolt

Spreader Beams

Spreader beams are used to support long, hard to handle loads. The use of these beams eliminates load tipping, sliding or bending. They also decrease the possibility of low sling angles. See illustration #250.

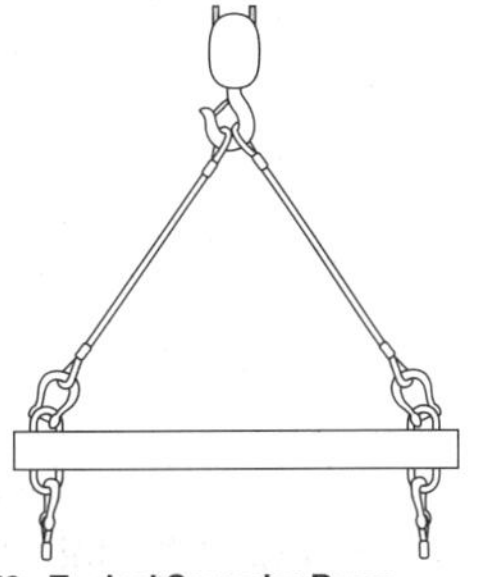

Illustration #250 - Typical Spreader Beam

Note: Custom fabricated lifting beams, or any other lifting device, must be designed by an engineer and have its capacity clearly stamped. It should be test lifted at 125% of rated capacity.

Sheaves

Always check the condition and dimensions of sheave grooves before a new wire rope is placed in service.

The bottom of the groove should have an arc of support of at least 120° to 150°, as indicated in illustration #251.

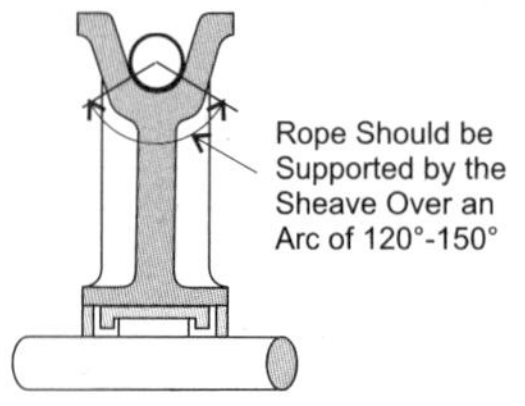

Illustration #251 - Area of Sheave Supporting Wire Rope

To ensure a long and efficient rope life, the grooves should be smoothly contoured, free of surface defects and have rounded edges.

When the groove diameter wears to less than the values for minimum conditions, re-groove or replace the sheave. This is indicated in table #96.

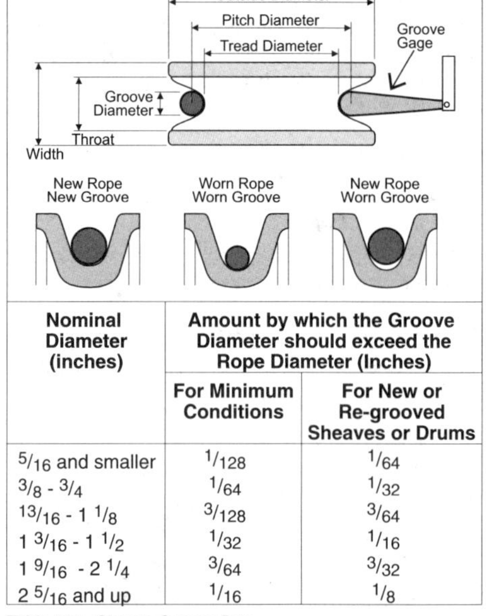

Nominal Diameter (inches)	Amount by which the Groove Diameter should exceed the Rope Diameter (Inches)	
	For Minimum Conditions	For New or Re-grooved Sheaves or Drums
5/16 and smaller	1/128	1/64
3/8 - 3/4	1/64	1/32
13/16 - 1 1/8	3/128	3/64
1 3/16 - 1 1/2	1/32	1/16
1 9/16 - 2 1/4	3/64	3/32
2 5/16 and up	1/16	1/8

Table #96 - Sheave Groove Data

Sheave Diameter

Do not operate wire rope over sheaves which are too small in diameter. When using small diameter sheaves, the excessive and repeated bending and straightening of the wires leads to premature failure from fatigue. The recommended "critical" diameters are shown in table #97.

Sheave Diameter Table		
Rope Construction	**Minimum Diameter**	**Critical Diameter**
6 x 9 Seale	34 x d	20 x d
6 x19 Filler Wire	30 x d	16 x d
6 x 19 Warrington	30 x d	16 x d
8 x 19 Seale	26 x d	16 x d
8 x 19 Filler Wire	26 x d	16 x d
6 x 22 Filler Wire	23 x d	16 x d
8 x 19 Warrington	21 x d	14 x d
8 x 19 Filler Wire	21 x d	14 x d
6 x 37 Seale	18 x d	14 x d

Table #97 - Sheave Diameter Data

Note: The sheave diameter rule of thumb requires the critical diameter of a sheave to be at least 20 times the diameter of the wire rope.

Snatch Blocks

Snatch blocks are used to change the direction of a wire rope. The block can open so all the rope does not have to feed through. See illustration #252.

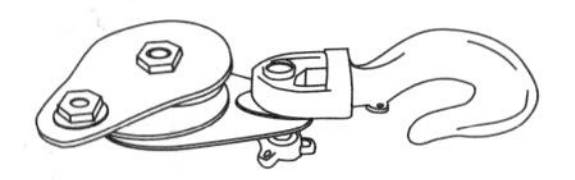

Illustration #252 - Typical Snatch Block

The load on the snatch block varies with the angle between the lead and load lines. To determine the stress on a hook, multiply the pull on the lead line by the correct factor from table #98.

Note: When the lead and load lines are parallel, the load on the block hook is double the weight of the load, therefore the rigging holding the block must support at least double the load weight.

Snatch Blocks

The angles between the lead and load lines are indicated in illustration #253.

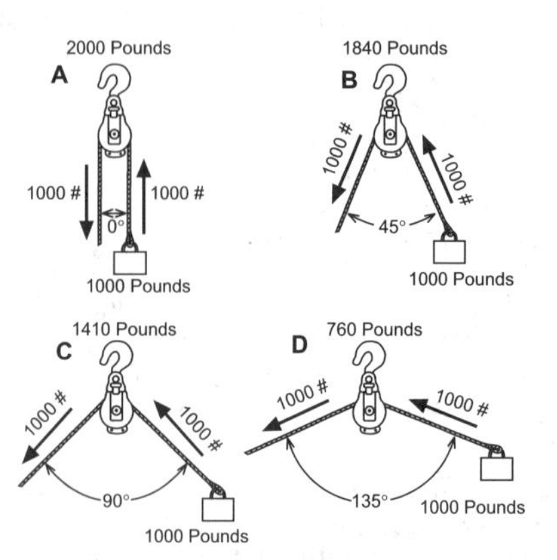

Illustration #253 - Lifting Tension on Snatch Block Hook

Note: On a straight vertical pull (A) the block rigging must hold double the load weight.

Multiplication Factors For Snatch Block Loads	
Angle Between Lead and Load Lines	**Multiplication Factor**
10°	1.99
20°	1.97
30°	1.93
40°	1.87
50°	1.81
60°	1.73
70°	1.64
80°	1.63
90°	1.41
100°	1.29
110°	1.15
120°	1.00
130°	.84
140°	.68
150°	.52
160°	.35
170°	.17
180°	.00

Table #98 - Snatch Block Factors

Slings and Chokers

Slings and chokers; are continually subjected to abnormal abuse due to overloading, abrasion, crushing, kinking and impact loading. Extra stress is often put on chokers when the sling angle is decreased to unacceptable levels.

There is no consistent design factor used for slings by wire rope manufacturers and the authors of safety books. The design factors of charts and formulas vary from a low of 5: 1 to a high of 8 : 1.

The rule of thumb formula to find the working load limit (WLL) for a sling or choker is:

Diameter in inches squared x 6 = WLL in tons.

$D^2 \times 6 = WLL$

WLL Example:

3/4 inch wire rope

3/4" x 3/4" x 6 = 9/16 x 6 = 3.4 tons WWL

Note: Based on a 6:1 design factor. Formula does not apply to metric sizes.

Note: All slings SHALL be tagged to show the manufacturer, and the rated load, including the lifting angle for each type of hitch.

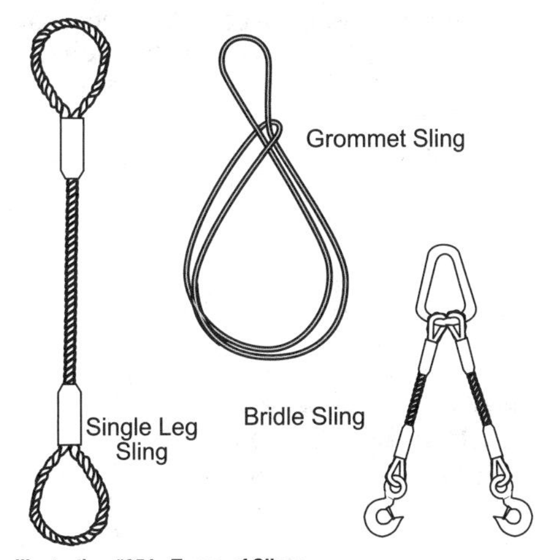

Illustration #254 - Types of Slings

Single and Double Basket Hitch

This is a method of supporting a load by wrapping the sling around the load and securing both ends of the sling on the hook. Make sure this hitch is used on loads that balance and cannot shift and slip out of the sling, see illustration #255. Do not place slings too far apart, as this could produce extra stress on the legs.

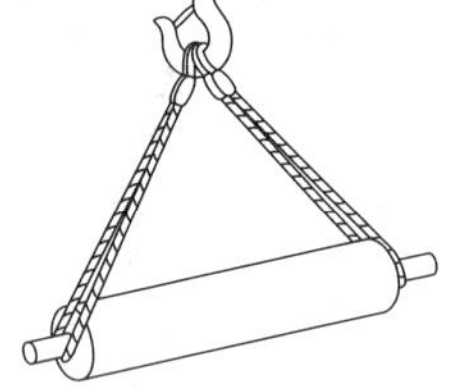

Illustration #255 - Double Basket Hitch

Double Wrap Basket Hitch

This basket hitch has one extra wrap completely around the load. It is excellent for pipe and tubes. The hitch exerts a full 360 degree contact and pulls the load together, see illustration #256.

Note: Sling charts rate a basket hitch at 200% of a single vertical hitch. To get the full basket rating, the load diameter to sling diameter ratio must be 25:1. Lifting a rectangular load will not allow the 200% basket rating.

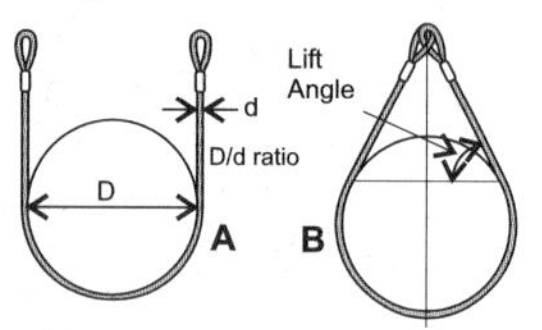

Illustration #256A - Basket Hitch

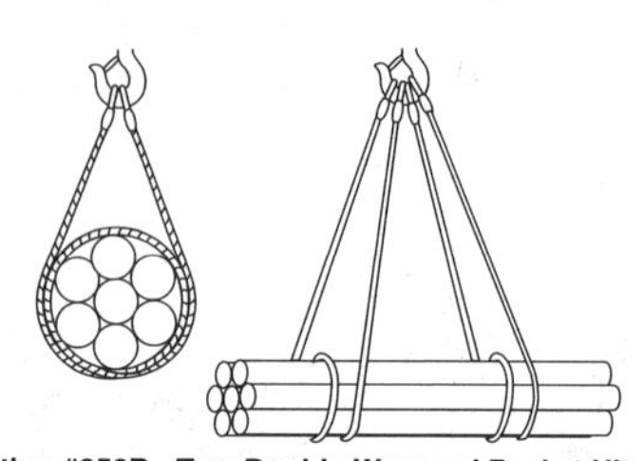

Illustration #256B - Two Double Wrapped Basket Hitches

Single and Double Wrap Choker Hitches

The choker hitch tightens as the load is lifted due to the noose formed at the point of choke, see illustration #257. The single choker hitch does not provide full 360° contact with the load and should not be used to lift loose bundles. A double wrap choker hitch is in full contact with the load because the choker is wrapped completely around the load before it is hooked into the vertical part of the sling, and should be used when lifting loose material such as pipe.

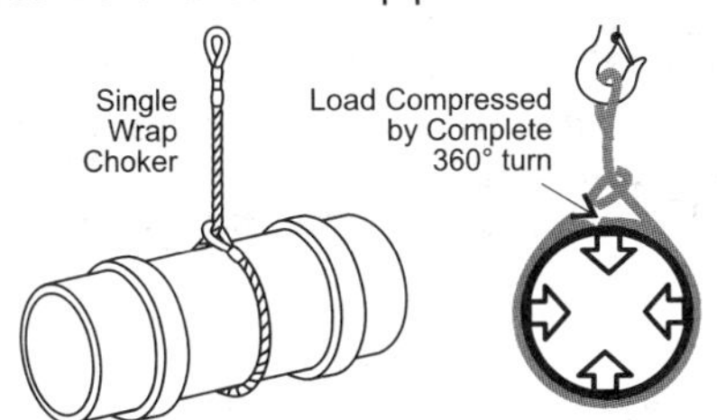

Illustration #257 - Single and Double Wrap Choker Hitch

Double Choker Hitches

The double choker hitch is made up of two single chokers that are spread and attached to the load. The hooks must be pointing out, as indicated in illustration #258. The double choker hitch does not provide full 360° contact with the load and should not be used to lift loose bundles. An extra wrap should be used when lifting pipe.

Note: A choker hitch is rated at 75% of a single vertical hitch.

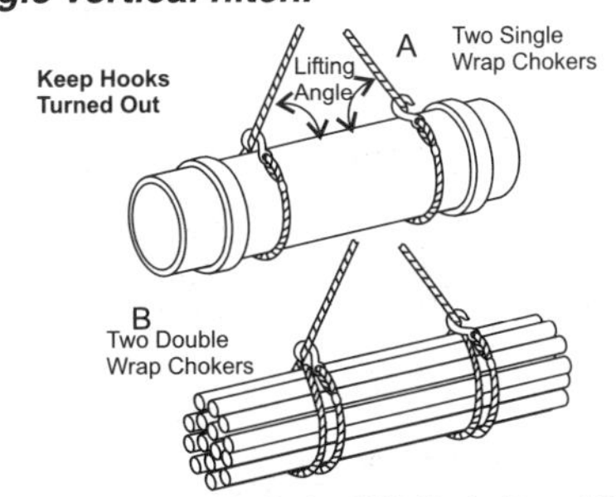

Illustration #258 - Double Choker With Hooks Turned Out

Synthetic Slings

Synthetic Web Slings: offer good protection for machined parts, are non-sparking, light weight and flexible. It is a flexible flat sling that has the ability to hug the load and keep it more secure from slippage, as indicated in illustration #259. This sling is usually made of nylon or polyester, however other materials are available.

Illustration #259 - Synthetic Web Sling

Nylon resists most alcohols, aldehydes, alkalis, and hydrocarbons, but it is not recommended for use around acids. Polyester can be used in acidic conditions. Nylon has a 10% stretch factor, while polyester has a lower stretch factor.

Both nylon and polyester can be used in temperatures up to 200 degrees F.

Rigging hardware, for example rings and shackles, is available to use with web slings (see the hardware section). The flat shape of the hardware is more suited to these slings than the usual curved types. Any folding, bunching or pinching in a standard shackle will reduce the rated load of a synthetic sling.

Note: Check with a reputable safety systems distributor for the proper sling material used with specific hazardous products.

Note: Synthetic slings must be removed from service if any of the following conditions are present: acid or caustic burns; melting or charring of any surface part; snags, punctures, tears, or cuts; broken stitches; distorted fittings; or the colored core warning yarns are showing.

Note: All sling types, including synthetic web sling must have the proper identification to show the name of the manufacturer, the rated load, and the type of material used. See illustration #260.

Note: Synthetic slings are manufactured in single and double ply. The double ply capacity ranges from 140% to 200% that of a single ply, depending upon sling type and hook up.

Edge protectors are available. They are the sewn-in, sliding, and replaceable types.

Synthetic web slings are identified by a Type Number, and they are available in a number of eye configurations. See illustration #261.

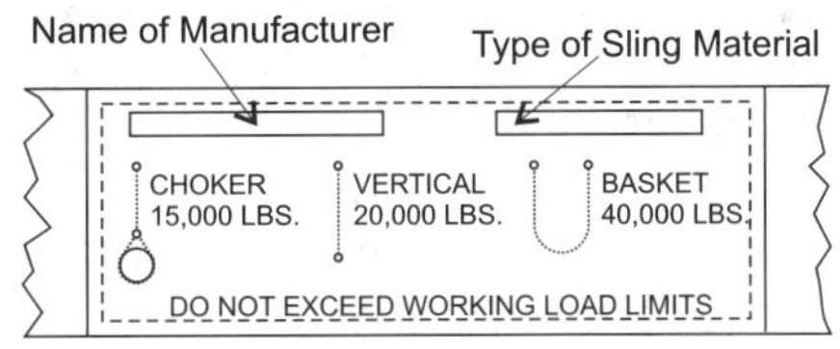

Illustration #260 - Synthetic Sling Identification Example

Synthetic Web Sling Grades

Although most web slings look similar to the untrained eye, they can vary in capacity for several reasons.

- Number of plies
- Grade of webbing material
- Efficiency of sewing

Be aware that a double ply web sling may not necessarily have twice the capacity of a single ply.

Web slings are available in three grades of webbing, and their rating is based on the efficiency of the sewing.

The WLL of a sling is based on the formula: Material breaking strength x efficiency of sewing ÷ design (safety) factor.

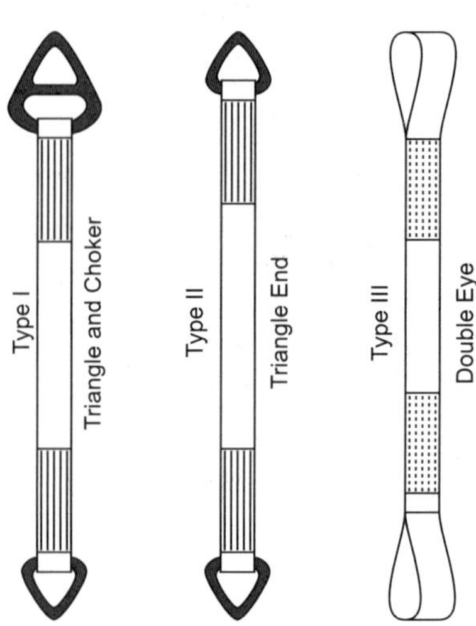

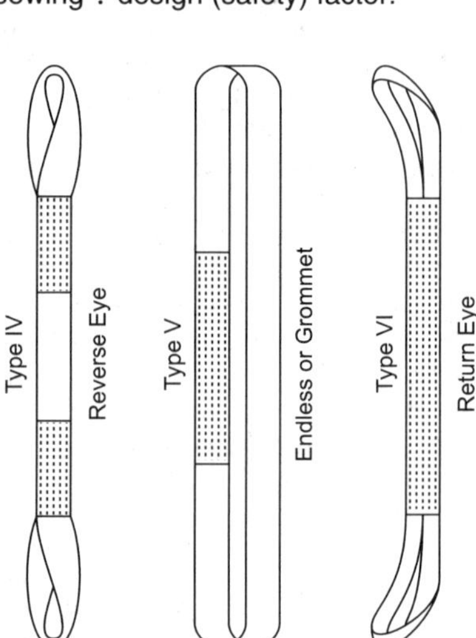

Illustration #261 - Synthetic Sling Types

Rated Loads for Single Ply, Class 5, Synthetic Web Slings (in pounds)

Hitch Type	Types I, II, III, and IV Single Leg			Two Leg or Single Basket Horizontal Angle				Type IV Endless
Width (inches)	Vertical	Choker	Basket	Vertical	60 deg	45 deg	30 deg	Vertical
1	1100	880	2200	2200	1900	1600	1100	2200
1 1/2	1600	1280	3200	3200	2800	2300	1600	3200
1 3/4	1900	1520	3800	3800	3100	2700	1900	1800
2	2200	1760	4400	4400	3800	3100	2200	4400
3	3300	2640	6600	6600	5700	4700	3300	6600
4	4400	3520	8800	8800	7600	6200	4400	8800
5	5500	4400	11,000	11,000	9500	7800	5500	11,000
6	6600	5280	13,200	13,200	11,400	9300	6600	13,200

Notes:
(a) The rated loads are based on stuffer weave construction webbing with a minimum certified tensile strength of 6,800 pounds per inch of width of webbing.
(b) Rated loads for Types III and IV slings apply to both tapered and non-tapered eye constructions. Rated loads for type V slings are for non-tapered webbing.
(c) For Type VI slings, consult the manufacturer for rated loads.
(d) For choker hitch, the angle of choke shall be 120 degrees or greater.

Table 99 - Synthetic Sling Capacities

Note: Always refer to the sling identification tag for the sling capacity. If the tag is missing, do not use the sling.

Synthetic Roundslings: are flexible lightweight slings made up of load carrying fibers covered with a tough non-load carrying cover. They are very flexible with limited stretch. Normal cover wear does not affect the strength, and wear points can be moved around. A safety feature is that the cover will rip when the sling is overloaded and over-stretched. See illustration #262.

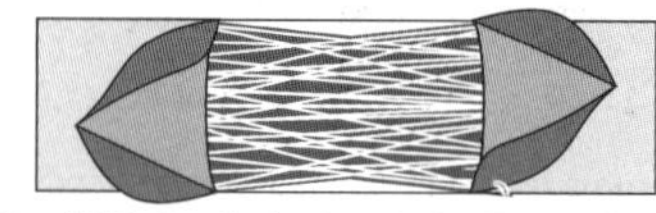

Illustration #262 - Synthetic Roundsling Example

Note: Always refer to the roundsling ID tag. Due to the wide number of manufacturers, there may be discrepancies in sizes and capacities, and color coding of round slings may not be standardized.

New Synthetic Sling Types

Twin Type: is one of the newest designs in slings. It is constructed using two roundsling types encased in an outer cover. See illustration #263 for an end view example.

This sling type is made of fibers from a kevlar type material which has a better weight to strength ratio than steel. For comparison purposes, one example is a type of 35 foot endless sling used in a basket configuration that has a capacity of 140,000 pounds.

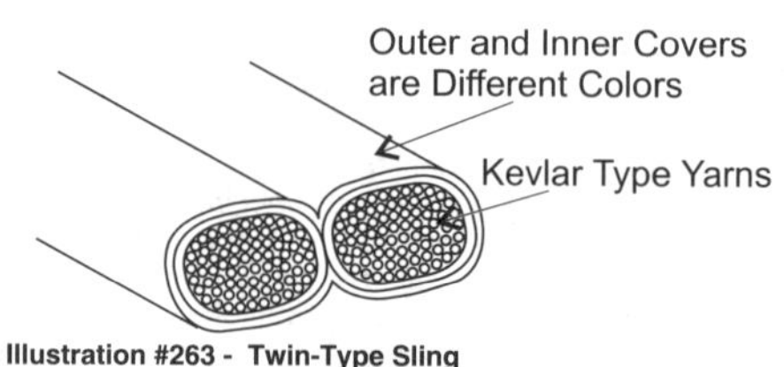

Illustration #263 - Twin-Type Sling

This sling only weighs about 60 pounds while a wire rope sling of equal capacity weighs approximately 600 pounds.

Some of the other features of this sling type include:

- Fiber-optic tell-tales: installed with both ends near the identification tag. A light shining through one end is visible at the other end. No light coming through could indicate damage and it should be removed from service.
- Load carrying yarns never come in contact with load. Outer cover material about 4 times more durable than polyester.
- Two inner bundles are covered in a different colored material than the outer cover to provide an instant alert for damage.
- Has overload indicator "tails" that are readily seen, but shrink and disappear when the sling is overloaded (illustration #264).

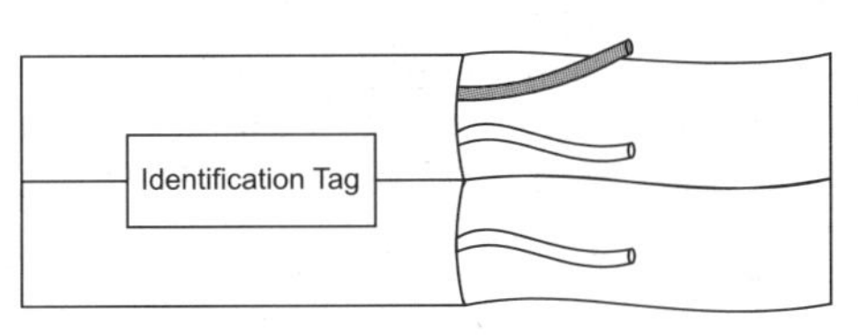

Illustration #264 - Twin-Type Sling

- Should not recoil at break, thereby reducing or eliminating the whiplash effect characteristic of chain, wire rope, and synthetic rope.
- The sling is actually two separate slings in one, with each making its own hook to load connection.
- Available in a continuous loop sling, two leg bridle sling with hardware, and eye and eye sling.

Sling Angle

The safe working load of any sling configuration depends on the size of the sling and the angles formed by the legs to the horizontal of the load.

A two legged sling with an angle of 90° on each sling would have exactly 50% of the load on each leg, see illustration #265A.

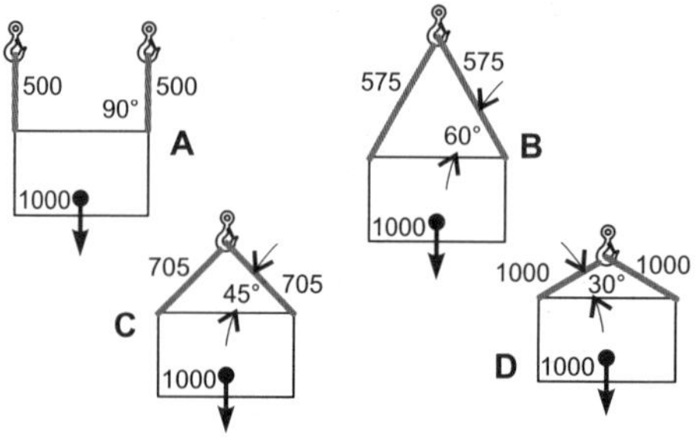

Illustration #265 - Sling Angle Affecting Tension

As the angle decreases, the load on each leg will increase. At 30°, the load on each leg will equal the load itself, as indicated in illustration #265D.

Note: The recommended safe lifting angle is 60°. See illustration #266.

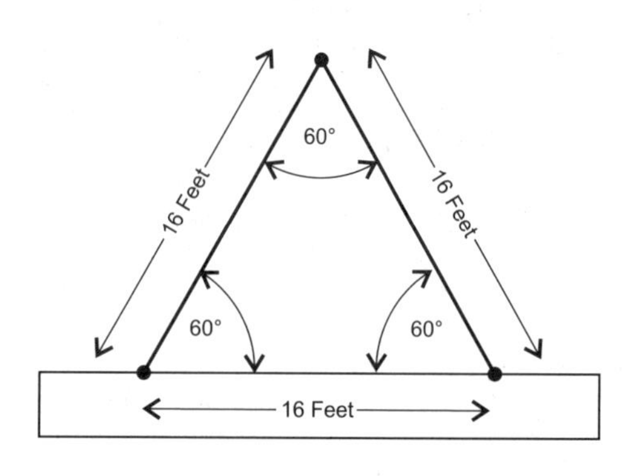

Illustration #266 - 60 Degree Safe Lifting Angle

Sling Leg Loading

Illustration #267 shows a 2,000 pound (908 kg) load carried by two slings (half the load and one sling is illustrated).

The loading on one sling leg as it moves out to different angles is shown. In a vertical lift, it is 1,000 pounds (454 kg), and at a 50 angle, it is 11,470 pounds (5214 kg), or nearly 6 times the load weight.

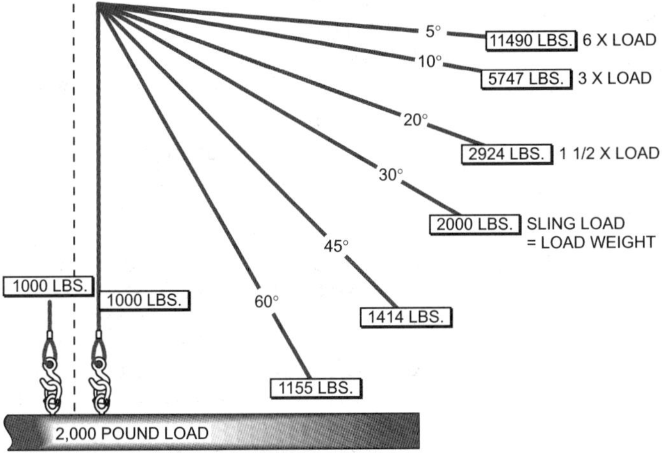

5° 11490 LBS. 6 X LOAD
10° 5747 LBS. 3 X LOAD
20° 2924 LBS. 1 1/2 X LOAD
30° 2000 LBS. SLING LOAD = LOAD WEIGHT
45° 1414 LBS.
60° 1155 LBS.
1000 LBS.
1000 LBS.
2,000 POUND LOAD

Illustration #267 - Sling Leg Stress Chart

Sling Angle	Load Angle Factor
5°	11.490
10°	5.747
15°	3.861
20°	2.924
25°	2.364
30°	2.000
35°	1.742
40°	1.555
45°	1.414
50°	1.305
55°	1.221
60°	1.155
65°	1.104
70°	1.064
75°	1.035
80°	1.015
85°	1.004
90°	1.000

Sling Load Formula

This formula is accurate on symmetrical loads. A tape measure and basic math can be used for calculations rather than degrees and angles. All slings must be of equal length and the center of gravity must be at load center. The formula is: $T = \frac{W}{N} x \frac{L}{V}$

T = tension per leg (pounds or kg)
W = load weight (pounds or kg)
N = number of sling legs
L = sling length (feet or meters)
V = vertical height (feet or meters)

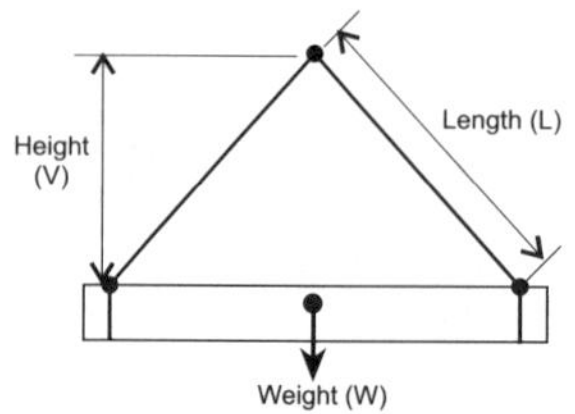

Illustration #268 - Calculating Sling Stress by Formula

Sling Stress Formula Example

W = weight = 5,250 pounds (2386.4 kg)
N = number of legs 2
V = vertical height 5 feet (1.524 m)
L= sling length = 8 feet (2.438 m)
T = W/N x L/V

Imperial Solution:

T = 5250/2 x 8/5 = 42000/10 = 4200 lbs.

Metric Solution:

T = 2386.4/2 x 2.438/1.524
= 5818.04/3.048 = 1908.8 kg

Tension per leg = 4,200 lbs. (1908.8 kg)

Note: This formula is for any number of sling legs. However, it is strongly recommended that the formula be applied using only two legs, as there is no way of knowing whether each leg of a three or four leg bridle is carrying its share of the load. With an inflexible load and more than two legs, it is possible to have all the weight on only two legs, while the others merely balance the load.

Centering a Load

When preparing a lift, the main considerations are weight, size and the center of gravity of the object. Estimate the center of gravity and spot the hook directly over it. See illustration #269.

A load rigged with the hook not positioned above the center of gravity will always swing when lifted. This changes the load distribution on the slings.

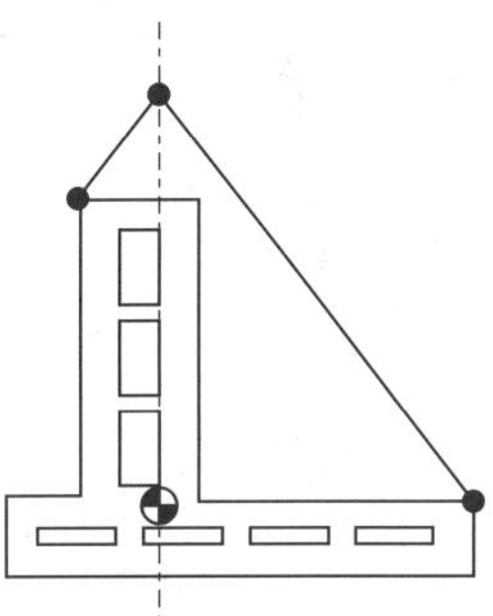

Illustration #269 - Hook-Up With Hook Over Center of Gravity

Softeners

All sharp corners should be covered by pads or softeners to prevent the sling from being bent or cut. These softeners can be purchased, or made from a split pipe section, padding or blocking, see illustration #270.

A good rule to follow is to make sure that the length of contact of the wire rope is equal to one rope lay, or seven times the rope diameter.

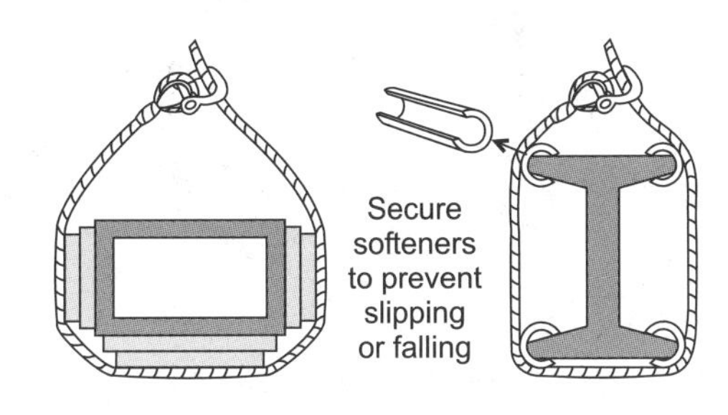

Illustration #270 - Use Softeners on All Sharp Corners

Sling WLL

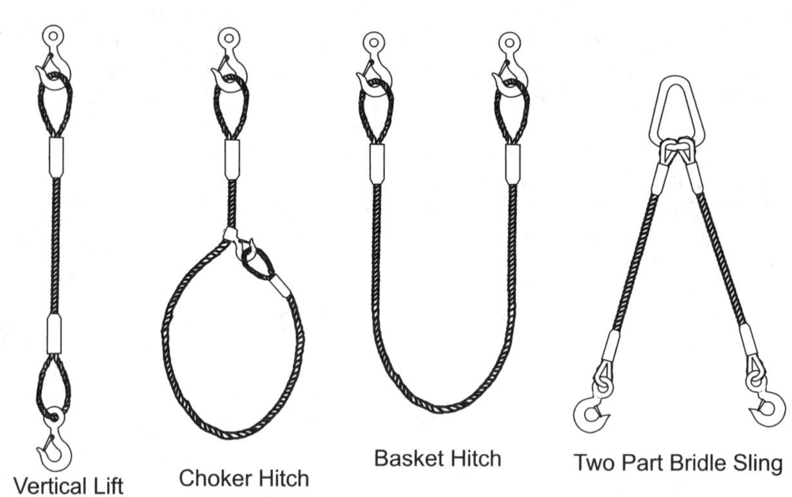

Table #100A - Sling Configurations for Table #100A

Sling WLL

		SLING HITCH TYPE						ANGLE FROM HORIZONTAL							
Rope Size		Vertical Lift		Choker Hitch		Basket Hitch		75°		60°		45°		30°	
ins	mm	lbs	kq	lbs	kq	lbs	kq	lbs	kq	lbs	kq	lbs	kq	lbs	kq
1/4	6.4	920	417	700	318	1,840	835.0	1,780	807	1,600	726	1,300	590	920	417
3/8	9.5	2,020	916	1,520	690	4,040	1833.0	3,900	1769	3,500	1587	2,860	1297	2,020	916
1/2	12.7	3,740	1696	2,800	1270	7,480	3393.0	7,220	3275	6,480	2939	5,280	2395	3,740	1696
5/8	16.0	5,600	2540	4,200	1905	11,200	5080.0	10,840	4917	9,700	4400	7,920	3592	5,600	2540
3/4	19.0	8,080	3665	6,060	2748	16,160	7330.0	15,640	7094	14,020	6359	11,460	5198	8,080	3665
7/8	22.2	10,920	4953	8,180	3710	21,840	9906.0	21,100	9571	18,920	8582	15,480	7022	10,920	4953
1	25.4	14,180	6432	10,680	4844	28,360	12864.0	27,400	12430	24,600	11158	20,060	9099	14,180	6432
1 1/8	28.6	16,660	7557	12,500	5670	33,320	15114.0	32,300	14651	28,900	13109	23,500	10660	16,660	7557
1 1/4	31.8	20,740	9407	15,540	7049	41,480	18815.0	40,120	18198	36,000	16330	29,300	13290	20,740	9407
1 3/8	35.0	25,340	11500	19,000	8618	50,680	22988.0	49,000	22226	43,880	19904	35,840	16257	25,340	11494
1 1/2	38.0	30,620	13889	22,960	10415	61,240	27778.0	59,500	29989	53,040	24058	43,300	19640	30,620	13889
1 5/8	41.3	35,900	16284	26,920	12211	71,800	32568.0	70,000	31751	62,180	28204	50,760	23025	35,900	16284
1 3/4	44.5	41,160	18670	30,860	13998	82,320	37340.0	80,000	36287	71,280	32332	58,200	26400	41,160	18670
1 7/8	47.6	48,320	21917	36,240	16438	96,640	43833.0	93,600	42456	83,680	37956	68,320	30990	48,320	21918
2	50.8	52,760	23930	39,560	17944	105,520	47863.0	102,400	46448	91,380	41450	74,600	33840	52,760	23932

Note: Safe Working Load Calculated on Improved Plow Steel 6 x 19 IWRC for sizes 1/4 inch (6.4 mm) to 1 inch (25.4 mm), 6 x 37 IWRC for sizes 1 1/8 inch (28.6 mm) to 2 inch (50.8 mm) DESIGN FACTOR = 6 : 1

Table #100 - Wire Rope Sling Capacities

Set-Up On Outriggers

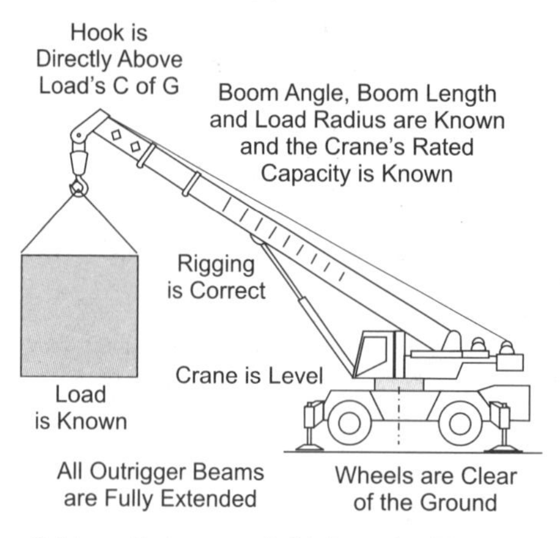

Illustration #271 - Mobile Crane Set-Up Summary

The majority of all crane accidents happen when lifting smaller loads because too much is taken for granted and the job is not planned.

A safe set-up similar to those shown above in illustration #271 will help to reduce crane accidents, whether it is crane tipping or a structural failure.

Note: ANSI B30.5a-2002 concerning Load Indicators, Rated Capacity Indicators, Rated Capacity Load Limiters. Effective 2003, all cranes with a maximum capacity of 3 tons or more SHALL have a load indicator, rated capacity indicator, or rated capacity load limiter.

Note: Do not use the load indicator to test weigh a load.

Load Pick and Carry

All crane capacity ratings are based on the crane being used in a stationary position. This also applies to models which are used for pick and carry operations. Traveling with suspended loads entails many variables, including:

- Type of terrain.
- Boom length.
- Stopping and starting momentum.

It is impossible to establish a single standard rating procedure with any assurance of safety. Therefore, when traveling with a load, the prevailing conditions must be evaluated to determine the applicable safety precautions.

Precautions to be followed during pick and carry operations:

- Keep the load as close to the carrier as possible.
- Keep the boom as short as possible.
- Keep the boom as low as possible.
- Load and boom to be carried in line with the direction of travel and in line with the axis of the crane.
- Rough terrain crane - boom and load must be over the front.
- Carrier mounted crane - (engine at rear) boom and load over the rear.
- Crawler - boom and load over the idler end.
- Loads - should be carried close to the ground and tied back to the carrier or controlled with tag lines.

Traveling With a Load

Use extreme caution if the load is behind the direction of travel (illustration #272). If the load is not snubbed to the crane, the load could swing out when the crane starts, thus allowing the crane to walk out from the load and tip over.

Never make any sudden starts or stops. In illustration #273, a sudden start could push the boom over backwards.

Note: Snub the load to the crane to avoid swingout.

The Crane Can Walk Out
From Under The Load
=== **USE CAUTION** ===

Snub
Off Load

Illustration #272 - Snub Load to Crane When Traveling

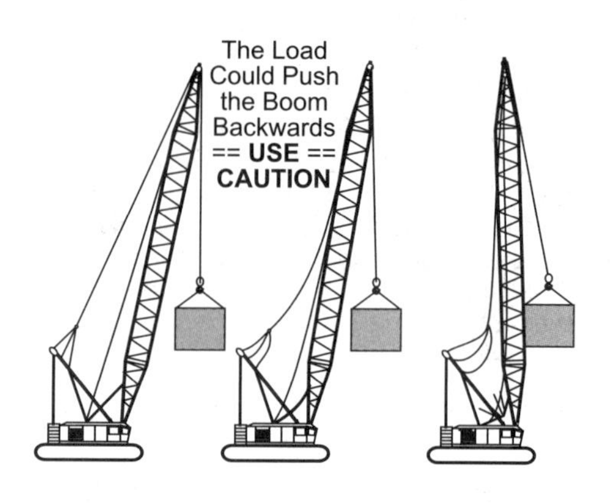

Illustration #273 - Do Not Make Sudden Starts or Stops

Power Line Clearances

Accidental electrocutions are among the most frequently repeated crane accidents. Most of these are caused when the boom contacts or approaches too close to overhead power lines. The fatality rate is high, particularly among riggers guiding the load.

While the danger is greater from high voltage transmission lines, where flashover can occur without actual contact, fatal accidents have resulted from contact with 440 volts and 220 volt service lines and strut lighting systems.

The safest procedure is to request the local electrical authority to cut off the power.

If, for any reason this is not possible or practical, and it is necessary for cranes to be under or near hot power lines, see table #101 for guidelines.

Electrical Hazards Clearance Guide

OPERATING NEAR HIGH VOLTAGE POWER LINES	
Normal Voltage (Phase to Phase)	**Minimum Required Clearance**
to 50 kV	10 ft. (3,05 m)
Over 50 to 200 kV	15 ft. (4.60 m)
Over 200 to 350 kV	20 ft. (6.10 m)
Over 350 to 500 kV	25 ft. (7.62 m)
Over 500 to 750 kV	35 ft. (10.67 m)
Over 750 to 1 000 kV	45 ft. (13.72 m)
IN TRANSIT WITH NO LOAD AND BOOM LOWERED	
Normal Voltage (Phase to Phase)	**Minimum Required Clearance**
to 0.75 kV	4 ft. (1 .22 m)
Over 0.75 to 50 kV	6 ft. (1.83 m)
Over 50 to 345 kV	10 ft. (3.05 m)
Over 345 to 750 kV	16 ft. (4.87 m)
Over 750 to 1 000 kV	20 ft. (6.10 m)

Table #101 - Minimum Powerline Clearances

Absolute Limit of Approach

Every live powerline has an area around it called the limit of approach. A crane boom, loadline, or load cannot operate in this area without the power being cut off. This is an absolute, no exception rule.

The absolute limit of approach area will vary somewhat with provincial, state, federal, or other regulating bodies. However the guidelines shown in table #101 and illustration #274 are close to those guidelines.

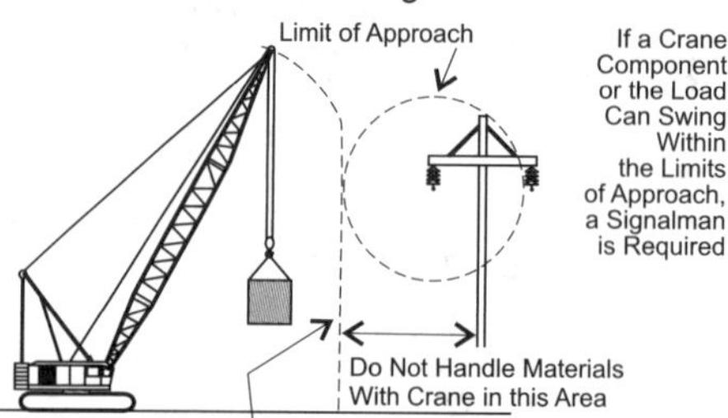

Illustration #274 - Use Signalman in Limit of Approach Area

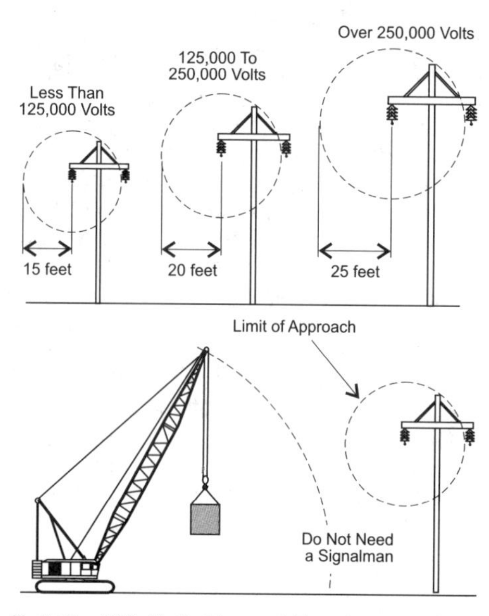

Illustration #275 - Limit of Approach Examples

Powerline Contact

The operator should remain in the cab after powerline contact until the power has been disconnected. If this is not practical, the operator must not step from the crane. He must jump clear with both feet together, being careful not to touch the crane. See illustration #276.

Illustration #276 - Jump Clear With Both Feet Together

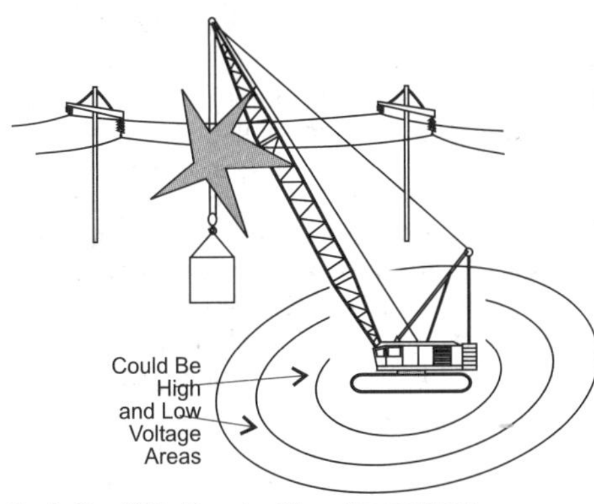

Illustration #277 - Energized Zone Around Crane

Crane and Hoist Signals

The person giving signals and the crane operator must work as a team. The lift should be discussed with both parties knowing what the steps will be and what signals will be used.

If there are several people in the ground crew, only one person shall give signals, although anyone can give a stop signal. It is recommended that the person giving the signals wear a fluorescent vest. Always stand in view of the operator and give clear signals. Try to avoid signals which will cause jerky operation movements. See illustration #278A and #278B for standard crane signals.

CONTROL OF PLATFORMS OR SKIPS

One bell or light........to STOP
Two bells or lights......to RAISE
Three bells or lights....to LOWER
Four bells or lights......ALL CLEAR

HORN SIGNALS FOR TRAVELLING AND MOBILE CRANES
(and as a warning for travel direction for Crawler Machines)

One blast............STOP
Two blasts.........FORWARD
Three blasts........... REVERSE

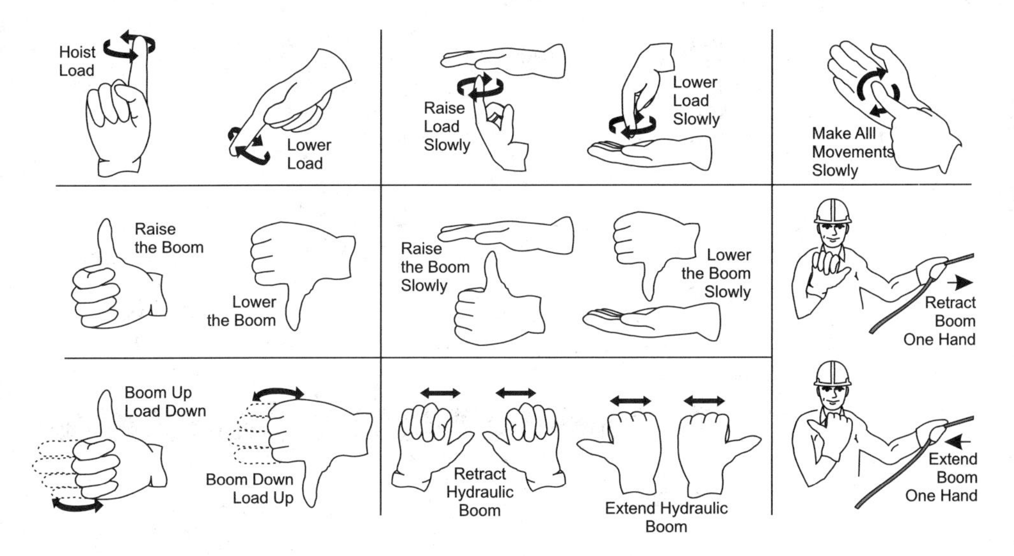

Illustration #278A - Crane and Hoist Signals

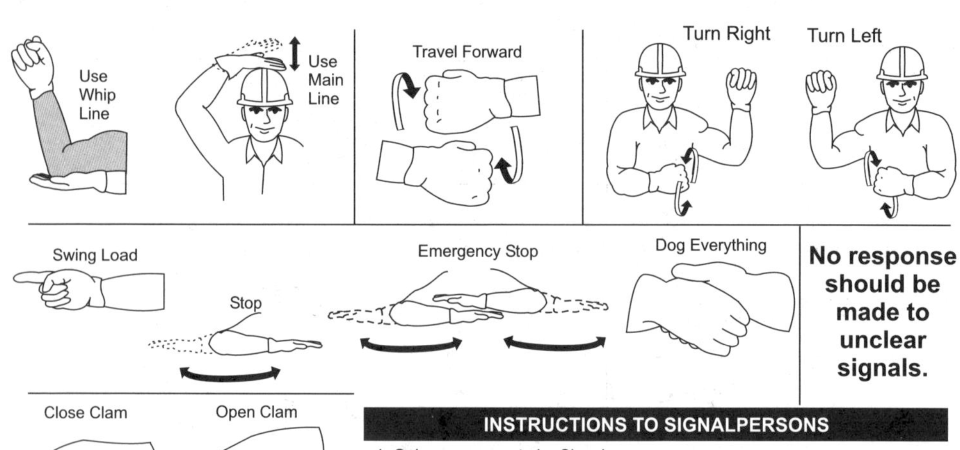

INSTRUCTIONS TO SIGNALPERSONS

1. Only one person to be Signalperson
2. Make sure the Operator can see you and acknowledges the signal given
3. Signalperson must watch the load - the Operator is watching you
4. Don't swing the load over other workers; warn them to keep out of the way

Illustration #278B - Crane and Hoist Signals

SECTION TWELVE

APPENDICES

Appendices Index

Safety and Accident Prevention

Each and every worker has the obligation to work safely with consideration for everyone on the jobsite. Accidents don't just happen without a reason, and usually that reason is a result of a mistake. Therefore if accidents are to be eliminated or reduced, the only way is by exercising caution at all times. Each worker has the responsibility to follow safety rules and use common safety practices at all times. Safety practices are developed by training, consciously practicing safe work habits, and by considering consequences before acting. Careless, unsafe actions without foresight inevitably result in accidents, injuries and death.

General Safety Requirements Include:

Dress Properly For The Job:

- Clothing, including boots, must fit properly and be in good condition.
- Do not wear rings and other jewelry items that may catch or become entangled on objects or moving parts.
- Long hair should be covered or confined.

Use The Appropriate Personal Protective Equipment Needed For The Job:

- Wear approved head and eye protection when required.
- Wear gloves for hand protection. Gloves should not be used when operating power tools if they interfere with the control operation or may catch on revolving parts of the power tool.
- If needed, use proper ear protection, breathing apparatus, and safety harness or nets.

Use Hand and Power Tools in a Safe and Proper Manner:

- Keep power tool guards in place and in working order.
- Keep tools in good condition and repair or replace when needed.
- Do not use extensions or "cheaters" on pipe wrenches, open end wrenches, or come-a-longs.
- Always pull on wrench handles and adjust stance to prevent a fall if something slips or lets go.
- Know the proper use, limitation and possible hazards of the tool before using it.

Keep the Work Area Clean:

- Keep material and tools properly secure and out of the way when not in use.
- Clean up any spilled water or oil from the work area immediately.

Establish Personal Sate Working Habits and Standards:

- Avoid all horseplay and practical jokes.
- Avoid any substance abuse which may impair safety and endanger co-workers.
- Avoid strains and back injuries; use the leg muscles, not the back muscles when lifting. If the load is too heavy, get help or use a proper lifting device.
- Think safety while performing any job and if not sure of the safe way of doing a job, ask.
- Help the less experienced worker in safe job practice.
- Learn first aid to cope with emergency injuries and illness.

Know Job Site Safety and Emergency Procedures: Find out and follow all job site safety rules and regulations.

- Find out where the jobsite First Aid Kit or First Aid Station is located and who is responsible for first aid treatment.

Jobsite Emergency Procedures

- Find out to whom hazardous or unsafe situations are reported.
- Know fire regulations and procedures, as well as the location of nearest fire extinguisher and alarm.
- Know the nearest emergency exit route and area evacuation procedures.

Know the Hazards of Material and Equipment:

- Follow and understand manufacturer's instructions and warnings dealing with any hazardous material or equipment.
- Know the hazardous product symbols.
- Take the proper precautions when handling dangerous, toxic or poisonous material.

Ladders

Many accidents are associated with the use of ladders. Most of them can be predicted and prevented with proper planning, proper selection of ladders, good work procedures, and adequate maintenance. Some of the more prevalent hazards in the use of ladders are: instability, electrical shock, falls from one level to another, and leaning too far out to the side.

To minimize the chance of accidents, these precautionary measures should be followed.

- Inspect ladders before using. Use only undamaged, unpainted ladders with non-slip feet.
- The ladder should extend 3 feet (1 m) beyond any landing.
- The ladder should be tied or wired at the top to prevent movement.
- The base should be placed 1/4 of the vertical height out from the wall.

- Never stand on the top two rungs of a step ladder.
- Ladders built on the job should be made from first grade materials.
- Don't use ladders around doorways or other blind entrance areas.
- Use both hands and always face the ladder when ascending or descending.

Scaffolds

Scaffolds make up a very important part of many industrial jobs. They require careful planning and construction to avoid serious accidents.

To help minimize accidents, the following precautionary measures should be followed.

- The height of rolling scaffolds must not exceed three times the smallest dimension of the base in height, unless securely tied off to a structure, or outriggers are used.
- Scaffolds 10 feet (3 m) or more in height must have perimeter guardrails around all open sides.
- Scaffold planks must be a minimum of 2" x 10" (50 mm x 250 mm) and be free of defects.
- All sections, including cross braces, must be used when erecting tubular scaffolds. Secure the scaffold to the structure no more than 20 feet (6 m) up, and every 15 feet (4.5 m) thereafter.
- Wheels must be locked or blocked if the scaffold is higher than twice the width.

Trenching

Trenches five feet (1.5 m) or more in depth shall have cave-in protection by installing temporary protective structures or by cutting back the walls of the trench, see illustration #279.

When the cutback method is used in trenches over 5 feet (1.5 m) in depth, the walls of the trench should be cut back as follows:

- In hard compact soil, not less than 30 degrees from vertical.
- In all other soils, not less than 45 degrees from vertical.
- The excavated material should be piled at least three feet (1 m) from the edge and not less than 45 degrees from the vertical. In trenches over 10 feet (3 m) in depth, soil piles should be placed in accordance with the specifications of a professional engineer.

Note: Each state or province could have different regulations for the shoring cut back or for temporary protective structures. Contact the applicable OSHA/ OH&S Department for the standards. See table #102A and #102B for suggested minimum size and spacing for trench shoring.

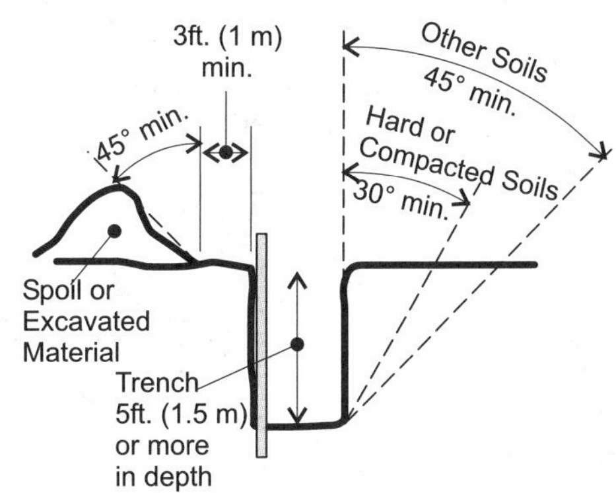

Illustration #279 - Cut Back or Use Shoring When Trenching

SUGGESTED MINIMUM SIZES AND SPACING FOR TRENCH SHORING						
Use For Hard Solid Soil Conditions						
	Uprights		**Stringers**		**Trench Width & Brace Size**	
Trench Depth	**Min. Size**	**Max. Spacing**	**Min. Size**	**Max. Spacing**	**Up to 6 ft. (1.8 m)**	**6 to 12 ft. (1.8 to 3.7 m)**
5 to 10 ft. (1.5 to 3 m)	2" x 10" (38 x 235 mm)	6 ft. (1.8 m)	4" x 6" (89 x 140 mm)	4 ft. (1.2 m)	4" x 4" (89 x 89 mm)	6" x 6" (140 x 140 mm)
10 to 15 ft. (3 to 4.5 m)	2" x 10" (38 x 235 mm)	4 ft. (1.2 m)	4" x 6" (89 x 140 mm)	4 ft. (1.2 m)	4" x 6" (89 x 140 mm)	6" x 6" (140 x 140 mm)
15 to 20 ft. (4.5 to 6 m)	2" x 10" (38 x 235 mm)	Close Sheeting	6" x 6" (140 x 140 mm)	4 ft. (1.2 m)	6" x 6" (140 x 140 mm)	6" x 8" (140 x 184 mm)
Use For Crumbling and Cracking Soil Conditions						
	Uprights		**Stringers**		**Trench Width & Brace Size**	
Trench Depth	**Min. Size**	**Max. Spacing**	**Min. Size**	**Max. Spacing**	**Up to 6 ft. (1.8 m)**	**6 to 12 ft. (1.8 to 3.7 m)**
5 to10 ft. (1.5 to 3 m)	2" x 10" (38 x 235 mm)	4 ft. (1.2 m)	4" x 6" (89 x 140 mm)	4 ft. (1.2 m)	4" x 6" (89 x 140 mm)	6" x 6" (140 x 140 mm)
10 to 15 ft. (3 to 4.5 m)	2" x 10" (38 x 235 mm)	3 ft. (0.9 m)	6" x 6" (140 x 140 mm)	4 ft. (1.2 m)	6" x 6" (140 x 140 mm)	6" x 8" (140 x 184 mm)
15 to 20 ft. (4.5 to 6 m)	2" x 10" (38 x 235 mm)	Close Sheeting	6" x 8" (140 x 184 mm)	4 ft. (1.2 m)	6" x 8" (140 x 184 mm)	6" x 8" (140 x184 mm)

Table #102A - Trench Shoring and Safety

SUGGESTED MINIMUM SIZES AND SPACING FOR TRENCH SHORING						
Use For Sand and Free Running Soil Conditions						
	Uprights		Stringers		Trench Width & Brace Size	
Trench Depth	**Min. Size**	**Max. Spacing**	**Min. Size**	**Max. Spacing**	**Up to 6 ft. (1.8 m)**	**6 to 12 ft. (1.8 to 3.7 m)**
5 to10 ft. (1.5 to 3 m)	2" x 10" (38 x 235 mm)	Close Sheeting	6" x 6" (140 x 140 mm)	4 ft. (1.2 m)	6" x 6" (140 x 140 mm)	6" x 8" (140 x l84 mm)
10 to 15 ft. (3 to 4.5 m)	2" x 10" (38 x 235 mm)	Close Sheeting	6" x 8" (140 x 184 mm)	4 ft. (1.2 m)	6" x 8" (140 x 184 mm)	8" x 8" (184 x 184 mm)
15 to 20 ft. (4.5 to 6 m)	2" x 10" (38 x 235 mm)	Close Sheeting	8" x 8" (184 x 184 mm)	4 ft. (1.2 m)	6" x 8" (140 x 184 mm)	8" x 10" (184 x 235 mm)

1. Vertical maximum spacing for all cross braces shall be 4 ft. (1.2 m).
2. Horizontal maximum spacing for all cross braces shall be 6 ft. (1.8 m).
3. Approved trench jacks may be used instead of, or in combination with cross braces, stringers or shoring.

Table #102B - Trench Shoring and Safety

Confined Space Entry

Safe work orders should be requested and received before any work is begun on containers which previously contained flammable gases, explosive substances or purge gases.

A confined space hazard and assessment program, specific for the work being conducted, should be written for the work in each and every confined space pertaining to the applicable workplace (as per OSHA/OH&S regulations).

Confined spaces include vessels, bins, tanks, tankers, tunnels, silos, sewers, utility vaults or chambers and pipelines. The hazards encountered in these confined spaces may include:

- Toxic gases.
- Flammable or explosive vapors.
- Oxygen deficiency.

Note: See table #103 for hazard information on common gases.

Only proper planning and testing can eliminate the risk of serious injury or death while entering, working, or leaving confined spaces.

Flammable and explosive substances may be present in a container because it was previously used for one of the following substances:

- Gasoline, light oil or other volatile liquid that releases potentially hazardous vapors at atmospheric pressure.
- An acid that reacts with metals to produce hydrogen.
- A non-volatile oil or a solid that, at ordinary temperature, will not release potentially hazardous vapors, but will release such vapors if the container is exposed to heat.
- A combustible solid, finely divided particles of which may still be present in the form of an explosive dust cloud.

All containers previously used for combustible, explosive or toxic substances should be cleaned prior to starting repair work. Containers should be gas tested before each entering.

- Empty and drain the container including all internal piping, traps and stand pipes.
- Insert a blind flange at the first outside joint of any pipe entering the vessel or container.
- Provide adequate protection for personnel cleaning the container. Air supply respirators may be necessary.
- Clean all compartments in containers having two or more compartments.
- Solution cleaning by using water, chemicals or steam should be decided upon after determining the properties of the substance.
- Maintain an alert person outside the manway as a safety watch.
- It is critical to realize that 60% of deaths in confined spaces are would-be rescuers who are not fully trained and/or adequately equipped for the rescue.

Dust and Fiber Hazards

There are many jobs where exposure to asbestos, silica or coal dust exceeds levels set out in regulation by OH&S/OSHA.

Workers should have a Fibrosis check up every two years if working in any occupation involving these materials. The risk of disease is very real if working where levels of dust or fibers are in the air. If not detected early, dust diseases could disable for life, or kill.

HAZARD INFORMATION FOR COMMON GASES				
Gas	Flammable Limits(% of gas in an air/gas)	Odor	Compared To Air	Major Hazards
Acetylene	2.5 to 81	Yes	Heavier	Flammable Asphyxiant
Ammonia	15 to 28	Yes	Lighter	Toxic
Argon	None	No	Heavier	Asphyxiant
Carbon Dioxide	None	No	Heavier	Asphyxiant
Carbon Monoxide	12.5 to 74	No	Lighter	Flammable Toxic
Chlorine	None	Yes	Heavier	Toxic Corrosive
Ethane	3.0 to 12.5	No	Heavier	Flammable Asphyxiant
Gasoline	1.4 to 7.4	Yes	Heavier	Flammable Asphyxiant
Helium	None	No	Lighter	Asphyxiant
Hydrogen	4.0 to 75	No	Lighter	Flammable Asphyxiant
Hydrogen Sulfide	4.3 to 45	Yes	Heavier	Toxic, Corrosive,Flammable
Methane	5.0 to 15	No*	Lighter	Flammable Asphyxiant
Nitrogen	None	No	Lighter	Asphyxiant
Oxygen	None	No	Heavier	Highly Reactive; Supports Combustion
Propane	2.2 to 9.5	No*	Heavier	Flammable Asphyxiant
Sulphur Dioxide	None	Yes	Heavier	Toxic, Corrosive
* Propane and Methane are usually given a distinctive odorant by the producer or gas processor to ensure that they are readily detectable.				

Table #103 - Common Hazardous Gas Information

Identification of Piping Systems (ANSI A13.1-1981)

This is a common system of labeling for piping containing hazardous materials. It consists of a legend and color format.

Information Found on Label or Legend:

- Legend or wording is used as the primary method of identification of piping contents, with pressure or temperature given as necessary to identify hazards.
- Arrows are used to indicate flow direction.
- Color field and letters used to identify the characteristic hazards which include: Flammable or Explosive, Chemically Active or Toxic, Temperatures or Pressures, Radioactive, Materials of Inherently Low Hazard, and Fire Quenching Material.

Note: See table #104 for color and size designations for legend and field, and illustration #280 for legend positioning.

Designation of Colors Used in Field and Lettering for Piping System Identification		
Classification	**Color of Field**	**Color of Ltrs. for Legend**
Materials Inherently Hazardous		
Flammable or Explosive	Yellow	Black
Chemically Active or Toxic	Yellow	Black
Extreme Temperature or Pressures	Yellow	Black
Radioactive	Yellow	Black
Materials of Inherently Low Hazard		
Liquid or Liquid Admixture	Green	White
Gas or Gaseous Admixture	Blue	While
Fire Quenching Materials		
Water, Foam, C0$_2$, Halon etc.	Red	White

Table #104 - Piping Identification Color Code

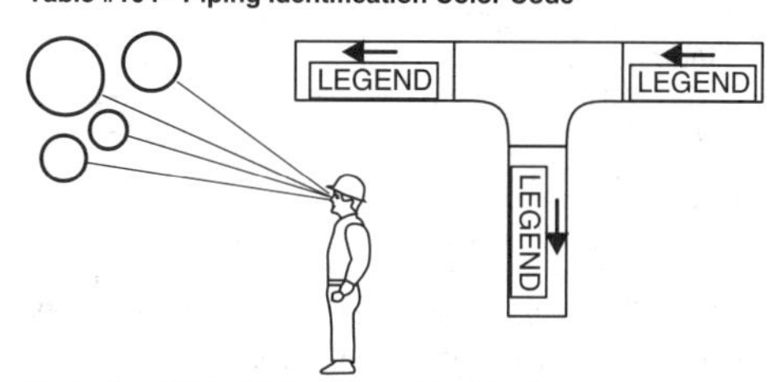

Illustration #280 - Piping Legend Positioning

STEAM TRAP SELECTION GUIDE

A = First Choice B = Alternate Choice

Application	FT Range	FT/TV/ SLR	FT/SLR	TD Range	BPT	SM	Ther- moton	IB Range
Canteen Equipment								
Boiling Pans - Fixed	A	B	B[1]	B	B			
Boiling Pans - Tilting	–	A	B	–	B			
Boiling Pans - Pedestal	B	B	B[1]	–	A[2]			
Steaming Ovens	–	–	–	–	A[2]			
Hot Plates	B	B	B[1]	–	A[2]			
Fuel Oil Heating								
Bulk Oil Storage Tanks	–	–	–	A	–	–	–	B[1]
Line Heaters	A	–	–	–	–	–	–	B[1]
Outflow Heaters	A	–	–	–	–	–	–	B[1]
Tracer Lines & Jacketed Pipes	–	–	–	B	A[3]	B	B	–
Hospital Equipment								
Autoclaves; and Sterilizers	B	B	B[1]	–	A	–	–	B
Industrial Dryers								
Drying Coils (continuous)	A	–	–	–	B	B	–	B
Drying Coils (grid)	–	–	–	–	B	A	–	B[1]
Drying Cylinders	B	A	B[1]	–	–	–	–	B[1]
Multi Bank Pipe Dryers	A	–	–	–	B	–	–	B[1]
Multi Cylinder Sizing Machines	B	A	B[1]	–	–	–	–	B[1]

(1) With air vent in parallel. (2) At end of cooling leg - minimum length 3 ft.(1m) (3) Use special tracing traps which offer fixed temperature discharge option. Note: Steam trap information courtesy Spirax Sarco Canada Ltd.

Table #105A - Steam Trap Selection Guide

STEAM TRAP SELECTION GUIDE

A = First Choice B = Alternate Choice

Application	FT Range	FT/TV/ SLR	FT/SLR	TD Range	BPT	SM	Ther-moton	IB Range
Laundry Equipment							–	
Garment Presses	B	–	–	A	–	–	–	B
Ironers and Calendars	B	A	B[1]	B[1]	B	–	–	B[1]
Solvent Recovery Units	A	–	–	B	–	–	–	B
Tumbler Dryers	A	B	B[1]	–	–	–	–	B[1]
Presses								
Multi Platen Presses (parallel connections)	B	–	–	A	–	–	–	B
Multi Platen Presses (series connections)	–	–	–	A[1]	–	–	–	B[1]
Tire Molds	B	–	–	A	B	–	–	B
Process Equipment								
Boiling Pans - Fixed	A	B	B[1]	B[1]	B	–	–	–
Boiling Pans - Tilting	–	A	B		–	–	–	–
Brewing Coppers	A	B	B[1]		–	–	–	B[1]
Digesters	A	–		B[1]	–	–	–	B[1]
Evaporators	A	B	B[1]	–	–	–	–	B[1]
Hot Tables	–	–	–	B	A	–	–	–
Retorts	A	–	–	–	–	–	–	B[1]
Bulk Storage Tanks	–	–	–	A[1]	–	–	–	B[1]
Vulcanizers	B	–	–	A[2]	–	–	–	B[1]

(1) With air vent in parallel. (2) At end of cooling leg - minimum length 3 ft.(1m) (3) Use special tracing traps which offer fixed temperature discharge option. Note: Steam trap information courtesy Spirax Sarco Canada Ltd.

Table #105B - Steam Trap Selection Guide

STEAM TRAP SELECTION GUIDE A = First Choice B = Alternate Choice								
Application	FT Range	FT/TV/ SLR	FT/SLR	TD Range	BPT	SM	Ther-moton	IB Range
Space Heating Equipment								
Shell & Tube Exchangers	A	B	B[1]	–	–	–	–	B[1]
Heating Coils & Unit Heaters	A	B	B[1]	–	–	–	–	B[1]
Radiant Panels & Strips	A	B	B[1]	B[1]	–	–	–	B[1]
Cabinet Heaters	B	–	–	–	A	B	–	–
Overhead Pipe Coils	B	–	–	–	A	–	–	B[1]
Steam Mains								
Horizontal Runs	B	–	–	A	B[2]	–	–	B
Separators	A	–	–	B	B[2]	–	–	B
Terminal Ends	B	–	–	A[1]	B[2]	–	–	B[1]
Shut Down Drain (Frost Protection)	–	–	–	–	B[3]	–	A	–
Tanks And Vats								
Process Vats (Rising Discharge Pipe)	B	–	–	A	B	–	–	B
Process Vats (Discharge Pipe at Base)	A	–	–	B	B	–	–	B
Small Coil Heated Tanks (quick boiling)	A	–	–	–	B	–	–	B
Small Coil Heated Tanks (slow boiling)	–	–	–	–	–	–	A	–

(1) With air vent in parallel. (2) At end of cooling leg - minimum length 3 ft.(1m) (3) Use special tracing traps which offer fixed temperature discharge option. Note: Steam trap information courtesy Spirax Sarco Canada Ltd.

Table #105C - Steam Trap Selection Guide

KEY TO STEAM TRAP SELECTION TABLE	
FT Range	Float/Thermostatic
FT/TV/SLR	Float/Thermostatic with Steam Lock Release
FT/SLR	Float/Steam Lock Release
TD Range	Thermodynamic
BPT	Balanced Pressure Thermostatic
SM	Bi-metallic
Thermoton	Liquid Expansion
IB Range	Inverted Bucket

Table #105D - Steam Trap Type List

Steam Traps Quick Sizing Guide

Need To Know:

I. ***The steam pressure at the trap:*** after any pressure drop through control valves or equipment.

2. ***The Lift:*** if any, after the trap.
Rule of Thumb: 2 ft. = 1 psi back pressure (approximate).

3. ***Possible Sources of Backpressure:*** in the condensate return system.

Example:

a. Condensate taken to a pressurized de-aerator tank.

b. Local back pressure due to discharges of numerous traps close together into small sized return.

4. ***Quantity of Condensate:*** to be handled, obtained from:

a. Measurement

b. Calculation of heat load (see General Usage Formula)

c. Manufacturer's Data

5. ***Safety Factor:*** these factors depend upon particular applications, typical examples being as follows:

	General	With Temp. Control
Mains Drainage	x2	–
Storage Heaters	x2	–
Space Unit Heaters	x2	x3
Air Heating Coils	x2	x4
Submerged Coils (low level drain)	x2	–

Submerged Coils (siphon drain)	x3	–
Rotating Cylinders	x3	–
Tracing Lines	x2	–
Platen Presses	x2	–

Rule of Thumb: Use factor of 2 on everything except Temperature Controlled Air Heater Coils and Converters, and Siphon applications.

How To Use:

The difference between the steam pressure at the trap, and the total black pressure, including that due to any lift after the trap, is the ***Differential Pressure***.

The quantity of condensate should be multiplied by the appropriate factor, to produce ***Sizing Load***. The trap may now be selected using the ***Differential Pressure*** and the ***Sizing Load***.

Example: A trap is required to drain 22 lb/hr of condensate from a 4 inch insulated steam main, which is supplying steam at 100 psig. There will be a lift after the trap of 20 feet.

Supply Pressure = 100 psig
Lift = 20 ft = 10 psi approx.

Therefore:

Differential Pressure = 100 - 10 = 90 psi
Quantity = 22 lb/hr
Mains Drainage Factor = 2
Therefore sizing load = 44 lb/hr

General Calculations for Condensate Loads

Heating water with steam

$$\text{lbs Condensate/hr} = \frac{GPM}{2} \times (1.1) \times Temp.Rise\ ^\circ F$$

Heating fuel oil with steam

$$\text{lbs Condensate/hr} = \frac{GPM}{4} \times (1.1) \times Temp.Rise\ ^\circ F$$

Heating air with steam coils

$$\text{lbs Condensate/hr} = \frac{GPM}{800} \times Temp.Rise\ ^\circ F$$

Steam Radiation

$$\text{lbs Condensate/hr} = \frac{Sq.Ft.E.D.R}{4}.$$

Steam Jacketed Dyers

$$lbsConcentrate / hr = \frac{100(Wi - Wf) + (Wi + \Delta T)}{L}$$

Wi = Initial weight of the material-pounds per hour
Wf = Final weight of the material-pounds per hour
ΔT= Temperature rise of the material °F
L = Latent heat of the steam BTU/lb

Steam Tracing Lines

Approximate load is 10 to 50 lb/hr for each 100 ft. of tracer.

Heating Liquids in Steam Jacketed Kettles and Steam Heated Tanks

$$lbs\,Concentrate / hr = \frac{G\,x\,s.g.\,x\,Cp\,x\,\Delta T\,x\,8.3}{L\,x\,t}$$

G = Gallons of liquid to be heated
s.g.= Specific gravity of the liquid
Cp = Specific heat of the liquid
ΔT = Temperature rise of the liquid °F
L = Latent heat of the steam BTLU/lb
t = Time in hours

Sterilizers, Autoclaves, Retorts Heating Solid Material

$$lbs\,Concentrate / hr = \frac{W\,x\,Cp\,x\,\Delta T}{L\,x\,t}$$

W = Weight of materials-lbs
Cp = Specific heat of the material
ΔT= Temperature rise of the material °F
L = Latent heat of steam BTU/lb
t = Time in hours

Heating Air with Steam; Pipe Coils and Radiation

$$lbs\,Concentrate / hr = \frac{A\,x\,U\,x\,\Delta T}{L}$$

A = Area of the heating surface in square feet
U = Heat transfer coefficient (2 for free convection)
ΔT= Steam temperature minus the air temperature °F
L = Latent heat of steam BTU/lb

Note: The condensate load to heat the equipment must be added to the condensate load for heating the material. Use same formula.

D-1 Temp. Conversion Formulas & Scales

Temperature Conversion

To Convert From:	Use the Formula:	To Obtain:
Degrees Celsius	(°C x 9/5) + 32	Degrees Fahrenheit
Degrees Celsius	°C + 273.16	Kelvin
Degrees Fahrenheit	(°F - 32) x 5/9	Degrees Celsius
Degrees Fahrenheit	°F + 459.69	Degrees Rankin

Temperature Scales

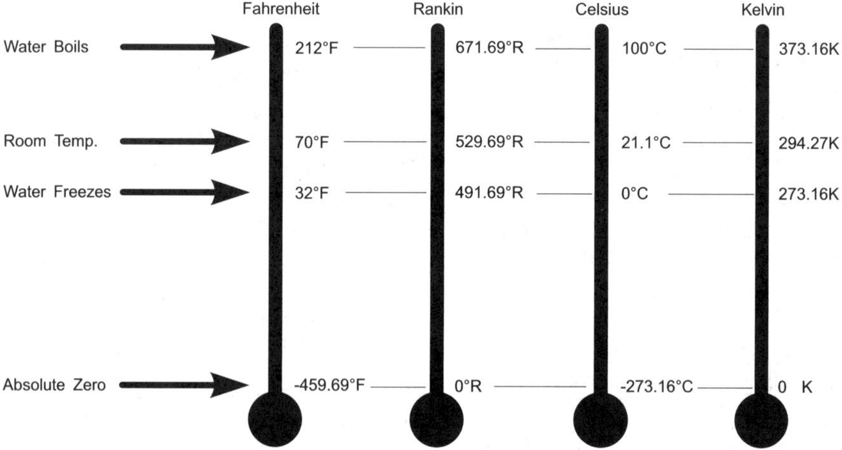

	Fahrenheit	Rankin	Celsius	Kelvin
Water Boils	212°F	671.69°R	100°C	373.16K
Room Temp.	70°F	529.69°R	21.1°C	294.27K
Water Freezes	32°F	491.69°R	0°C	273.16K
Absolute Zero	-459.69°F	0°R	-273.16°C	0 K

Table #106 - Temperature Conversion and Scales

Temperature Conversion Table

To use the following table, the known temperature in either degrees Fahrenheit (°F) or degrees Celsius (°C) is first located in the bold center column. The corresponding temperature in degrees Fahrenheit is found to the right and degrees Celsius to the left.

°C		°F
-73.3	**-100**	-148
-67.8	**-90**	-130
-62.2	**-80**	-112
-56.7	**-70**	-94
-51.1	**-60**	-76
-45.6	**-50**	-58
-40.0	**-40**	-40
-34.4	**-30**	-22
-28.9	**-20**	-4
-23.3	**-10**	14
-17.8	**0**	32.0
-17.2	**1**	33.8
-16.7	**2**	35.6
-16.1	**3**	37.4
-15.6	**4**	39.2
-15.0	**5**	41.0
-14.4	**6**	42.8
-13.9	**7**	44.6
-13.3	**8**	46.4
-12.8	**9**	48.2

°C		°F
-12.2	**10**	50.0
-11.7	**11**	51.8
-11.1	**12**	53.6
-10.6	**13**	55.4
-10.0	**14**	57.2
-9.4	**15**	59.0
-8.9	**16**	60.8
-8.3	**17**	62.6
-7.8	**18**	64.4
-7.2	**19**	66.2
-6.7	**20**	68.0
-6.1	**21**	69.8
-5.6	**22**	71.6
-5.0	**23**	73.4
-4.4	**24**	75.2
-3.9	**25**	77.0
-3.3	**26**	78.8
-2.8	**27**	80.6
-2.2	**28**	82.4
-1.7	**29**	84.2

°C		°F
-1.1	**30**	86.0
-0.6	**31**	87.8
0.0	**32**	89.6
0.6	**33**	91.4
1.1	**34**	93.2
1.7	**35**	95.0
2.2	**36**	96.8
2.8	**37**	98.6
3.3	**38**	100.4
3.9	**39**	102.2
4.4	**40**	104.0
5.0	**41**	105.8
5.6	**42**	107.6
6.1	**43**	109.4
6.7	**44**	111.2
7.2	**45**	113.0
7.8	**46**	114.8
8.3	**47**	116.6
8.9	**48**	118.4
9.4	**49**	120.2

Table #107A - Temperature Conversions

D-2 Temperature Conversion

°C		°F
10.0	**50**	122.0
10.6	**51**	123.8
11.1	**52**	125.6
11.7	**53**	127.4
12.2	**54**	129.2
12.8	**55**	131.0
13.3	**56**	132.8
13.9	**57**	134.6
14.4	**58**	136.4
15.0	**59**	138.2
15.6	**60**	140.0
16.1	**61**	141.8
16.7	**62**	143.6
17.2	**63**	145.4
17.8	**64**	147.2
18.3	**65**	149.0
18.9	**66**	150.8
19.4	**67**	152.6
20.0	**68**	154.4
20.6	**69**	156.2
21.1	**70**	158.0
21.7	**71**	159.8
22.2	**72**	161.6
22.8	**73**	163.4

°C		°F
23.3	**74**	165.2
23.9	**75**	167.0
24.4	**76**	168.8
25.0	**77**	170.6
25.6	**78**	172.4
26.1	**79**	174.2
26.7	**80**	176.0
27.2	**81**	177.8
27.8	**82**	179.6
28.3	**83**	181.4
28.9	**84**	183.2
29.4	**85**	185.0
30.0	**86**	186.8
30.6	**87**	188.6
31.1	**88**	190.4
31.7	**89**	192.2
32.2	**90**	194.0
32.8	**91**	195.8
33.3	**92**	197.6
33.9	**93**	199.4
34.4	**94**	201.2
35.0	**95**	203.0
35.6	**96**	204.8
36.1	**97**	206.6

°C		°F
36.7	**98**	208.4
37.2	**99**	210.2
37.8	**100**	212.0
43	**110**	230
49	**120**	248
54	**130**	266
60	**140**	284
66	**150**	302
66	**160**	320
66	**170**	338
66	**180**	356
88	**190**	374
93	**200**	392
99	**210**	410
100	**212**	413
104	**220**	428
110	**230**	446
116	**240**	464
121	**250**	482
127	**260**	500
132	**270**	518
138	**280**	536
143	**290**	554
149	**300**	572

Table #107B - Temperature Conversions

°C		°F
154	**310**	**590**
160	**320**	608
166	**330**	626
171	**340**	644
177	**350**	662
182	**360**	680
188	**370**	698
193	**380**	716
199	**390**	734
204	**400**	752
210	**410**	770
216	**420**	788
221	**430**	806
227	**440**	824
232	**450**	842
238	**460**	860
243	**470**	878
249	**480**	896
254	**490**	914
260	**500**	932
266	**510**	950
271	**520**	968
277	**530**	986
282	**540**	1004
288	**550**	1022
293	**560**	1040
299	**570**	1058
304	**580**	1076
310	**590**	1094
316	**600**	1112
321	**610**	1130
327	**620**	1148
332	**630**	1166
338	**640**	1184
343	**650**	1202
349	**660**	1220
354	**670**	1238
360	**680**	1256
366	**690**	1274
371	**700**	1292
377	**710**	1310
382	**720**	1328
388	**730**	1346
393	**740**	1364
399	**750**	1382
404	**760**	1400
410	**770**	1418
416	**780**	1436
421	**790**	1454
427	**800**	1472
432	**810**	1490
438	**820**	1508
443	**830**	1526
449	**840**	1544
454	**850**	1562
460	**860**	1580
466	**870**	1598
471	**880**	1616
477	**890**	1634
482	**900**	1652
488	**910**	1670
493	**920**	1688
499	**930**	1706
504	**940**	1724
510	**950**	1742
516	**960**	1760
521	**970**	1778
527	**980**	1796
532	**990**	1814
538	**1000**	1832

Table #107C - Temperature Conversions

HANGERS AND SUPPORTS Recommended Maximum Spacing					
Nominal Pipe Size		Steel Pipe Systems		Copper Tubes	
inches	mm	feet	m	feet	m
1/4	8	7	2.1	5	1.5
3/8	10	7	2.1	5	1.5
1/2	15	7	2.1	5	1.5
3/4	20	7	2.1	5	1.5
1	25	7	2.1	6	1.8
1 1/4	32	7	2.1	7	2.1
1 1/2	40	9	2.7	8	2.4
2	50	10	3.4	8	2.4
2 1/2	65	11	3.4	9	2.7
3	80	12	3.7	10	3.0
3 1/2	90	13	4.0	11	3.4
4	100	14	4.3	12	3.7
5	125	16	4.9	13	4.0
6	150	17	5.2	14	4.3
8	200	19	5.8	16	4.9
10	250	20	6.1	18	5.5
12	300	23	7.0	19	5.8
14	350	25	7.6	-	-
16	400	27	8.2	-	-
18	450	28	8.5	-	-
20	500	30	9.1	-	-

Table #108 - Pipe Hangers and Supports

Hanger Spacing For Other Material

Asbestos Cement: as per manufacturer's recommendations.

Cast Iron Pressure Pipe: 12 ft. (3.7 m) maximum.

Cast Iron Soil Pipe: 10 ft. (3.0 m) maximum. Fiberglass Reinforced Pipe: as per manufacturer's recommendations.

Fiberglass Reinforced Pipe: as per manufacturers' recommendations.

Glass Pipe: 8 ft. (2.4 m) maximum.

Plastic: depends on material and temperature, as per manufacturer's recommendations.

Note:

1. Extra support or hangers are required when heavy valves or fittings are placed in a piping system, at changes of direction, and/or as per code or job specifications.
2. Flexible plastic tube or similar material should be supported continuously.
3. Rod diameter may be reduced one size smaller when double rods are used (minimum reduction 3/8 of inch (9.6 mm) (table #109).
4. Maximum safe load taken at rod temperature of 610°F (321°C) conforming to ASTM A107 (table #109).

Minimum Rod Sizes					
Nominal Pipe Size		Rod Size		Max. Safe Load	
inches	mm	inches	mm	lbs	kg
2" and smaller	50 and smaller	3/8	9.6	610	276
2 1/2 to 3 1/2	65 to 90	1/2	12.7	1,130	512
4 to 5	100 to 125	5/8	15.8	1,810	821
6	150	3/4	19.1	2,710	1229
8 to 12	200 to 300	7/8	22.2	3,770	1710
14 to 18	350 to 450	1	25.4	4,960	2249
20 to 24	500 to 600	1 1/4	31.8	8,000	3628

Table #109 - Single Rod Diameter

Rod Length Calculations for Fabrication of U-Bolts

If a U-bolt is to be fabricated from rod for a particular pipe size, the length of rod needed can be calculated easily using the following formula:

Rod Length = A + B + C
(see illustration #281)

Where:

A = 1.571 x (Pipe O.D. + Rod Diameter)

B = Pipe O.D.

C = 2 x (Plate Thickness + Washer Thickness + Nut Thickness + Length of Thread Beyond End of Nut)

Example:

Find the length of 3/8 in. (10 mm) rod needed to fasten a 4 in. (100 mm) pipe to a 1/4 in. (6 mm) steel plate using 1/8 in. 3 mm) washers, 3/8 in. (10 mm) thick hex nuts; with 3/8 in. (10 mm) of thread needed beyond end of the nuts.

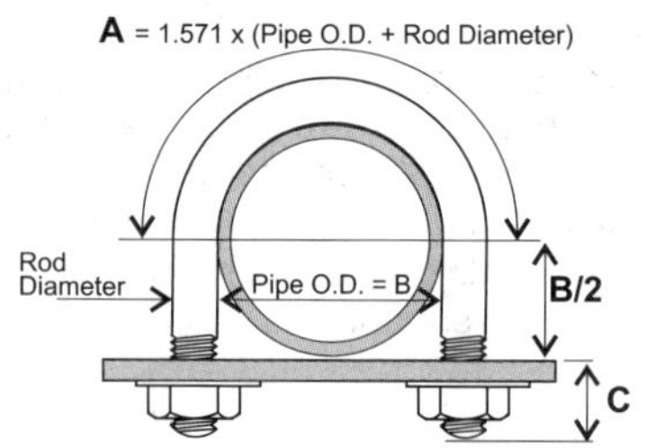

Illustration #281 - U-bolt Rod Length Calculation

Solution Steps:

Imperial Calculation:

1. Use the rod length calculation formula:
   Rod Length = A + B + C
2. Calculate A Length:
   A = 1.571 x (Pipe O.D. + Rod Diameter)
   A = 1.571 x (4.5 in. + 0.375 in.)
   A = 7.659 inches
3. Calculate B Length:
   B = Pipe O.D.
   B = 4.5 inches

4. Calculate C Length:
   C = 2 x (Plate Thickness + Washer Thickness + Nut Thickness + Length of Thread Beyond End of Nut)
   C = 2 x (0.25 in. + 0. 125 in. + 0.375 in. + 0.375 in.)
   C = 2.25 inches
5. Add together A + B + C to find Rod Length:
   Rod Length
   = A + B + C
   = 7.659 in. + 4.5 in. + 2.25 in.
   = 14.409 (14 3/8) inches

Note: Decimals of an inch and fractional equivalents can be found in the Equivalent Chart in the Appendices

Metric Calculation:

1. Use the rod length calculation formula:
   Rod Length = A + B + C
2. Calculate A Length:
   A = 1.571 x (Pipe O.D. + Rod Diameter)
   A = 1.571 x (114.3 mm + 10 mm)
   A = 195.275 mm
3. Calculate B Length:
   B = Pipe O.D.
   B = 114.3 mm
4. Calculate C Length:
   C = 2 x (Plate Thickness + Washer Thickness + Nut Thickness + Length of Thread Beyond End of Nut)
   C = 2 x (6 mm + 3 mm + 10 mm + 10 mm)
   C = 58mm
5. Add together A + B + C to find Rod Length:
   Rod Length
   = A + B + C
   = 195.275 mm + 114.3 mm + 58 mm
   = 367.575 mm

SHEET METAL AND WIRE GAGE - INCHES								
Gage Number	U.S. Standard Gage (1)	American Standard Wire Gage (A.W.G.) or Brown & Sharpe Gage (2)	U.S. Manufac-turer Standard Gage (M.S.G.) (3)	Twist Drill and Steel Wire Gage (4)	Birmingham or Stubs Iron Wire Gage (B.W.G.) (5)	Standard Wire Gage (S.W.G.) (British Imperial Wire Gage) (6)	American National Screw Gage Machine Screw (7)	American National Standard Gage Wood Screw (8)
0000	0.4063	0.4600	0.3886		0.454	0.400		
000	0.3750	0.4096	0.3587		0.425	0.372		
00	0.3438	0.3648	0.3288		0.380	0.348		
0	0.3125	0.3249	0.2990		0.340	0.324	0.0600	0.060
1	0.2813	0.2893	0.2690	0.2280	0.300	0.300	0.0730	0.073
2	0.2656	0.2576	0.2541	0.2210	0.284	0.276	0.0860	0.086
3	0.2500	0.2294	0.2391	0.2130	0.259	0.252	0.0990	0.099
4	0.2344	0.2043	0.2242	0.2090	0.238	0.232	0.1120	0.112
5	0.2188	0.1819	0.2092	0.2055	0.220	0.212	0.1250	0.125
6	0.2031	0.1620	0.1943	0.2040	0.203	0.192	0.1380	0.138
7	0.1875	0.1442	0.1793	0.2010	0.180	0.176		0.151
8	0.1719	0.1284	0.1644	0.1990	0.165	0.160	0.1640	0.164
9	0.1563	0.1144	0.1495	0.1960	0.148	0.144		0.177
10	0.1406	0.1018	0.1345	0.1935	0.134	0.128	0.1900	0.190
11	0.1250	0.0907	0.1196	0.1910	0.120	0.116		0.203
12	0.1094	0.0808	0.1046	0.1890	0.109	0.104	0.2160	0.216
13	0.0938	0.0719	0.0897	0.1850	0.095	0.092		

Table #110A - Sheet Metal & Wire Gage - Inches

SHEET METAL AND WIRE GAGE - INCHES								
Gage Number	U.S. Standard Gage (1)	American Standard Wire Gage (A.W.G.) or Brown & Sharpe Gage (2)	U.S. Manufacturer Standard Gage (M.S.G.) (3)	Twist Drill and Steel Wire Gage (4)	Birmingham or Stubs Iron Wire Gage (B.W.G.) (5)	Standard Wire Gage (S.W.G.) (British Imperial Wire Gage) (6)	American National Screw Gage Machine Screw (7)	American National Standard Gage Wood Screw (8)
14	0.0781	0.0640	0.0747	0.1820	0.083	0.080		0.242
15	0.0703	0.0570	0.0673	0.1800	0.072	0.072		
16	0.0625	0.0508	0.0598	0.1770	0.065	0.064		0.268
17	0.0563	0.0452	0.0538	0.1730	0.058	0.056		
18	0.0500	0.0403	0.0478	0.1695	0.049	0.048		0.294
19	0.0438	0.0358	0.0418	0.1660	0.042	0.040		
20	0.0375	0.0319	0.0359	0.1610	0.035	0.036		0.320
21	0.0344	0.0284	0.0329	0.1590	0.032	0.032		
22	0.0313	0.0253	0.0299	0.1570	0.028	0.028		
23	0.0281	0.0225	0.0269	0.1540	0.025	0.024		
24	0.0250	0.0201	0.0239	0.1520	0.022	0.022		0.372
25	0.0219	0.0179	0.0209	0.1495	0.020	0.020		
26	0.0188	0.0159	0.0179	0.1470	0.018	0.018		
27	0.0172	0.0141	0.0164	0.1440	0.016	0.0164		
28	0.0156	0.0126	0.0149	0.1405	0.014	0.0148		
29	0.0141	0.0112	0.0135	0.1360	0.013	0.0136		
30	0.0125	0.0100	0.0120	0.1285	0.012	0.0124		

Table #110B - Sheet Metal & Wire Gage - Inches

SHEET METAL AND WIRE GAGE - INCHES								
Gage Number	U.S. Standard Gage (1)	American Standard Wire Gage (A.W.G.) or Brown & Sharpe Gage (2)	U.S. Manufac-turer Standard Gage (M.S.G.) (3)	Twist Drill and Steel Wire Gage (4)	Birmingham or Stubs Iron Wire Gage (B.W.G.) (5)	Standard Wire Gage (S.W.G.) (British Imperial Wire Gage) (6)	American National Screw Gage Machine Screw (7)	American National Standard Gage Wood Screw (8)
31	0.0109	0.0089	0.0105	0.1200	0.100	0.0116		
32	0.0102	0.0079	0.0097	0.1160	0.009	0.0108		
33	0.0094	0.0070	0.0090	0.1130	0.008	0.0100		
34	0.0086	0.0063	0.0082	0.1110	0.007	0.0092		
35	0.0078	0.0056	0.0075	0.1100	0.005	0.0084		
36	0.0076	0.0050	0.0067	0.1065	0.004	0.0076		
37	0.0068	0.0044	0.0064	0.1040		0.0068		
38	0.0060	0.0039	0.0060	0.1015		0.0060		
39	0.0052	0.0035	0.0056	0.0995		0.0052		
40	0.0048	0.0031		0.0980		0.0048		

NOTE:
1. United States Standard Gage is an older gage method used for measuring sheet and plate iron and steel.
2. Used to measure some tubing wall thickness and for wires in all metals excluding iron or steel wire.
3. Newer gage method replacing the U.S. Standard Sheets. Based on a weight of 41.83 pounds per square foot per inch of thickness.
4. Extended number sizes, letters, inches and millimetres equivalent can be found in the Appendices under Standard Twist Drill Sizes.
5. Used for iron wire, types of seamless tubing, steel pressure tube and steel mechanical tube.
6. The Standard Wire Gage (S.W.G.) formerly known as the Imperial Wire Gage is the one wire gage that has been legally recognized in Great Britain.
7. Machine Screw size, customarily change to fractions of an inch after gage #12.

Table #110C - Sheet Metal & Wire Gage - Inches

SHEET METAL AND WIRE GAGE - MILLIMETRES								
Gage Number	U.S. Standard Gage (1)	American Standard Wire Gage (A.W.G.) or Brown & Sharpe Gage (2)	U.S. Manufacturer Standard Gage (M.S.G.) (3)	Twist Drill and Steel Wire Gage (4)	Birmingham or Stubs Iron Wire Gage (B.W.G.) (5)	Standard Wire Gage (S.W.G.) (British Imperial Wire Gage) (6)	American National Screw Gage Machine Screw (7)	American National Standard Gage Wood Screw (8)
0000	10.320	11.684	9.8704		11.531	10.160		
000	9.525	10.404	9.1110		10.795	9.4488		
00	8.733	9.265	8.3515		9.652	8.8392		
0	7.938	8.251	7.5946		8.636	8.2296	1.5240	1.5240
1	7.145	7.348	6.8326	5.7912	7.620	7.6200	1.8542	1.8542
2	6.746	6.543	6.4541	5.6134	7.2136	7.0104	2.1844	2.1844
3	6.350	5.827	6.0731	5.4102	6.5786	6.4008	2.5146	2.5146
4	5.954	5.189	5.6947	5.3086	6.0452	5.8928	2.8448	2.8448
5	5.558	4.620	5.3137	5.2197	5.5880	5.3848	3.1750	3.1750
6	5.159	4.115	4.9352	5.1816	5.1562	4.8768	3.5052	3.5052
7	4.763	3.663	4.5542	5.1054	4.5720	4.4704		3.8354
8	4.366	3.261	4.1758	5.0546	4.1910	4.0640	4.1656	4.1656
9	3.970	2.906	3.7973	4.9784	3.7592	3.6576		4.4958
10	3.571	2.586	3.4163	4.9149	3.4036	3.2512	4.8260	4.8260
11	3.175	2.304	3.0378	4.8514	3.0480	2.9464		
12	2.779	2.052	2.6568	4.8006	2.7686	2.6416	5.4864	5.4864
13	2.383	1.826	2.2784	4.6990	2.4130	2.3368		

Table #110D - Sheet Metal & Wire Gage - Millimetres

SHEET METAL AND WIRE GAGE - MILLIMETRES								
Gage Number	U.S. Standard Gage (1)	American Standard Wire Gage (A.W.G.) or Brown & Sharpe Gage (2)	U.S. Manufacturer Standard Gage (M.S.G.) (3)	Twist Drill and Steel Wire Gage (4)	Birmingham or Stubs Iron Wire Gage (B.W.G.) (5)	Standard Wire Gage (S.W.G.) (British Imperial Wire Gage) (6)	American National Screw Gage Machine Screw (7)	American National Standard Gage Wood Screw (8)
14	1.984	1.626	1.8974	4.6228	2.1082	2.0320		6.1468
15	1.786	1.448	1.7094	4.5720	1.8288	1.8288		
16	1.588	1.290	1.5189	4.4958	1.6510	1.6256		6.8072
17	1.430	1.148	1.3665	4.3942	1.4732	1.4224		
18	1.270	1.024	1.2141	4.3053	1.2446	1.2192		7.4676
19	1.113	0.909	1.0617	4.2164	1.0668	1.0160		
20	0.953	0.810	0.9119	4.0894	0.8890	0.9144		8.128
21	0.874	0.721	0.8357	4.0386	0.8128	0.8128		
22	0.795	0.643	0.7595	3.9878	0.7112	0.7112		
23	0.714	0.572	0.6833	3.9116	0.6350	0.6096		
24	0.635	0.511	0.6071	3.8608	0.5588	0.5588		9.4488
25	0.556	0.455	0.5309	3.7973	0.5080	0.5080		
26	0.478	0.404	0.4547	3.7338	0.4572	0.4572		
27	0.437	0.358	0.4166	3.6576	0.4064	0.4166		
28	0.396	0.320	0.3785	3.5687	0.3556	0.3759		
29	0.358	0.285	0.3429	3.4544	0.3302	0.3454		
30	0.318	0.254	0.3048	3.2639	0.3048	0.3150		

Table #110E - Sheet Metal & Wire Gage - Millimetres

SHEET METAL AND WIRE GAGE - MILLIMETRES

Gage Number	U.S. Standard Gage (1)	American Standard Wire Gage (A.W.G.) or Brown & Sharpe Gage (2)	U.S. Manufacturer Standard Gage (M.S.G.) (3)	Twist Drill and Steel Wire Gage (4)	Birmingham or Stubs Iron Wire Gage (B.W.G.) (5)	Standard Wire Gage (S.W.G.) (British Imperial Wire Gage) (6)	American National Screw Gage Machine Screw (7)	American National Standard Gage Wood Screw (8)
31	0.277	0.226	0.2667	3.0480	0.2540	0.2946		
32	0.259	0.259	0.2464	2.9464	0.2286	0.2743		
33	0.239	0.178	0.2286	2.8702	0.2032	0.2540		
34	0.218	0.160	0.2083	2.8194	0.1778	0.2337		
35	0.198	0.142	0.1905	2.7940	0.1270	0.2134		
36	0.193	0.127	0.1702	2.7051	0.1016	0.1930		
37	0.173	0.112	0.1626	2.6416		0.1727		
38	0.152	0.099	0.1524	2.5781		0.1522		
39	0.132	0.089	0.1422	2.5273		0.1321		
40	0.122	0.079		2.4892		0.1219		

NOTE:
1. United States Standard Gage is an older gage method used for measuring sheet and plate iron and steel.
2. Used to measure some tubing wall thickness and for wires in all metals excluding iron or steel wire.
3. Newer gage method replacing the U.S. Standard Sheets. Based on a weight of 41.83 pounds per square foot per inch of thickness.
4. Extended number sizes, letters, inches and millimetres equivalent can be found in the Appendices under Standard Twist Drill Sizes.
5. Used for iron wire, types of seamless tubing, steel pressure tube and steel mechanical tube.
6. The Standard Wire Gage (S.W.G.) formerly known as the Imperial Wire Gage is the one wire gage that has been legally recognized in Great Britain.
7. Machine Screw size, customarily change to fractions of an inch after gage #12.

Table #110F - Sheet Metal & Wire Gage - Millimetres

METAL PROPERTIES								
Metal	Symbol	Specific Gravity	Specific Heat (Btu/lb °F)	Melting Point °C	°F	Weight lb/ft²	Mass kg/m³	Expansion Coefficient
Aluminum (Cast)	Al	2.56	0.2185	658	1217	159.67	2557.7	12.5
Aluminum (Rolled)	Al	2.71	-		-	169.00	2707.15	
Antimony	Sb	6.71	0.051	630	1166	418.87	6709.73	6.3
Bismuth	Bi	9.8	0.031	271	520	611.71	9798.77	7.5
Boron	B	2.3	0.3091	2300	4172	143.60	2300.28	1.1
Brass	-	8.51	0.094	-	-	531.36	8511.67	11.0
Cadmium	Cd	8.6	0.057	321	610	536.89	8600.25	16.6
Calcium	Ca	1.57	0.17	810	1490	97.98	1569.51	13.9
Carbon	C	2.22	0.165	-	-	138.59	2220.02	0.67
Chromium	Cr	6.8	0.12	1510	2750	424.57	6801.04	4.5
Cobalt	Co	8.5	0.11	1490	2714	530.67	8500.61	6.7
Copper	Cu	8.89	0.094	1083	1982	555.03	8890.83	9.1
Columbium		8.57	-	1950	3542	534.99	8569.82	4.0
Gold	Au	19.32	0.032	1063	1945	1205.97	19,318.01	8.0
Iridium	Ir	22.42	0.033	2300	4170	1399.51	22,418.26	3.5
Iron	Fe	7.86	0.11	1520	2768	455.16	7291.05	6.6
Iron (Cast)	Fe	7.218	0.1298	1375	2507	450.14	7210.63	
Iron (Wrought)	Fe	7.7	0.1138	1500 -1600	2732 -2912	480.21	7692.31	
Lead	Pb	11.37	0.031	327	621	709.86	11,370.99	16.4
Lithium	Li	0.057	0.941	186	367	36.81	589.65	31.0
Magnesium	Mg	1.74	0.25	651	1204	108.69	1741.07	14.3
Manganese	Mn	8	0.12	1225	2237	499.39	7999.55	12.8

Table #111A - Metal Properties

METAL PROPERTIES								
Metal	Symbol	Specific Gravity	Specific Heat (Btu/lb °F)	Melting Point °C	°F	Weight lb/ft²	Mass kg/m³	Expansion Coefficient
Mercury	Hg	13.59	0.032	38.7	37.7	848.28	13588.30	
Molybdenum	Mo	10.2	0.0647	2620	4748	635.9	10186.25	3.05
Monel Metal	-	8.87	0.127	1360	2480	552.96	8857.67	
Nickel	Ni	8.8	0.13	1452	2646	551.23	8829.96	7.6
Phosphorus	P	1.82	0.177	43	111.4	113.53	1818.60	69.0
Platinum	Pt	21.5	0.033	1755	3191	1342.14	21499.27	4.3
Potassium	K	0.87	0.17	62	144	54.26	869.17	46.0
Selenium	Se	4.81	0.084	220	428	300.67	4816.32	20.6
Silicon	Si	2.4	0.1762	1427	2600	150.34	2408.24	
Silver	Ag	10.53	0.056	961	1761	657.5	10532.26	
Sodium	Na	0.97	0.29	97	207	60.48	968.81	39.5
Steel	-	7.858	0.1175	1330 -1378	2372 -2532	490.58	7858.43	6.5
Strontium	Sr	2.54	0.074	-	-	158.63	2541.04	
Sulphur	S	2.07	0.175	115	235.4	129.6	2076.02	
Tantalum	Ta	10.8	-	2850	5160	674.27	10800.89	3.6
Tin		7.29	0.056	232	450	455.16	7291.05	12.7
Titanium	Ti	5.3	0.13	1900	3450	330.91	5300.73	3.9
Tungsten	W	19.1	0.033	3000	5432	1192.32	19099.35	2.2
Uranium	U	18.7	-	-	-	1167.26	18697.92	-
Vanadium	V	5.5	-	1730	3146	343.35	5500.00	-
Zinc	Zn	7.19	0.094	419	786	448.93	7191.25	16.5

** Note: Expansion Coefficient per °F x 10^{-6}

Table #111B - Metal Properties

H-1 Standard Twist Drill Sizes

Inch	mm	Wire Gage	Decimals of an inch
		80	0.0135
		79	0.0145
1/64			0.0156
	0.4		0.0157
		78	0.0160
		77	0.0180
	0.5		0.0197
		76	0.0200
		75	0.0210
	0.55		0.0217
		74	0.0225
	0.6		0.0236
		73	0.0240
		72	0.0250
	0.65		0.0256
		71	0.0260
	0.7		0.0276
		70	0.0280
		69	0.0293
	0.75		0.0295
		68	0.0310
1/32			0.0313
	0.8		0.0315
		67	0.0320
		66	0.0330
	0.85		0.0335
		65	0.0350
	0.9		0.0354
		64	0.0360
		63	0.0370
	0.95		0.0374
		62	0.0380
		61	0.0390
	1		0.0394
		60	0.0400
		59	0.0410
	1.05		0.0413
		58	0.0420
		57	0.0430
	1.1		0.0433
	1.15		0.0453
		56	0.0465
3/64			0.0469
	1.2		0.0472
	1.25		0.0492
	1.3		0.0512
		55	0.0520
	1.35		0.0531
		54	0.0550
	1.4		0.0551
	1.45		0.0571
	1.5		0.0591
		53	0.0595
	1.55		0.0610
1/16			0.0625
	1.6		0.0630
		52	0.0635
	1.65		0.0650
	1.7		0.0669
		51	0.0670
	1.75		0.0689
		50	0.0700
	1.8		0.0709
	1.85		0.0728
		49	0.0730
	1.9		0.0748
		48	0.0760
	1.95		0.0768
5/64			0.0781
		47	0.0785
	2		0.0787
	2.05		0.0807
		46	0.0810
		45	0.0820
	2.1		0.0827
	2.15		0.0846
		45	0.0860
	2.2		0.0866
	2.25		0.0886
		43	0.0890
	2.3		0.0906
	2.35		0.0925
		42	0.0935
3/32			0.0938
	2.4		0.0945
		41	0.0960
	2.45		0.0966
		40	0.0980
	2.5		0.0984
		39	0.0995
		38	0.1015
	2.6		0.1024
		37	0.1040
	2.7		0.1063
		36	0.1065
	2.75		0.1083
7/64			0.1094
		35	0.1100
	2.8		0.1102
		34	0.1110

Table #112A - Standard Twist Drill Sizes

Inch	mm	Wire Gage	Decimals of an inch
		33	0.1130
	2.9		0.1142
		32	0.1160
	3		0.1181
		31	0.1200
	3.1		0.1220
1/8			0.1250
	3.2		0.1260
	3.25		0.1280
		30	0.1285
	3.3		0.1299
	3.4		0.1339
		29	0.1360
	3.5		0.1378
		28	0.1405
9/64			0.1406
	3.6		0.1417
		27	0.1440
	3.7		0.1457
		26	0.1470
	3.75		0.1476
		25	0.1495
	3.8		0.1496
		24	0.1520
	3.9		0.1535
		23	0.1540
5/32			0.1563
		22	0.1570
	4		0.1575
		21	0.1590
		20	0.1610
	4.1		0.1614
	4.2		0.1654
		19	0.1660
	4.25		0.1673
	4.3		0.1693
		18	0.1695
11/64			0.1719
		17	0.1730
	4.4		0.1732
		16	0.1770
	4.5		0.1772
		15	0.1800
	4.6		0.1811
		14	0.1820
		13	0.1850
	4.7		0.1850
	4.75		0.1870
3/16			0.1875
	4.8		0.1890
		12	0.1890
		11	0.1910
	4.9		0.1929
		10	0.1935
		9	0.1960
	5		0.1969
		8	0.1990
	5.1		0.2008
		7	0.2010
13/64			0.2031
		6	0.2040
	5.2		0.2047
		5	0.2055
	5.25		0.2067
	5.3		0.2087
		4	0.2090
	5.4		0.2126
		3	0.2130
	5.5		0.2165
7/32			0.2188
	5.6		0.2205
		2	0.2210
	5.7		0.2244
	5.75		0.2264
		1	0.2280
	5.8		0.2283
	5.9		0.2323
		A	0.2340
15/64			0.2344
	6		0.2362
		B	0.2380
	6.1		0.2402
		C	0.2420
	6.2		0.2441
		D	0.2460
	6.25		0.2461
	6.3		0.2480
1/4		E	0.2500
	6.4		0.2520
	6.5		0.2559
		F	0.2570
	6.6		0.2598
		G	0.2610
	6.7		0.2638
17/64			0.2656
	6.75		0.2657
		H	0.2660
	6.8		0.2677
	6.9		0.2717
		I	0.2720

Table #112B - Standard Twist Drill Sizes

H-1 Standard Twist Drill Sizes

Inch	mm	Wire Gage	Decimals of an inch
	7		0.2756
		J	0.2770
	7.1		0.2795
		K	0.2810
9/32			0.2812
	7.2		0.2835
	7.25		0.2854
	7.3		0.2874
		L	0.2900
	7.4		0.2913
		M	0.2950
	7.5		0.2953
19/64			0.2969
	7.6		0.2992
		N	0.3020
	7.7		0.3031
	7.75		0.3051
	7.8		0.3071
	7.9		0.3110
5/16			0.3125
	8		0.3150
		O	0.3160
	8.1		0.3189
	8.2		0.3228
		P	0.3230

Inch	mm	Wire Gage	Decimals of an inch
	8.25		0.3248
	8.3		0.3268
21/64			0.3281
	8.4		0.3307
		Q	0.3320
	8.5		0.3346
	8.6		0.3386
		R	0.3390
	8.7		0.3425
11/32			0.3438
	8.75		0.3345
	8.8		0.3465
		S	0.3480
	8.9		0.3504
	9		0.3543
		T	0.3580
	9.1		0.3583
23/64			0.3594
	9.2		0.3622
	9.25		0.3642
	9.3		0.3661
		U	0.3680
	9.4		0.3701
	9.5		0.3740
3/8			0.3750

Inch	mm	Wire Gage	Decimals of an inch
		V	0.3770
	9.6		0.3780
	9.7		0.3819
	9.75		0.3839
	9.8		0.3858
		W	0.3860
	9.9		0.3898
25/64			0.3906
	10		0.3937
		X	0.3970
		Y	0.4040
13/32			0.4063
		Z	0.4130
	10.5		0.4134
27/64			0.4219
	11		0.4331
7/16			0.4375
	11.5		0.4528
29/64			0.4531
15/32			0.4688
	12		0.4724
31/64			0.4844
	12.5		0.4921
1/2			0.5000
	13		0.5118

Inch	mm	Wire Gage	Decimals of an inch
33/64			0.5156
17/32			0.5313
	13.5		0.5315
35/64			0.5469
	14		0.5512
9/16			0.5625
	14.5		0.5709
37/64			0.5781
	15		0.5906
19/32			0.5938
39/64			0.6094
	15.5		0.6102
5/8			0.6250
	16		0.6299
41/64			0.6406
	16.5		0.6496
21/32			0.6563
	17		0.6693
43/64			0.6719
11/16			0.6875
	17.5		0.6890
45/64			0.7031
	18		0.7087
23/32			0.7188
	18.5		0.7283

Table #112C - Standard Twist Drill Sizes

Inch	mm	Wire Gage	Decimals of an inch
47/64			0.7344
	19		0.7480
3/4			0.7500
49/64			0.7656
	19.5		0.7677
25/32			0.7812
	20		0.7874
51/64			0.7969
	20.5		0.8071
13/16			0.8125
	21		0.8268
53/64			0.8281
27/32			0.8438
	21.5		0.8465
55/64			0.8594
	22		0.8661
7/8			0.8750
	22.5		0.8858
57/64			0.8906
	23		0.9055
29/32			0.9063
59/64			0.9219
	23.5		0.9252
15/16			0.9375
	24		0.9449
61/64			0.9531
	24.5		0.9646
31/32			0.9688
	25		0.9843
63/64			0.9844
1			1.0000
	25.5		1.0039
1 1/64			1.0156
	26		1.0236
1 1/32			1.0313
	26.5		1.0433
1 3/64			1.0469
1 1/16			1.0625
	27		1.0630
1 5/64			1.0781
	27.5		1.0827
1 3/32			1.0938
	28		1.1024
1 7/64			1.1094
	28.5		1.1220
1 1/8			1.1250
1 9/64			1.1406
	29		1.1417
1 5/32			1.1562
	29.5		1.1614
1 11/64			1.1719
	30		1.1811
1 3/16			1.1875
	30.5		1.2008
1 13/64			1.2031
1 7/32			1.2188
	31		1.2205
1 15/64			1.2344
	31.5		1.2402
1 1/4			1.2500
	32		1.2598
1 17/64			1.2656
	32.5		1.2795
1 9/32			1.2813
1 19/64			1.2969
	33		1.2992
1 5/16			1.3125
	33.5		1.3189
1 21/64			1.3281
	34		1.3386
1 11/32			1.3438
	34.5		1.3583
1 23/64			1.3594
1 3/8			1.3750
	35		1.3780
1 25/64			1.3906
	35.5		1.3976
1 13/32			1.4063
	36		1.4173
1 27/64			1.4219
	36.5		1.4370

Table #112D - Standard Twist Drill Sizes

INCH AND MILLIMETRE EQUIVALENT OF LETTER TWIST DRILL SIZE					
Drill Size	Inch	mm	Drill Size	Inch	mm
A	0.234	5.944	N	0.302	7.671
B	0.238	6.045	O	0.316	8.026
C	0.242	6.147	P	0.323	8.204
D	0.246	6.248	Q	0.332	8.433
E	0.250	6.350	R	0.339	8.611
F	0.257	6.528	S	0.348	8.839
G	0.261	6.629	T	0.358	9.093
H	0.266	6.756	U	0.368	9.347
I	0.272	6.909	V	0.377	9.576
J	0.277	7.036	W	0.386	9.804
K	0.281	7.137	X	0.397	10.084
L	0.290	7.366	Y	0.404	10.262
M	0.295	7.493	Z	0.413	10.490

Table #113 - Equivalent Letter Table

Tap Size	Pitch	Form	75%
0	80	NF	3/64
1	64	NC	53
	72	NF	53
	56	NS	54
2	56	NC	50
	64	NF	50
3	48	NC	47
	56	NF	45
4	40	NC	43
	48	NF	42
	32	NS	45
	36	NS	44
5	40	NC	38
	44	NF	37
6	32	NC	36
	40	NF	33
	36	NS	34
8	32	NC	29
	36	NF	29
	40	NS	28
10	24	NC	25
	32	NF	21

Tap Size	Pitch	Form	75%
	30	NS	22
12	24	NC	16
	28	NF	14
	32	NS	13
14	20	NS	10
	24	NS	7
1/4	20	NC	7
	28	NF	3
5/16	18	NC	F
	24	NF	I
3/8	16	NC	5/16
	24	NF	Q
7/16	14	NC	U
	20	NF	25/64
1/2	13	NC	27/64
	20	NF	29/64
9/16	12	NC	31/64
	18	NF	33/64
5/8	11	NC	17/32
	18	NF	37/64
11/16	11	NS	19/32
	16	NS	5/8

Tap Size	Pitch	Form	75%
3/4	10	NC	21/32
	16	NF	11/16
7/8	9	NC	49/64
	14	NF	13/16
1	8	NC	7/8
	12	NF	59/64
	14	NS	15/16
1 1/8	7	NC	63/64
	12	NF	1 3/64
1 1/4	7	NC	1 7/64
	12	NF	1 11/64
1 3/8	6	NC	1 7/32
	12	NF	1 19/64
1 1/2	6	NC	1 11/32
	12	NF	1 27/64
1 5/8	5 1/2	NS	1 29/64
1 3/4	5	NC	1 9/16
1 7/8	5	NS	1 11/16
2	4 1/2	NC	1 25/32
1/16	27	NPT	R
1/8	27	NPT	R

Tap Size	Pitch	Form	75%
1/4	18	NPT	7/16
3/8	18	NPT	37/64
1/2	14	NPT	23/32
3/4	14	NPT	59/64
1	11 1/2	NPT	1 5/32
1 1/4	11 1/2	NPT	1 1/2
1 1/2	11 1/2	NPT	1 47/64
2	11 1/2	NPT	2 7/32

Self Tapping Screws		
Size	Decimal	Drill
No. 4	0.112	5/64
No. 6	0.138	3/32
No. 7	0.155	7/64
No. 8	0.165	1/8
No. 10	0.191	9/64
No. 12	0.218	5/32
No. 14	0.251	3/16

Table #114A - Imperial Tap Drill and Self Tapping Screw Sizes

H-3 Tap Drill Sizes (Imperial)

Tap Size	Thread Pitch mm	Tap Drill mm	Alt. Tap Drill
I.S.O. Metric Coarse			
1.6	0.35	1.25	3/64
1.7	0.35	1.35	55
1.8	0.35	1.45	54
2	0.40	1.60	1/16
2.2	0.45	1.75	50
2.3	0.40	1.90	49
2.5	0.45	2.05	46
2.6	0.45	2.15	44
3	0.50	2.50	40
3.5	0.60	2.90	33
4	0.70	3.30	30
4.5	0.75	3.70	27
5	0.80	4.20	19
5.5	0.90	4.60	15
6	1.00	5.00	9
7	1.00	6.00	15/64
8	1.25	6.80	H
9	1.25	7.80	5/16
10	1.50	8.50	Q
11	1.50	9.50	3/8

Tap Size	Thread Pitch mm	Tap Drill mm	Alt. Tap Drill
12	1.75	10.20	Y
14	2.00	12.00	15/32
16	2.00	14.00	35/64
18	2.50	15.50	39/64
20	2.50	17.50	11/16
22	2.50	19.50	49/64
24	3.00	21.00	53/64
27	3.00	24.00	61/64
30	3.50	26.50	1 3/64
33	3.50	29.50	1 5/32
36	4.00	32.00	1 1/4
36	4.00	35.00	1 3/8
I.S.O. Metric Fine			
3	0.35	2.65	37
4	0.35	3.65	27
4	0.50	3.50	29
4.5	0.45	4.05	21
5	0.50	4.50	16
5	0.70	4.30	18
5	0.75	4.25	18

Tap Size	Thread Pitch mm	Tap Drill mm	Alt. Tap Drill
5.5	0.50	5.00	9
6	0.50	5.50	7/32
6	0.75	5.25	5
7	0.75	6.25	D
8	0.5	7.50	M
8	1.00	7.00	J
9	0.50	8.50	Q
9	1.00	8.00	O
10	0.50	9.50	3/8
10	0.75	9.25	U
10	1.00	9.00	T
10	1.25	8.75	11/32
11	1.00	10.00	X
12	1.00	1/8 1.00	7/16
12	1.25	10.75	27/64
12	1.50	10.50	Z
13	1.50	11.50	29/64
13	1.75	11.25	7/16
14	1.25	12.75	1/2
14	1.50	12.50	31/64
15	1.50	13.50	17/32

Tap Size	Thread Pitch mm	Tap Drill mm	Alt. Tap Drill
16	1.00	15.00	19/32
16	1.25	14.75	37/64
16	1.50	14.50	9/16
18	1.00	17.00	43/64
18	1.25	16.75	21/32
18	1.50	16.50	41/64
18	2.00	16.00	5/8
20	1.00	19.00	3/4
20	1.50	18.50	47/64
20	2.00	18.00	45/64
22	1.00	21.00	53/64
22	1.50	20.50	13/16
22	2.00	20.00	25/32
24	1.00	23.00	29/32
24	1.50	22.50	7/8
24	2.00	22.00	55/64
24	2.50	21.50	27/32

Table #114B - Metric Tap Drill Sizes

Pipe Tap Sizes (NPT)		
Nom. Pipe Size inches	Threads Per Inch (TPI) Pitch	Tap Drill Size inches
1/8	27	11/32
1/4	18	7/16
3/8	18	19/32
1/2	14	23/32
3/4	14	15/16
1	11 1/2	1 5/32
1 1/4	11 1/2	1 11/32
1 1/2	11 1/2	1 23/32
2	11 1/2	2 3/16
2 1/2	8	2 9/16
3	8	3 3/16
4	8	4 3/16
5	8	5 5/16
6	8	6 5/16

Table #115 - Pipe Tap Sizes

Note:

Screw sizing of #0 = 0.06 inches

Each # (number) increases the size by 0.013 inches

Example:

#10 screw = 10 x 0.013 + 0.06 = 0.190 inches

Fractions	Decimals Inch	mm
	0.00394	0.1
	0.00787	0.2
	0.01	0.254
	0.01181	0.3
1/64	0.015625	0.3969
	0.01575	0.4
	0.01969	0.5
	0.02	0.508
	0.02362	0.6
	0.02756	0.7
	0.03	0.762
1/32	0.03125	0.7938
	0.0315	0.8
	0.03543	0.9
	0.03937	1.0
	0.04	1.016
3/64	0.046875	1.1906
	0.05	1.27
	0.06	1.524
1/16	0.0625	1.5875
	0.07	1.778
5/64	0.078125	1.9844
	0.07874	2.0
	0.08	2.032
	0.09	2.286

Fractions	Decimals Inch	mm
3/32	0.09375	2.3812
	0.1	2.54
7/64	0.109375	2.7781
	0.11	2.794
	0.11811	3.0
	0.12	3.048
1/8	0.125	3.175
	0.13	3.302
	0.14	3.556
9/64	0.140625	3.5719
	0.15	3.810
5/32	0.15625	3.9688
	0.15748	4.0
	0.16	4.064
	0.17	4.318
11/64	0.171875	4.3656
	0.18	4.572
3/16	0.1875	4.7625
	0.19	4.826
	0.19685	5.0
13/64	0.2	5.08
	0.203125	5.1594
	0.21	5.334
7/32	0.21875	5.5562
	0.22	5.588

Fractions	Decimals Inch	mm
	0.23	5.842
15/64	0.234375	5.9531
	0.23622	6.0
	0.24	6.096
1/4	0.25	6.35
	0.26	6.604
17/64	0.265625	6.7469
	0.27	6.858
	0.27559	7.0
	0.28	7.112
9/32	0.28125	7.1438
	0.29	7.366
19/64	0.296875	7.5406
	0.30	7.62
	0.31	7.874
5/16	0.3125	7.9375
	0.31496	8.0
	0.32	8.128
21/64	0.32815	8.3344
	0.33	8.382
	0.34	8.636
11/32	0.34375	8.7312
	0.35	8.89
	0.35433	9.0
23/64	0.359375	9.1281

Fractions	Decimals Inch	mm
	0.36	9.144
	0.37	9.398
3/8	0.375	9.525
	0.38	9.652
	0.39	9.906
25/64	0.390625	9.9219
	0.39370	10.0
	0.40	10.16
13/32	0.40625	10.3188
	0.41	10.414
	0.42	10.668
27/64	0.421875	10.7156
	0.43	10.922
	0.43307	11.0
7/16	0.4375	11.1125
	0.44	11.176
	0.45	11.430
29/64	0.453125	11.5094
	0.46	11.684
15/32	0.46875	11.9062
	0.47	11.938
	0.47244	12.0
	0.48	12.192
31/64	0.484375	12.3031
	0.49	12.446

Table #116A - Millimetres, Fractions, Decimal Inch Equivalents

Fractions	Decimals Inch	mm
1/2	0.50	12.7
	0.51	12.954
	0.51181	13.0
33/64	0.515625	13.0969
	0.55	13.970
	0.55118	14.0
	0.56	14.224
9/16	0.5625	14.2875
	0.57	14.478
37/64	0.578125	14.6844
	0.58	14.732
	0.59	14.986
	0.59055	15.0
19/32	0.59375	15.0812
	0.60	15.24
39/64	0.609375	15.4781
	0.61	15.494
	0.62	15.748
5/8	0.625	15.875
	0.62992	16.0
	0.63	16.002
	0.64	16.256
41/64	0.640625	16.2719

Fractions	Decimals Inch	mm
	0.65	16.510
21/32	0.65625	16.6688
	0.66	16.764
	0.66929	17.0
	0.67	17.018
43/64	0.671875	17.0656
	0.68	17.272
11/16	0.6875	17.4625
	0.69	17.526
	0.70	17.78
45/64	0.703125	17.8594
	0.70866	18.0
	0.71	18.034
23/32	0.71875	18.2562
	0.72	18.288
	0.73	18.542
47/64	0.734375	18.6531
	0.74	18.796
	0.74803	19.0
3/4	0.75	19.050
	0.76	19.304
49/64	0.765625	19.4469
	0.77	19.558

Fractions	Decimals Inch	mm
	0.78	19.812
25/32	0.78125	19.8438
	0.78740	20.0
	0.79	20.066
51/64	0.796875	20.2406
	0.80	20.320
	0.81	20.574
13/16	0.8125	20.6375
	0.82	20.828
	0.82677	21.0
53/64	0.828125	21.0344
	0.83	21.082
	0.84	21.336
27/32	0.84375	21.4312
	0.85	21.590
55/64	0.859375	21.8281
	0.86	21.844
	0.86614	22.0
	0.87	22.098
7/8	0.875	22.225
	0.88	22.352
	0.89	22.606
57/64	0.890625	22.6219

Fractions	Decimals Inch	mm
	0.90	22.860
	0.90551	23.0
29/32	0.90625	23.0188
	0.91	23.114
	0.92	23.368
59/64	0.921875	23.4156
	0.93	23.622
15/16	0.9375	23.8125
	0.94	23.876
	0.94488	24.0
	0.95	24.130
61/64	0.953125	24.2094
	0.96	24.384
31/32	0.96875	24.6062
	0.97	24.638
	0.98	24.892
	0.984375	25.0031
	0.99	25.146
1	1.00000	25.4000

Table #116B - Millimetres, Fractions, Decimal Inch Equivalents

Useful Formulas

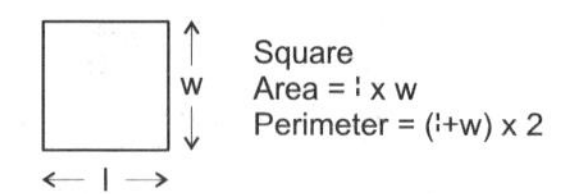

Square
Area = l x w
Perimeter = (l+w) x 2

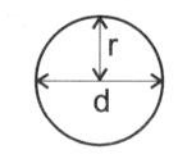

Circle
Area = $\pi \times r^2$ or $\frac{\pi \times d^2}{4}$
Perimeter = π x d

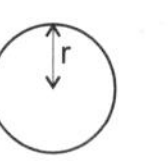

Sphere
Volume = 4/3 x πr^3

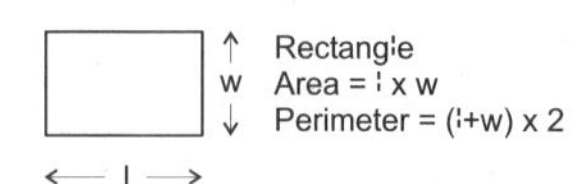

Rectangle
Area = l x w
Perimeter = (l+w) x 2

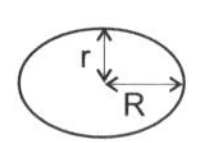

Ellipse
Area = π x R x r
Perimeter = $6.283 \times \sqrt{\frac{R^2 + r^2}{2}}$

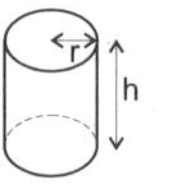

Cylinder
Volume = $\pi r^2 \times h$

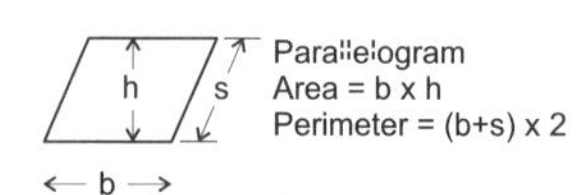

Parallelogram
Area = b x h
Perimeter = (b+s) x 2

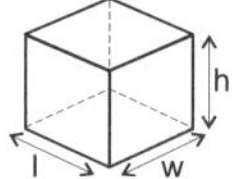

Cube or Rectangular solid
Volume = l x w x h

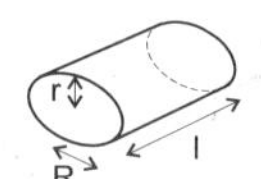

Elliptical Tank
Volume = π x r x R x l

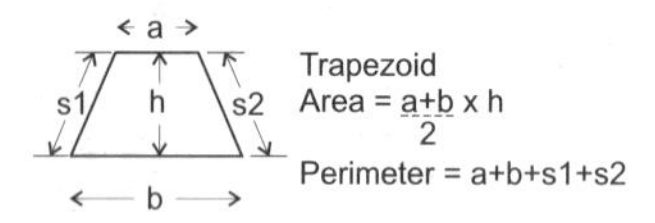

Trapezoid
Area = $\frac{a+b}{2} \times h$
Perimeter = a+b+s1+s2

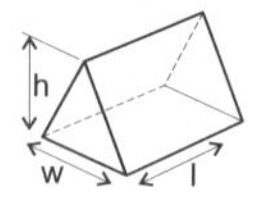

Prism
Volume = $\frac{(l \times w \times h)}{2}$

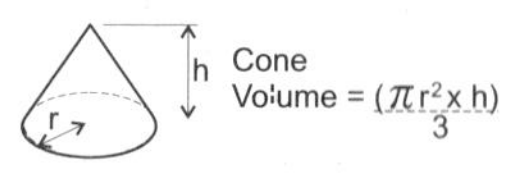

Cone
Volume = $\frac{(\pi r^2 \times h)}{3}$

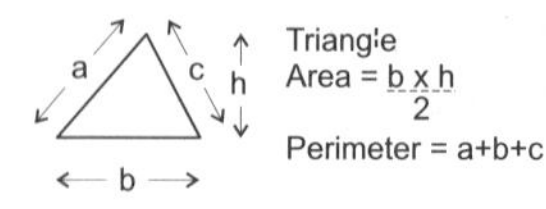

Triangle
Area = $\frac{b \times h}{2}$
Perimeter = a+b+c

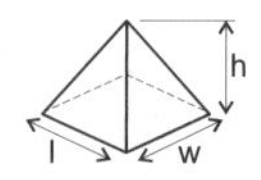

Pyramid
Volume = $\frac{(l \times w \times h)}{3}$

π = 3.1416

Table 117 - Useful Mathematical Formulas

Use of Equivalent Tables

In using the equivalent tables, units in the left hand column are equivalent to the number under each unit across the top of the table.

Example: In the length equivalent table, 1 yard in the left hand column is equivalent to 36 inches located under the Inches units column.

When a number of each unit must be converted to another unit, multiply the number of the particular unit times the number found in the table to obtain the needed unit on the top of the table.

Example: 5 yards x 36 inches = 180 inches

Scientific Notation

In the equivalent tables, scientific notation is used to express a number when the amount of zeros needed are excessive (very small or very large in magnitude).

To change a number from scientific notation to ordinary terms, simply move the decimal point either left or right as dictated by the exponent.

Positive exponents reflect the number of places the decimal point moves to the right.

Example: 9.53×10^{4} to ordinary terms, move the decimal point 4 places to the right and add the needed zeros:

$9.53 \times 10^{4} = 95,300$

Negative exponent reflects the number of places the decimal point moves to the left.

Example: 9.53×10^{-4} to ordinary terms, move the decimal point 4 places to the left and add the needed zeros:

$9.53 \times 10^{-4} = 0.000953$

K-1 Unit Equivalent Tables

	mm	cm	m	km	Inches	Feet	Yards	Miles
mm	1	0.1	0.001	1×10^{-6}	39.37×10^{-3}	3.2808×10^{-3}	1.0936×10^{-3}	6.214×10^{-7}
cm	10	1	0.01	1×10^{-5}	39.37×10^{-2}	3.2808×10^{-2}	1.0936×10^{-2}	6.214×10^{-6}
m	1000	100	1	0.001	39.37	3.2808	1.0936	6.214×10^{-4}
km	1×10^{6}	1×10^{5}	1000	1	39370	3281	1093.613	0.6214
Inches	25.4	2.54	0.0254	2.54×10^{-5}	1	8.33×10^{-2}	2.778×10^{-2}	1.578×10^{-5}
Feet	304.8	30.48	0.3048	3.048×10	12	1	0.3333	1.894×10^{-4}
Yards	914.4	91.44	0.9144	9.144×10^{-4}	36	3	1	5.682×10^{-4}
Miles	1609344	160934.4	1609.344	1.609344	63,360	5280	1760	1

Table #118A - Unit Equivalents (Length)

	mm^2	cm^2	m^2	km^2	Square Inches	Square Feet	Square Yards	Square Miles
mm^2	1	0.01	1×10^{-6}	1×10^{-12}	1.550×10^{-3}	1.076×10^{-5}	1.1×10^{-6}	3.861×10^{-13}
cm^2	100	1	0.0001	1×10^{-10}	0.155	10.76×10^{-4}	1.196×10^{-4}	3.861×10^{-11}
m^2	1×10^{6}	1×10^{5}	1	1×10^{-6}	1550	10.764	1.196	3.861×10^{-7}
km^2	1×10^{12}	1×10^{10}	1×10^{6}	1	155×10^{7}	10764×10^{-3}	1196×10^{3}	0.3861021
Sq. Inches	645.16	6.4516	6.451×10^{-4}	6.451×10^{-10}	1	6.944×10^{-3}	7.716×10^{-4}	2.491×10^{-10}
Sq. Feet	92903	929	0.09290	9.29×10^{-8}	144	1	0.1111	3.587×10^{-8}
Sq. Yards	836127	8.361	0.8361	8.361×10^{-7}	1296	9	1	3.2283×10^{-7}
Sq. Miles	2590×10^{9}	2590×10^{7}	2.590×10^{6}	2.590	4.01×10^{9}	27.88×10^{6}	3.098×10^{6}	1

Table #118B - Unit Equivalents (Area)

	Cubic cm	litres	Cubic m	Cubic Inches	Cubic Feet	U.S. Gallon	Imperial Gallon	U.S. Barrels
Cubic cm	1	1×10^{-3}	1×10^{-6}	61.024×10^{-3}	0.035315×10^{-3}	0.2642×10^{-3}	0.21997×10^{-3}	0.00629×10^{-3}
litres	1000	1	1×10^{-3}	61.024	0.0353	0.2642	0.22	0.00629
Cubic m	1×10^{6}	1000	1	61023.744	35.3147	264.2	219.9694	6.29
Cubic Inches	1.63871	0.0164	1.638706×10^{-5}	1	5.787×10^{-4}	4.329×10^{-3}	3.606×10^{-3}	1.03×10^{-4}
Cubic Feet	2.8317×10^{4}	2.8317×10	2.8317×10^{-2}	1728	1	7.48055	6.22888	0.1781
U.S. Gallon	3.785×10^{3}	3.78541	3.785×10^{-3}	231	0.1337	1	0.833	2.38×10^{-2}
Imperial Gallon	4.5461×10^{3}	4.54609	4.5461×10^{-3}	277.4193	0.1605	1.20095	1	2.877×10^{-2}
U.S. Barrels	158.98×10^{3}	158.98	0.15898	9701.856	5.6145	42	34.973	1

Table #118C - Unit Equivalents (Volume)

	litres/minute	Cubic m/hour	U.S. Gallons/minute	Can. Gallons/minute	Cubic in/minute	Cubic ft/minute	U.S. Barrels/Day
litres/minute	1	0.06	0.2642	0.220	61.024	0.0353	9.057
Cubic m/hour	16.667	1	4.403	3.6616	1017.062	0.5886	151
U.S. Gallons/min	3.7854	0.22712	1	0.8327	231.0	0.1337	342.72
Can. Gallons/min.	4.5461	0.2728	1.2009	1	277.4193	0.1605	414.288
Cubic in/min.	0.0164	9.832×10^{-4}	4.329×10^{-3}	3.604×10^{-3}	1	5.787×10^{-4}	1.4832
Cubic ft/min.	28.3169	1.699	7.4805	6.2288	1728	1	256.464
U.S. Barrells/Day	0.1104	6.624×10^{-3}	0.02917	0.02429	6.7374	3.899×10^{-3}	1

Table #118D - Unit Equivalents (Volume Flow)

K-1 Unit Equivalent Tables

	BTU	Foot Pounds	Horsepower Hours	Joules	Calorie	kilowatt Hour
BTU	1	777.9	3.929×10^{-4}	1055	252	2.93×10^{-4}
Foot Pounds	1.285×10^{-3}	1	5.051×10^{-7}	1.356	0.3239	3.766×10^{-7}
Horsepower Hrs.	2545	1.98×10^{6}	1	2.685×10^{6}	6.414×10^{5}	0.7457
Joules	9.48×10^{-4}	0.7376	3.725×10^{-7}	1	0.2389	2.778×10^{-7}
Calorie	3.968×10^{-3}	3.087	1.559×10^{-6}	4.186	1	1.163×10^{-6}
kilowatt Hr.	3413	2.655×10^{6}	1.341	3.6×10^{6}	8.60×10^{5}	1

Table #118E - Unit Equivalents (Energy)

	Ounce	Pound	Milligram	Gram	Kilogram	Short Ton	Long Ton	Metric Tonne
Ounce	1	0.0625	28344.67	28.3447	0.02835	3.125×10^{-5}	2.79×10^{-5}	2.835×10^{-5}
Pound	16	1	453257.8	453.2578	0.4536	0.0005	4.464×10^{-4}	4.536×10^{-4}
Milligram	35.28×10^{-6}	2.205×10^{-6}	1	0.001	0.001×10^{-3}	1.102×10^{-9}	9.842×10^{-10}	1×10^{-9}
Gram	35.28×10^{-3}	2.205×10^{-3}	1000	1	0.001	1.102×10^{-6}	9.842×10^{-7}	1×10^{-6}
Kilogram	35.28	2.205	1×10^{6}	1000	1	1.102×10^{-3}	9.842×10^{-4}	0.001
Short Ton	32000	2000	907.2×10^{6}	907.2×10^{3}	907.2	1	0.8929	0.907
Long Ton	35840	2240	1016×10^{6}	1016×10^{3}	1016	1.12	1	1.016
Metric Tonne	35280	2205	1×10^{9}	1×10^{6}	1000	1.103	0.9842	1

Table #118F - Unit Equivalents (Mass)

	kPa	Bar	Atmo-sphere	In W.C.	Ft. W.C.	m W.C.	In. Hg	mm Hg	PSI
kPa	1	0.01	9.869 x 10^{-3}	4.01463	0.334553	0.10197	0.2953	7.50064	0.14504
Bar	100	1	0.9869	401.463	33.4553	10.197	29.52998	750.0615	14.5038
Atmosphere	101.325	1.01325	1	406.78	33.8985	10.3323	29.92	760	14.696
In. W.C.	0.24909	2.49089 x 10^{-3}	0.00246	1	0.08333	0.0254	73.5559 x 10^{-3}	1.86833	36.12729 x 10^{-3}
Ft. W.C.	2.98907	29.8907 x 10^{-3}	0.0295	12	1	0.3048	0.88267	22.4199	0.43353
m W.C.	9.80665	0.09807	0.09678	39.3701	3.28084	1	2.8959	73.5561	1.42233
In. Hg	3.3864	33.86389 x 10^{-3}	0.03342	13.5951	1.132925	0.34532	1	25.4	0.49115
mm Hg	0.1333	1.33322 x 10^{-3}	1.31579 x 10^{-3}	0.53524	0.0446	0.0136	39.36996 x 10^{-3}	1	19.33672 x 10^{-3}
PSI	6.8947	0.06895	0.06805	27.6799	2.30666	0.70307	2.0360	51.7151	1

Table #118G - Unit Equivalents (Pressure)

Note:
See page 523 for an explanation of Scientific Notation for tables #118A to #118G.

Pressure Terms

Pressure is defined as force acting upon a unit area of surface. The relationship between pressure, force and area in both the Imperial and Metric systems are shown in table #119.

Related Pressure Terms

1. Absolute Pressure - is pressure above absolute zero or a perfect vacuum. It can be determined by:

- Absolute Pressure = Gage Pressure + Local Atmospheric Pressure
- Absolute Pressure = Local Atmospheric Pressure - Vacuum

$$Pressure = \frac{Force}{Area}$$

Imperial Units	Metric Units
$\frac{Pounds}{Square\ Inch} = Pounds\ per\ Square\ Inch$	$\frac{Newton}{Square\ Metre} = Pascal\ (Pa)$
or	or
$\frac{lbs}{in^2} = psi$	$\frac{N}{m^2} = Pa$
	or
	$1000\ Pa = KiloPascal\ (kPa)$

Table #119

2. Atmospheric Pressure is the pressure exerted by the atmosphere or the weight of air above the earth.

3. Barometric or Local Atmospheric Pressure is the pressure of the atmosphere above absolute zero value. This pressure will vary depending on altitude and climatic conditions.

4. Differential Pressure is the pressure difference between two pressures. It can be determined by:

- Differential Pressure = Pressure #1 - Pressure #2

5. Gage Pressure is pressure above local atmospheric pressure or the difference between absolute pressure and local atmospheric pressure. It can be determined by:

- Gage Pressure = Absolute Pressure - Local Atmospheric Pressure

6. Vacuum is pressure below atmospheric pressure.

It is important when referring to pressure that it be specified to a reference.

Standard Atmospheric Pressure

Standard atmospheric pressure or standard atmosphere can be stated as:

- 14.696 pounds per square inch (psi)
- 101.325 kilopascals (kPa)
- 29.92 inches of mercury (Hg)
- 760 millimetres of mercury (mm Hg) or torrs (Torr)
- 407 inches of water column (WC)
- 1.01325 bars or 1013.25 millibar

K-2 Pressure Terms

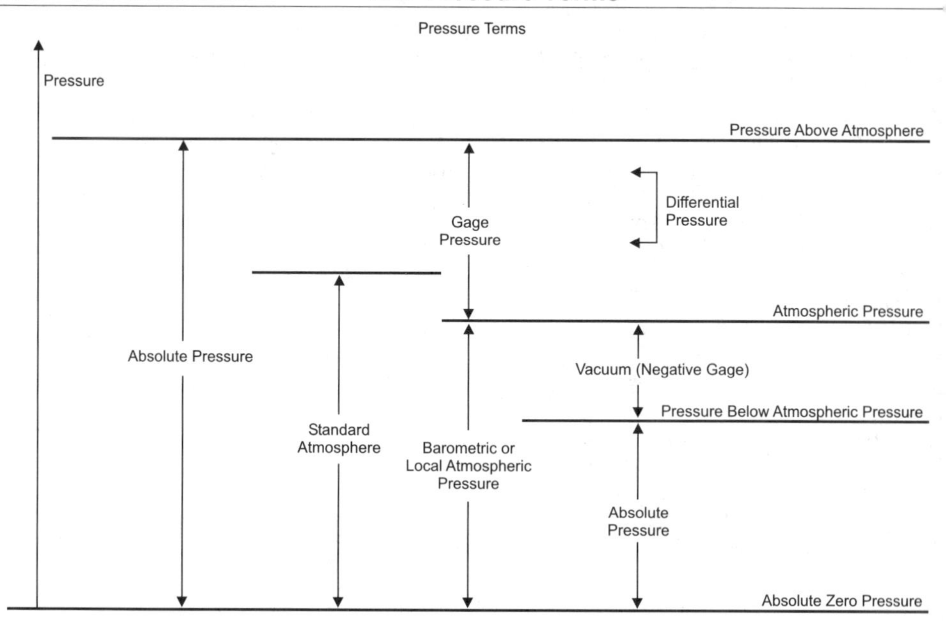

PSI to kPa Conversion Table

To use the following table, the known pressure in pounds per square inch (psi) is found in the left-hand column and the equivalent pressure in kilopascals (kPa) is found directly to the right.

To convert pressures greater than 100 psi or less than 1 psi, simply move the decimal point in the kPa unit either to the left or right as dictated by the psi unit.

Example:

To find the equivalent kPa unit that corresponds to 0.1 psi, simply move the decimal point one place to the left in the 1 psi column and one place left in the corresponding kPa column, thus 0.1 psi = 0.68948 kPa.

To find the equivalent kPa unit that corresponds to 1000 psi, simply move the decimal point three places to the right in the 1psi column and three places to the right in the corresponding kPa column, thus 1000 psi = 6894.8 kPa.

To convert units not found in the table, simply add the known units from the table together.

Example:

To find the equivalent kPa unit that corresponds to 178.54 psi

100.00 psi = 689.4757
78.00 psi = 537.7911
0.54 psi = 3.7232

178.54 psi = 1230.99 kPa

psi	kPa	psi	kPa	psi	kPa	psi	kPa	psi	kPa
1	6.8948	21	144.7899	41	282.685	61	420.5802	81	558.4753
2	13.7895	22	151.6847	42	289.5798	62	427.475	82	565.3701
3	20.6843	23	158.5794	43	296.4746	63	434.3697	83	572.2649
4	27.579	24	165.4742	44	303.3693	64	441.2645	84	579.1596
5	34.4738	25	172.3689	45	310.2641	65	448.1592	85	586.0544
6	41.3685	26	179.2637	46	317.1588	66	455.054	86	592.9491
7	48.2633	27	186.1584	47	324.0536	67	461.9487	87	599.8439
8	55.1581	28	193.0532	48	330.9484	68	468.8435	88	606.7386
9	62.0528	29	199.948	49	337.8431	69	475.7383	89	613.6334
10	68.9476	30	206.8427	50	344.7379	70	482.633	90	620.5282
11	75.8423	31	213.7375	51	351.6326	71	489.5278	91	627.4229
12	82.7371	32	220.6322	52	358.5274	72	496.4225	92	634.3177
13	89.6318	33	227.527	53	365.4221	73	503.3173	93	641.2124
14	96.5266	34	234.4217	54	372.3169	74	510.212	94	648.1072
15	103.4214	35	241.3165	55	379.2117	75	517.1068	95	655.0019
16	110.3161	36	248.2113	56	386.1064	76	524.0016	96	661.8967
17	117.2109	37	255.106	57	393.0012	77	530.8963	97	668.7915
18	124.1056	38	262.0008	58	399.8959	78	537.7911	98	675.6862
19	131.0004	39	268.8955	59	406.7907	79	544.6858	99	682.581
20	137.8951	40	275.7903	60	413.6854	80	551.5806	100	689.4757

Table 120 - psi to kPa Conversion

Pressure Head Conversion Table

To use the following table, the known pressure in pounds per square inch (psi) or height in feet of water column is first located in the **Bold** center column. The corresponding head in feet of water is found to the right and pressure in psi to the left.

Note: Table based on water @ 60°F (15.6 °C).

- To convert feet head of any liquid to pounds per square inch (psi):
  psi = H x 0.433 x R.D.

Where:

psi = Pressure in pounds per square inch.

H = Height of liquid in feet.

0.433 = Constant based on the height of one foot of water will exert a pressure of 0.433 psi.

R.D. = Relative Density or Specific Gravity.

- To convert psi to equivalent height in feet of liquid:

$$Height\ ft. = \frac{psi}{0.433 \times R.D. of\ liquid}$$

Pressure (psi)		Head (ft. H_2O)
0.433	**1**	2.309
0.866	**2**	4.618
1.299	**3**	6.927
1.732	**4**	9.236
2.165	**5**	11.545
2.598	**6**	13.854
3.031	**7**	16.162
3.464	**8**	18.471
3.897	**9**	20.78
4.33	**10**	23.089
4.763	**11**	25.398
5.196	**12**	27.707
5.629	**13**	30.016
6.062	**14**	32.325
6.495	**15**	34.634
6.928	**16**	36.943
7.361	**17**	39.252
7.794	**18**	41.561
8.227	**19**	43.87
8.66	**20**	46.179
9.093	**21**	48.487
9.526	**22**	50.796
9.959	**23**	53.105
10.392	**24**	55.414
10.825	**25**	57.723
11.258	**26**	60.032
11.691	**27**	62.341

Table #121A - Pressure Head Conversion

K-4 Pressure Head Conversion

Pressure (psi)		Head (ft. H_2O)
12.124	**28**	64.65
12.557	**29**	66.959
12.99	**30**	69.268
13.423	**31**	71.577
13.856	**32**	73.886
14.289	**33**	76.195
14.722	**34**	78.504
15.155	**35**	80.812
15.588	**36**	83.121
16.021	**37**	85.43
16.454	**38**	87.739
16.887	**39**	90.048
17.32	**40**	92.357
17.753	**41**	94.666
18.186	**42**	96.975
18.619	**43**	99.284
19.052	**44**	101.593
19.485	**45**	103.902
19.918	**46**	106.211
20.351	**47**	108.52
20.784	**48**	110.829
21.217	**49**	113.137
21.65	**50**	115.446
22.083	**51**	117.755
22.516	**52**	120.064

Pressure (psi)		Head (ft. H_2O)
22.949	**53**	122.373
23.382	**54**	124.682
23.815	**55**	126.991
24.248	**56**	129.3
24.681	**57**	131.609
25.114	**58**	133.918
25.547	**59**	136.227
25.98	**60**	138.536
26.413	**61**	140.845
26.846	**62**	143.154
27.279	**63**	145.462
27.712	**64**	147.771
28.145	**65**	150.08
28.578	**66**	152.389
29.011	**67**	154.698
29.444	**68**	157.007
29.877	**69**	159.316
30.31	**70**	161.625
30.743	**71**	163.934
31.176	**72**	166.243
31.609	**73**	168.552
32.042	**74**	170.861
32.475	**75**	173.17
32.908	**76**	175.479
33.341	**77**	177.787

Pressure (psi)		Head (ft. H_2O)
33.774	**78**	180.096
34.207	**79**	182.405
34.64	**80**	184.714
35.073	**81**	187.023
35.506	**82**	189.332
35.939	**83**	191.641
36.372	**84**	193.95
36.805	**85**	196.259
37.238	**86**	198.568
37.671	**87**	200.877
38.104	**88**	203.186
38.537	**89**	205.495
38.97	**90**	207.804
39.403	**91**	210.112
39.836	**92**	212.421
40.269	**93**	214.73
40.702	**94**	217.039
41.135	**95**	219.348
41.568	**96**	221.657
42.001	**97**	223.966
42.434	**98**	226.275
42.867	**99**	228.584
43.3	**100**	230.893

Table #121B - Pressure Head Conversion